Formal Description Techniques VII

Formal Description Techniques VII

Proceedings of the 7th IFIP WG 6.1 international conference on formal description techniques

Edited by

Dieter Hogrefe

and

Stefan Leue

Department of Computer Science and Applied Mathematics
University of Berne
Switzerland

Sponsored by IFIP WG 6.1 in cooperation with ACM SIGCOMM and the IEEE Computer Society Technical Committee on Distributed Processing

Published by Chapman & Hall on behalf of the International Federation for Information Processing (IFIP)

CHAPMAN & HALL

London · Glasgow · Weinheim · New York · Tokyo · Melbourne · Madras

Published by Chapman & Hall, 2–6 Boundary Row, London SE1 8HN, UK

Chapman & Hall, 2–6 Boundary Row, London SE1 8HN, UK

Blackie Academic & Professional, Wester Cleddens Road, Bishopbriggs, Glasgow G64 2NZ, UK

Chapman & Hall GmbH, Pappelallee 3, 69469 Weinheim, Germany

Chapman & Hall USA, One Penn Plaza, 41st Floor, New York NY 10119, USA

Chapman & Hall Japan, ITP-Japan, Kyowa Building, 3F, 2-2-1 Hirakawacho, Chiyoda-ku, Tokyo 102, Japan

Chapman & Hall Australia, Thomas Nelson Australia, 102 Dodds Street, South Melbourne, Victoria 3205, Australia

Chapman & Hall India, R. Seshadri, 32 Second Main Road, CIT East, Madras 600 035, India

First edition 1995

Printed in Great Britain by TJ Press Ltd, Padstow, Cornwall

ISBN 0 412 64450 9

A catalogue record for this book is available from the British Library

Printed on permanent acid-free text paper, manufactured in accordance with ANSI/NISO Z39.48-1992 and ANSI/NISO Z39.48-1984 (Permanence of Paper).

CONTENTS

PART SIX Semantics

Preface

FORTE'94 was the Seventh International Conference on Formal Description Techiques (``FDTs"). The six previous FORTE conferences were held in Stirling, Scotland (1988), Vancouver, Canada (1989), Madrid, Spain (1990), Sydney, Australia (1991), Perros-Guirec, France (1992), and Boston, USA (1993). The next FORTE conference will be held in Montreal, Canada, in 1995. FORTE'94 was sponsored by IFIP WG 6.1 in cooperation with ACM SIGCOMM and the IEEE Computer Society Technical Committee on Distributed Processing.

The FORTE series of conferences concentrates on formal techniques applicable to distributed systems and protocols. Initially, the focus was on the techniques standardized by ISO and ITU-T (formerly CCITT), namely Estelle, LOTOS, and SDL, but the scope has widened considerably, and recent FORTE conferences have also included papers on other techniques. Much of the importance of these formal techniques is the contribution that their use can make in improving the quality of the software of distributed systems.

In addition to a strong theoretical component, FORTE always includes a practical component, attracting practitioners in the field by actively soliciting papers from industry and including demonstrations of electronic tools for manipulating formal descriptions of systems. FORTE'94 placed additional emphasis on this aspect by offering a session entitled *FDTs go commercial*, initiated and chaired by Harry Rudin, which comprised the presentation of three industrial usage reports and a related panel discussion.

The conference provided a forum for presentation of the state of the art in theory, application, tools and industrialization of formal techniques and provided an excellent orientation for newcomers, in particular during the tutorial day. By bringing together both researchers and practitioners, FORTE conferences open communication between these groups. This helps keep the research vital and improves the state of the practitioner's art. To promote discussions about interesting ongoing research FORTE'94 also included the presentation of position statements.

In addition to 24 full research papers and 12 position statements, FORTE'94 had 3 invited speakers: Nancy Lynch, Leslie Lamport, and Gerard Holzmann. There were also 6 tutorials for which supporting material appeared in a tutorial notes volume handed out to participants of the tutorials.

There were a total of 121 articles submitted: 112 research papers and 9 industrial usage reports. From these, 24 (approximately 21%) full research papers and 3 (33%) industrial usage reports were accepted. Table 1 summarizes the contributions and acceptances according to geographic location. In the case of multiple authors, the author who was identified as contact author was used to determine the geographic origin of a paper.

A conference such as this could not have taken place without the effort of a great many individuals and organizations. The editors wish to thank all of them. We gratefully acknowledge the sponsorship of IFIP WG 6.1, in cooperation with ACM SIGCOMM and the IEEE Computer Society Technical Committee on Distributed Processing. The University of Berne provided the conference facilities as well as technical support and equipment. We thank the Swiss National Science Foundation and the Beer-Brawand-Funds of the University of Berne for the generous financial support they provided. Further financial support was contributed by Alcatel STR, Ascom Tech, Siemens, Sun Microsystems, Swiss Telecom, and Swissair (as official conference carrier).

The program committee provided ideas and helped obtain the many (approximately 320) reviews that are crucial to the high quality of the conference. We owe our gratitude to all reviewers for their help. Special thanks are due to Ana Cavalli, Reinhard Gotzhein, Jan Kroon, Elie Najm, Tommaso Bolognesi, Harry Rudin, Jean-Pierre Courtiat, Jean-Pierre Hubaux, Claude Petitpierre, and Jan de Meer, who joined the editors at the program committee meeting in July, 1994, to complete the selection process. Ana Cavalli helped us with the dissemination of the call for papers. Richard Tenney offered us his editorial advice on this and a few other texts related to FORTE'94.

We also owe our gratitude to Manuela Beck, Lucia Engel, Urs Gasser, Heike Horn, Sylvia Schaad, and Patrick Zumbrunn for their effort in organizing this conference.

Berne, October 1994

Dieter Hogrefe
Stefan Leue

Table 1

Country	Research Papers			Industrial Reports	
	Submitted	Full Accepted	Position Statement	Submitted	Accepted
Australia	3	0	1	0	0
Brazil	1	0	0	0	0
Bulgaria	1	0	0	0	0
Canada	11	2	1	0	0
China	1	0	0	0	0
Denmark	2	0	0	0	0
France	12	5	1	2	1
Germany	15	3	3	0	0
Greece	1	0	0	1	0
Hong-Kong	2	0	0	0	0
Hungary	2	0	0	0	0
Italy	11	1	2	1	0
Japan	6	2	1	1	0
Lithuania	1	0	0	0	0
Norway	1	0	0	0	0
Poland	2	0	0	0	0
Russia	1	0	0	0	0
Scotland	1	0	1	0	0
Spain	3	1	0	2	1
Sweden	2	1	0	0	0
Switzerland	3	0	0	0	0
Taiwan, R.O.C.	1	0	0	0	0
The Netherlands	7	3	1	2	1
U.S.A.	13	3	0	0	0
United Kingdom	9	3	1	0	0
Total	112	24	12	9	3

Conference Organizers

Conference chairperson: Dieter Hogrefe (University of Berne, CH)
Conference organization chairperson: Stefan Leue (University of Berne, CH)

Program Committee

Paul Amer (University of Delaware, USA)
Gregor v. Bochmann (University of Montreal, CDN)
Tommaso Bolognesi (CNUCE, I)
Ed Brinksma (University of Twente, NL)
Ana Cavalli (INT, F)
Jean-Pierre Courtiat (LAAS-CNRS, F)
Piotr Dembinski (Polish Academy of Science, PL)
Ove Faergemand (EURESCOM, D)
Reinhard Gotzhein (University of Kaiserslautern, D)
Dieter Hogrefe (University of Berne, CH)
Gerard Holzmann (AT&T Bell Laboratories, USA)
Jean-Pierre Hubaux (EPF Lausanne, CH)
Toshihiko Kato (KDD, J)
Jan Kroon (PTT Research, NL)
Stefan Leue (University of Berne, CH)
Luigi Logrippo (University of Ottawa, CDN)
Nancy Lynch (MIT, USA)
Lynn Marshall (BNR, CDN)
Jan de Meer (GMD Fokus, D)
Elie Najm (ENST, F)
Linda Ness (Bellcore, USA)
Ken Parker (Telecom Research Laboratory, AUS)
Bjorn Pehrson (SICS, S)
Claude Petitpierre (EPF Lausanne, CH)
Juan Quemada (DIT ETSIT UPM, E)
Harry Rudin (IBM Research, CH)
Deepinder Sidhu (University of Maryland, USA)
Richard Tenney (University of Massachusetts, USA)
Ken Turner (University of Stirling, UK)
Ümit Uyar (CUNY, USA)
Son Vuong (University of British Columbia, CDN)

Organization Committee

General organization: Manuela Beck (University of Berne, CH)
General organization: Lucia Engel (University of Berne, CH)
Technical support: Heike Horn (University of Berne, CH)
Finances: Sylvia Schaad (University of Berne, CH)
General organization: Patrick Zumbrunn (University of Berne, CH)

List of Referees

V. S. Alagar
Rajeev Alur
Paul D. Amer
Flemming Andersen
Charles Andre
Pierre Azema

Cinzia Bernadeschi
Ramesh Bharadwaj
Gregor v. Bochmann
Tommaso Bolognesi
Martin Brachwitz
Jan Bredereke
Ed Brinksma
Stan Budkowski
Jörg Burmeister

Thomas Cain
Ana R. Cavalli
Nam Hong Cheng
Kong Eng Cheng
Giuseppe Ciaccio
Andrzej Cichocki
Robert G. Clark
John Colton
Jean-Pierre Courtiat

Piotr Dembinski
Michel Diaz
Abel Divin
Stephen Donaldson
A. J. M. Donaldson
Khalil Drira

Henk Eertink
Pierra-Alain Etique

Alessandro Fantechi
Kazi Farooqui
Arnaud Fevrier
Ove Færgemand

Anders Gammelgaard
Jean-Paul Gaspoz
Stefania Gnesi

Reinhard Gotzhein
Jean-Charles Grégoire

Abdelhakim Hafid
A. G. Hamilton
Dieter Hogrefe
Gerard J. Holzmann
Robert Huis in 't Veld

Misbah Islam

Stan Jedrysiak

Günter Karjoth
Pim Kars
Toshihiko Kato
Joost-Pieter Katoen
Ferhat Khendek
Ahmed Khoumsi
Tim Kleingeld-MacKenzie
A. S. Krishnakumar
Finn Kristoffersen
Jan Kroon

Peter Ladkin
Rom Langerak
David Larrabeiti Lopez
Diego Latella
George Leih
Jacek Leszczylowski
Stefan Leue
Jing-Yue Lin
Luis Llana
Luigi Logrippo
Nancy A. Lynch

Patrick Maigron
Lynn S. Marshall
Juan Martins
Mieke Massink
Antoni Mazurkiewicz
José A. Mañas
Jan de Meer
Laurent Mounier
Bruno Müller-Clostermann

Elie Najm
Carlos Miguel Nieto
A. Nymeyer
Bo Bichel Norbæk

Philippe Oechslin
Anders Olsen
Daniel Ouimet

Fabio Paternó
Santiago Pavon
Anton M. Peeters
Björn Pehrson
Doron A. Peled
Kim Dam Petersen
Claude Petitpierre
Alexandre Petrenko
Jimmi S. Pettersson
Simon Pickin
Luis Ferreira Pires

Juan Quemada

Gualberto Rabay
Charles Rattray
Nathalie Rico
Joaquin Salvachua Rodriguez
Frank Roessler
Rudolf Roth
Anne Rouger
Harry Rudin

Djamel-Eddine Saidouni
Pierre de Saqui-Sannes

Patrick Senac
Deepinder P. Sidhu
Jacques Sincennes
Richard O. Sinnott
Stephane Somé
Marian Srebrny
Marek Sredniawa
Silke Storp

Richard L. Tenney
M. J. A. Tesselaar
Jeron Thierry
Martti Tienari
Daniel Toggweiler
Kenneth J. Turner

Ümit Uyar

Rüdiger Valk
Ferance Vallo
Raghu Vallurupalli
Tran Huy Vu
Son T. Vuong

H. M. K. Warkentyne
Ishan Weerakoon
Alex Wijesinha
Jozef Winkowski
Mark F. Witteman

George M. Yee

Invited Talk

Proving Performance Properties (even Probabilistic Ones)

Nancy Lynch
MIT Laboratory for Computer Science
Cambridge, MA 02139, USA
`lynch@theory.lcs.mit.edu`

September 2, 1994

Abstract: This paper surveys some new tools and methods for formally verifying time performance properties of systems that satisfy timing assumptions. The techniques are potentially of practical benefit in the validation of real-time process control and communication systems. The tools and methods include nondeterministic timed automaton models, invariant assertion and simulation techniques for proving worst-case time bounds, probabilistic timed automaton models, and Markov-style techniques for proving probabilistic time bounds. All of these techniques are well suited for (partial) mechanization.

1 Introduction

A rich collection of formal methods have become well established for proving correctness properties – usually, safety and liveness properties – for asynchronous concurrent systems. The most important of these techniques are invariant assertion methods and simulation (refinement) methods for proving safety properties, and temporal logic methods for proving liveness properties. Other important techniques include algebraic and partial ordering methods.

But basic correctness properties are not the only ones that must be proved. Performance properties – especially time performance properties – are often nearly as important as correctness. In fact, for some systems, timing properties of system components are required to ensure the correct operation of the system as a whole. Yet formal methods have not been used very much to prove performance properties; nearly all of the proofs that have been carried out for such properties are ad hoc.

The goal of this paper is to show some ways in which proofs of timing properties can be formalized and systematized. We discuss the methods in three groups: those that are used for proving worst-case time bounds for asynchronous and timing-based algorithms, those for proving timing properties for hybrid systems, and those for proving probabilistic time bounds for probabilistic systems. Another paper in this proceedings, by Luchangco, Soylemez, Garland and Lynch [12], shows how proofs using the methods in the first group can be verified mechanically.

2 Time Bounds for Asynchronous and Timing-Based Algorithms

2.1 Theory

The basic model that we use for modelling asynchronous and timing-based systems is the nondeterministic *timed automaton* model of Lynch and Vaandrager [18, 17, 14]. We base our work on an automaton model rather than any particular specification language, programming language or proof system in order to obtain the greatest flexibility in choice of specification and proof methods.

Timed automata. A *timed automaton* A consists of a set $states(A)$ of states, a nonempty set $start(A) \subseteq states(A)$ of start states, a set $acts(A)$ of actions, including a special *time-passage* action ν, a set $steps(A)$ of steps (transitions), and a mapping $now_A : states(A) \rightarrow \mathsf{R}^{\geq 0}$. ($\mathsf{R}^{\geq 0}$ denotes the nonnegative reals.) The actions are partitioned into *external* and *internal* actions, where ν is considered external; the *visible* actions are the non-ν external actions; the visible actions are partitioned into *input* and *output* actions. The set $steps(A)$ is a subset of $states(A) \times acts(A) \times states(A)$. We write $s \xrightarrow{a}_A s'$ as shorthand for $(s, \pi, s') \in steps(A)$, and usually write $s.now_A$ in place of $now_A(s)$. We sometimes suppress the subscript or argument A.

A timed automaton must satisfy five axioms: [A1] If $s \in start$ then $s.now = 0$. [A2] If $s \xrightarrow{\pi} s'$ and $\pi \neq \nu$ then $s.now = s'.now$. [A3] If $s \xrightarrow{\nu} s'$ then $s.now < s'.now$. [A4] If $s \xrightarrow{\nu} s''$ and $s'' \xrightarrow{\nu} s'$, then $s \xrightarrow{\nu} s'$. The statement of [A5] requires the auxiliary definition of a *trajectory*, which describes restrictions on the state changes that can occur during time-passage. Namely, if I is any interval of $\mathsf{R}^{\geq 0}$, then an *I-trajectory* is a function $w : I \rightarrow states$, such that $w(t).now = t$ for all $t \in I$, and $w(t_1) \xrightarrow{\nu} w(t_2)$ for all $t_1, t_2 \in I$ with $t_1 < t_2$. That is, w assigns a state to each time t in interval I, in such a way that time-passage steps can connect any pair of states in the range of w. If w is an I-trajectory and I is left-closed, then let $w.fstate$ to be the first state of w, while if I is right-closed, then let $w.lstate$ denote the last state of w. If I is a closed interval, then an I-trajectory w is said to *span* from state s to state s' if $w.fstate = s$ and $w.lstate = s'$. The final axiom is: [A5] If $s \xrightarrow{\nu} s'$ then there exists a trajectory that spans from s to s'.

Timed Executions and Timed Traces. A *timed execution fragment* is a finite or infinite alternating sequence $\alpha = w_0\pi_1 w_1 \pi_2 w_2 \cdots$ where:

1. Each w_j is a trajectory and each π_j is a non-time-passage action.
2. If α is a finite sequence, then it ends with a trajectory.
3. If w_j is not the last trajectory in α then its domain is a closed interval. If w_j is the last trajectory then its domain is left-closed (and either right-open or right-closed).
4. If w_j is not the last trajectory then $w_j.lstate \xrightarrow{\pi_{j+1}} w_{j+1}.fstate$.

A *timed execution* is a timed execution fragment for which $w_0.fstate$ is a start state. We focus mainly on the *admissible* timed executions, i.e., those in which the *now* values occurring in the states approach ∞. A state of a timed automaton is defined to be *reachable* if it is the final state of the final trajectory in some finite timed execution.

In order to describe the problems to be solved by timed automata, we require a definition for their visible behavior. We use *timed traces*, where the *timed trace* of any timed execution is just the sequence of visible events that occur in the timed execution, paired with their times of occurrence. The *admissible timed traces* of the timed automaton are just the timed traces that arise from all the admissible timed executions.

Composition. Let A and B be timed automata satisfying the following *compatibility* conditions: A and B have no output actions in common, and no internal action of A is an action of B, and vice versa. Then the *composition* of A and B, written as $A \times B$, is the timed automaton defined as follows.

- $states(A \times B) = \{(s_A, s_B) \in states(A) \times states(B) : s_A.now_A = s_B.now_B\}$;
- $start(A \times B) = start(A) \times start(B)$;
- $acts(A \times B) = acts(A) \cup acts(B)$; an action is *external* in $A \times B$ exactly if it is external in either A or B, and likewise for *internal* actions; a visible action of $A \times B$ is an *output* in $A \times B$ exactly if it is an output in either A or B, and is an *input* otherwise;
- $(s_A, s_B) \xrightarrow{\pi}_{A \times B} (s'_A, s'_B)$ exactly if
  1. $s_A \xrightarrow{\pi}_A s'_A$ if $\pi \in acts(A)$, else $s_A = s'_A$, and
  2. $s_B \xrightarrow{\pi}_B s'_B$ if $\pi \in acts(B)$, else $s_B = s'_B$;
- $(s_A, s_B).now_{A \times B} = s_A.now_A$.

Then $A \times B$ is a timed automaton. If α is a timed execution of $A \times B$, we write $\alpha|A$ and $\alpha|B$ for the projections of α on A and B, respectively. For instance, $\alpha|A$ is defined by projecting all states in α on the state of A, removing actions that do not belong to A, and collapsing consecutive trajectories.

MMT automata. An important special case of the timed automaton model, describable in a particularly simple way, is the MMT timed automaton model [20], developed by Merritt, Modugno and Tuttle. We use a special case of their definition from [19, 14].

An MMT automaton is an I/O automaton [13] together with upper and lower bounds on time. An I/O automaton A consists of a set $states(A)$ of states, a nonempty set $start(A) \subseteq states(A)$ of start states, a set $acts(A)$ of actions, (partitioned into *external* and *internal* actions; the external actions are further partitioned into *input* and *output* actions), a set $steps(A)$ of steps, and a partition $part(A)$ of the locally controlled (i.e., output and internal) actions into at most countably many "tasks". The set $steps(A)$ is a subset of $states(A) \times acts(A) \times states(A)$; An action π is said to be *enabled* in a state s provided that there exists a state s' such that $(s, \pi, s') \in steps(A)$, i.e., such that $s \xrightarrow{\pi}_A s'$. It is required that the automaton be *input-enabled*, by which is meant that π is enabled in s for every state s and input action π. The final component, *part*, is sometimes called the *task partition*. Each class in this partition groups together actions that are supposed to be part of the same "task".

An MMT automaton is obtained by augmenting an I/O automaton with certain upper and lower time bound information. Let A be an I/O automaton with only finitely many tasks. For each task C, define lower and upper time bounds, $lower(C)$ and $upper(C)$, where $0 \leq lower(C) < \infty$ and $0 < upper(C) \leq \infty$; that is, the lower bounds cannot be infinite and the upper bounds cannot be 0.

A timed execution of an MMT automaton A is defined to be an alternating sequence of the form $s_0, (\pi_1, t_1), s_1, \cdots$ where the π's are input, output or internal actions (but not time-passage actions). For each j, it must be that $s_j \xrightarrow{\pi_{j+1}} s_{j+1}$. The successive times are nondecreasing, and are required to satisfy the given *lower* and *upper* bound requirements. Finally, *admissibility* is required: if the sequence is infinite, then the times of actions approach ∞. Each timed execution of an MMT automaton A gives rise to a *timed trace*, which is just the subsequence of external actions and their associated times. The *admissible timed traces* of the MMT automaton A are just the timed traces that arise from all the timed executions of A.

It is not hard to transform any MMT automaton A into a naturally-corresponding timed automaton A'. First, the state of the MMT automaton A is augmented with a *now* component, plus *first*(C) and *last*(C) components for each task. The *first*(C) and *last*(C) components represent, respectively, the earliest and latest time in the future that an action of task C is allowed to occur. The time-passage action ν is also added. The *first* and *last* components get updated in the natural way by the various steps, according to the *lower* and *upper* bounds specified in the MMT automaton A. The time-passage action has explicit preconditions saying that time cannot pass beyond any of the *last*(C) values, since these represent deadlines for the various tasks. Restrictions are also added on actions of any task C, saying that the current time *now* must be at least equal to *first*(C). The resulting timed automaton A' has exactly the same admissible timed traces as the MMT automaton A.

Invariants and simulations. We define an *invariant* of a timed automaton to be any property that is true of all reachable states.

The definition of a simulation is paraphrased from [18, 17, 14]. We use the notation $f[s]$, where f is a binary relation, to denote $\{u : (s, u) \in f\}$. Suppose A and B are timed automata and I_A and I_B are invariants of A and B, respectively. Then a *simulation* from A to B with respect to I_A and I_B is a relation f over *states*(A) and *states*(B) that satisfies:

1. If $u \in f[s]$ then $u.now = s.now$.
2. If $s \in start(A)$ then $f[s] \cap start(B) \neq \emptyset$.
3. If $s \xrightarrow{\pi}_A s'$, $s, s' \in I_A$, and $u \in f[s] \cap I_B$, then there exists $u' \in f[s']$ such that there is a timed execution fragment from u to u' having the same timed visible actions as the given step.

Note that π is allowed to be the time-passage action in the third item. The most important fact about simulations is that they imply admissible timed trace inclusion:

Theorem 2.1 *If there is a simulation from timed automaton A to timed automaton B, with respect to any invariants, then every admissible timed trace of A is an admissible timed trace of B.*

In practice, the simulations often have an interesting form: a system of inequalities [19].

2.2 Applications

We sketch several examples of time bounds for asynchronous and timing-based concurrent algorithms, proved using the tools and methods described above. The MMT model is sufficient to describe all of these. The examples are summarized from [14, 15, 11].

Counting Process. The first and simplest example involves an automaton that counts down from some fixed positive integer k and then reports its completion. The time between successive steps is assumed to be in the range $[c_1, c_2]$.

Count:

Actions:
Output: *report*
Internal: *decrement*

State components:
count, initially $k > 0$
reported, Boolean, initially *false*

Transitions:

decrement
Precondition:
$count > 0$
Effect:
$count := count - 1$

report
Precondition:
$count = 0$
$reported = false$
Effect:
$reported := true$

Tasks and bounds:
{*report*}, bounds $[c_1, c_2]$
{*decrement*}, bounds $[c_1, c_2]$

Informally, it is easy to see that the time until a *report* occurs is in the range $[(k + 1)c_1, (k+1)c_2]$. In order to prove this formally, we express these time bound assumptions by a trivial high-level reporting automaton called *Report*.

Report:

Actions:
Output: *report*

State components:
reported, Boolean, initially *false*

Transitions:

report
Precondition:
$reported = false$
Effect:
$reported := true$

Tasks and bounds:
{*report*}, bounds $[(k+1)c_1, (k+1)c_2]$.

We can show that *Count* implements *Report*, that is, that every admissible timed trace of *Count* is an admissible timed trace of *Report*. Specifically, we define $(s, u) \in f$ provided that the following hold:

- $u.now = s.now$,
- $u.reported = s.reported$,
- $u.last(report) \geq$
$$\begin{cases} s.last(decrement) + s.count \cdot c_2 & \text{if } s.count > 0, \\ s.last(report) & \text{otherwise.} \end{cases}$$
- $u.first(report) \leq$
$$\begin{cases} s.first(decrement) + s.count \cdot c_1 & \text{if } s.count > 0, \\ s.first(report) & \text{otherwise.} \end{cases}$$

The *now* and *reported* conditions are straightforward. The *last*(*report*) component is constrained to be at least as large as a calculated upper bound on the latest time until a *report* action is performed by *Count*. If $count > 0$, then this calculated bound is the last time at which the first *decrement* can occur, plus the additional time required to do $count - 1$ *decrement* steps, followed by a *report*; since each of these *count* steps could take at most time c_2, this additional time is at most $count \cdot c_2$. On the other hand, if $count = 0$, then this calculated bound is the last time at which the *report* can occur. The interpretation of the condition for *first*(*report*) is symmetric – the lower bound to be proved should be no larger than a calculated lower bound on the earliest time until a *report* action is performed by *Count*.

Lemma 2.2 *The relation f is a forward simulation from Count to Report.*

Proof: The proof involves verifying the three properties in the definition of a forward simulation (after proving some trivial invariants). The correspondence between *now* values is immediate, and it is easy to check that the unique start states of the two automata are related by f. The interesting part of the proof is the third condition – the step correspondence. For this, we consider cases based on types of transitions.

For example, consider a transition $s \xrightarrow{decrement}_{Count} s'$, where $u \in f[s]$. Then $s.count > 0$. Thus, $u.now = s.now$, $u.reported = s.reported$, $u.last(report) \geq s.last(decrement) + s.count \cdot c_2$, and $u.first(report) \leq s.first(decrement) + s.count \cdot c_1$. It suffices to show that $u \in f[s']$. The first two conditions in the definition of f carry over immediately.

For the third condition, consider first the case where $s'.count > 0$. Then the left side of the inequality, *last*(*report*), does not change, while on the right side, *last*(*decrement*) cannot increase by more than c_2, and the second term decreases by exactly c_2. (The reason why *last*(*decrement*) cannot increase by more than c_2 is as follows: The construction of the timed automaton from the MMT automaton for *Count* — captured by invariants — implies that $s.now \leq s.last(decrement)$. On the other hand, $s'.last(decrement) = s'.now + c_2$ and $s'.now = s.now$.) So the inequality still holds after the step.

Similar arguments can be made for the case where $s'.count = 0$, and for the lower bound. ∎

This lemma implies the following theorem, which says that *Count* satisfies the timing requirements.

Theorem 2.3 *Every admissible timed trace of Count is an admissible timed traces of Report.*

It is possible to use an equational theorem prover to verify proofs of simulations such as this one, as well as proofs of the needed invariants. In fact, such a theorem prover can fill in some of the proof steps. Work along these lines, using the Larch theorem prover [6], is described in [12].

The *Count* algorithm is trivial; in the rest of this section, we sketch more interesting time bound results that can be proved using the same strategy. Without such a stylized method, proving such time bounds would be a difficult task.

Fischer mutual exclusion algorithm. The Fischer mutual exclusion algorithm [5] is a popular test case for formal methods for verifying real-time algorithms. In this algorithm, a collection of user processes compete for control of a resource, using a single shared variable x that they can only access using *read* and *write* operations. We model the entire assembly of processes and x as an MMT automaton, where the tasks correspond to the different types of steps performed by the processes (several tasks per process). A high-level description of the algorithm is:

Fischer, process i:
L: wait for $x = 0$
    $x := i$
    if $x \neq i$ then go to L
    (Critical Region)
    $x := 0$

As described so far, this algorithm can violate mutual exclusion. For example, two processes, i and j, might *both* test x and find its value to be 0. Then i might set $x := i$ and immediately check and see its own index, and then j might do the same. In order to avoid this bad interleaving, we introduce time restrictions: for each process, the time for setting x, once x has been seen equal to 0, is in the range $[0, a]$, while the time to check x after it has been set is in the range $[b, \infty]$, for some constants a, b, where $a < b$. This rules out the bad interleaving above, since any process i that sets $x := i$ is made to wait long enough before checking to ensure that any other process j that tested x before i set x (and therefore might subsequently set $x := j$) has already set $x := j$. That is, there should be no processes left at the point of setting x, when i finally checks.

It is easy to translate this code into an MMT automaton *Fischer* (say, in precondition-effect style). The input actions are of the form try_i, by which a process i tries to get access to the resource, and $exit_i$, by which a process returns the resource, and the outputs are $crit_i$, by which the algorithm grants access to the resource (critical region) and rem_i, by which the algorithm gives permission to the user program to return to the remainder of its processing. There are also internal actions $test_i$, set_i, $check_i$ and $reset_i$. The tasks correspond to the individual actions. Then mutual exclusion can be expressed as an invariant on the algorithm's state (more precisely, on the state of a composed *Fischer* system automaton consisting of the *Fischer* automaton and and a collection of nondeterministic *user* automata). It can be proved as a consequence of the following invariant, which is in turn proved easily by induction:

Lemma 2.4 *If* $pc_i = check$ *and* $x = i$ *and* $pc_j = set$ *then* $first(check_i) > last(set_j)$.

That is, if i is about to check x successfully, then any other process that is about to set x is scheduled to do so before i can check x.

Mutual exclusion is the property that is usually proved for this algorithm. But since this paper is about proving time bounds, we take the example further. We assume an upper bound of c, $c \geq b$, for each *check* action, and an upper bound of a for all the other actions, and we show that the time from when *any* process is trying to obtain the resource until *some* process has it is at most $2c + 5a$.

Following the same strategy as for the *Count* example, we formulate the specification of the time bound as an MMT automaton *Mutex* with the same external actions as the *Fischer* algorithm. The *Mutex* automaton is very nondeterministic; it expresses only the mutual exclusion property plus the given time bound. Then we use the simulation method to show that the *Fischer* system implements the *Mutex* system (i.e., the composition of the *Mutex* automaton and the users).

In this example, the time bound is most easily understood in terms of reaching certain *milestones*. In particular, once the critical region is empty, the first important event is for the shared variable x to be "seized" by some contending process. The next important event is for the value of x to "stabilize" so that it can no longer be modified by any process until someone reaches the critical region. It turns out that the first milestone is reached within at most time $c + 3a$, then the second milestone is reached within additional time a, and finally some process reaches the critical region within additional time $c + a$; the total is $2c + 5a$.

Formally, we can express these milestones as the events of another intermediate MMT automaton I. A trivial simulation can be used to show that the I system implements the *Mutex* system, and then a more complicated simulation can be used to show that the *Fischer* system implements the I system. For example, a key piece of the simulation from *Fischer* to I is the set of inequalities that involve calculated bounds on time for some process to seize x:

1. $u.last(seize) \geq s.last(reset_i) + c + 2a$ if $s.pc_i = reset$.
2. If $s.x = 0$ then $u.last(seize) \geq \min_i\{h(i)\}$,

$$\text{where } h(i) = \begin{cases} s.last(check_i) + 2a & \text{if } s.pc_i = check, \\ s.last(test_i) + a & \text{if } s.pc_i = test, \\ s.last(set_i) & \text{if } s.pc_i = set, \\ \infty, & \text{otherwise.} \end{cases}$$

Thus, if some process i has just exited, the calculated bound is the maximum time until process i resets $x := 0$, plus $c + 2a$. If $x = 0$, the calculated bound is the minimum of a collection of individual process bounds, where the bound for process i measures the time until i sets x (if no other process does so in the meantime).

Theorem 2.5 *Every admissible timed trace of the Fischer system is an admissible timed trace of the Mutex system.*

Again, the proof can be checked mechanically [12].

Dijkstra mutual exclusion algorithm. Dijkstra's asynchronous algorithm [4] for mutual exclusion using read/write shared memory is one of the earliest published distributed algorithms.

```
Dijkstra, process i:
L: flag(i) := 1
   while x ≠ i do
      if flag(x) = 0 then x := i
   flag(i) := 2
   for j ≠ i do:
      if flag(j) = 2 then goto L
   (Critical Region)
   flag(i) := 0
```

Unlike the *Fischer* algorithm, the *Dijkstra* algorithm guarantees mutual exclusion regardless of the relative speeds of processes. For a time bound, we can assume an upper bound of a for all process steps and show that the time from when any process is trying until some process is critical is at most $(3n + 11)a$. (Here and elsewhere, n is the number of processes in the system.)

The proof strategy is the same as for the Fischer algorithm, using an intermediate automaton with *seize* and *stabilize* milestones, plus an additional *dropback* milestone. The *dropback* event indicates a point by which all but the process whose index is in x have dropped back to the first stage of the algorithm, where $flag = 1$ (or by which someone goes critical, it this happens first). I uses upper bounds $(n + 5)a$, $2a$, $(n + 1)a$ and $(n + 3)a$ for the milestones *seize*, *stabilize*, *dropback* and *critical*, respectively. Using I, we can provide two stylized simulations to show the needed time bound.

LeLann-Chang-Roberts leader election algorithm. The LeLann-Chang-Roberts leader election algorithm for ring networks [10, 3] works as follows:

> **LCR:**
> Every process sends its process identifier clockwise. Smaller identifiers that encounter larger identifiers are discarded. If a node receives its own identifier in a message, it elects itself as leader.

If we assume an upper bound of ℓ on the step time for each process, and an upper bound of d on the time to deliver the oldest message in each channel, then we can show that the time until a leader is elected is at most $(n+1)\ell+nd$. The difficulty of the proof involves the possible pile-up of messages in channels if some processes and channels operate faster than others. The proof again uses an intermediate automaton I, this time with n milestones. Milestone i, $1 \leq i \leq n$, is reached when the *slowest* token has progressed distance i around the ring. The bounds for the successive milestones are all $\ell + d$, and the bound for the final leader report is ℓ. The simulation from the *LCR* system to I determines how many milestones have been reached based on the least progress made by any identifier.

Discussion. Timed automata, MMT automata, simulations and invariants have been successful for verifying time bounds for a substantial collection of asynchronous and timing-based concurrent algorithms. These have not all been toy examples, but include algorithms for which time bounds are otherwise difficult to obtain. The method provides information (invariants, simulations) that help in documenting the operation of the algorithms. The method is systematic, and lends itself to mechanical assistance.

The proofs have more common structure than just the use of simulations and invariants. In several of the cases, an intermediate milestone automaton is used. In every case, an important part of the simulation is a set of inequalities involving calculated upper and/or lower bounds. These calculated bounds can usually be expressed as "progress functions" that measure the time until the specification-level goals are reached. If progress functions satisfy certain properties (as detailed in [19]), then a relation derived from them in a systematic way is guaranteed to be a simulation. Identifying such common structure should help in further systematizing the proofs.

3 Timing Properties for Hybrid Systems

The results of Section 2.2 show how simulation and invariant methods can be used to prove time performance properties for asynchronous and timing-based algorithms. Now we show how the same methods can be used to prove properties of "hybrid" systems containing both real world and computer components, for example, real-time process control systems. The real world components typically include physical quantities that change continuously as time passes, as well as quantities that are changed by discrete events.

3.1 Theory

It turns out that the timed automaton and MMT automaton models introduced in Section 2.1 for representing algorithms are also adequate for representing most real-time systems with real world components. The main difference is that the trajectory mechanism provided by the general timed automaton model now becomes important for modelling continuous changes in the real-world components. The simulation and invariant methods are still used in much the same way as before.

3.2 Applications

We discuss one example: the Generalized Railroad Crossing (GRC) problem [7].

Generalized railroad crossing. The problem involves real world trains and gates, interacting with a computer system via sensors and actuators. There are several parallel tracks on which trains travel through a road intersection I. Before arriving in I, each train arrives in the general region R, and triggers a sensor that notifies the computer system of its arrival. When a train leaves I, it also triggers a sensor. The computer system is capable of lowering and raising a crossing gate. Parameters ϵ_1 and ϵ_2 describe lower and upper bounds on the time from when a train triggers the arrival sensor until it reaches I, and parameters γ_{down} and γ_{up} describe upper bounds on the time to lower and raise the gate completely. There are two requirements:

1. A *safety property*, which says that if a train is in I then the gate is down.
2. A *utility property*, which says that the gate must be up unless there is a reason it should not be: either within following time ξ_1 or preceding time ξ_2 (two additional parameters), some train will be in I.

The problem is to formalize these requirements and to design and verify a computer system that satisfies them. This problem was originally proposed as a challenge problem for comparing the effectiveness of various formal methods for handling such problems.

We present the approach from [8, 9]. We begin by defining separate timed automata for the trains and gate components, plus a placeholder automaton to represent the interface between the real world components and the computer system. Initially, we give discrete automata *Trains* and *Gate*, and later infer results about the corresponding continuous components using general properties of timed automaton composition. For example, *Trains* is the following MMT automaton.

Trains:

Actions:
Output: $enterR(r)$, r a train
$enterI(r)$, r a train
$exit(r)$, r a train

State components:
for each train r:
$r.status \in \{not\text{-}here, P, I\}$, initially *not-here*

Transitions:

$enterR(r)$
Precondition:
$r.status = not\text{-}here$
Effect:
$r.status := P$

$enterI(r)$
Precondition:
$r.status = P$
Effect:
$r.status := I$

$exit(r)$
Precondition:
$r.status = I$
Effect:
$r.status := not\text{-}here$

Tasks and bounds:
Each action is a separate task; all bounds are trivial except each $enterI(r)$ has $[\epsilon_1, \epsilon_2]$.

Similarly, *Gate* is modelled as a discrete MMT automaton, with four states, *up*, *going-down*, *down* and *going-up*. It has inputs *lower* and *raise*, which place the gate in state *going-down* or *going-up*, respectively, and outputs *down* and *up*, which signal the arrival of the gate at the extreme positions.

For the specification, there is also a trivial *CompSpec* component describing the computer system's interface. The complete specification *AxSpec* consists of the composition of these three automata, plus two axioms corresponding to the safety and utility properties, which constrain the set of executions.

To describe an implementation, we replace *CompSpec* by *CompImpl* – a timed automaton that decides when to raise and lower the gate. It lowers the gate when there is a train that might reach I within time γ_{down}, and raises it when there is no train that could reach I within time $\gamma_{up} + \gamma_{down}$. The complete system *SysImpl* is just the composition of the three components, with no extra constraints. We must show that *SysImpl* satisfies the requirements, specifically, that each admissible timed execution of *SysImpl* projects on the real world components to look like an admissible timed execution of *AxSpec*. This is the same as saying that *SysImpl* satisfies the two axioms.

The safety property is proved as an invariant of *SysImpl*. We are more interested here in the utility property, however, since it can be classified as a performance property. As for the previous performance properties, we would like to prove this using a simulation. But to do this, we need a specification automaton, which is not the same as an axiomatic specification. Thus, we give a second specification, *OpSpec*, in the form of a timed automaton. Instead of having an axiom about time intervals, we put in explicit *last* components as before, as deadlines for certain specification-level goals. For example, *OpSpec* includes the provision that whenever the gate is lowered, some train will be in the crossing within time ξ_1. This is expressed by means of a *last* component that gets set when *CompImpl* lowers the gate to time ξ_1 in the future, and gets disabled when some train enters I.

An ad hoc argument is used to prove that *OpSpec* implements *AxSpec* (in the sense that each admissible timed execution of *OpSpec* projects on the real world to look like an admissible timed execution of *AxSpec*). (The implementation relationship does not need to go the other way.) Then the more systematic simulation methods are used as before to show that *SysImpl* implements *OpSpec*. Technically, the simulation proof only shows inclusion of sets of admissible timed traces, and not the preservation of the complete view of the real world. But general properties of timed automaton composition yield the stronger correspondence as a corollary. We obtain:

Theorem 3.1 *For every admissible timed execution α of SysImpl, there is an admissible timed execution α' of AxSpec such that $\alpha' | Trains \times Gate = \alpha | Trains \times Gate$.*

Also, as noted earlier, we can define more realistic models of the real world components, such as a *Gate'* automaton with a specific function describing the position of the gate while it is being lowered or raised. We can then infer results about the continuous models from those for the discrete models, using general properties of timed automaton composition.

Discussion. Timed automata, simulations and invariants have proved able to cope successfully with the Generalized Railroad Crossing problem, in particular, with its performance properties. The results obtained are very general – e.g., they apply to arbitrary values of the various parameters. This is in contrast, say, to model-checking methods which (usually) only work for specific constants. As before, the method provides useful information in the form of invariants and simulations, is systematic, and lends itself to mechanical assistance.

It appears that applications experts for process control systems prefer axiomatic to operational specifications. Furthermore, the axioms should be statements about the real world, and should (as far as possible) be independent of each other. On the other hand, an operational, automaton-style specification is needed for the application of invariant and simulation proof methods. We conclude that both types of specifications should be provided, and a proof that the operational specification implements the axiomatic specification should be given. In the GRC example, this proof of correspondence between the two specifications is ad hoc. It remains to develop a user-friendly language for axiomatic specification of real-time systems, and a systematic way of relating (and perhaps translating) such specifications to operational specifications.

It is useful to work initially with simplified, discrete models of the real world components, and then to use general properties of timed automaton composition to extend the results to more realistic, continuous models.

The GRC example is small. It remains to see how well these methods scale up to realistic size process control systems. We are currently looking at some problems arising in transportation systems such as the Personal Rapid Transit system currently being designed by Raytheon.

4 Probabilistic Time Bounds for Probabilistic Algorithms

We finish with a collection of new formal tools and methods for proving time bound properties for probabilistic systems. The motivating examples are the many randomized distributed algorithms that have been developed in the theory of distributed computing. The appropriate time bounds for such algorithms are generally *probabilistic time bounds*, i.e., assertions that some event occurs within a certain amount of time t, with at least a certain probability p.

It is notoriously difficult to obtain correct proofs for randomized distributed algorithms, and so it is worthwhile to develop systematic proof methods. The methods we use to prove probabilistic time bounds are somewhat different from the simulation methods discussed in the previous sections. However, the development of these methods represents work in progress; eventually, the methods used for probabilistic algorithms ought to be better unified with those used for non-probabilistic algorithms.

4.1 Theory

The theory we used is taken from [22, 16, 21].

Probabilistic timed automata. We modify the timed automaton model to incorporate probabilistic state transitions; specifically, we allow non-time-passage steps to be of the form (s, π, Π), where Π is a probability distribution whose domain is a subset of the states of the algorithm. The meaning is that when the step occurs, a probabilistic choice according to distribution Π is used to determine the new state. A probabilistic timed automaton (pta) has a combination of nondeterministic and probabilistic branching in its timed executions, where nondeterminism is involved in the choice of the next step.

Meaningful probabilistic statements about the behavior of a pta must be based on a probability distribution on its timed executions. In order to obtain such a distribution, we must resolve all nondeterministic choices, making all the branching probabilistic. We do this by hypothesizing an entity called an *adversary* that determines the next step; formally, an adversary is a function that, given the finite execution performed so far, returns the following (possibly probabilistic) step. A fixed adversary $\mathcal{A}$ gives rise to an execution tree with only probabilistic branch points, and thereby to a probability distribution on timed executions.

An adversary is said to be *admissible* provided that all the timed executions that appear in its distribution are admissible, i.e., provided that it always allows time to pass to ∞. When we claim that an algorithm (pta) guarantees some property P with at least probability p, we mean that for every admissible adversary $\mathcal{A}$, the probability of P in the probability distribution of timed executions generated by $\mathcal{A}$ is at least p.

Markov rules. One practical method for proving probabalistic time bound properties is based on formulating Markov-style statements of the form $\mathcal{U} \xrightarrow[p]{t} \mathcal{U}'$, where $\mathcal{U}$ and $\mathcal{U}'$ are sets of states. This statement means that for every admissible adversary $\mathcal{A}$, if the algorithm is started in a state in $\mathcal{U}$, then in the probability distribution generated by $\mathcal{A}$, the probability that a state in $\mathcal{U}'$ is reached within time t is at least p. Such statements can be combined:

Lemma 4.1 *If* $\mathcal{U} \xrightarrow[p]{t} \mathcal{U}'$ *and* $\mathcal{U}' \xrightarrow[p']{t'} \mathcal{U}''$ *then* $\mathcal{U} \xrightarrow[pp']{t+t'} \mathcal{U}''$.

Coin lemmas. In most situations where a property of the form $\mathcal{U} \xrightarrow[p]{t} \mathcal{U}'$ is true, the probability p arises as the probability that certain designated random choices are made in a favorable way. In such a case, the statement can usually be proved by fixing the results of those choices in the favorable way and converting all the remaining random choices to nondeterministic choices. This gives rise to an ordinary (non-probabilistic) timed automaton A. Proving that the desired time bound holds *with certainty* for A implies that it holds with probability p in the original pta. The soundness of this strategy is justified by a series of *coin lemmas* in [21].

4.2 Applications

We have so far carried out two significant proofs of probabilistic time bound properties for randomized distributed algorithms.

Lehmann-Rabin Dining Philosophers algorithm. The randomized Dining Philosophers algorithm of Lehmann and Rabin works as follows.

LehmannRabin:

Each contending process executes a loop wherein it randomly chooses a direction, *left* or *right*, waits to obtain the fork in that direction, and then instantaneously examines the other fork. If the second fork is free, it takes it and proceeds to the critical region; otherwise, it drops its first fork and returns to the beginning of the loop.

The safety property, i.e., that no two processes with conflicting resource requirements ever reach their critical regions simultaneously, is easy to show. A more interesting property is the following probabilistic time bound claim: from any state where some process is trying to obtain its forks, within time 14ℓ, and with probability at least $\frac{1}{16}$, some process will reach the critical region.[1] This can be expressed in the form $\mathcal{T} \xrightarrow[\frac{1}{16}]{14\ell} \mathcal{C}$. where $\mathcal{T}$ is the set of states in which some process is trying to get its forks and $\mathcal{C}$ is the set of states in which some process is in the critical region.

This property is proved in [16] using Markov rules and coin lemmas. We use a chain of five Markov rules marking progress from $\mathcal{T}$ to $\mathcal{C}$, each with its own time and probability bounds, and combine them using Lemma 4.1. Some of these rules involve probability 1, and in fact describe results that hold with certainty; these can be proved by methods such as the simulation method already developed for proving time bounds for non-probabilistic timed automata. The others all involve the results of particular random coin tosses, generally the first coin tosses by a small number of neighboring processes. These can be proved by using the coin lemmas to reduce the problem to one involving non-probabilistic timed automata, then using methods for proving time bounds for non-probabilistic timed automata. In [16], operational arguments are used to prove the claims for non-probabilistic timed automata, but it is clear in principle that the arguments could be redone more systematically using simulations.

For an example of an argument that is done for this proof, consider a situation involving a process i and its right neighbor $i+1$. Suppose that i has already obtained its left fork and is about to test its right fork (the one shared with $i+1$), while process $i+1$ is about to toss its coin. This situation is formally described as a set $\mathcal{U}$ of states of the algorithm. We claim that $\mathcal{U} \xrightarrow[\frac{1}{2}]{\ell} \mathcal{U}'$, where $\mathcal{U}'$ is the set of states in which some process has both its forks. The probability of $\frac{1}{2}$ is just the probability that $i+1$ chooses *right*. In that case, we claim that the given situation leads to $\mathcal{U}'$ with absolute certainty, within time ℓ. To see this, note that i must test its right fork within time ℓ. If it succeeds in getting it, then we are done. If not, then it must be that in the meantime, $i+1$ has obtained it. But if $i+1$ has obtained it, then it must be $i+1$'s second fork, since $i+1$'s next coin toss is *right*. Thus in this case, $i+1$ must have both its forks and again we are done.

Aggarwal-Kutten spanning tree algorithm. A larger and more complex example is the Aggarwal-Kutten randomized algorithm for finding a spanning tree in a network based on an unknown undirected graph G [2]. Processes are assumed to be identical (e.g., they do not have unique identifiers), and so the only way to break symmetry in the construction of the tree is using randomness. The algorithm is too complicated to explain quickly, but the basic idea is that each process tries to form a spanning tree with itself as the root, using a broadcast-convergecast strategy. In doing this, each process

[1] Again, ℓ is an upper bound on process step time.

uses a randomly-chosen identifier, and if it discovers the existence of another root (by discovering a different identifier) it takes steps to combine with the other root.

The problem is that there might be more than one root with the same identifier, preventing them from discovering each other's existence. To address this problem, the roots make repeated random choices of new identifiers, according to a careful discipline, and propagate these throughout their trees. Then it turns out that within time proportional to the diameter of the network, and with at least a nonzero constant probability, a unique spanning tree will result.

This algorithm is a randomized distributed algorithm of typical difficulty; the state of the art in the distributed algorithms area has till now not permitted careful proofs of such algorithms. Nevertheless, in [1] this bound is proved, using Markov rule and coin lemma methods. There are two interesting points to be noted about the proof. First, as is common when careful proofs are carried out for complicated algorithms, a mistake was found in the original code. (It was easy to fix.) Second, although the final proof is fairly lengthy, the parts of it that deal with probabilities constitute only a few pages. One effect of our work on systematizing such proofs has been the reduction of reasoning about probabilistic systems to reasoning about non-probabilistic systems.

Discussion. Systematic proofs of probabilistic time bounds for probabilistic algorithms are possible. The methods we have used so far – Markov rules and coin lemmas – are somewhat different from those used for non-probabilistic algorithms. Mechanical assistance is clearly possible for the non-probabilistic parts of the reasoning; we do not know if additional assistance is possible for the probabilistic parts.

Intuitively speaking, there seems to be some redundancy in the methods we have applied so far to probabilistic systems. Both the strategy of combining Markov rules and the techniques for proving individual Markov rules involve measuring progress toward a goal. Perhaps there is a way of unifying these methods.

It should also be possible to extend simulation methods directly to probabilistic systems, without first removing the probabilistic choices. The simulation relations themselves might be probabilistic, corresponding states with probability distributions on states rather than with single states. It remains to define such correspondences, verify their soundness, and see how they can be used in verification.

5 Conclusions

There are now a wide range of practical, systematic methods for verifying time performance properties for concurrent systems. These include timed automata, simulations, invariants, probabilistic timed automata, Markov rules and coin lemmas. Computer assistance is possible for most of these, using equational theorem provers such as Larch.

Remaining work includes developing the formal methods for probabilistic systems further, and unifying them with the simulation and invariant methods. It also includes further attempts to automate these methods, and to apply them to many more examples.

References

[1] Sudhanshu Aggarwal. Time optimal self-stabilizing spanning tree algorithms. Master's thesis, MIT Electrical Engineering and Computer Science, May 1994.

[2] Sudhanshu Aggarwal and Shay Kutten. Time optimal self stabilizing spanning tree algorithms. In R.K. Shyamasundar, editor, *13th International Conference on Foundations of Software Technology and Theoretical Computer Science*, volume 761 of *Lecture Notes in Computer Science*, pages 400–410, Bombay, India., December 1993. Springer-Verlag.

[3] Ernest Chang and Rosemary Roberts. An improved algorithm for decentralized extrema-finding in circular configurations of processes. *Communications of the ACM*, 22(5):281–283, May 1979.

[4] E.W. Dijkstra. Solution of a problem in concurrent programming control. *Communications of the ACM*, 8(9):569, September 1965.

[5] Michael Fischer. Re: Where are you? E-mail message to Leslie Lamport. Arpanet message number 8506252257.AA07636@YALE-BULLDOG.YALE.ARPA (47 lines), June 25, 1985 18:56:29EDT.

[6] Stephen J. Garland and John V. Guttag. A guide to LP, the Larch Prover. Technical report, Digital Systems Research Center, 130 Lytton Avenue, Palo Alto, California 94301, December 1991. Research Report 82.

[7] C. L. Heitmeyer, R. D. Jeffords, and B. G. Labaw. A benchmark for comparing different approaches for specifying and verifying real-time systems. In *Proc., Tenth Intern. Workshop on Real-Time Operating Systems and Software*, May 1993.

[8] Constance Heitmeyer and Nancy Lynch. The generalized railroad crossing: A case study in formal verification of real-time systems. In *Proceedings of the 15th IEEE Real-Time Systems Symposium.*, San Juan, Puerto Rico, December 1994. To appear.

[9] Constance Heitmeyer and Nancy Lynch. The generalized railroad crossing: A case study in formal verification of real-time systems. Technical Memo MIT/LCS/TM-511, Laboratory for Computer Science, Massachusetts Institute of Technology, Cambridge, MA, 1994. To appear.

[10] G. LeLann. Distributed systems, towards a formal approach. In *IFIP Congress*, pages 155–160, Toronto, 1977.

[11] Victor Luchangco. Using simulation techiniques to prove timing properties. Master's thesis, MIT Electrical Engineering and Computer Science, 1994. In progress.

[12] Victor Luchangco, Ekrem Söylemez, Stephen Garland, and Nancy Lynch. Verifying timing properties of concurrent algorithms. In *Proceedings of the Seventh International Conference on Formal Description Techniques for Distributed Systems and Communications Protocols (FORTE'94)*, IFIP Transactions, Berne, Switzerland, October 1994. IFIP WG6.1, Elsevier Science Publishers B. V. (North Holland).

[13] N. Lynch and M. Tuttle. An introduction to Input/Output automata. *CWI-Quarterly*, 2(3):219–246, September 1989. Centrum voor Wiskunde en Informatica, Amsterdam, The Netherlands.

[14] Nancy Lynch. Simulation techniques for proving properties of real-time systems. In *A Decade of Concurrency: Reflections and Perspectives*, Lecture Notes in Computer Science, pages 375–424, REX School/Symposium, Noordwijkerhout, the Netherlands, June 1993. Springer-Verlag.

[15] Nancy Lynch. Simulation techniques for proving properties of real-time systems. In Sang H. Son, editor, *Principles of Real-Time Systems.* Prentice Hall, 1994. To appear.

[16] Nancy Lynch, Isaac Saias, and Roberto Segala. Proving time bounds for randomized distributed algorithms. In *Thirteenth Annual ACM Symposium on the Principles of Distributed Computing*, Los Angeles, CA, August 1994.

[17] Nancy Lynch and Frits Vaandrager. Forward and backward simulations – Part II: Timing-based systems. Submitted for publication.

[18] Nancy Lynch and Frits Vaandrager. Forward and backward simulations for timing-based systems. In *Proceedings of REX Workshop "Real-Time: Theory in Practice"*, volume 600 of *Lecture Notes in Computer Science*, pages 397–446, Mook, The Netherlands, June 1991. Springer-Verlag.

[19] Nancy A. Lynch and Hagit Attiya. Using mappings to prove timing properties. *Distrib. Comput.*, 6:121–139, 1992.

[20] Michael Merritt, Francemary Modugno, and Mark R. Tuttle. Time constrained automata. In J. C. M. Baeten and J. F. Goote, editors, *CONCUR'91: 2nd International Conference on Concurrency Theory*, volume 527 of *Lecture Notes in Computer Science*, pages 408–423, Amsterdam, The Netherlands, August 1991. Springer-Verlag.

[21] Roberto Segala. PhD thesis, MIT Dept. of Electrical Engineering and Computer Science, 1993. In progress.

[22] Roberto Segala and Nancy Lynch. Probabilistic simulations for probabilistic processes. In *Proceedings of the 5th International Conference on Concurrency Theory - CONCUR'94*, Lecture Notes in Computer Science, Uppsala, Sweden, August 1994. Springer-Verlag.

PART ONE

New Techniques

1

Semantic-Oriented Description Techniques for Network Management Information Modeling

Deh-Min Wu

Technical University of Berlin, Institute of Open Communication Systems,
Department of Computer Science, ZAZ 520, Hardenbergplatz 2, 10623 Berlin, FR Germany,
e-mail: wu@fokus.gmd.de

Abstract

This paper proposes a semantic-oriented approach to classifying network management information so that the formal description of network states can be supported. We describe the classification concept which is slightly different from the approaches like *rule-based* or specification language *object-Z*. We apply the situation classification capability extended from those found in *KL-ONE* or *Description Logic*, in order to organize the state and event for network management in a hierarchical structure. This approach is applied in order to *formally describe information* for network management purposes.

Keyword Codes: K6.4; F4.3; H1.1

Keywords: System Management; Formal Languages; Systems and Information Theory

1. INTRODUCTION

Today the integrated network is getting larger and larger, causing the information overhead to increase and hence making common understanding more difficult than before. One example of this sort of problem is specifying the network state and events. Without any help of formal description techniques, it is difficult to understand the real state of the network, and to enable an efficient deduction of network management actions. Therefore, the first step in managing a network is to specify these monitoring data for the network manager.

Classification of information is applied to determine whether management information possesses particular characteristics so that the specification of information corresponding to these properties can be carried out. The function of ***defining an information model*** is to abstract data so that the semantics of raw data from network resources is presented to the manager. The trend for the management of future broadband networks is object-oriented. Therefore, our concept of information classification has integrated object-oriented analysis with *term classification facilities* [1], like those found in the knowledge representation language *KL-ONE* and its successor *description logics*.

A general question is how to model world information. Different proposals have been made in the last decade for solving this problem, e.g. Chen's *entity-relationship* model, Sowa's *conceptual graph*, and *object-oriented* concepts [6,7,8]. For distributed applications like network management we prefer the ideas set out in situation theory, which is based on *cognitive science* and *decentralized artificial intelligence* [2,3,4,5], for the following reasons:

- A situation is the partial information that is always combined with spatial, temporal data. Network events, which are recognized and reported by agents, are passed onto the network manager with their location of occurrence and sequence.

- A situation provides information about system configuration and state changes; a situation is applied as the discrete information exchanged by intelligent autonomous agents to describe the network occurrence and its environment in order to derive an action.

The most important differences between the approach as INFOMOD information exchange for open distributed processing [9] and our approach is that our analysis attempts to support complex information structures and consistent semantics. As will be discussed later, the structure of information modeled to represent a network situation is similar to the relational model. In such a model the relationships between network elements are specified by the participants associated in a relationship. Our approach differs from *Object-Z* [10] in that a situation is a description and not a schema definition. A situation may contain an individual, a variable or another situation within it. These characteristics make our proposal more flexible with complex description cases.

The application of situations to represent the network configuration is proposed and their classification is discussed. Our approach is designed for descriptive management information, where information has to be inductively defined to ensure a consistent interpretation. Such an approach is suitable for defining new specific concepts by more generic primitive concepts like those in [1]. The objectives and also the major advantages of our proposal include:

- Situation classification provides a complex structural description. We specify a network situation by a relation and objects, e.g., a relation and its participants are a data structure to be classified. Suppose a certain network configuration is represented by a concrete real situation. Then we can apply our classification algorithm to find out its situation type.

- In addition, our proposal specifies information semantically based on the embedded situational information inductively defined by more general information. Each parameter in a situation is translated to its most generic definition and these are then compared. This characteristic allows the information exchange to be achieved with a common understanding of most generic, primitive definitions, but with no obligation to the specific classes.

2. SITUATION DESCRIPTION

In the management information models all network resources are abstracted as objects and the relation between them is modeled by a relationship. We extend the understanding of object classes as concept type in order to apply the object classes and instances to represent the concepts and individuals in a situation. In description logic, concepts are assertions or descriptions about the state of the world. Individuals are particular instantiations of concepts, and roles provide a way to relate individuals. We accordingly obtain a mapping between class and concept, instance and individual, attribute and role.

Based on this management information model, we apply the situation theory of Barwise and Perry [4,5] and Devlin [3] to describe network states and events. In a situation the world is viewed as a collection of objects, properties, and relations. An ***infon*** is a unit (discrete) information item, and a ***situation*** is modeled as a set of infons f_i. An infon is defined as an associated object set, formally represented as a tuple of $f_i = <r_i, o_{i1}, o_{i2},..., o_{in}, p_i>$, where r_i is n-ary relationship, o_{ij} are objects, and p_i is the polarity of the infon. If the polarity is 1, then the infon is a positive state of affairs. The value 0 identifies it as negative. A situation is represented as a set of infons $\{f_1, f_2,..., f_m\}$.

A composite situation description like

> ***Situation A:*** "A line has cable failure at 8 p.m." ⇔ *<has, line, cable failure,* l_a*, 20, 1>*

it describes a real situation happened in network at 8 p.m.. We use the letters l_a for an arbitrary location. This situational information is exchanged between agents and managers which classify this information into defined situation types, which are known or understood in a particular management domain. For instance,

> ***Situation B:*** "A connection has at least one failure at night" ⇔ *<has, connection, <atleast, 1, failures>,* l_b*, night, 1>*

is a defined situation type. We will say that situation *B* subsumes situation *A*, because a cable or a line is a physical element of a connection, and 8 p.m. is intuitively defined as time at night. In this example, the real situation *A* is categorized by a network manager with the situation class *B*.

The description logic uses concepts as the description of a data structure. This language is a set of terms built up from a set of primitive concepts and roles, using the operators *and, or, not, at-least, at-most, all* to construct more specific terms. A concept is used as an object class and a role is as a binary relationship. For instance, all terms used to represent a situation can be modeled by more generic terms in description logic in order to define their semantics.

The technique underlying it is term subsumption. This is the capability to recognize whether a term is more special than another one. This is simply the task of a network manager to classify information so that descriptive information acquires their semantics class. The situation subsumption is therefore used to classify situation descriptions and to specify them with subsumption relations.

The classification problem of a situation can be supported by the introduction of the subsumption. A is subsumed by B if the extension of A to the real world domain is contained in B. The notation $\subset$ is used for set inclusion. Mathematically, an extension function ξ is defined as

Definition 1: ξ : *description class* $\rightarrow$ *instances*

$\xi(A) \subset \xi(B)$, *for all subordinate description class A of B*

A partial order subsumption relation $\leq$ is defined as a binary relation between situation classes *A and B*:

Definition 2: $A \leq B$, *iff* $\xi(A) \subset \xi(B)$

The problem of network management is to determine the class of a situation description, e.g. to test whether the received information can be classified by defined situation classes or not. The proposed subsumption algorithm applied to solve this problem is recursively defined to decompose a situation description. Since a situation is a set of infons, this algorithm tests all infon subsumption between two situations. In turn, the subsumption for each object/relation in infons are tested for the comparison of infons. Finally, the algorithm checks whether an inheritance specialization relation exists between two classes or not.

3. MANAGED RELATIONSHIPS AND OBJECTS

Object-oriented modeling is employed for situation description while inheriting the characteristics of the conceptual model. If objects are used to represent a concept such as a failure or a connection, we call them *conceptual managed objects* (CMO). The classification problem of these conceptual objects is similar to the term subsumption in description logic.

As defined previously, an infon consists of a relation and objects. The situation subsumption is not solved if the relation is not included into the subsumption algorithm. For instance, the relation "has" in our previous example could have the same semantics as "own" or "get". For this reason, it is necessary to define a class to group all relations with the same semantics. According to relationship classes, we can apply specialization to declare more specific relationship classes. In this manner, relation is an entity with roles referring to other objects. We call such relationships *conceptual relationship objects* (CRO).

Any CROs and CMOs used in a situation are inductively defined information to ensure a consistent interpretation. Such an approach is suitable for defining new specific concepts by more generic primitive concepts. If a relationship is added to the *Guidelines for the Definition of Managed Objects* (GDMO), a relationship in a situation is defined as an n-ary association which has to be mapped to binary relationships allowed in description logics. That is, a relationship is referred to as an entity containing referential attributes.

A relationship is specified by a relationship class, and a relationship referring to an object/ properties with roles associates with other objects/properties. An attribute in a CMO definition is replaced by a role in that of a relationship. Applying the relationship as a special type of MO, a situation can be defined and classified for exchanging in a distributed environment. As is at present conceived by a joint CCITT-ISO group [7], a template for the structure of a relationship class may be as follows:

<relationship-class-label> RELATIONSHIP CLASS
[DERIVED FROM <relationship-class-label>
[,<relationship-class-label>];]*
[BEHAVIOUR <behaviour-definition-label>
[,<behaviour-definition-label>];]*
[ROLE <role-label> role-properties
*[REGISTERED AS object-identifier];] **
REGISTERED AS object-identifier;

To identify CMO and CRO in different open management systems with common understanding of their values, their classes are specified and managed in inheritance, containment and registration trees. Inheritance helps us derive new object classes from an existing one, and a containment tree declares object dependency/containing relations. A registration tree provides a global unique object identifier for naming an object in open systems. We identify a CMO/CRO by $i{::}\ C$, where C is the class of an object i. We apply $*{::}C$ (or simply C) to denote an arbitrary element in class C.

4. Subsumption Algorithm

We apply set theory for the extension function in the description logic to the definition of the subsumption rules. We use the letters C to denote a CMO/CRO class (a set), f an infon (information unit) and i an object instance (a set element). Before we introduce subsumption, we define semantics equivalence for class/instance of conceptual objects:

Definition 3: Two classes C, C' are equivalent ($C \cong C'$) iff $\xi(C) = \xi(C')$.

Definition 4: If two instances $i{::}C$ and $i'{::}C'$ have the same instance values, and $\xi(C) \subset \xi(C')$ or $\xi(C') \subset \xi(C)$, then i and i' are equivalent ($i \cong i'$).

To check if two object instances are semantically equivalent, we only need to compare the values involved in these instances. This can be demonstrated as the case of having information sent twice to the manager. How should we define a case where two situation descriptions are semantically equivalent for network managers so that duplicated events for one state change can be ignored. From these two definitions we have:

Proposition 1: CMO/CRO $O = *{::}\ C$ and $O' = *{::}\ C'$, O and O' are equivalent iff $C \cong C'$. * is a wild-card.

Proposition 2: CMO/CRO $O = i{::}\ C$ and $O' = i'{::}\ C'$, O and O' are equivalent iff $i \cong i'$.

The subsumption of infons will be carried out by the following steps. Two relation notations are used:

- relation $X :< Y$, where X and Y are classes, is true when class X is a direct subordinate class of Y in an inheritance hierarchy.
- relation $i := Y$ is true when i is an instance of class Y.

We apply the following propositions to subsume conceptual objects C, C' and C''. The intersection operation is denoted as $\cap$ and implication as $\Rightarrow$.

Proposition 3: $C :< C' \Rightarrow C \leq C'$

Proof: Because C is derived from the class C', the extension of C in the real world is also an extension of C', i.e. $\xi(C) \subset \xi(C')$.

Proposition 4: $C :< C'$ *and* $C' \leq C'' \Rightarrow C \leq C''$

Proof: Applying Proposition *3*, $C :< C' \Rightarrow C \leq C'$. When we apply the transitive property of the relation $\leq$, then $C \leq C'$ and $C' \leq C''$ implies $C \leq C''$.

Proposition 5: $i := C$ *and* $C \leq C' \Rightarrow i := C'$

Proof: $i := C \Rightarrow i \in \xi(C)$,

$C \leq C' \Rightarrow \xi(C) \subset \xi(C')$, i.e., $i \in \xi(C')$

Proposition 6: $C \leq C'$ *or* $C \cong C' \Rightarrow i{::}\ C \subset *{::}\ C'$, where * denotes a wild-card variable.

Proof: Suppose $\exists\ i \in \xi(C)$, we know that $\xi(C) \subset \xi(C')$

that implies $i \in \xi(C')$

Proposition 7: $C \leq C' \Rightarrow *{::}\ C \subset *{::}\ C'$

Proof: Applying proposition *5*, $\forall\ i \in \xi(C)$ and $\xi(C) \subset \xi(C') \Rightarrow i \in \xi(C')$

In our proposed subsumption algorithm, the set operations intersection and union are applied to support the description operators such as *and, or, atleast,* etc. Its theoretical background is outside the scope of this paper and can be referenced in object-centered modeling such as [1]. Our proposal applies the subsumption concept to the situation classification. Notice that the classification of a situation is based on the comparison of set inclusion between the extension of two descriptions.

5. SITUATION SUBSUMPTION

We apply the propositions described in the previous section for the subsumption propositions for CMO/CRO. Suppose given CRO a_1, a_1' and CMO a_i, a_i' associated in infons f and f', we know that there is an equivalence or a subsumption relation between f and f' if one of the following conditions is satisfied:

Proposition 8: For infons, $f = <a_1, a_2, ..., a_n>$ and $f' = <a_1', a_2', ..., a_n'>$, $f \cong f'$ iff $a_i \cong a_i'$, where $i = 1, ..., n$.

Proof: Two infons are equivalent when all their contained CMOs/CROs are equivalent.

For instance, a controller C_i reports that the observation *situation* A is detected, which is given in section 2. Then the composite situation C

Situation C:: *"Controller C_i observes situation A at 8 p.m."* $\Leftrightarrow$ *<observes, C_i::Controller, A::Situation, l_c, 20, 1)*

is equivalent to the situation D, iff the situation A is replaced by its definition.

Situation D: *"Controller C_i observes "A line has cable failure at 8 p.m." "* $\Leftrightarrow$ *<observes, C_i::Controller, <has, line, cable failure, l_a, 20, 1>, l_c, 20, 1)*

Proposition 9: For infons, $f = <a_1, a_2, ..., a_n>$ and $f' = <a_1', a_2', ..., a_n'>$, $f \leq f'$ iff there exists at least one i with $a_i \leq a_i'$ and for all others $a_j \cong a_j'$, where $j = 1, ..., n, j \neq i$.

Proof: Infon f is more special than infon f' if the relation and all object participants in f are either (at least once) more special than or equivalent to them in f'.

Proposition 10: For infons, $f = <a_1, ..., a_i, b_1, b_m, a_{i+1}, ..., a_n>$ *and* $f' = <a_1', ..., a_i', a_{i+1}' ..., a_n'>$, $f \leq f'$ iff either $a_i \leq a_i'$ *or* $a_i \cong a_i'$, where $1 \leq i \leq n$, and b_j arbitrary objects, $1 \leq j \leq m$.

Proof: Infon f is more special than infon f' if the n objects $a_1, a_2, ..., a_n$ in infon f are more special than or equivalent to that part $a_1', a_2', ..., a_n'$ of f', and the roles participants in f, i.e. $b_1, ..., b_m$, are additional restrictions to f.

Now we have to make some propositions normally allowed in situation theory, but conventionally not permitted in the *object-oriented model* and *object-Z*. Infons may involve infons as their parameters, e.g. a conceptual object may occur buried in one or more levels inside an infon.

Proposition 11: For infons f and f', $f \leq f'$ iff $f = <a_1, a_2, ..., f', ..., a_n>$

Proof: An infon f' is more general than an infon f when f' is contained in f.

In *Proposition 11*, f' is an embedded situation which is specified by an indirect description, as the example in C that $C \leq A$ satisfies.

Proposition 12: For infons $f=<a_1,a_2,..., a_n, a_{n+1}, ..., a_m>$ *and* $f' = <a_1',a_2',...,a_n',s>$, and a situation $s = \{f_1, ..., f_k\}$, $f' \leq f$ iff both of the following are satisfied:

- $\forall\, a_i, 1 \leq i \leq n$, either $a_i' \leq a_i$ *or* $a_i \cong a_i'$
- $\forall a_i, n+1 \leq i \leq m, \exists f_j = <...,b_l, ...>$ that $b_l \cong a_i$ or $b_l \leq a_i$, where $1 \leq j \leq k$

Proof: The first condition checks the subsumption relation of their n objects $a_1,a_2,..., a_n$ according to Proposition *10*. The second condition ensures that all the remaining objects in f are contained in any infons of s, i.e. for all CMOs in f, we always find a more specific CMO in s corresponding to it.

To show an application to the *proposition 12* a situation E is given,

Situation E: *"A controller observes a failure at 8 p.m."* $\Leftrightarrow$ *<observes, *::Controller, *::failure,* l_c*, 20, 1>*

The situation E subsumes the situation C as situation C specifies a more complete description than that in E. The situation E only states that a failure is observed but the situation C specifies exactly where this failure is manifested. We further give the following propositions for situation subsumption, however, they are only used to ensure that all situation infons are subsumed by infons in another situation:

Proposition 13: $S = \{f_1,..., f_n\}$ and $S' = \{f_1',..., f_n'\}$, $S \cong S'$ iff $\forall f_i$ in S, $\exists f_j'$ in S' that satisfies $f_i \cong f_j'$, and vice versa.

Proposition 14: $S = \{f_1,..., f_n\}$ and $S' = \{f_1', ..., f_m'\}$, $S \leq S'$ iff $\forall f_j'$ in S', $\exists f_i$ in S that satisfies either $f_i \cong f_j'$ or $f_i \leq f_j'$

Proof: We apply the inductive approach to prove this proposition. In the case of m=1: Suppose $S' = \{f_1'\}$ and $\exists f_l$ in S that satisfies either $f_l \cong f_1'$ or $f_l \leq f_1'$, then $S=\{..., f_l, ...\} \leq \{..., f_1,...\} \leq \{f_1\} = S'$ is satisfied.

In the case of $m=k$-1, *Proposition 14* satisfies: $S \leq \{f_1', ..., f_{k-1}'\} = S'$. We try to prove whether $m=k$ is also satisfied. That is to prove $S \leq S'' = \{f_1', ..., f_{k-1}', f_k'\}$. As the pre-condition of *Proposition 14,* for f_k' in S'' we find a f_l in S that $f_l \leq f_k'$, and $S=\{f_1,\ldots,f_l,\ldots,f_n\} \leq \{f_1', \ldots, f_{k-1}', f_l\} \leq \{f_1', \ldots, f_{k-1}', f_k'\} = S''$.

By induction, for all $k=1,\ldots,m$ that *Proposition 14* satisfies.

Notice that the following case is in general not decidable: Suppose there is one infon f_k' in S', and we cannot find any f_l in S that $f_l \leq f_k'$ *or* $f_l \cong f_k'$, then $S \leq S'$ cannot be determined automatically in our study. As an example of the above propositions, we give two situations $F = \{A, C\}$, $G = \{B, E\}$, i.e.

Situation F: A=*<has, line, cable failure,* l_a*, 20, 1> and C=<observes,* C_i*::Controller, A::Situation,* l_c*, 20, 1>*

Situation G: B=*<has, connection, <atleast, 1, failures>,* l_b*, night, 1> and E=<observes, *::Controller, *::Failure,* l_c*, 20, 1>*

To recognize if the situation G has a more general class than that of situation F, we check whether G subsumes F. This means proving the fact that both infons B and E in G subsume one infon in F. Since $A \leq B$ *and* $C \leq E$, hence the situation G subsumes the situation F is obtained.

Suppose the domain of all situations in the real world is the set SIT, then the situations S and S' are elements from this domain. As defined in the previous section, situation S is subsumed by S' iff $S \leq S'$. We define a function SSUB for situation subsumption:

Definition 5: SSUB: SIT $\times$ SIT $\rightarrow$ { *true, false* }

$$\text{SSUB}(S, S') = \begin{cases} \text{true, iff } S \leq S', \\ \text{false, } S \nleq S' \end{cases}$$

If a real situation S is subsumed by a situation class S', then S' contains more generic information than S or S is more specific. We can specify all descriptions for network states and events using situation classes. In our previous example, a real situation C is subsumed by a more general situation E. The type of a real situation description is then classified, e.g. a real situation C has the situation description class E.

Suppose a real situation S is classified by classes in a situation hierarchy, then more general situation classes, say σ, are found as classes of this real situation description S. The most specific one in σ is defined as the (most specific) situation class of S.

In other words, all descriptions subsumed by a situation description class are defined as real situations for that class. In this way, the network information is organized by the situation subsumption function SSUB, i.e. the proposed algorithm provided by the SSUB function compares the extensions of two given situations. All description classes are structured into a hierarchy according to their subsumption relations.

The ISO/OSI structure of management information employs an object-oriented concept to model managed information. According to this object-oriented concept, classification of information is the pre-condition to define a model for information exchange in open systems. If the network situation descriptions have to be included into CMIP (*Common Management Information Protocol* [12]), we have to provide a classification method for organizing them in an object-oriented approach. In this case, the proposed classification of situation descriptions using the subsumption function SSUB proposed in this paper can be applied to structure description information for CMIP.

6. EVALUATION OF APPLICATIONS

This proposal begins with the conventional approach GDMO used by CMIP and tries to integrate description facilities into it. Situation theory and description logics which are mathematically well-founded are applied to support the proposal. The situation theory abstracts the state of affairs with mathematical formula. We apply the description logics to describe a situation so that situation theory can be represented in CMOs and CROs. That is to say, the realization of situation theory is solved by description logic.

The propositions provide the subsumption algorithms for the situation classification. Because a situation is described by recursive composition of CMOs and CROs, it makes the classification more difficult than that of CMOs alone. As described, a relationship is formally specified with a CMO-like structure and the classification of CROs is also defined. Therefore, the CRO can be integrated into GDMO and the similarity of their structure simplifies the classification of situations.

For open information modeling the classification of information is the most basic pre-condition. The situation theory has still no open information model for its application in network management. This proposal supports us to construct one for situation theory. Because CRO may be considered as an extension to GDMO, and the classification of composition of CMOs and CROs can be clarified by the subsumption algorithms, hence the classification of information in situation description is achieved. The model of situational information is therefore constructed for the pre-condition of information exchange in open environment.

For problem solving using current GDMO, we may solve a problem in the framework of an object-oriented method. For instance, suppose an object is defective, the handling of this fault has to be stored in the class information of this object. A current GDMO solution is to specify the behavior and actions in text for the handling of managed objects. No formal description techniques are applied for the specification of such information. The major benefits of such situational modelling for a managing application is to support the definition of the behavior and actions. One possible approach is to specify the behavior by a rule.

As an example for a behavior rule we may specify it formally as:

Situation H: *"if the situation H and situation A are true, then apply action T"* ⇔ *<then, <and, H::Situation, A::Situation,* l_h, t_h*::Time, 1>, <apply, managing system, T::Action,* l_h, t_h*::Time, 1>,* l_h, **::Time, 1>*

Action T: *"go step* S_1 *and* S_2*, if successful, then go step* S_5*, else go step* S_9*"* ⇔*<go, <then, <are, <go,* S_1*::Step,* S_2*::Step,* l_t, t_t*, 1>, successful,* l_t, t_t*, 1>,* S_5*::Step,* S_9*::Step,* l_t, t_t*, 1>,* l_t, t_t*, 1>*

7. SCENARIO EXAMPLES AND IMPLEMENTATION

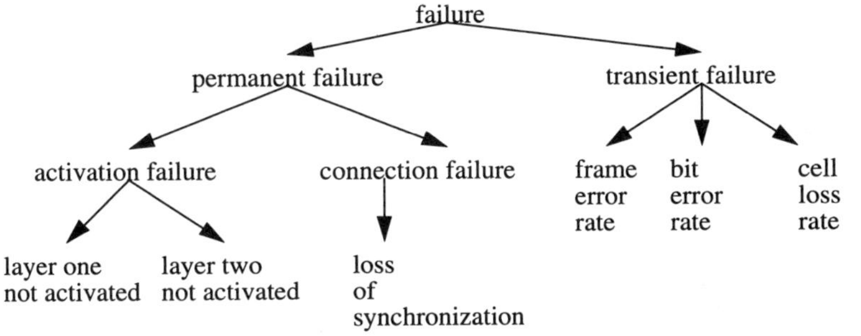

Figure 1 : CMO class inheritance hierarchy for failures

In this section we depict a part of our management information model as an example to illustrate how this proposal is applied in practice to specify monitoring information for network management. As described, we model the network elements as CMOs. CMO classes specified for network failures are illustrated in Figure 1, and for network elements and their configurations are depicted in Figure 2. In these two figures, a subordinate node is a more specific CMO class than its superior.

The CROs are also modeled like CMOs. The relations representing a situation are structured in an inheritance hierarchy. For instance, the relation class *has-parameter* is more specific than *has*. After defining these conceptual objects, we derive a situation hierarchy according to the subsumption function as depicted in Figure 4. For better understanding, we describe the situations in Figure 4 in the text. Each subordinate situation is subsumed by its superior. For example, the situation descriptions $S_{1,1,1}$ and $S_{1,1,2}$ are represented formally as

$S_{1,1,1}$ = *<has,*
networkElement,
failure>

$S_{1,1,2}$ = *<has,*
<hasParameter, subscriberLine,
<is, lineOperationalState, X> and
<is, lineAdministrativeState, Y>>,
activationFailure>

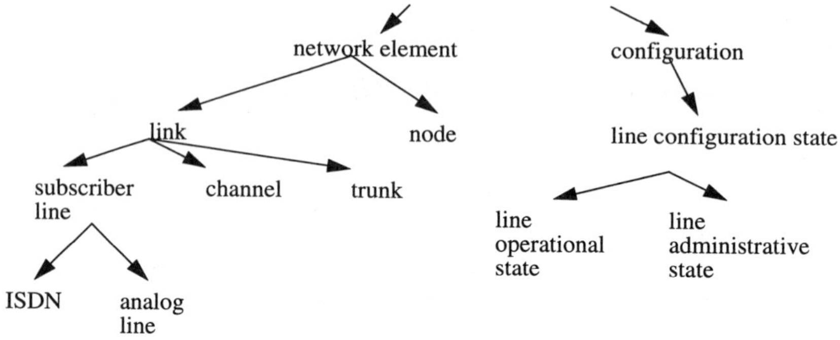

Figure 2 : CMO class inheritance hierarchy for network physical elements and logical parameters

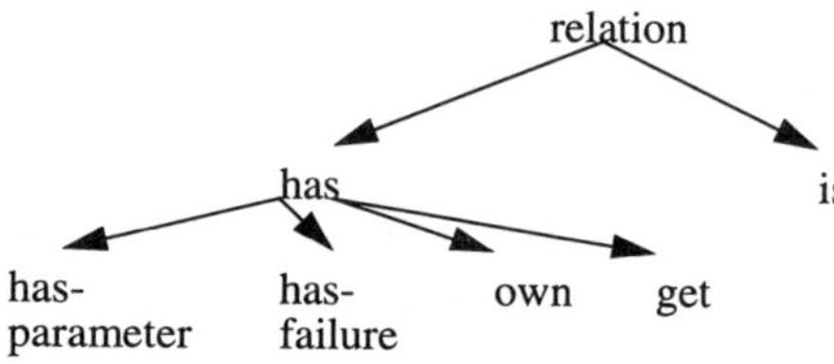

Figure 3 : CRO inheritance hierarchy

In addition, we apply SSUB to classify a network real situation. For example in Figure 4 a real situation $S_{1,1,2}$ is subsumed by situation $S_{1,1,1}$, i.e. $S_{1,1,2} \leq S_{1,1,1}$. The SSUB tests if the *network element* class and *failure* class in $S_{1,1,2}$ are more specific than those in $S_{1,1,1}$. The following subsumption relationships are proved in order to reason the validity of $S_{1,1,2} \leq S_{1,1,1}$.

subscriberLine $\leq$*networkElement.*

<hasParameter, subscriberLine, lineConfigurationState>
$\leq$ *subscriberLine.*
<is, lineOperationalState, X> and <is, lineAdministrativeState, Y>
$\leq$ *lineConfigurationState.*
activationFailure $\leq$ *failure.*

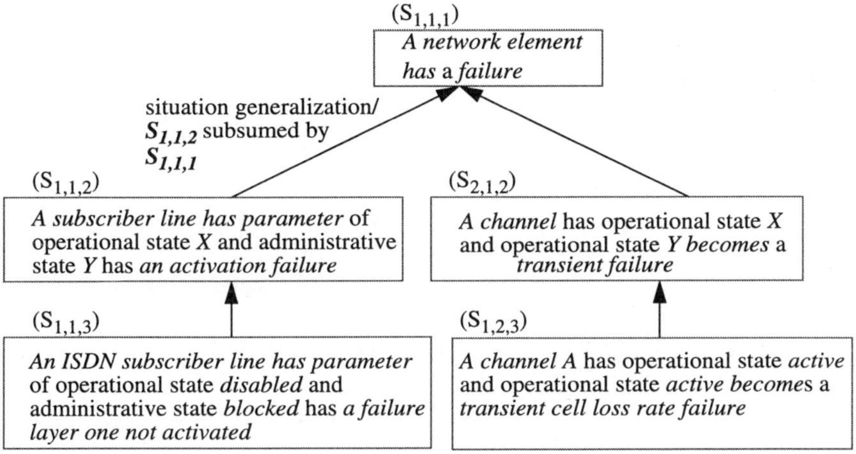

Figure 4 : An example in a digital switching system for situation hierarchy organized with the help of the subsumption function

We have used the tool *DAMOCLES* [13], which is being developed by GMD Fokus within the *Berkom* project, to maintain all conceptual object classes. *DAMOCLES* enables us to define and to administer managed object classes interactively. It can be used to check the syntactical correctness and the consistency of an information model according to the *GDMO* [11]. For the subsumption of conceptual objects, all managed objects defined in *DAMOCLES* are translated into concept definitions in the *BACK* system [14], which is developed at the TU Berlin within the framework of an *ESPRIT* project for concept management. *BACK* is based on *description logic*, which is realized in Prolog and supports the concept subsumption facilities.

8. CONCLUSION

This paper describes an extended definition of the term subsumption capability in order to describe network situations formally. This proposal puts forward the situational modeling realized by [15] so that these subsumption propositions are suitable for situational description in an open environment. The most important characteristic of our proposal is: Situation classification supports structuring of situation hierarchy. In addition, our proposal specifies information semantically, based on the embedded situational information recursively defined by a more general description.

REFERENCES

1. Bernhard Nebel, "Reasoning and Revision in Hybrid Representation Systems", Lecture Notes in Artificial Intelligence 422, Edited by J. Siekmann, Springer-Verlag, 1990
2. Eric Werner, "Distributed Cooperation Algorithms", Decentralized A.I. Editors Y. Demazeau J.-P. Müller, North-Holland, 1990
3. Devlin, Keith J., "Logic and Information", Cambridge University Press, 1991
4. Jon Barwise, John Perry, "Situationen und Einstellungen: Grundlagen d. Situationssemantik", de Gruyter, 1987
5. Jon Barwise, "The Situation in Logic", CSLI, Lecture Notes 17, 1989
6. Peter Pin-Shan Chen, "The Entity-Relationship Model - Toward a Unified View of Data", ACM Transactions on Database Systems, Vol. 1, No. 1, March 1976, pp. 9-36
7. ISO/IEC N7126, "General Relationship Model - Third Working Draft", ISO/IEC JTC 1/SC 21/WG 4 Meeting, Ottawa, May 1992
8. J. F. Sowa, "Conceptual Structures - Information Processing in Mind and Machine" Reading/MA: Addison Wesley, 1984
9. J.J. van Griethuysen, "Enterprise modelling, a necessary basis for modern information systems", MMJ Information Engineering, Casteren - The Netherlands, Convener ISO/IEC JTC1/SC21/WG7 - ODP
10. David Carrington, David Duke, Roger Duke, Paul King, Gordon Rose, and Graeme Smith, "Object-Z: An Object-Oriented Extention to Z", Formal Description Techniques, II, Ed. S.T. Vuong, Elsevier, IFIP, 1990
11. ISO/IS 10165-4, "Information processing systems - Open systems interconnection (OSI) - Structure of Management Information", Part 4: Guidelines for the Definition of Managed Objects, May 1992
12. Information Processing Systems - Open Systems Interconnection - Common Management Information Protocol - Part 1', 1990
13. Marcus Wittig, Martin Pfeiler, "A tool supporting the Management Information Modeling process", IFIP Transactions Integrated Network Management III, Editor H.-G. Hegering and Y. Yemini, pp739- 750, 1993
14. Thomas Hoppe, Carsten Kindermann, J. Joachim Quantz, Albrecht Schmiedel and Martin Fischer, "BACK V5: Tutorial & Manual", Technische Universität Berlin, Projekt KIT-BACK, March 1993
15. Deh-Min Wu, "TMN Performance Assurance using Situation Modelling on ATM-Networks", Technical Report 1993/19 of the Department of Computer Science, Technical University of Berlin, July 1993

2

A Service Request Description Language

Claudia Popien and Bernd Meyer

Aachen University of Technology, Department of Computer Science IV, Ahornstr. 55
D-52056 Aachen, GERMANY. E-mail: popien@informatik.rwth-aachen.de

Abstract

To meet the new requirements of growing computer networks the client-server model has to be improved by new concepts with encreased functionality. Therefore, a service management in distributed systems is realized by a service trading concept.

This paper presents a language given in Backus-Naur Form which enables specification of service requests. First, a matching criteria is specified as conditioned Boolean expression in order to evaluate appropriate service offers. Additionally, a selection criteria can be used to yield the most suitable service. To limit the scope of a search it is possible to specify policies and search constraints on service properties which restrict the search on particular trading objects. The Service Request Description Language presented here is related to the ANSAware trader to increase its functionality by giving means for service searching and selecting.

Keyword Codes: F.4.3, D.2.1, D.3.1
Keywords: ODP, Service Trading, SRDL

1. INTRODUCTION

The current development of information processing systems leads to global network of interconnected computers. Therefore, new services for managing resources of distributed systems and networks are needed. A growing number of components which offer services leads to a new situation which enforces an open service market. The problem of the client's need to select a server arises. Therefore, the client-server model is extended to a three-party model. A so-called trader enables binding between client and server at run time. To realize trading a service directory is used, which stores service types and their associated service property values. Different trading realizations are compared in [Ke 93].

In this paper, trading is considered according to the Reference Model of Open Distributed Processing (ODP) [ODP P1-P4], [Ste 90]. The tasks of the ODP trader are devided into domain and type management, respectively. Domain management supports finding a service, type management supports interface matching. It is the aim of this paper to improve trading concepts given in the ODP Reference Model by presenting matching criteria and search constraints in a language with formal syntax and semantics. In addition, the concept of a policy will be elaborated formally and applied to trading.

The following section introduces elementary ODP trading concepts. The notions of a service and its properties are explained. After defining the structure of a service offer, a query language for service import is defined in section three. Both, matching and selection criteria are given in Backus-Naur Form, accompanied by example queries. In order to restrict the search space or scope of an importing query, search constraints are defined in section four. In addition, the scope for imports of the requested trader has to be considered, which is defined by a trader search policy. A notation for specifying policies is given, too. Whereas the previous sections deal with syntactical constructs, their semantics is given in section five. It consists of four steps for import operations on a given service database. Section six is concerned with service management, especially with the evaluation of dynamic service properties. Finally, this paper concludes with a brief summary of contributions made and a view on future work.

2. SERVICE TRADING IN DISTRIBUTED SYSTEMS

A trader is an object that performs trading, primarily satisfying identification requirements [Gei 92]. In the following traders based on ODP are considered [ODP Tr], [MaBl 92]. A trader can be seen as an object from which another object can buy (import) its needs and sell (export) its services in a distributed environment, see Fig. 1.

In order to study the process of service trading in more detail some notions are introduced. A *service* is a function provided by an object at a computational interface. It is a set of capabilities available at an interface of an object. Every service is an instance of a *service type*. Associated with each service type is an *interface type*, which determines the computational behaviour. Interfaces of the same type provide the same functionality. However, instances of the same service type may differ in some noncomputational aspects. These additional aspects are called *service properties*.

For trading, service properties can be classified as static or dynamic, see Fig. 2. A *static service property* is a property that changes infrequently. The values are asserted by an exporter in a service offer and are stored in the *service directory* (SD) of the trader. A *dynamic service property* is a property whose values change much faster than the rate at which the trader is asked to perform a match based on that property value. Values of dynamic properties are obtained on demand by the trader.

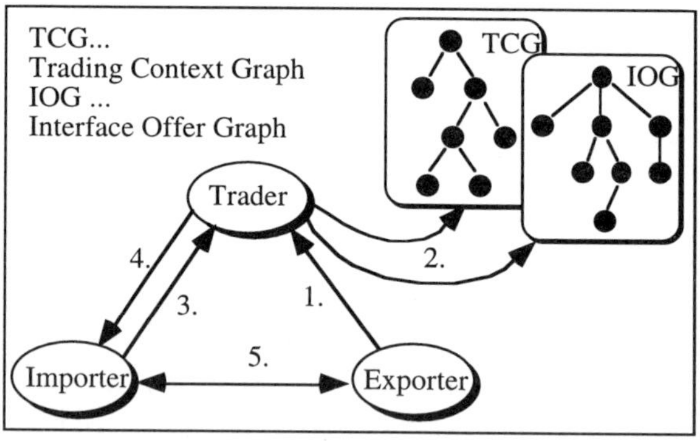

Fig. 1: The process of trading in ODP

A *service offer* describes a service that is being traded. It is an assertion made by an exporter about a service that is offered for use by other objects at a computational interface. In addition to service type and service property values, a service offer can also have *service offer property values*. Considering the trader in more detail, it performs two major functions: the *domain management function* and *type management function.*

The type management function realizes the management of the subtype relationship between types. The service type hierarchy is represented as a directed acyclic graph, called *Interface Offer Graph* (IOG), in which each node is a service type and each directed edge represents supertype to subtype relationships.

The domain management function manages the *service offer space*. The service offer space within a trader may be structured into sets called contexts which are defined by a containment relationship between contexts. A trader context structure can be represented as a directed acyclic graph, called *Trading Context Graph* (TCG), where nodes represent trading contexts and arcs represent a containment relationship. A service offer is a member of a trader context, and of any super context of that trader context. A trader service offer space can reflect administrative structures and organizations, or service types.

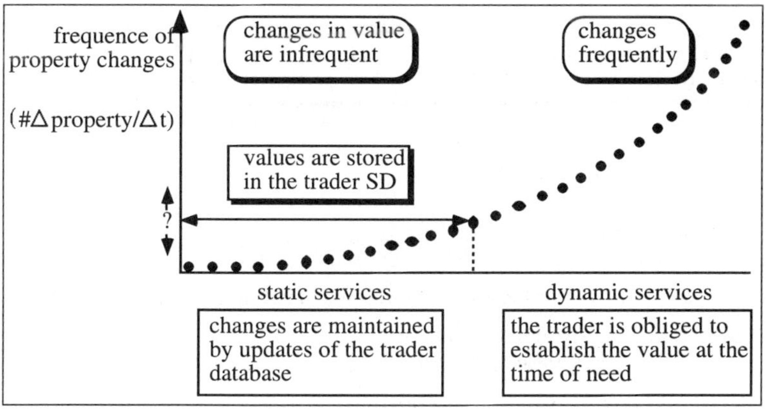

Fig. 2: Static vs. dynamic service properties

3. DESCRIBING MATCHING CRITERIA AND SELECTION CRITERIA IN SERVICE REQUESTS

Considering the description of a service offer, it is possible to derive a structure for service offers of an importer. A service offer consists of "Exporter Id", "Service Interface Id", "Service Type Description" and "Service Offer Properties", where a "Service Type Description" can be further refined into an "Interface Type Description" and a "Service Property Type Description". Service properties are "Static Service Properties" or "Dynamic Service Properties", respectively described as ordered pairs

<service property name, service property value>.

The structure of a service offer without service offer properties is described in Fig. 3.

The importer has two different functions which are used to find suitable services. These functions are `search` and `select`. Search is the behaviour which searches the service offers of a trader and returns a set of service offers which satisfy some matching criteria. Its inputs are the service type of the desired service and a *matching criteria*. The requirement specification comprises two parts:

- the required computation service behaviour, given by a service type description and
- the required properties, expressed by a matching criteria.

To perform a `search` operation, the matching criteria must be studied in more detail. A matching criteria is a set of rules which are applied to the total set of service offers in order to yield a smaller set of acceptable service offers. The result of a `search` operation is the set of service offers with the correct service type. These service offers also satisfy the matching constraint. In this sense, a matching criteria is a criteria on service properties against which the service offers are evaluated. The successful execution of this operation will return a list of interface identifiers, service properties and service offer properties.

Until now, no further recommendations have been given, in which way these functions should be performed. ANSAware [ANSA] only offers a reduced syntax for matching constraints. Therefore, a concept is presented that enables specifying parameters of these functions. In this concept, a matching criterion is specified as a Boolean expression. The criterion is specified in a Backus-Naur Form (BNF) as followes.

Exporter Identifier	Service Interface Identifier	Service Type Id	Static Service Properties	Dynamic Service Properties

Fig. 3: The detailed structure of a service offer

The language presented here, is called Service Request Description Language (SRDL). SRDL specifications given according to the BNF are called SRDL-texts.

service-request ::= <service_request_operation> <search-constraint> "**END SERVREQ**"

service_request_operation ::= <search_operation>
| <select_operation>

search-operation ::= "**SEARCH**" <service-type-identifier> "**WITH**" <conditional-matching-criteria>
select-operation ::= **"SELECT"** <service-type-identifier> "**WITH**" <conditional-selection-criteria>

Therefore, a conditional_matching_criteria is defined based on a basic_matching_criteria and a simple_matching_criteria:

basic-matching-criteria ::= <service-property-identifier> <general-relation> <real-number>
| <service-property-identifier> <equal-or-not-symbol> <service-property-value-identifier>
| <service-property-value-identifier> <element-or-not-symbol> <service-property-identifier>.
simple-matching-criteria ::= <basic-matching-criteria>
| "**NOT**" <simple-matching-criteria>
| <simple-matching-criteria> <and-or-or-symbol> <simple-matching-criteria>.
conditional-matching-criteria ::= <simple-matching-criteria>
| **"IF SUCCESS"** <conditional-matching-criteria> "**THEN**" <conditional-matching-criteria> "**ELSE**" <conditional-matching-criteria> "**END**"
| **"IF SUCCESS"** <conditional-matching-criteria> "**THEN**" "**TAKE THAT**" "**ELSE**" <conditional-matching-criteria> "**END**".

The used notions and identifiers are explained in the following:

letter ::= "a" | "b" | ... | "z" | "A" | "B" | ... | "Z".
digit ::= "0" | "1" | ... | "9".
normal-character ::= <letter> | <digit>.
identifier ::= <letter> [{<normal-character> | "_"} normal-character].
service-property-identifier ::= <identifier>
| <identifier> "/d".

service-property-value-identifier ::= <identifier>.

service-type-identifier ::= <identifier>.

real-number ::= {<digit>} <digit> ["." <digit> {<digit>}].
relation-symbol ::= "<" | "≤" | ">" | "≥".
equal-or-not-symbol ::= "=" | "≠".
general-relation ::= <relation-symbol> | <equal-or-not-symbol>.
element-or-not-symbol ::= "∈" | "∉".
and-or-or-symbol ::= "**AND**" | "**OR**".

As an example a SRDL-text is given so that all services of service type PRINTER with special static service properties "location" and "cost_per_page" and a special value for the dynamic service property "queue_length" should be searched which fulfill the following matching constraint:

SEARCH PRINTER **WITH**
IF SUCCESS location = CompCent **AND**
cost_per_page ≤ 0.10
THEN **TAKE THAT**
ELSE **IF SUCCESS** cost_per_page < 0.10 **AND** queue_length/d ≤ 3
THEN cost_per_page < 0.10 **AND** queue_length/d ≤ 5
ELSE location = CompCent **END** **END**

A matching criterion needs no specific values for all properties of an exported service offer. The non-specified properties are essentially "don't care" values and are "true" when matched with any values for those properties in an exported service offer.

The `select` function searches a set of service offers and returns a single service offer which satisfies some selection criteria. It chooses the most suitable service offer that satisfies the importer's matching and selection criteria. This operation can be regarded as a simple case of a search behaviour.

To study the selection function it is necessary to consider *selection criteria*. A selection criterion is a set of rules which are applied to the total set of service offers to yield a single service offer. It is a criterion which will be applied to select one service offer from this set of service offers resulting from the matching criteria. Based on the Backus-Naur Form the selection criteria can be specified as follows.

select-operation ::= **"SELECT"** <service-type-identifier> "**WITH**" <conditional-selection-criteria>

Therefore, a conditional_selecion_criteria is defined based on a basic_selection_criteria and a simple_selecion_criteria:

basic-selection-criteria ::= <min-or-max-symbol> "(" <service-property-identifier> ")"
| <choice-symbol> "(" <conditional-matching-criteria> ")".
simple-selection-criteria ::= <basic-selection-criteria> {"**AND**" <simple-matching-criteria>}.
conditional-selection-criteria ::= <simple-selection-criteria>
| "**IF SUCCESS**" <conditional-selection-criteria>
"**THEN**"<conditional-selection-criteria> "**ELSE**"
<conditional-selection-criteria> "**END**"
| "**IF SUCCESS**" <conditional-selection-criteria>
"**THEN**" "**TAKE THAT**" "**ELSE**"
<conditional-selection-criteria> "**END**".

The additionally used notions and identifiers are explained in the following:

min-or-max-symbol ::= "**MINIMUM**" | "**MAXIMUM**".
choice-symbol ::= "**FIRST**" | "**LAST**" | "**RANDOM**".

An example for selecting a service of type PRINTER is given in the following:

SELECT PRINTER **WITH**
IF SUCCESS
RANDOM (type = Laser **AND** cost_per_page < 0.10 **AND** A3 ∈ paper_size)
THEN FIRST (**MINIMUM** (cost_per_page) **AND** type = Laser **AND** A3 ∈ paper_size)
ELSE MINIMUM (cost_per_page) **AND** A3 ∈ paper_size **END**

This SRDL language to describe matching and selection criteria presented here, has a formal semantics. By mapping Boolean expressions onto the values "true" and "false" it is possible to get corresponding services.

4. CONSIDERING POLICIES AND SEARCH CONSTRAINTS

A search constraint specifies a restriction for matching import requests. This object is a constraint on service and link properties that result in the set of traders in which the search is performed. A search constraint includes criteria for limiting the scope of a search and focusing a search to particular trader objects. When evaluated at a particular starting trader object, a search constraint may be resolved to a set of trader objects. If a local trader fails to return a service offer to its clients it can establish an import contract with a remote trader. This trader acts as exporting trader and has to establish an exporting contract, see Fig. 4.

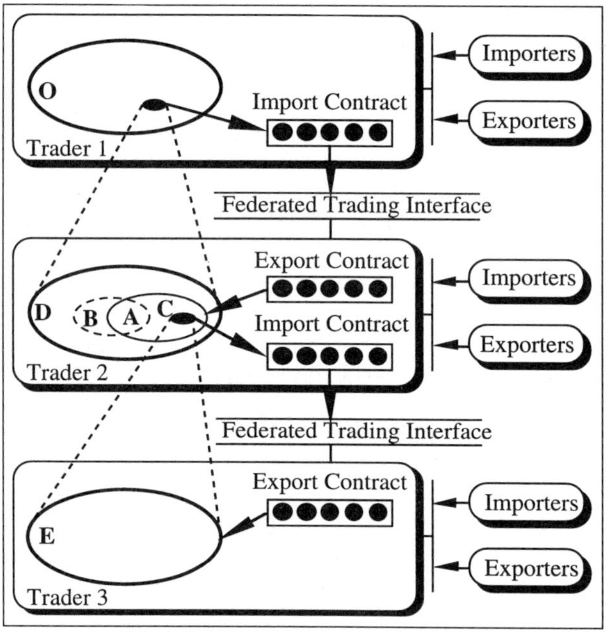

Fig. 4: A search path within a trading federation

The new system is called *trading federation* [BeRa 91], [PoMe 93]. The *trading offer domain* (TOD) is extended. It contains the set of all service offers that describe services available to objects in a federated trading offer domain, including those service offers that have been made available through federation.

Links to federated operations' service offers are set up in the appropriate search spaces of the importing trader when an import contract is established. When a trader receives a search request, it looks up its database to find suitable exports by accessing its relevant search spaces. If one of the entries in the search space is a link to an import contract, this contract is used to generate an equivalent federated operation, i.e. a federated-search, using the information in this import contract. On the other hand, when an exporting trader receives a federated search-request, it associates the search with an export contract identified by the requesting trader. The search is restricted to the search spaces of the exporting trader specified in the export contract. To describe the restricted search space, a context relative naming model is used in the following. Each name is a path which describes the route to a particular context.

4.1 Defining Search Constraints

Each context can have multiple names provided that the names are unique. In a context relative naming model, unique names are guaranteed if each edge in the graph has a unique label relative to the source of the edge.

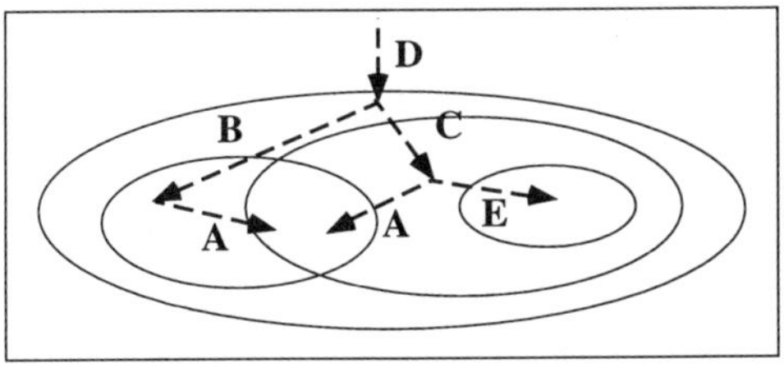

Fig. 5: An example for a context-relative naming of a context structure

In our example in Fig. 5, trading context A is named both "/D/B/A" and "/D/C/A". It is not necessary to assign the label "A" to both of the edges (TC-B, TC-A) and (TC-C, TC-A). The edge (TC-B, TC-A) could be given any label while the edge (TC-C, TC-A) could be given any label that is different from the label of the edge (TC-C, TC-E). The subset relationships can be derived from the names. The context named "/D/B/A" is a subset of the context named "/D/B", which is a subset of the context named "/D".

The search constraint can be described as a search path in the context structure. The BNF given before is extended to:

search-constraint ::= **"IN"** <search-expression>.
search-expression ::= "/" <trading-context> {"/" <trading-context>}.
trading-context ::= <identifier>.

As an example, a service offer should be searched in trading context E. This search constraint is specified as followes.

SELECT PRINTER **WITH**
MINIMUM (queue_length/d) **AND** cost_per_page < 0.10 **AND** location = CompCent
IN /D/C/E **END SERVREQ**

This concept for specifying matching criteria and search constraints is included into the global ODP service management concept.

4.2 Defining Policies

In large distributed systems, changes of configuration are likely to appear quite often. Therefore, the behaviour of management and administration services must be flexible to cope with dynamicalLy changing configurations and requirements. In order to achieve a flexible object behaviour, it has to be determined by a policy. Changing the policy must enforce a change in the objects´ behaviour.

Referring to the definition given by the RM ODP, a policy is defined as a prescriptive relation between one or more objects and some reference behaviour. It can be expressed as an obligation, a permission and a prohibition. In the following, it will be focussed on restrictions of behaviour, interface and operations. Valid behaviour, operations and parameter value ranges are specified. A policy is valid as long as its policy maker, a role of an object defining policies, changes it.

```
<policy> ::= "POLICY" <name>
             "FOR AGENTS" <object_name> {<object_name>}
             ["ENFORCED BY MANAGER" <object_name>]
             "IS" <behaviour_specification>
             ["WHERE" <behaviour_specification> {<behaviour_specification>}]
             "END POLICY" .
```

A policy consists of a unique name, one or more objects to which it is applied called agents, and the prescribed behaviour for these objects. Optionally, another object can be determined as the manager, enforcing the policy on the agents. If an object with manager role is determined, the following describes the manager behaviour. If the manager role is omitted whereas it defines the agent behaviour. This behaviour specification might contain names of additional specifications, that are specified after the (optional) "where" statement. Thus behaviour can be split into blocks, increasing the readability of the specification and giving the designer a means for structuring.

```
<behaviour_specification> ::= "BEHAVIOUR" <behaviour_name> ":"
                               <behaviour>"END BEHAVIOUR".

<behaviour> ::= <behaviour_combination>
              | <restriction> | <activity> .

<behaviour_combination> ::= <behaviour> "AND" <behaviour>
                          | <behaviour> "OR" <behaviour>.

<restriction>      ::= "RESTRICT" <restriction_type>.
<restriction_type> ::= <behaviour_restriction>
                     | <interface_restriction>
                     | <operation_restriction> .

<behaviour_restriction> ::= "BEHAVIOUR TO" <behaviour>
                            ["AT" (<object_name>)] .
<interface_restriction> ::= "INTERFACE TO"
                            <operation_name> {<operation_name>}
                            ["AT" <object_name> {<object_name>}] .
<operation_restriction> ::= "SCOPE OF" <operation_name> {<operation_name>}
                            "TO" <scope_description>
                            ["AT" <object_name> {<object_name>}]
                            |
                            "ATTRIBUTE RANGE OF" <attribute_name>
                            "IN" <operation_name> {<operation_name>}
                            "TO" <range_description>
                            [" <object_name> {<object_name>}] .
```

With a policy specification, restrictions on behaviour, on interface and on operations of objects can be made. Restrictions on operations refer to their scope and parameter value ranges. If restrictions do not contain the object, they refer to, the object or objects determined by the policy header are assumed. In the following we focus on restricting the scope of operations. The next two production rules determine the way the scope of an operation can be described.

<scope_description> ::= "{" <scope> "}" .
<scope> ::= <scope> ";" <scope>
| "**SUBGRAPH OF**" <path> ["**TO DEPTH**" <depth>]
| <path> .

Range descriptions can be expressed as a single value or a set of valid values.

<range_description> ::= <value> | "{"<value_set>"}" .
<value_set> ::= <value> ";" <value_set> | ε .

The scope of an operation is expressed by a set of paths to objects that can be used during operation processing. Each TOD of a federated trader will result in one or more path or subgraph descriptions. Within the following text, a path leads to a context containing a number of service offers.

<activity> ::= "**BEHAVIOUR**" <behaviour_name>
| <action> {";" <action>} "." .
<action> ::= <interaction>
| <conditional_action> | <action_name> .
<interaction> ::= <operation_invocation>
| <operation_reception>
| <notification_sending>
| <notification_reception> .
<conditional_action> ::= "**IF**" <condition> "**THEN**" <activity>
["**ELSE**" <activity>] "**END IF**" .

<operation_invocation> ::= "**INVOKE OPERATION**" <operation_name>
"**AT**" (<object_name>)
["**WITH**" (<value>)] ["**RESULTS IN**" (<variable>)] .
<operation_reception> ::= "**ON INVOCATION OF**" <operation_name>
"**FROM**" <object_name>
["**WITH**" (<variable>)]
"**DO**" <activity>
"**RETURNING**" (<value>) .
<notification_sending> ::= "**SEND NOTIFICATION**" <notification_name>
"**TO**" (<object_name>)
["**WITH**" (<value>)] .
<notification_reception> ::= "**ON NOTIFICATION**" <notification_name>
"**FROM**" <object>
["**WITH**" (<variable>)]
"**DO**" <activity> .

The behaviour specification language should at least contain constructs to describe operation invocation and reception, notification sending und reception, conditional action and linear temporal ordering of actions, as given by the above described production rules. The following policy specification gives an example for the use of the above defined notation. The name structure refers to the context-relative naming scheme used in Fig. 5.

POLICY TraderSearchPolicy
FOR AGENTS LocalTrader

```
IS BEHAVIOUR LocalServiceImport :
        RESTRICT INTERFACE TO search select
    AND
        RESTRICT SCOPE OF search select TO {SUBGRAPH OF "/D/C"}
    AND
        RESTRICT ATTRIBUTE RANGE OF location
            IN search select TO rwth_aachen
    END BEHAVIOUR
END POLICY
```

With this policy, the search space for service import performed by the local trader is restricted to the context "/D/C" and all subcontexts. Service import includes the search and select operation of the trader service interface. In addition, all exporters must be located at the RWTH Aachen. Related work on the policy concept can be found in e.g. [RPB 93], where examples of policy specifications are provided, but nothing is said on syntax nor semantics for their specification language.

5. THE FOUR-STEP-CONCEPT FOR A FORMAL SEMANTICS

Now, the aim is to consider the formal semantics of a service request. This formal semantics is devided into four steps which have well-defined interfaces. Starting point of our semantics is the data base of the trader, i.e. the service direcotry, called service offer set (SOS), together with the service request and a policy specified formally in our Service Request Description Language, presented above. The result is the interpretation of a SRDL-text in zero, one or more service offers according to our service offer set, see Fig. 6.

The first step consists of limiting the search space of a service request with respect to a corresponding policy, and is performed by the function `restricting_scope`. The operations' search space is defined in a SRDL-text and will be restricted to the search space the requested trader is allowed to access, which is determined by the trader search policy. During the computation of this function, the SOS remains unchanged.

The second step of the formal semantics is denoted by the function `restricting_ TOD`. It limits the whole scope of service offers by mapping the search constraints onto a smaller set of service offers which are of interest. This step considers only the SRDL-text and the service offer set. The result is a new service offer set, which is the interface to the third phase of this semantics.

After having obtained a set of service offers which is of interest, the third step applies the `IT&SSP_check` function. During this phase the interface type is considered, and static service properties are evaluated in order to reduce the set of appropriate services. The result of `IT&SSP_check` is a subset of SOS' which includes services with demanded static service properties. This function requires the service offer set SOS' which is the result of the previous step and it demands the SRDL-text of the given service request. These two values are mapped by the `IT&SSP_check` function onto the power set of SOS', i.e. onto $\wp$(SOS'). The SRDL-text remains unchanged.

The fourth step of this formal semantics differents between `search` and `select` service requests. In case of a search service request it is important to prove if there are dynamic service properties included in the given SRDL-text. If no dynamic service properties are considered, the result of the third phase is equal to the result of our fourth phase. Otherwise, it is necessary to apply the `search` function onto an element of $\wp$(SOS') and the SRDL-text in order to get a subset of the previously considered service offer set. In case of a `select` function it is always necessary to apply this function. If there are no dynamic service properties requested by the given selection criteria, the most suitable service is selected. Otherwise, the corresponding dynamic service properties are evaluated and the most suitable service offer is selected. Result of the selection process is a single service offer.

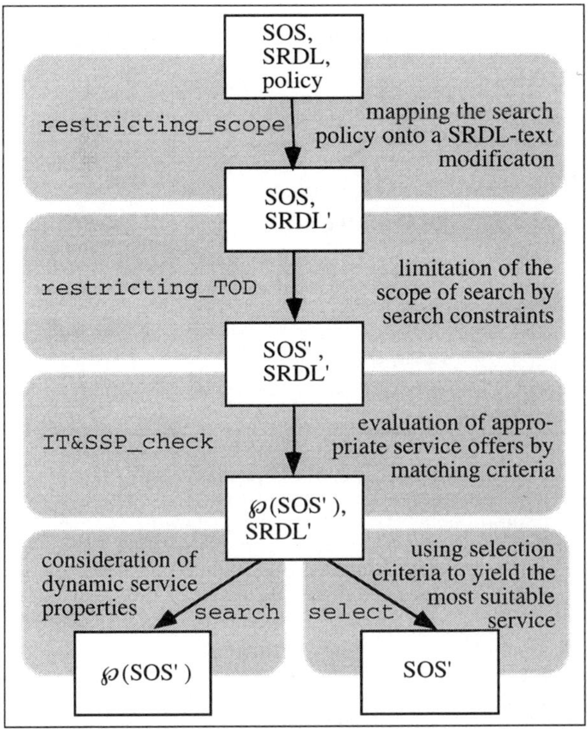

Fig. 6: The four-step semantics for searching and selecting a service

The following subchapters explain these four steps of semantics in more detail. An exact formal description of each of these four steps is given.

5.1 The First Step: Restricting the Search Space

The first step of the semantics of service requests restricts the search space defined by the invoker, forming a part of the importer policy, in accordance with the search policy valid for the requested trader. Generally, the trader search policy has a higher priority than the importer policy. Thus,the search space used for a service request is an intersection of the search spaces of the trader search policy and the importer policy.

`restricting_scope`: policy, SRDL-text-> SRDL-text

The function `restricting_scope` extracts the operation_restriction for the SRDL-text from the policy specification. Since the scope_description of a policy and the search-constraint of a service-request do not have the same syntax, it also has to perform an adaption. Path specifications do not need to be changed, whereas subgraph specifications must be flatted to a set including a path for each context in a subgraph. Both, the operation_restriction and the unchanged SRDL-text serve as input for the function $[[\cdot]]_{ro}$. The semantics of an operation_restriction is given by a function $[[\cdot]]_{ro}$.

$[[\cdot]]_{ro}$: operation_restriction, SRDL-text-> SRDL-text

It maps a service request and an operation restriction onto a modified service request taking the restrictions into account. The result is a modification of the search space. It is out of the scope of this paper to present this semantics, for more detail see [PMK 94].

5.2 The Second Step: Mapping Search Constraints onto a Limited Service Offer Set

The second step of formal semantics performs a function

`restricting_TOD`: SOS, SRDL' →SOS'.

The formal semantics is given in the following. First, the SRDL-text is projected onto the search-expression.

if	SRDL-text = service-request
with	service-request = <service_request_operation> <search-constraint> **"END SERVREQ"**,
	search-constraint = "**IN**" <search-expression>
then	RESULT = <search-expression>

A second step is to map the RESULT, i.e. the search-expression of our SRDL-text onto the corresponding set of service offers.

In our context relative naming model presented in chapter 4, unique names are guaranteed if each edge in the graph has a unique label relative to the source of the edge. For example, a search-expression "/D/B/A" denotes the subset obtained by taking the set D which contains the subset B which includes subset A. The subset relationships can be derived from the names. The context named "/D/B/A" is a subset of the context named "/D/B", which is a subset of the context named "/D". The search path is easy to derive and describes the set SOS' which is used in the following.

5.3 The Third Step: Formal Interpretation of Search and Select

In the following, we consider the third phase of our four-step-model. This step maps a set of service offers onto a smaller set of service offers according to special matching criteria.

The basic set of service offers which has to be considered is denoted by

$$SOS' = (SO_1, SO_2, ..., SO_g),$$

SOS denotes the service offer set, and SO is a service offer. All together, we have a set of g elements. Considering the i-th service offer, it can be described as a 8-tupel

$$SO_i = (EId_i, SId_i, IT_i, SSP_i = \{sp_{i1}, sp_{i2}, ..., sp_{in_i}\}, DSP_i = \{dp_{i1}, dp_{i2}, ..., dp_{im_i}\}, SOP_i = \{so_{i1}, so_{i2}, ..., so_{ik_i}\}; tog_i, iog_i),$$

where sp_{ij} := (<spn_{ij}> "=" <spv_{ij}>), dp_{ij} := (<dpn_{ij}> "=" <dpv_{ij}>), and so_{ij} := (<son_{ij}> "=" <sov_{ij}>).

EId_i denotes the exporter identifier of the i-th service offer, SId_i denotes the service interface of the i-th service offer, IT_i describes the interface type of i-th service offer, SSP_i is the set of all static service properties and DSP_i is the set of dynamic service properties. The third phase of formal semantics maps all service offers in a suitable set of service offers SOS' and a matching criteria described as Service Request Description Language text into the power set of SOS', i.e., $\wp$(SOS'). That means:

`IT&SSP-check`: SRDL'-text, SOS' → $\wp$(SOS').

Beginning a search or selection process, a variable *res* denoting a result set has the value empty. The function KEEP <search> which is used in the following, stores simple BNF expressions which include dynamic service properties of type '<identifier>"/d"' for the 4th phase of formal semantics. Now, we have all means to describe the 3rd semantics phase of the SRDL.

if SRDL-text = search,
sti is a service-type-identifier,
cmc is basic-matching-criteria
with search = **SEARCH** sti **WITH** cmc,
cmc = <service-property-identifier><general-relation><real-number>
then <service-property-identifier> = <identifier>"/d" → KEEP <search>,
<service-property-identifier> ≠ <identifier>"/d" →
∀ i∈SOS: sti = SId_i ∧ ∃ x: service-property-identifier = spn_{ix} ∧ spv_{ix} ∈ real
∧ spv_{ix} <general-relation><real-number>
→ res = res ∪ SO_i

if SRDL-text = search,
sti is a service-type-identifier,
cmc is basic-matching-criteria
with search = **SEARCH** sti **WITH** cmc,
cmc = <service-property-identifier> <equal-or-not-symbol> <service-property-value-identifier>
then <service-property-identifier> = <identifier>"/d" → KEEP <search>,
<service-property-identifier> ≠ <identifier>"/d" →
∀ i∈SOS: sti = SId_i ∧ ∃ x: service-property-identifier = spn_{ix} ∧ spv_{ix} <equal-or-not-symbol><service-property-value-identifier>
→ res = res ∪ SO_i

if SRDL-text = search,
sti is a service-type-identifier,
cmc is a simple-matching-criteria
with search = **SEARCH** sti **WITH NOT** cmc,
cmc = <service-property-identifier><general-relation><real-number>
then <service-property-identifier> = <identifier>"/d" → KEEP <search>,
<service-property-identifier> ≠ <identifier>"/d" →
∀ i∈SOS: sti = SId_i ∧ ∃ x: service-property-identifier = spn_{ix} ∧ spv_{ix} ∈ real
∧ ¬ spv_{ix} <general-relation><real-number>
→ res = res ∪ SO_i

if SRDL-text = search,
sti is a service-type-identifier,
cmc is simple-matching-criteria
cmc_1 and cmc_2 are simple-matching-criteria
with search = **SEARCH** sti **WITH** cmc,
cmc = cmc_1 **AND** cmc_2 .
then ∀ i∈SOS: if ∃ i: (**SEARCH** sti **WITH** cmc_1 → res = res ∪ SO_i
∧ **SEARCH** sti **WITH** cmc_2 → res = res ∪ SO_i)
→ res = res ∪ SO_i

if SRDL-text = search,
sti is a service-type-identifier,
cmc is a conditional-matching-criteria
cmc_1, cmc_2 and cmc_3 are conditional-matching-criteria
with search = **SEARCH** sti **WITH** cmc,
cmc = **IF SUCCESS** cmc_1 **THEN** cmc_2 **ELSE** cmc_3 **END**
then ∀ i∈SOS: if ∃ i: (**SEARCH** sti **WITH** cmc_1 → res = res ∪ SO_i)
then search = **SEARCH** sti **WITH** cmc_2
else search = **SEARCH** sti **WITH** cmc_3

Note: So far, the "IF SUCCESS"-condition considers only static service properties.

if SRDL-text = search,
sti is a service-type-identifier,
cmc is a conditional-matching-criteria
cmc_1 and cmc_2 are conditional-matching-criteria
with search = **SEARCH** sti **WITH** cmc,
cmc = **IF SUCCESS** cmc_1 **THEN TAKE THAT ELSE** cmc_2 **END**
then $\forall$ i$\in$SOS: if $\exists$ i: (**SEARCH** sti **WITH** $cmc_1 \rightarrow res = res \cup SO_i$)
then search = **SEARCH** sti **WITH** cmc_1
else search = **SEARCH** sti **WITH** cmc_2

If SRDL-text = select with select = "**SELECT**" <service-type-identifier> "**WITH**" <conditional-selection-criteria> <search-constraint> "**END SERVREQ**", so the same semantics operations as above are valid, accepting the modification, that search is substituted by select. Until there is no rule which could be performed, these rules are applied, then the next step is performed.

Going in more detail, the above described structure is taken for all BNF-expressions, but here, only the structure of this semantics phase should be explained. The whole description could be found in an internal paper [PMK 94].

Finally, *res* = SOS'' $\in \wp$(SOS') and denotes the set of all service offers coresponding to the given matching criteria. If SRDL-text is a search request and there is a dynamic service-property-identifier so this service property is evaluated in the next step of formal semantics; if there is no dynamic service property involved in the request description, so *res* denotes the final result of the search. If SRDL-text is a select request then it is in every case necessary to consider the following step of formal semantics.

5.4 The Fourth Step: Performing `Search` and `Select`

First, the case of a search request should be considered. If KEEP <search> stores no expressions, so *res* denotes the final result. Otherwise, the same rules as above are applied to SOS'', accepting the extension, that $spn_i \rightarrow dpn_i$ and $spv_{ix} \rightarrow dpv_i$ are substituted, i.e.:

4th-semantics-phase(`search`): SRDL-text, $\wp$(SOS') $\rightarrow$ $\wp$(SOS').

The result of applying these rules is an equal or smaller set SOS''' $\in \wp$(SOS'). The values for the dynamic service properties are updated.

If, in the second case, we have a select operation, it is possible to combine the rules given above and the following rules. The fourth step uses the preselected service offers included into SOS' and applies the whole set of rules, i.e.:

4th-semantics-phase(`select`): SRDL-text, $\wp$(SOS') $\rightarrow$ SOS'

Similar, the variable *res* denoting a result set is set empty when applying all rules once more including the following rules:

if SRDL-text = select,
sti is a service-type-identifier,
csc is basic-selection-criteria
with select = **SELECT** sti **WITH** csc,
csc = **MINIMUM** "(" <service-property-identifier> ")"
then c=0, $\forall$ i$\in$SOS: (sti = $SId_i \wedge \exists$ x: service-property-identifier = $spn_{ix} \wedge spv_{ix} \in$ real
$\wedge$ ($c = 0 \vee spv_{ix} < s$)
$\rightarrow res := SO_i \wedge c := c+1 \wedge s := spv_{ix}$)

if SRDL-text = select,
sti is a service-type-identifier,
csc is basic-selection-criteria
with select = **SELECT** sti **WITH** csc,

csc = <choice-symbol> "(" <conditional-matching-criteria> ")"
then (∀ i∈SOS: (sti = SId_i ∧ ∃ x: service-property-identifier = spn_{ix} ∧ spv_{ix} ∈ real
→ res := res ∪ SO_i));
res := <choice-symbol> (res)

if SRDL-text = select,
sti is a service-type-identifier,
csc is a conditional-selection-criteria
csc_1, csc_2 and csc_3 are conditional-selection-criteria
with select= **SELECT** sti **WITH** cmc,
cmc = **IF SUCCESS** csc_1 **THEN** csc_2 **ELSE** csc_3 **END**
then ∀ i∈SOS: if ∃ i: (**SELECT** sti **WITH** cmc_1 → res = res ∪ SO_i)
then select = **SELECT** sti **WITH** cmc_2
else select = **SELECT** sti **WITH** cmc_3

if SRDL-text = select,
sti is a service-type-identifier,
csc is a conditional-selection-criteria
csc_1 and csc_2 are conditional-selection-criteria
with select= **SELECT** sti **WITH** cmc,
cmc = **IF SUCCESS** csc_1 **THEN TAKE THAT ELSE** csc_2 **END**
then ∀ i∈SOS: if ∃ i: (**SELECT** sti **WITH** cmc_1 → res = res ∪ SO_i)
then select = **SELECT** sti **WITH** cmc_1
else select = **SELECT** sti **WITH** cmc_2

The other semantics expressions could be presented in a similar way. Now, *res* denotes the service offer, which is choosen by the selection criteria. This is the endpoint of our formal semantics.

6. CONCLUSIONS

To perform Open Distributed Processing it is important to enable a concept for searching and selecting service offers. The Service Request Description Language presented here in a BNF is a first step in that direction. It enables the specification of matching criteria, selection criteria, search constraints and policies. This language has not only a formal syntax, but also a formal semantics. Therefore it is possible to map the specification of a service request onto the offered entries given in the service directory. So far, the Service Request Description language does not consider service offer property values. It is difficult to find a general description for such properties corresponding to all exporters, but this subject is for further study. Separate examinations study several approaches to include Quality of Service (QoS)-aspects. However, at the moment there is no solution to cope with all requirements.

Another point for further work is to integrate openness into that approach. Therefore, the formal language has to be extended in a way, that the matching constraint also includes a transformation of data types. Going into more detail implies that the interface type of a service is further subdevided into signature, behaviour, constraints, and the role of the client, i.e. either client/server or provider/consumer. The signature of such an interface type includes data types. By mapping these data types it is possible to achieve openness within the well known transformations.

Further possibilities to enhance this formal language arise from the value-added service of service combining [PoHe 94]. By sequential performance of several services it could be possible to get a new service which has the interface type or properties which are requested. This possiblity should be includend into the language.

Currently, the ODP Working Group of our Computer Science Department is busy with implementing this language in order to support the ANSAware trading realization. Based on tools supporting BNFs we have implemented a syntax checker for that language and are now working on the interface between the service directory and the processing of that language. Thus, one or more services can be searched or selected by specifying the conditions which are important for the user of distributed environments.

REFERENCES

[ANSA] Architecture Projects Management Ltd.: *An Overview of ASNAware 4.1.* Document RM.099.02 February 1993

[BeRa 91] Bearman, M.; Raymond, K.: *Federating Traders: An ODP Adventure.* In: IFIP Workshop on Open Distributed Processing, Berlin 1991, North Holland

[Gei 92] Geihs, K.: *Trader Interaction Models and Infrastructure Implications.* In: Proceedings of IFIP TC6 International Conference on Information Networks and Data Communication IV, Espoo SF, North Holland, 1992

[Ke 93] Keller, L.: *From Name Server to the Trader - An Overview about Trading in Distributed Systems.* In : PIK 16(1993)3, pp. 122 - 133

[LOTOS] Information processsing systems - Open systems interconnection - LOTOS - A formal description technique based on the temporal ordering of observational behaviour. International Standard , ISO 1988

[MaBl 92] Macartney, A. J.; Blair, G.S.: *Flexible Trading in distributed multimedia systems.* In: Computer Networks and ISDN Systems 25 (1992) 145-157, Elsevier Science Publishers B.V., North Holland, 1992

[ODP P1] ISO/IEC JTC1/SC21/WG 7 N885(preliminary): *Basic Reference Model of Open Distributed Processing - Part 1: Overview and User Model*, Nov. 1993

[ODP P2] ISO/IEC JTC1/SC21 Nxxxx: *WG7 DIS Basic Reference Model of Open Distributed Processing - Part 2: Descriptive Model*, Feb. 1994

[ODP P3] ISO/IEC JTC1/SC21 Nxxxx: *WG7 DIS Basic Reference Model of Open Distributed Processing - Part 3: Prescriptive Model*, Feb. 1994

[ODP P4] ISO/IEC JTC1/SC21 N7056: *Working Draft for the Basic Reference Model of Open Distributed Processing - Part 4: Architectural Semanitcs*, Aug. 1993

[ODP Tr] ISO/IEC JTC1/SC21 N8409: *Working Document - ODP Trading Function* Jan. 1994

[Pa 91] Park, H. J. et al: *Configuration management of object groups.* In: The Australian Computer Journal, Vol. 23, No. 4, Dezember 1991

[PoHe 94] Popien, C.; Heineken, M.: *Trading Enhancement by Service Combination in ODP.* In: Proceedings of the IFIP International Conference on Open Distributed Processing. North Holland Amsterdam London New York 1994 pp. 384 - 387

[PMK 94] Popien, C.; Meyer, B.; Kuepper, A.: *A Formal Approach to Service Import in ODP Trader Federations.* In: Aachener Informatik-Berichte 94-6, ISSN 0935-3232, Aachen 1994, pp. 1-24

[PoMe 93] Popien, C.; Meyer, B.: *Federating ODP Traders: An X.500 Approach.* In: Proceedings of the IEEE International Conference on Communications ICC'93, May 1993, Geneva, Switzerland, pp. 313 - 318

[RPB 93] Roos, J.; Putter, P.; Bekker, C.: Modelling Management Policy using Enriched Managed Objects. Proceedings of IFIP International Symposium on Integrated Network Management. North Holland 1993, pp. 207 - 215

[Ste 90] Stefani, J.: *Open Distributed Processing - The next Target for the Application of Formal Description Techniques.* In: Proceedings of FORTE III, Madrid 1990, North Holland

PART TWO

Applications

3

Modelling and Verification of a Multiprocessor Realtime OS Kernel

Thierry CATTEL[1,2]

[1]Laboratoire de Téléinformatique, Ecole Polytechnique Fédérale
CH-1015 Lausanne, Switzerland

This paper reports the experience of the Software Engineering Laboratory of the National Research Council of Canada with the modelling and verification of the kernel of Harmony, a portable real-time multitasking multiprocessor operating system. In this paper we explain the aim of this study and give the first results. We use a modelling approach and formalize the models of the system, the scenarios and the properties that are to be checked in PROMELA using the SPIN tool. Several models of the systems were produced with various degrees of abstraction and completeness. The most recent is a tractable one that enables the expression, simulation and verification of any scenario that consists of a bounded number of tasks that may use all the services of the kernel. An exhaustive verification of the intertask communication features of Harmony was carried out by model-checking. It revealed a bug that has been in the system for more than ten years. The first verifications of the dynamic task management primitives lead to the discovery of other bugs and serious critical races. This paper shows that it is possible to detect more than deadlocks when using formal methods for the study of a real medium-sized operating system that encompasses complex internal management.

Introduction

Formal specifications of application software are appearing in many areas [8]. However it is not yet a common practice for operating systems. Some attempts have been made, in particular for UNIX [5], but they mainly have provided an unambiguous documentation. Some complete verifications of limited parts of systems [4-22-23-24] were done. Some attempts of abstract architectural specification have also been done [21] but have not led to significant results yet.

When we began this work, our objective was clearly to illustrate of the applied use of formal methods by proving or verifying some interesting properties of the Harmony kernel. The sole production of a specification was not considered sufficient.

The complexity of concurrent systems led us to choose a framework favoring proof or verification power over expressive power. We would not only address parts of the Harmony kernel but, as much as possible, its totality. Manual proof techniques, even supported by proof environments, did not seem to be realistic tools to use because of the complexity of the task and because of the very error prone nature of such activities, even for the smallest systems. An automatic or semiautomatic approach seemed to be a better choice. We initially studied the state of the art and could see the common points with the domain of communication protocol design and validation [25]. Specification and verification of real-sized communication protocols is a common practice [17]. Communicating finite state machines appeared as the obvious choice to model our system. Two kinds of approaches tend to be used in studying finite state machines. Transition-oriented techniques [11] avoid the issue of state explosion,

---

[2] This research was done at the Software Engineering Laboratory, National Research Council of Canada, Ottawa, Ontario, Canada K1A OR6, and is registered as technical report NRC 38309.

but usually require manual proofs. State-oriented techniques such as reachability analysis or model-checking may rely on automated tools. State explosion is always an issue, but we found a tool that addressed this problem. We decided to use PROMELA as a specification language, and to use the SPIN[14-15] tool to check the large state spaces for different kinds of properties. PROMELA's expressive power is rather low, since it provides only a few basic data types and only very primitive type constructors, but it offers the possibility for expressing safety, liveness or precedence properties by logical assertions, valid end states, progress states, or linear temporal logic expressions. A model may be animated by the SPIN simulator, or submitted for verification. Full state or partial state verification may be used, depending on the problem size. During simulation or verification, the properties specified are checked, and if an error is found, it is possible to run the simulator again and look at the state values at each step.

In the next sections we briefly present the problem, we explain the approach used to produce the models of Harmony and we show how these models were used to verify some interesting properties. In this paper we mainly focus on deadlocks, livelocks, and other safety properties for the primitives of Harmony kernel.

1. Harmony Kernel

Harmony is a portable real-time multitasking multiprocessor operating system being developed at the NRC since the early 80's. Version 4.1 was released in 1994. For details about Harmony see [6-7]. Basically, Harmony is an operating system based on a micro-kernel and system servers. The features of the kernel are interrupt management, task scheduling, intertask communication, dynamic task management, memory management and I/O.

There are three interrupt management primitives. *_Disable* masks all the interrupts on the current processor and *_Enable* reenables those interrupts that can occur while the calling task is executing. *_Await_interrupt* puts a task to sleep until a given interrupt is raised.

Task scheduling is based on a priority preemptive scheme. There are *N* priority levels. On a given processor, the ready task having the highest priority runs until it blocks, or until it is preempted by a higher priority task that has become ready to run, or until it is killed. Priorities are fixed for the whole life of a task; only when it dies does it change temporarily. Four functions implement task scheduling. *_Dispatch* finds the task of the highest priority among the ready queues and gives the processor to it. *_Add_ready* makes a task ready and inserts it in the ready queues according to its priority. *_Block* removes the running task from the ready queues and dispatches another, *_Block_signal_processor* has the same effect as *_Block*, but it also signals the processor of the task given as a parameter. These four functions are not part of the kernel interface; they are private to the kernel.

The intertask communication primitives of Harmony, implement the *send-receive-reply* scheme. *_Send* is a blocking primitive. It sends a message to a specified task and returns when the correspondent has not only received the message but has replied to it. *_Receive* receives a message either from a specific task or from any task. It blocks until a message is sent to the receiver. *_Try_receive* has the same functionality as *_Receive* but is non blocking. *_Reply* is a non blocking primitive that returns a message in response to a sender. In comparison to the Ada rendez-vous that may be nested according to the lexical scope and that offers an implicit reply, Harmony is not a language but a collection of OS services whose reply is explicit and may be postponed indefinitely. This allows simpler tasking structures.

Dynamic task management is provided by the *_Create*, *_Destroy* and *_Suicide* primitives. Tasks can be created dynamically on any processor but cannot migrate during their life. A task may commit suicide or kill any other task even over processor boundaries or among its ancestors. Destruction is immediate even for a task that is waiting on another task. This leads to uniform semantics at the cost of a more intricate implementation. Tasks waiting on the victim must be unblocked with an adequate return code. Before a task is destroyed, its offspring are destroyed and the resources it possesses are released. Task management is implemented thanks to a system server called the local task manager (LTM) on each

processor. The Harmony kernel provides cache and non cache memory management and I/O facilities that we do not consider here.
Fig.1.1 shows a logical view of an application running under Harmony on a multiprocessor. Application tasks communicate with other tasks over processor boundaries or with local system servers, or may wait for an interrupt.

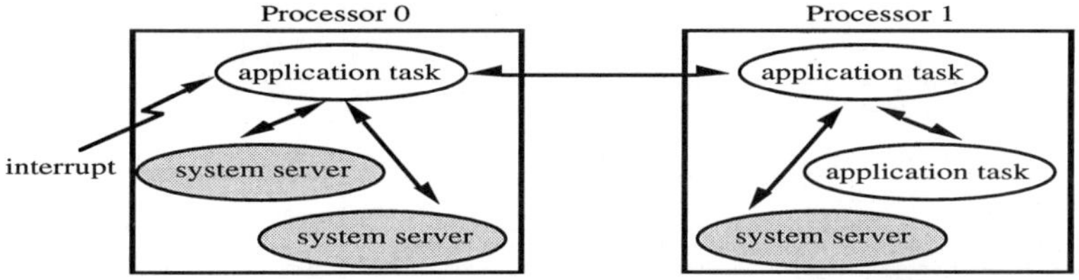

Fig.1.1 — Logical view of a multiprocessor application

Mutual exclusion across processor boundaries is implemented in the kernel by an ownership protocol rather than by locks. The implementation of this protocol is distributed among the communication primitives, the LTMs, and a procedure, *_TD_service* , that is invoked by the interprocessor interrupt handler. At most one processor can own a task descriptor[1] (TD) at a given time. Only the processor that owns a TD may access the fields that are shared between processors. Interprocessor interrupts are used to signal the transfer of ownership of a TD.

Several interprocessor interrupts coming from remote processors may arrive before the current one is treated. Therefore, a way is needed to memorize the pending interrupts. On some platforms, this is achieved by a hardware device called a hardware FIFO. On other platforms, software mailboxes [19] are used. The actual communication between tasks is done by copying the message from the sender's memory space to the receiver's.

2. Tools and modelling approach

PROMELA [14] models communicating sequential processes which may be created dynamically and communicate asynchronously or synchronously by messages through channels. The available data types are booleans and integers with the usual logical and arithmetic operators. Type constructors are restricted to channels and one dimension arrays. Control statements are the sequence, non deterministic choices, loops, unconditional branchings, and atomic sequence. Structuring facilities do not exist except the ones provided by the C preprocessor.

PROMELA models may be simulated with SPIN. It is possible to insert trace messages in the model, and to look at the model state after each step, namely the variables values and the current instruction. From a PROMELA model, SPIN generates a validator that may either attempt a complete space search or a partial one using the supertrace algorithm[12]. The validator detects all deadlock or livelock situations, violations of logic or linear temporal logic propositions[2-9-10], and unreachable code.

PROMELA has a very restricted expressive power. In particular, there is no direct means of representing data of type real, multiple dimensions arrays, pointers, structures and union type. Nor is it possible to structure models with subprograms. But the experience confirmed that this is not a problem and that it is always possible to address such issues.

A model is a simplification of a real system. The model only needs to be equivalent to the full system with respect to the properties that are being validated [13]. Verifying some properties of a system assumes two preliminaries: the modelling of the system and the modelling of the

[1] A task descriptor (TD) or task control block (TCB) is a record that holds information about a task: its name or id, its current state, its descendants, its ancestor, the memory blocks it has allocated, etc.

properties. The absence of deadlocks and livelocks was our minimal requirement, but we knew that some other properties would have to be checked. For instance task destruction is known to suffer from a possible but highly unlikely critical races.

To write the models, we took into account interrupts, intertask communication, task management and scheduling by producing partial models. We separately modelled each of these concerns in an abstract way and in a more concrete one. We thus had an incremental approach based on a *"divide to conquer"* principle that moved along two axes, as shown in Fig.2.1, horizontally from abstract to more detailed, and vertically from partial to more complete, the basic features appearing first.

Abstraction / **Features**		*abstract*	*concrete*
Interrupts	***interprocessor***	hardware FIFO	software mailboxes
	physical	several priorities	several priorities
Intertask communication		channels	_TD_service FSM
Task mngt	***creation***	synch. + offsprings	synch. + offspr. +queues
	destruction	synch. + offsprings	synch. + offspr. +queues
Scheduling		nondeterministic	priority preemptive

Fig.2.1 — Criterias for incremental modelling approach

The best abstraction we found for interprocessor interrupts management is to have a FIFO on each processor that memorizes the pending interrupts; this is exactly what a hardware FIFO does. Communication through PROMELA channels is the obvious abstraction for intertask communication. Apart from looking at a particular aspect of it, (e.g. synchronization and offspring-tree management), it is difficult to produce an abstraction for task management. Non deterministic scheduling is a generalization of priority preemptive scheduling; it does not determine which task to run among the ready ones, but let any choice possible.

We successively produced several types of models. From abstract-partial to complete-detailed. With the first type, we attempted to produce a very abstract specification of the *send-receive-reply* scheme in terms of communications through PROMELA channels. We obtained satisfying results only for some restricted communication situations and topologies (e.g. *_Send-_Receive_any-_Reply*) but if we tried to produce a model addressing all situations, it became artificially complicated. We started using some overall specifications of Harmony written with Petri nets [18]. It was quite easy to simulate these nets or even to produce equivalent and clearer models using PROMELA channels. We did not gain very much with this first type of models for two reasons. First it could not provide an abstract specification for the detailed models we produced later, thus preventing any benefit from the specification refinement methodology [1-16]. Second the obtained model was not suitable for proving properties, in particular deadlock absence in the system. Another type of model focused on dynamic task management with some restrictions (e.g. a task cannot kill one of its ancestors). Since the communication situations involved in task creation and destruction occurs always on the one hand between the creator/killer and the local task manager (LTM) and between the LTM and the killed task (there is no communication between the LTM and the created), we could use PROMELA channels for abstracting intertask communications. This type of model relies on a nondeterministic scheduling totally implicit, namely no code is written for it, it relies on the scheduling of the simulator/validator. This type of model was very useful to better understand Harmony task management. Another type, described in §3., addresses the intertask communication as implemented in *_TD_service* and relies on a hardware FIFO and an explicit nondeterministic scheduling that lets run any ready task. The last type, described in §4., addresses intertask communication and dynamic task management with no restriction and relies on a hardware FIFO and a priority preemptive scheduling, namely each task has a priority and is allowed to run according to it when it is ready. This type of model really corresponds to the concrete system and is used in our current work for the actual verification of the whole kernel. The same type of model with software mailboxes instead of hardware FIFO, we did not write yet, but could serve in the future for performing more specific verifications.

3. Intertask communication

Before looking at the model used for the verification of intertask communication, we explain how it was obtained. In order to increase the confidence one has of a real system, it is possible to produce a model of this system, to express and check formally some of its properties. That means that if we succeed in checking or proving a property for the model, we are sure its true for the model, but for the real system it is hard to say, since unfortunately there exists no way of proving that a model corresponds to a real system. In the case of Harmony, the real system was studied through its C and assembler code. We tried to keep the detailed model as close as possible to the code, only abstracting by pruning irrelevant things for the verification we had in mind. We made this choice for easing model management and for convincing people that the model corresponds to the real system. A drawback is that it probably limits the abstraction one can produce of the system, perhaps disabling the actual verification of complex properties. Nevertheless, we succeeded in producing a detailed model using the full intertask communication and dynamic task management primitives, running according to a priority preemptive scheduling. Before reaching this point, we had already produced partial models for detailed intertask communication and roughly validated them. However, we found a more sensible practice to consist in taking the complete detailed model, pruning the irrelevant details out of it (task management, scheduling) in order to get a consistent restricted model suitable for the validation of intertask communication. We use this model in §3.1. It only addresses intertask communication synchronization, the actual data exchange through message is irrelevant for the properties we want to check, but could easily be taken into account if needed.

3.1. Modelling

The model of Harmony intertask communication is decomposed into several modules. At the top level, module *kernel* includes parameters, such as the number of processors, the definition of each part of the system and scenarios. It initializes the system by starting the interrupt handlers modelled as independent processes, and the user program. Module *td* models task descriptors (TD), defining data such as the state of a task, its correspondent, its processor and its send queue. It defines a function *_I_td* to initialize the TD of task, and *_Convert_to_td* to convert a task identifier of an existing task into its TD. Since we do not have pointers, in fact we do not manage task descriptors but only task identifiers. The send queue is limited to one-element queue, because the scenarios we will validate do not require longer queues. The receive and reply queues serve only in the case of task destruction, so we forget them here. Module *taskstates* is a set of constant definitions for all the possible states of a task. Module *processor* models interrupt masking primitives. We should note that in Harmony, there is no global mutual exclusion mechanism over processor boundaries, only local mutual exclusion mechanism relying on interrupt masking. A solution consisting in using the PROMELA *atomic* statement would be too strong, and could lead to forget some problematic interleavings. We thus implemented this mechanism at the model level thanks to spin-locks – but there is no use of spin-locks in Harmony though. When a process runs on its processor, and wants to mask, it takes the processor for itself only. The treatment of interprocessor interrupts is modelled into module *ipinterrupt*. It defines the hardware FIFOs as PROMELA channels, the primitives used to signal processors, *_Block_signal_processor* and *_Signal_processor*, and the interrupt handlers modelled as processes. The handling consists in calling *_Td_service* masked and then dispatching. Most of the kernel mechanics is encoded in the *_Td_service* procedure. This model uses a quite straightforward nondeterministic *scheduling*. Any task is allowed to run provided its state is READY. Module *queues* provides usual functions for queue manipulation. *_Send*, *_Receive* and *_Reply* almost exactly look like the ones in the real system, without all what concerns things we do not take into consideration, in particular cache memory management.

We do not consider dynamic task management in this model, but we have to *_Create* the application tasks. This initializes the task descriptor, makes it *READY* and actually runs the corresponding proctype. We implemented a very pragmatic startup that statically creates the application tasks of the scenario in the *init* process of the PROMELA model. For instance a sender with identifier 1, a receiver with identifier 2 running both on processor 0 :

```
#define _Start_up()\
    _Create(Sender,1,0);\
    _Create(Receiver,2,0)
```

The modelling of interrupt masking imposes a few adaptations. Any sequence of the system of the form *_Disable();S;_Enable();* may be left as it is in the model, except if *S* includes a blocking primitive such as *_Block* or *_Block_signal_processor*, it must be rewritten by unmasking just before calling the blocking primitive and by remasking right after. For instance *_Disable();S1;Block();S2;_Enable();* needs to be rewritten in the following manner *_Disable();S1;_Enable();Block();_Disable();S2;_Enable().* This could strongly shock any Harmony designer who demands to call *_Block_signal_processor* or *_Block* disabled. The task possessing the processor, through the lock, when calling the blocking primitive, may be preempted but has not the possibility to give back the lock, which leads always to a deadlock on the processor. Our solution could certainly encompass an issue if *_Block_signal_processor* or *_Block* would modify some critical data, but it is not the case in this model. *_Block_signal_processor* sends a message to the channel FIFO, which is an atomic operation and blocks till the current task gets *READY*.

We now dispose of all what is necessary for writing scenarios that model application programs. We will see some examples of them in the section dedicated to verification.

3.2. Verification

Having the model, the problem it to know what to verify. We are first interested in checking the absence of deadlock or livelock . We could do more specific verifications such as checking that in a situation of successful communication the two communicating tasks return with an appropriated return code. We are certainly interested in checking that all the code in used too. Now the problem is to know how to verify the previous properties. This may be done by trying to find a set of scenarios as complete as possible that would exhibit all possible communication schemes. We can enumerate a list of all the possible situations that a communication task may face :

A task tries to communicate with :

(1) a non existing task
(2) an existing task that does not attempt to communicate with any task
(3) an existing task that communicates with this particular task in the right way (e.g. receive then reply)
(4) an existing task that communicates with this particular task but in a wrong way (e.g. no receive before a reply, a receive instead of a send, ...)
(5) an existing task that does not communicate with this particular task but another one

Cases (1) to (4) involves only two distinct tasks that may run on one or two processors. Case (5) involves three distinct tasks that may run on one, two or three processors. A scenario composed of *T* tasks may thus run on one to *T* distinct processors. Taking into account, obvious symmetry considerations based on the fact that we consider all the processors identical, there are *P* distinct affectations of *T* distinct tasks on *T* distinct processors, where *P* is the number of partitions of a set of *T* elements. For instance if two tasks t_1 and t_2 may run on p_1 and p_2, there are two distinct affectations: $\{[t_1,t_2]\}$ and $\{[t_1], [t_2]\}$. That means that either both tasks may run on a single processor (either *p1* or *p2*, equivalently) or that t_1 runs on p_1 and t_2 on p_2 (or equivalently t_1 runs on p_2 and t_2 on p_1). If three tasks may run on three processors there are five distinct affectations.

To be exhaustive, we should draw a three dimensions array, with all the possible relevant behaviors of the first task in columns, all the possible behaviors of the second task in raw and all the possible behaviors of the third task in the third dimension. We then have to analyze each cell by anticipating the different possible states each task could reach, e.g. first task returns with code 0 and second tasks gets *SEND_BLOCKED*,... (See Fig.3.2.1).

		1	2	3
	Sender (id=1) **Receiver (id=2)**	***_Send(2)***	***_Send(99/0)***	∅
1	***sender=_Receive(1) [;_Reply(sender)]***	2-1	0-RSB	_-RSB
2	***sender=_Receive(0) [;_Reply(sender)]***	2-1	0-RB	_-RB
3	***sender=_Try_receive(1) [;_Reply(sender)]***	SB-0 | 2-1	0-0	_-0
4	***sender=_Try_receive(0) [;_Reply(sender)]***	SB-0 | 2-1	0-0	_-0
5	***_Receive(99)***	SB-0	0-0	_-0
6	***_Try_receive(99)***	SB-0	0-0	_-0
7	***_Reply(1)***	SB-0	0-0	_-0
8	***_Reply(99/0)***	SB-0	0-0	_-0
9	∅	SB-_	0-_	_-_

Legend :
- ***1*** *task identifier of Sender*
- ***2*** *task identifier of Receiver*
- ***0*** *dummy task identifier for (try)receive ANY*
- ***99*** *task identifier of a non existent task*
- ***∅*** *empty task*
- ***SB*** *SEND_BLOCKED*
- ***RSB*** *RCV_SPECIFIC_BLOCKED*
- ***RPB*** *REPLY_BLOCKED*
- ***RB*** *RCV_BLOCKED*
- ***_*** *non significant*
- ***x-y*** *Sender reaches state x and Receiver reaches state y*
- ***a | b*** *a or b depending on relative timing of Sender and Receiver*

Fig.3.2.1 — Elementary scenarios with two tasks

It is easy to write the code related to a given cell, submit it to verification, and interpret the results. In each error case we need to check if its a predictable deadlock, an assertion violation or something else. Do we need to exhaustively test the model this way ? Is the minimal complete set of scenarios smaller than the exhaustive one produced by this three dimensions array ? We believe so, though we cannot provide a completely formal justification. We can prune cases out of the cube that are redundantly treated. Taking into account the symmetry considerations argued above, we may reduce the cube to a pyramid. There are classes of cells having the same content. It is then possible to group the corresponding scenarios and verify them altogether. It chiefly saves time if the verification of the class produces no error. We are convinced that it is useless to test some particular two-task scenarios and most of the three tasks ones, if not all. We illustrate this point a bit further but first, we consider some two-task scenarios of Fig.3.2.1. We think that this table sums up all the interesting situations.

In cell 1-1 for instance, the sender whose identifier is 1, sends to the receiver whose identifier is 2. The receiver receives from this specific sender, and then replies to it. This is a regular case of successful communication, the sender should return with code 2 (its correspondent's identifier) and the receiver should return with code 1. The only difference in cell 2-1 with cell 1-1 is the *_Receive* any, but the resulting situation is the same. Both scenarios may then be grouped into a same class. In cell 4-1, depending on the relative timing of both processes, there may be two states reached. First, another regular case of successful communication, the sender has initiated its *_Send* before the receiver executes the successful *_Try_receive* and then the *_Reply*, both processes return with their correspondent's identifier as return code. Second, the receiver executes the *_Try_receive* before the sender has executed the *_Send*, the communication attempt fails and the receiver returns with code 0 and do not reply. When the sender does the *_Send* it gets *SEND_BLOCKED* and does not return.

Fig.3.2.2 shows the code for the class containing both scenarios of cells 1-1 and 2-1. Notice the *assert* clauses that identify impossible situations. Cells 5-1 to 9-1 (Fig.3.2.3) represent another class of scenarios. The sender always gets *SEND_BLOCKED* and thus never returns. The receiver returns with 0. Notice the assertion in *Sender* that *_Send* does not return.

```
proctype Sender(byte _Active){
   byte receiver;
   _Send(receiver,2);
   assert(receiver==2)}
proctype Receiver(byte _Active){
   byte c,sender,replyee;
   if :: c=0 :: c=1 fi;
   _Receive(sender,c);
   if
   :: sender==1 ->
      _Reply(replyee,sender);
      assert(replyee==1)
   :: sender!=1 -> assert(FALSE)
   fi}
```

Fig.3.2.2 — Scenario for cells 1-1 to 2-1

```
proctype Sender(byte _Active){
   byte receiver;
   _Send(receiver,2);
   assert(FALSE)}
proctype Receiver(byte _Active){
   byte c,sender,replyee;
   if
   :: _Receive(sender,99)
   :: _Try_receive(sender,99)
   :: if :: c=0 :: c=1 :: c=99 fi;
   _Reply(sender,c)
   :: goto FINISHED
   fi;
 CHECK:
   assert(!sender);
 FINISHED:
   skip}
```

Fig.3.2.3 — Scenario for cells 5-1 to 9-1

There are some *a priori* interesting scenarios that do not need to be verified because they may be decomposed into a set of elementary two-task scenarios. Let us take two examples. First, a two-task scenarios involving two senders that send to each other. Obviously the resulting situation is both senders *SEND_BLOCKED*. We think that verifying such a scenario or verifying the scenario corresponding to cell 9-1 of Fig.3.2.1 is equivalent, at least for the properties we check here. Second, a three-task scenario, a priori more problematic ; both first tasks correspond to cell 1-1 of Fig.3.2.1, the third attempts to reply to the first in place of the second. We think that it is equivalent to verify scenarios of cells 1-1 and 7-3. We claim that Fig.3.2.1 presents all the elementary scenarios. It even includes some redundancy. For instance it is certainly equivalent to verify only cell 9-1 instead of cells 5-1 to 9-1. We have coded and verified the scenarios of Fig.3.2.1 taking into account one or two processors, first separately and then grouped into classes. We tried some scenarios with three tasks too. In every cases, the results were consistent with what we have said above.

Another aspect is unreached code. Which scenario do we need to write to check if some code is never reached and thus is useless ? Intuitively, a scenario that verifies all the elementary situations at the same time would fit. We argued that Fig.3.2.1 contains all these elementary situations and it is even possible to suppress last raw and last column, it then is easy to write the corresponding scenario that verifies all the situations (Fig.3.2.4).

```
proctype _Sender(byte _Active){          proctype _Receiver(byte _Active){
    byte receiver,c;                         byte c,sender;
    if :: c=0 :: c=2 :: c=99 fi;             if :: c=0 :: c=1 :: c=99 fi;
    _Send(receiver,c)}                       if
                                             :: _Receive(sender,c)
                                             :: _Try_receive(sender,c)
                                             :: goto REPLY
                                             fi;
                                         REPLY:
                                             _Reply(rid,sender)}
```

Fig.3.2.4 — Scenario for identification of unreached code

Now that we have, model, properties and scenarios, how do we proceed to verification and how do we interpret the results ? First we separately validated all the scenarios of Fig.3.2.1. Before we added a *end:* label in front of the loop of *_Int_handling*, so that when both the sender and receiver finish, the handler is blocked on an interrupt reception but this is considered as a valid end state. For instance for cell 9-1 of Fig.3.2.1, verification confirmed that the only end states were the expected deadlocks where the sender is *SEND_BLOCKED*. It is possible to stop the validation after a given number *N* of errors encountered (*pan -cN*) and get a trace of the execution that lead to error number *N* ; it is thus possible to have a close look at each error. This was what we did, which confirmed the expected results of Fig.3.2.1. Then we grouped scenarios by classes and got the same results. We also tried a few scenarios with three tasks, for confirmation.

We found a bug during this validation. The problem concerns primitive *_Receive* that, on the opposite of *_Send*, *_Try_receive* and *_Reply*, does not check if the correspondent exists before signaling the correspondent's processor. This caused an error in the model, which could be interpreted as signaling a non existent processor. In the best case the result was a deadlock that left the receiver blocked, instead of returning with code 0, in the worst case an exception was raised leading to a system crash. We could fix it, implemented it in the model and returned to validation. This time no error was found. Namely there was no livelocks and only predictable deadlocks occurred, but none in case of predictable successful communication and no assertion was violated.

Regarding the identification of unreached code, we proceeded exactly the same way. In all the cases, we tried the scenario on one or two processors. The verification covered all the code of the model except at some places. There are two reasons to justify this very limited unreached code. First, when reabstracting the complete model, only keeping things related to intertask communication, we only suppressed entire cases of *_Td_service*, but never modified the remaining cases that contains some code useful to test if a task has died or if a message exchange was aborted. This code is obviously useless since we take task management out of consideration. We did not modify the primitive code neither and some similar code remained. Second, we use macros to simulate functions or procedures that have conditional bodies. If we use such a macro several times, we are ensured that some of its code will not be reached in at least one place. In order to get meaningful output from the verification, we need to use each macro once at the most. This constraint is rather easy to respect while using the macros that implement the primitives of Harmony, when writing scenarios (See *REPLY*, Fig.3.2.4) but it is quite impossible when writing the model. The fact that the validator indicates that all the code is reached does not prove that every situation in the code was reached, though. For a conditional statement such as :

if :: c1 && c2 -> S1 :: !c1 || !c2 -> S2 fi

it is enough that c_1 or c_2 be false to execute S_2, but, if we do not rewrite the code with exclusive conditions, we are not ensured that the case were both c_1 and c_2 were false was encountered. Obviously, we did not bother to write the model this way :

if :: c_1 && c_2 -> S_1 :: !c_1 && c_2 -> S_2 :: c_1 && !c_2 -> S_2 :: !c_1 && !c_2 -> S_2 *fi*

4. Task management

4.1. Modelling

The detailed model of Harmony complete kernel contains all the modules of the model related to intertask communication of §3.1, possibly completed or modified, plus specific modules that correspond to task management primitives, and local task managers.

Module *ltmmsg* in particular defines constants used by the LTM and the task management primitives *_Create*, *_Destroy*, *_Suicide*. Each primitive sends an appropriate message to the LTM in charge of the particular task type to be manipulated. Module *td* is much bigger than in the previous model, because a TD is composed of all the fields necessary to manage the offspring creation tree, the various queues, and some extra fields that are detailed further. Module *queues* is modified to treat real queues, module *trees* manages the offspring trees. Module *tdservice* is completed with the extra cases that correspond to task management plus an initial control for avoiding to treat several times a task being destroyed. Module *copymsg* takes into account exchange of messages of various lengths. Module *ltm* models the local task manager. This is a system server that runs on each processor and is in charge of managing the tasks that are running on the processor. It accepts three types of messages : *CREATE*, *DESTROY* and *SUICIDE*. The *CREATE* and *DESTROY* messages are sent by the primitives *_Create* and *_Destroy*. The *SUICIDE* message is sent by a task being killed as explained below. Task creation may be dynamic in PROMELA (*run*). It is okay for simulation, but it introduces infinite state space at validation time. The solution consists in prerunning a bounded number of application tasks at the initialization of the model, the actual creation of each task consists in synchronizing the beginning of its execution on the value of field *state* of its TD. This is slightly complicated by the possibility to create several tasks types (templates). This is solved by introducing an extra field in each TD, called *index*, that indicates the task type. We then only need a generic application task type (See Fig.4.1.1). By convention task of index TASK1 is the user program that initializes the application, it corresponds to the *main* task of a real system. Task identifier attribution is centralized in our model on the opposite of what is done in the real system, where each processor generates structured task identifiers, but this choice did not introduce any issue so far.

```
proctype Task(byte _Active){...
START:
 (state[_Active]==READY && valid[_Active]) ->
  do
  :: atomic{
    ( _Active==running[processor[_Active]] &&
     !treatment[processor[_Active]] &&
     (!len(fifo[processor[_Active]]) || masked[_Active])) ->
    if
    :: killed[_Active] ->
       /* _Infanticides() */
       ...
    :: !killed[_Active] ->
       if
      :: index[_Active]==TASK1 ->
#include "scenariotask1"
      :: index[_Active]==TASK2 -> ...
      fi
    fi}
  od}
```

Fig.4.1.1 — A generic application task

Task destruction imposes a difference in the control over application task execution. This control needs to be finer, to be able to take into account a task destruction as soon as possible. In the real system, when a task is killed, first, if it is involved in intertask communication, it is retrieved from its correspondent's queues. Then the LTM changes the task application code by the procedure *_Infanticide* and sets its priority to the same as the LTM's. This code serves for killing the victim's offspring and ends by sending a *SUICIDE* message to the LTM. The LTM then invalidates the victim's TD and unblocks all the possible victim's correspondents by walking through the victim's queues. When it is done the victim is really dead. The *_Send* in *_Infanticide* never returns, because the LTM never replies to this king of message. A call to *_Destroy* always return with a zero code. This is because it is implemented as the emission of a *DESTROY* message to the proper LTM. When the LTM receives the message, the requestor (killer) is enqueued on the LTM' s reply queue. The LTM retrieves it from its own reply queue and enqueues it on the victim's reply queue. When the victim reaches the very last step of its destruction, its LTM will unblock all its correspondents, the killer being normally the last one.

First, we have the issue of how to represent this context change between the task application code and the *_Infanticide* procedure. Second, how do we represent the fine execution control ? The solution imposes to regularly check if a task is killed or not, between the execution of its instructions. A boolean *killed* field is added to the TDs, which is set to true when the LTM changes the task procedure code.

Besides each application task has to manage a program counter (PC). Module *pc* provides the minimal features to execute a sequential instruction, a conditional instruction, an unconditional jump and also subprogram calls. Consider now an application task that calls a system primitive. The code of this primitive must include the fine control too.
That means that we have to write all the primitives to take this into consideration. The caller must take into account the length of the primitive called, in terms of instruction number, for managing the PC. This imposes to do kind of manual assembly.

Fig.4.1.2 shows the code of a two-tasks scenario. The main task creates a child and loops on sending messages to it, and the child receives messages from its father and replies to it.

```
/* main */
if
:: CAL2(0,     _Create,child,TASK2)
:: COND(20,    child==0,0,21)
:: SEQI(21,    request[SIZE]=MAXMSGLENGTH;request[RESULT]=99)
:: CAL4(22,    _Send,rid,request,response,child)
:: GOTO(35,    21)
fi
/* child */
if
:: SEQI(0,     request[SIZE]=MAXMSGLENGTH)
:: CAL3(1,     _Receive,requestor,request,_Father_id())
:: COND(55,    requestor==0,0,56)
:: SEQI(56,    reply[SIZE]=MAXMSGLENGTH;reply[RESULT]=127)
:: CAL3(57,    _Reply,rid,reply,requestor)
:: COND(88,    rid==0,56,0)
fi
```

Fig.4.1.2 — A two-tasks scenario

The priority preemptive scheduling is also introduced by adding data related to ready queues and running processes into module *processor*. A *priority* field is added to TDs too. The most differences appear in module *scheduling*, and *ipinterrupt*. The *_Add_ready* function non only sets the task state to READY but also does it insert the task into the ready queue of appropriate priority on the task's processor. *_Dispatch* on a given processor *p*, sets the running process for processor *p* by searching the task of highest priority in the ready queues of *p*. Only *_Block* and *_Block_signal_processor* change in module *ipinterrupt*, they take the executing task out of the ready queue.

Interruption masking is totally changed. Each task has an extra boolean field *masked* than *_Enable* sets to false and *_Disable* to true. All the tasks of the system are subject to priority scheduling, even LTMs and tasks executing *_Infanticide*. The overall control on application tasks needs to take that into account, *_Infanticide* and LTMs has to be written with a fine control but LTMs are not concerned by task destruction.

Besides interprocessor interrupt handling needs to be prioritary over the running tasks if there are not masked. The handling does not need the fine control over its execution since it is not interruptible. In summary, on a given processor, here are the running conditions for the various items :

Interprocessor Interrupt Handler :
an interprocessor interrupt is under treatment or
an interprocessor interrupt is pending and
the running task did not mask
LTMs and application tasks :
the task is the running task and
no interprocessor interrupt is in treatment and
no interprocessor interrupt is pending or
the task has masked

Fig.4.1.3 shows the skeleton of the LTMs and Fig.4.1.4 the skeleton of the interprocessor interrupt handlers.

```
proctype _Local_task_manager(byte _Active){
   byte requestor; message(request); ...
   do
   :: atomic{
          ( _Active==running[processor[_Active]] &&
          !treatment[processor[_Active]] &&
          (!len(fifo[processor[_Active]]) || masked[_Active])) ->
          if
          :: SEQI(0,          response[SIZE]=MAXMSGLENGTH)
          :: CAL3(1,          _Receive,requestor,request,0)
          :: COND(55,         (request[TYPE]==CREATE), 56, 200)
          ...
          :: COND(200,        (request[TYPE]==DESTROY), 201,400)
          ...
          :: COND(400,        (request[TYPE]==SUICIDE), 401,500)
          ...
          :: SEQI(500,        assert(FALSE))
          fi
   od}
```

Fig.4.1.3 — Local task manager

Startup now better conforms to reality. The variable *_Ltm_for_template* is a table that returns the identifier of the LTM in charge of a given task template. The variable *template_priority* returns the priority of a given template. In order to keep the startup pragmatic we prevent the user program from running before all the LTMs on all the processor are blocked on the reception of their first message. User program is started by initializing the first task, that will run on processor 0 (Fig.4.1.5).

```
proctype Int_handling(byte p){...
 do
 :: (len(fifo[p]) && !masked[running[p]]) ->
    /* _IP_int(p) */
           treatment[p]=TRUE;
           fifo[p]?candidate;
           /* _Td_service(candidate) */
              ...
           _Dispatch(p);
           treatment[p]=FALSE
 od}
```

Fig.4.1.4 — Interprocessor interrupt handler

```
init{...
   atomic{
   Proc_init(); ...
      _Avail_task_number = NB_PROC+1;
      _Ltm_for_template[TASK1]=...;...
      template_priority[TASK1]=...;...
      /* wait for all the LTMs be blocked on the reception of first message */
      (state[1]==RCV_BLOCKED && ...
         state[NB_PROC]==RCV_BLOCKED);
   /* start user program */
         _I_td(NB_PROC+1,0,0,template_priority[TASK1],TASK1);
      _Add_ready(NB_PROC+1)}}
```

Fig.4.1.5 — System startup

4.2. Verification

The verification of task management in Harmony is certainly the most interesting but the most challenging part of this work. With the model presented in §4.1, we are able to write, simulate and verify any scenario involving a bound numbers of application tasks that may use any primitives of the kernel, either for intertask communication or task management running according to a priority based scheduling. It is obvious that it is quite impossible to cover all the possible scenarios, but at least is it possible to write the most basic ones and verify them. We have defined a set of elementary application tasks and we have used them in combination to produce more or less complicated scenarios. Some were already presented in §4.1. As another example, here is a task that immediately suicides :

```
if
:: CALO(0,       _Suicide)
:: SEQI(18,      assert(FALSE))
fi
```

Fig.4.2.1 presents a two-tasks scenario. The main task creates a child and immediately destroys it. The child attempts to kill its father. When beginning this work, this kind of scenarios were suspected to encompass critical races.

These scenarios allowed us for finding several issues. First, the program restricted to the main task that immediately suicides, revealed that the termination of any application could be not very clean. When a victim has released all its children it sends a SUICIDE message to its LTM and this one requests from the father of the victim to remove it of its offspring tree, but the main task has no father and no control is done for that. At this point, the main task is the only remaining task of the application, all the children are killed, their resources released and it could seem not very important to end in a clean way or not, even crashing the machine. The problem is that the illegal treatment may imply wild stores, affect device registers and result in physical damage.

/* main task */	/* TASK2 */
if	*if*
:: CAL2(0, _Create,child,TASK2)	*:: CAL1(0, _Destroy,_Father_id())*
:: CAL1(20, _Destroy,child)	*:: ...*
:: ...	*fi*
fi	

Fig.4.2.1 — A two-task scenario for detecting critical races

Second, the scenario of Fig.4.2.1, revealed a lot of problems that were mostly detected by assertion violations. The *_Receive any* primitive should be atomic, namely it should always return when a message is received and with a non zero code, but in fact it was not the case. An assertion to check the requestor in the LTM code was violated, showing that *_Receive any* could return with zero. The problem was easily fixed and the new version of *_Receive any* proved to be atomic. Unfortunately the circumstances under which the bug was found suggest that it may exist much more serious issues inside the LTM. Study is underway.

Several other weird situations were found with this scenario, that are consequences of races between both simultaneous destructions, more precisely between the DESTROY or SUICIDE cases of the LTMs, the *_Infanticide* procedure and the part of *_Td_service* which retrieves the victim out of its queue. The cause was mostly a race between two concurrent executing items that corrupted the state of a task before it could be correctly treated by *_Td_service*, letting a pending interrupt that resulted either in a call of *_Td_service* for a task having five illegal possible states, or in progressing prematurately a task state. This could be considered as a light problem, but it leads to dramatic situations such as the call to *_Suicide* returns, a task is retrieved from a queue where it is no longer in, possibly causing wild stores, a task is queued in two different queues at the same moment, a victim task executing *_Infanticide* never succeeds in sending the SUICIDE message to the proper LTM and thus never really dies, or never sends a DESTROY message for killing one of its children, or never executes *_Infanticide* at all and thus never releases its offspring tree and never dies.

The explanation of these errors would have be very hard without a graphical debugger that we developped with Tcl/Tk [20] as an extension of the X interface of SPIN. During simulation, a window is created for every process of the model. The current instruction for each process is highlighted. Debugging may proceed step by step or in running mode. The execution stops if the simulation ends or if a breakpoint is encountered. There are unconditional, conditional breakpoints and stop conditions. A window allows for the print of variable values.

Conclusions

In this paper we have reported our experience in modelling and verifying a real size operating system kernel whose internal high complexity makes it an ideal candidate for formal methods. Most of the time was spent on producing the models, and only a little on verification. It takes a long time to get acceptable models for validation. Even at this point, we are not sure if they are too simplified or approximate in comparison to the real system, either hiding possible issues or bringing out artificial ones, or if some simplifications could be done, that would increase the model abstraction and thus allow a deeper validation. With this study we succeeded in producing a complete tractable model of Harmony kernel. We are thus able to write any scenario involving a bounded number of application tasks that may use the whole functionality of the system. We only touched upon the verification of the entire system but already succeeded in finding problems in the system. We have already exhaustively verified the intertask communication primitives of the kernel and found a bug that did not showed up during ten years of utilization. Some basic scenarios using the task management primitives revealed at least two other bugs and several critical races. The model we have written also provides a prototype of the system that may be simulated and could serve as a convenient pedagogical tool for Harmony users.

We had an reengineering approach, because the system existed already. We only studied it through its code. The first serious issue we met was to produce an adequate model. The ideal way would have consisted in proceeding by refinements from a very global and abstract model. Unfortunately we could not do that, first because that would have requested a total understanding of the system, which we had not, and second it is easier to apply such an approach to layered systems such as communication protocols. Harmony is a layered system but its kernel is the first layer and it is certainly impossible to apply this kind of decomposition here. For instance, task destruction, which is the most critical issue, uses the features of the whole kernel. It sometimes was possible to isolate concerns and model them separately, such as the relative priority between application tasks, LTMs and interprocessor handlings. We have the feeling that an approach by parts is easier than by refinement. Very often, we realized at validation time that the model was imperfect. It is certainly more realistic to abstract detailed models to check overall properties than produce detailed models form abstract ones. Only after a few iterations, might the ideal method be followed. The second serious issue was to design the tests. Is it possible to produce a complete set of separately validated scenarios that would not miss problematic situations ? When verifying the four intertask communication primitives, we were exhaustive. But it is certainly quite impossible to do an exhaustive verification when we take into account the task management primitives too. The only thing we can do is to define the most interesting scenarios and verify them.

How to validate big models like the ones produced for Harmony is a puzzling question. It is intuitively better to do complete validations of partial models instead of partial validations of complete models, but even with a partial model and a very restricted scenario, we may be faced with a forbidding state explosion. New techniques based on partial order reduction will be soon integrated into SPIN and they will help removing the limits. Nevertheless, to be efficient, such techniques require that the model to be checked present a minimum rate of communication interaction. Not surprisingly, since one of the basic services of Harmony is intertask communication, the resulting models presented a high rate of computation, thus nullifying the benefit of the reduction techniques. We certainly will gain in trying to abstract the detailed models obtained. Some parts of the system do not involve task destruction or only use restricted communication. It is thus possible to produce more abstract models using more interactions. Another serious issue arises when an error is found during validation. If the problem size is small, it is not too difficult to interpret the results. But if it is large, even if SPIN provides all the necessary information, it is really tough to follow the path that lead to the error, especially if it is long and if the model has a lot of variables. We had difficulties to explain the real interesting errors of the task management primitives, but we have begun the development of a graphical debugger on top of SPIN. It already helps in interpreting the validation output. Eventually we have a prototype of a really powerful tool.

Though it was not the initial aim of this study, it was proven that SPIN was very well suited for addressing this problem. Its low expressive power was only an apparent limitation that could always easily be circumvented. More important was the high verification power of validators produced by SPIN. We are convinced that designers would even gain more in using it as a design help and not only in *a posteriori* validations.

Future work should address some part of the questions raised above and some other areas of investigation such as the cache memory management in the kernel. We only used a restricted subset of SPIN's features for our validation and temporal logic should be taken into account for checking more elaborated properties. Most of the errors found tend to suggest they occur for processors with different relative speeds. Nevertheless they give good indications of possible errors for homogeneous processors. Since Harmony runs on boards with homogeneous processors, the errors may sometimes be discarded, but it is often tricky to say without the introduction of time. One could do queueing system modelling with QNAP2 [3] for instance, for doing not only qualitative but also quantitative validations.

Acknowledgments

I am very grateful to Charles-Antoine Gauthier and Morven Gentleman of Software Engineering Laboratory for their comments about this work and contribution to improve the first versions of this paper. I also would like to thank Gregory Duval who extended the X interface of SPIN to allow the graphical debugging of models.

References

1. Andrews D. & Ince D., *Practical Formal Methods with VDM* , McDrawhill Int., 1991.
2. Audureau-Enjalbert-Fariñas, *Logique temporelle, semantique et validation des programmes paralleles*, Masson, 1990.
3. Badel M, Duong D., Eyraud S., *The New Simulation Facilities in QNAP2*, Proc of ESS91, Het Pand, Belgium, November 1991.
4. Birrell A.D., Guttag J.V., Horning J.J. & Levin R., *Synchronization Primitives for a Multiprocessor*, A Formal Specification, Operating Systems Review, 21(5), 1987.
5. Doeppner T.W. & Giacalone A., *A Formal Description of the UNIX Operating System*, Proc. of 2nd ACM Symp. on Principles of Distrib. Syst., Montreal, Canada, August 1983.
6. Gauthier C., *Design and Implementation of Realtime Operating System on Thinwire Multiprocessor*, PhD candidacy paper, Ottawa University, June 1993.
7. Gentleman W.M. , MacKay S.A., Stewart D.A. & Wein M., *Using the Harmony Operating System, Release 3.0*, NRC/ERA-377, National Research Council of Canada, Ottawa, February 1989.
8. Gerhart S., Craigen D. & Ralston T., *Experience with Formal Methods in Critical Systems*, IEEE Software, January 1994, pp.21-39.
9. Glade B.B., *A temporal logic to Promela "never" clause converter*, EE594 Final Project, Cornell University, Electrical Engin. Department, 3 May 1991.
10. Gotzhein R., *Temporal Logic and applications - a tutorial*, Computer Networks and ISDN Systems 24(1992) 203-218.
11. Gouda M.G., *Protocol verification made simple : a tutorial*, Computer Networks, 25(9), April 93, pp. 969-980.
12. Holzmann G.J., *Algorithms for automated protocol verification*, AT&T Technical Journal Jan/Feb, 1990
13. Holzmann G.J., *Basic Spin Manual*, AT&T Bell Laboratories, March 1994.
14. Holzmann G.J., *Design and Validation of Computer Protocols*, 512 pgs, ISBN 0-13-539925-4, Publ. Prentice Hall, (c) 1991 AT&T Bell Laboratories.
15. Holzmann G.J., *Design and validation of protocols : a tutorial,* Computer Networks, 25(9), April 93, pp. 981-1017.
16. Jones C.B., *Systematic Software Development Using VDM*, Prentice Hall Inter., 1990
17. Lai R. & Jirachiefpattana A., *Verification of ISO ACSE protocol specified in Estelle,* Computer Communication 17(3), March 1994.
18. Li Y., *The Harmony operating system described by Petri nets*, Master thesis, Carleton University, 1987.
19. NRC, *Harmony Operating System, Release 4.0*, Application Note 18, National Research Council of Canada, Ottawa, Ont., March 1993.
20. Ousterhout J., *Tcl and the Tk toolkit*, Addison-Wesley, 1994.
21. Pêcheur C., *Using Lotos for specifying the CHORUS distributed operating system kernel*, Computer Communication 15(2), Mars 1992.
22. Pike R., Presotto D., Thompson K. & Holzmann G., *Process Sleep and Wakeup on Shared-memory Multiprocessor*, EurOpen'91 - Tromso
23. Ramsey N., *Correctness of Trap-Based Breakpoint Implementations*, Bell Communications Research, November 1993.
24. Spivey J.M., *Specifying a Real-Time Kernel*, IEEE Software, 7(5), 1990.
25. West C.H., *Protocol Validation - principles and applications*, Computer Networks and ISDN Systems 24, 1992, pp. 219-242.

4

Design of operating systems using the F.D.T. ESTELLE

O. González, V.G. García and M. Riesco
Department of Mathematics, University of Oviedo
C/ Calvo Sotelo, s/n, 33007 Oviedo, Spain

E. Vázquez
Department of Telematic Engineering, University of Madrid
ETS Ing. Telecomunicación, 28040, Madrid, Spain

Abstract - This paper describes the application of the Formal Description Technique ESTELLE to the modelling and design of Operating Systems. We describe the overall scheme of a multiprogramming operating system and then we design in a top-down way the operating system, proposing its specification in ESTELLE at the same time.

The use of ESTELLE allows not only the formal specification of an operating system like the one we describe, but also its simulation, and even its implementation, using development tools like the WorkStation EWS (developed in the Esprit project 1265: Sedos Estelle Demonstrator).

All of the above makes the known advantages of the usual application of Formal Description Techniques to Communication Protocols available to the modelling, formalization and design of other kind of systems, such as Operating Systems.

1. INTRODUCTION

Queue theory [1], Markov processes [1], Petri networks [2], etc., have been used in the modelling and study of the behaviour of Operating Systems. In this paper we present another choice: Formal Description Techniques (FDT's).

We have chosen ESTELLE among the different FDT's: ESTELLE [3], LOTOS [4], SDL [5] (described together in reference [6]), because we believe it is very suitable to the models and schema that define an operating system, basically for its concept of structure by means of modules and channels. There are other examples in [7] of the application of ESTELLE outside the field of communication protocols. In manufacturing and industrial applications, ESTELLE has been used for example in the development of software for flexible assembly devices and for a solar plant control system.

An operating system has several modes of operation, depending on how the processor (CPU), or processors, if they are present, executes the different processes in the system. So we speak of monoprogramming, multiprogramming, multiprocessing, etc.. In this paper, we are going to focus our attention on one of the possible modes of operation, multiprogramming. The main feature of multiprogramming is that the CPU does not need to finish execution of a process in order to execute another one. That is, the CPU will stop executing one process due to that process needing to wait (for an I/O operation, an event) or because the time (quantum) granted to that process is over. In any of these cases, the CPU will switch processes, that is, the CPU will execute any of the ready processes. Later on, the CPU will resume execution of the postponed process [8].

At the moment, most computer systems work following the multiprogramming model (minis and mainframes), which offers many advantages for the users and for the system itself over the monoprogramming model. Multiprogramming improves substantially the time parameters of the processes (response time, service time, variance) and the parameters that define the optimum use of the CPU (throughput) [9].

2. MULTIPROGRAMMING SCHEME - FIRST LEVEL

In Figure 1 we show the block diagram we have designed for the first level of a multiprogramming, multiuser system (with many user stations).

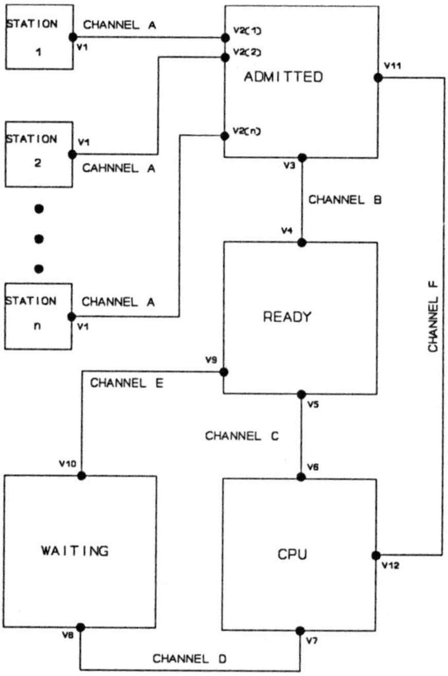

Figure 1 - Multiprogramming scheme

2.1 Module Descriptions

The different modules are:

STATION: This module represents every user connected to the computer. We assume that there are n user stations connected. We could have modelled different types of users of the system, but we chose the users to have the same model of behaviour, in order to keep the design simple. To model different behaviours we would just have to repeat for every type of user what we do here for the one type.

ADMITTED: This module manages the process submitted to execution by the n users. Management of secondary memory is also carried out this module. It also sends the results of the execution of processes to the corresponding users.

READY: It manages the processes in the ready state, that is, the processes which own all the resources needed to execute, with the exception of the CPU. This module also manages main memory. It is designed this way because all processes in the ready state have one thing in common: they are loaded into main memory.

WAITING: The processes move to this module when they are running and ask for an I/O operation or they must wait for an event to happen. When the I/O peripheral finishes the operation or the event arrives, the process will move to the ready state again (it only needs the processor to resume execution).

CPU: Processes are executed in this module. The model of operation of this module must be, obviously, multiprogramming.

2.2 Channel Descriptions

The design of the different communication channels shown in Figure1 must be specified. This design is very important in the modelling of the system, for it is through these channels where the messages flow between the specified modules.

The channels are:

Channel A

It connects the module STATION representing each user with the module ADMITTED. There are as many A channels as users (n). The messages flowing through this channel are shown in Figure 2.

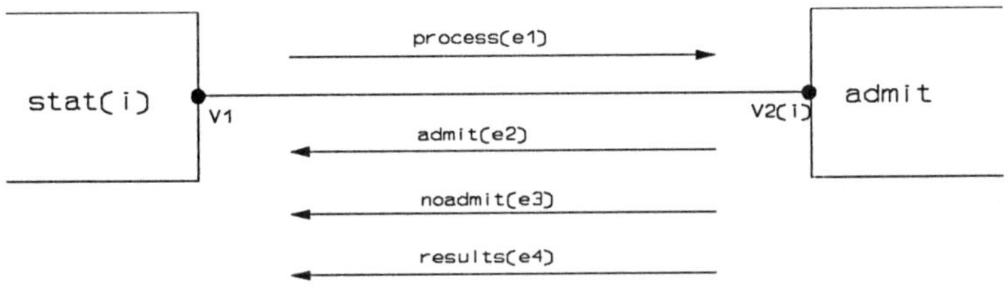

Figure 2 - Channel A

The meaning of each message is:

process (e1): it is a process that the user sends to the computer to run it. The e1 parameter specifies the process to run.

admit (e2): it informs the user that the computer (the module ADMITTED) accepts the execution of a previously sent process. The e2 parameter specifies the process admitted.

noadmit (e3): it is sent to the user when the module ADMITTED does not accept a process. The e3 parameter specifies the reasons of the rejection of the process: error, lack of resources, ..., so the user sending the process can act accordingly.

results (e4): it sends to the user partial results during the execution of a process, or final results after its execution. The results are sent in the parameter e4.

Channel B

It connects the module ADMITTED with the module READY. Figure 3 shows the messages flowing through this channel.

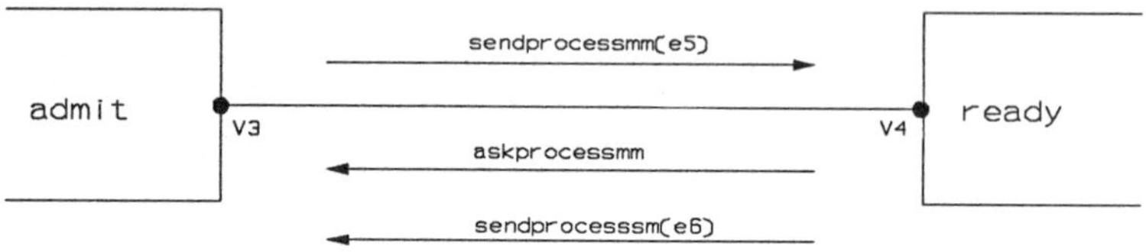

Figure 3 - Channel B

The meaning of each message is:

askprocessmm: using this message the module READY asks the module ADMITTED to send a process to it. That is, there is a need to load into main memory a process standing in secundary memory.

sendprocessmm (e5): it implies the sending of the process asked using the previous message. The module ADMITTED will choose a process among the ready processes, and will send it to main memory. The scheduling policy is called long term scheduling. It can be FCFS(First Come First Served), SJF(Shortest Job First), priority, etc..

sendprocesssm (e6): when the module READY, as an effect of the medium term scheduling chooses a process to leave main memory to the secondary memory, sends the process into parameter e6 to ADMITTED using this message. The reason is to try to light the load on the multiprogramming grade of the system (number of processes in main memory). This message is also sent whenever a processes ends execution, if it needs to be stored in secondary memory.

Channel C

It connects the READY module with the CPU module of the system. Figure 4 shows the messages of this channel.

The meaning of each message is:

askprocess: if the CPU has nothing to do (no process to run), it sends this message to READY asking it to send a process.

sendprocesstorun (e7): whenever the previous message asks READY to send a process, a short term scheduling is used to choose a ready process and send it to the CPU in the parameter e7 of this message. The short term scheduler will follow a known scheduling policy: FCFS, SJF, priority, ...

askmm (e8): a running process may need at any time more main memory. If this the case, this message will be sent to the module managing the main memory (READY). The parameter e8 specifies the features of this memory: amount, location, etc..

givemm (e9): the module READY, once a message asking for main memory is received, manages it and then tells the CPU processor to continue executing the process which needed that memory.

endprocess (e10): the CPU signals the module READY that the running process has finished execution. The memory used by this process must be freed and made available to other processes.

outprocess (e11): this message is needed whenever a running process must go directly to the READY state. For example, if we use a time sharing processor scheduling, the process running is discharged after an specified amount of time t (quantum) has expired. The process will resume execution in the future.

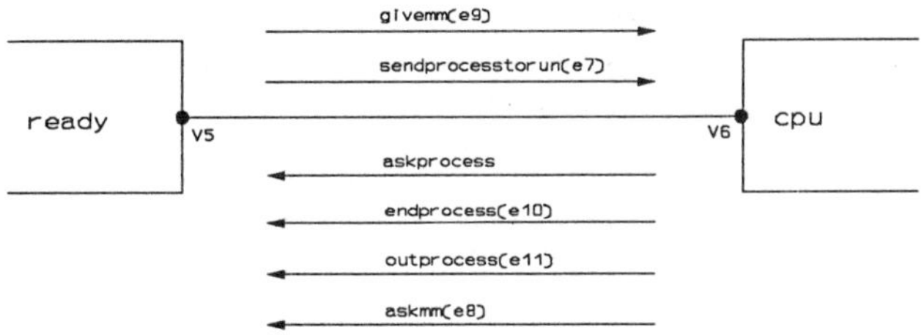

Figure 4 - Channel C

Channel D

It connects the CPU module with the WAITING module. Figure 5 shows the design:

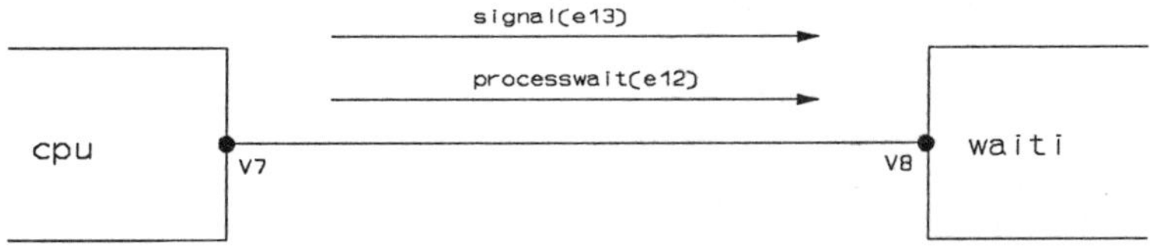

Figure 5 - Channel D

The meaning of each message of this channel is:

processwait (e12): if the running process reaches a point where it must wait for something, it must leave the CPU, allowing execution of another process. This message makes the process wait for the resources it needs to continue. Parameter e12 specifies the process and the wait causes.

signal (e13): a running process sends this message to signal another process, which was waiting, that the event it was waiting for is available. The process sending this message continues running, and the waiting process will move to the ready state. Parameter e13 contains information about the process leaving the waiting state.

Channel E

It connects the WAITING module with the READY module. Figure 6 shows this channel.

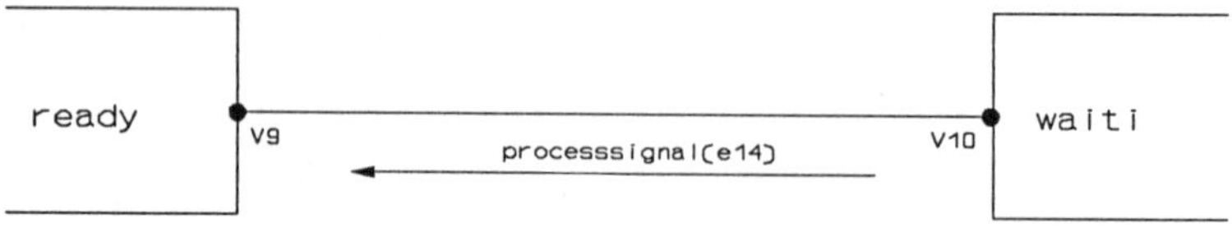

Figure 6 - Channel E

Its only message flowing is:

processsignal (e14): it signals the addition of a process to the ready state. Later it will enter execution when the short term scheduler chooses it.

Channel F

It connects the CPU module with the ADMITTED module. Figure 7 shows this channel.

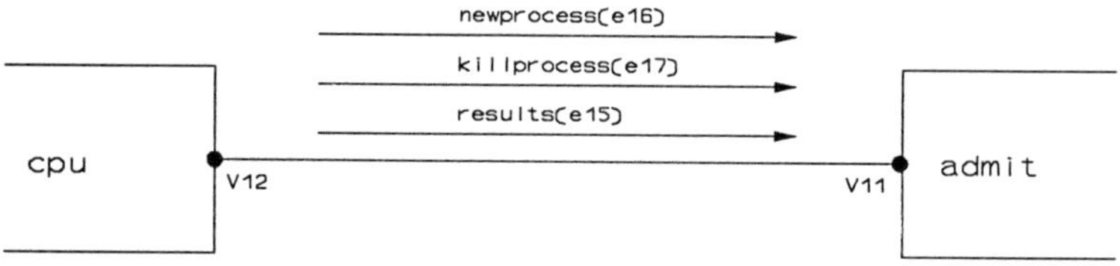

Figure 7 - Channel F

The messages are the following:

results (e15): whenever a process leaves the CPU (whether or not it has completed execution), the results are sent by the parameter e15 of this message to the ADMITTED module. Management of secondary memory is carried out by the ADMITTED module, so the processes results are sent to the users of the system.

newprocess (e16): a running process can create a child process. To achieve that, it will send this message to the ADMITTED process management module. Parameter e6 will carry information about the process to create.

killprocess (e17): a running process can kill a child process using this message. e17 signals the process to kill.

2.3 First Level ESTELLE Specification

The first level ESTELLE specification for the system here developed will be:

```
SPECIFICATION examplemultiprogramming ;
TIMESCALE  milliseconds ;
```

```
CONST
    n= ANY INTEGER ;  (* number of stations *) ...
TYPE   parameter1 = ... ;...
VAR ...

(* CHANNELS *)
CHANNEL A (stat,admit) ;
 BY  stat:  process (e1: parameter1);
 BY  admit: admit (e2: parameter2);
            noadmit (e3: parameter3);
            resultados (e4: parameter4);
CHANNEL B (admit,ready);
 BY  ready: askprocessmm;
            sendprocesssm(e6:parameter6);
 BY  admit: sendprocessmm (e5: parameter5);
CHANNEL C (ready,cpu);
 BY  ready: sendprocesstorun(e7:parameter7);
            givemm (e9: parameter9);
 BY  cpu:   askprocess;
            askmm (e8: parameter8);
            endprocess (e10: parameter10);
            outprocess (e11: parameter11);
CHANNEL D (cpu,waiti);
 BY  cpu:   processwait (e12: parameter12);
            signal (e13: parameter13);
CHANNEL E (waiti,ready);
 BY  waiti: processsignal (e14: parameter14);
CHANNEL F (cpu,admit);
 BY  cpu:   results (e15: parameter15);
            newprocess (e16: parameter16);
            killprocess (e17: parameter17);
(* END CHANNELS *)

(* MODULES OF THE SYSTEM *)
(* MODULE STATION *)
MODULE stationcab SYSTEMPROCESS ;
IP V1: A(stat) INDIVIDUAL QUEUE ;
END;
BODY stationbody FOR stationcab ;
.... (* SECOND LEVEL *)
END;
(* MODULE ADMITTED *)
MODULE admittedcab SYSTEMPROCESS;
IP V2: ARRAY [1..n] OF A(admit) INDIVIDUAL QUEUE;
   V3: B (admit) INDIVIDUAL QUEUE;
   V11:F (admit) INDIVIDUAL QUEUE;
END;
BODY admittedbody FOR admittedcab;
.... (* SECOND LEVEL *)
END;
(* MODULE READY *)
MODULE readycab SYSTEMPROCESS;
IP V4: B(ready) INDIVIDUAL QUEUE;
   V5: C(ready) INDIVIDUAL QUEUE;
   V9: E(ready) INDIVIDUAL QUEUE;
END;
BODY readybody FOR readycab;
.... (* SECOND LEVEL *)
END;
(* MODULE WAITING *)
MODULE waitingcab SYSTEMPROCESS;
IP V8: D(waiti) INDIVIDUAL QUEUE;
   V10:E(waiti) INDIVIDUAL QUEUE;
END;
```

```
BODY waitingbody FOR waitingcab;
.... (* SECOND LEVEL *)
END;
(* MODULE CPU *)
MODULE cpucab SYSTEMPROCESS;
IP V6: C(cpu) INDIVIDUAL QUEUE;
   V7: D(cpu) INDIVIDUAL QUEUE;
   V12:F(cpu) INDIVIDUAL QUEUE;
END;
BODY cpubody FOR cpucab;
.... (* SECOND LEVEL *)
END;
(* INITIALIZATION MODULE EXAMPLEMULTIPROGRAMMING *)
MODVAR
   station: ARRAY [1..n] OF stationcab ;
   admitted : admittedcab ;
   ready : readycab ;
   waiting : waitingcab ;
   cpu : cpucab ;
INITIALIZE
   BEGIN
   ALL i:1..n DO
     INIT station[i] WITH stationbody ;
   INIT admitted WITH admittedbody ;
   INIT ready WITH readybody ;
   INIT waiting WITH waitingbody ;
   INIT cpu WITH cpubody ;
   ALL i:1..n DO
     CONNECT station[i].V1 TO admitted.V2[i] ;
   CONNECT admitted.V3 TO ready.V4 ;
   CONNECT ready.V5 TO cpu.V6 ;
   CONNECT admitted.V11 TO cpu.V12 ;
   CONNECT ready.V9 TO waiting.V10 ;
   END ;
END.
(* END EXAMPLEMULTIPROGRAMMING *)
```

3. TOP-DOWN DESIGN

Once we have designed the first level of the system main modules and their communication channels we must go on designing the structure of each submodule (second level of design). This can be done in two different ways: going on dividing into submodules an so on (in a recursive way), or proposing a model for its behaviour with a finite extended automaton.

In this paper, due to lack of space, we are going to present the development of only two modules of the first level: STATION (that will be divided into submodules) and CPU (whose behaviour will be proposed with an automaton).

3.1 Module STATION

We can see the refinement of this module into submodules and its internal channels in Figure 8.

The three submodules of STATION are:

HANDLER: It performs the control of the module STATION.

TIMER: It performs statistics (response time, service time, performance, variance, etc.) and other parameters relative to the user of this STATION. We will not define it.

USER: It is the person who is seating at the STATION.

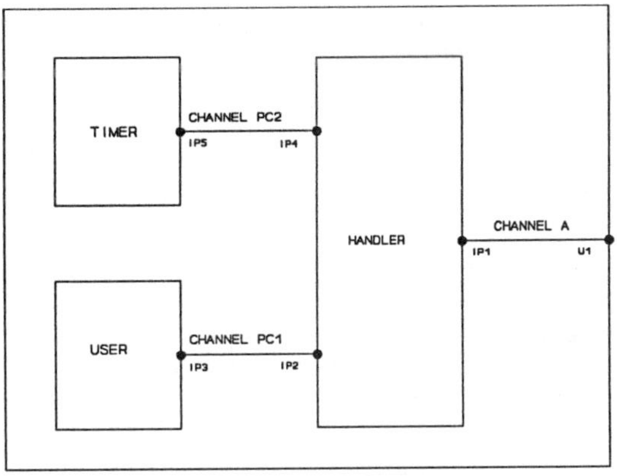

Figure 8 - Module STATION

Figure 9 shows a design for the internal channels PC1 and PC2. We will not explain in detail the meaning of each message that flows through both channels [10].

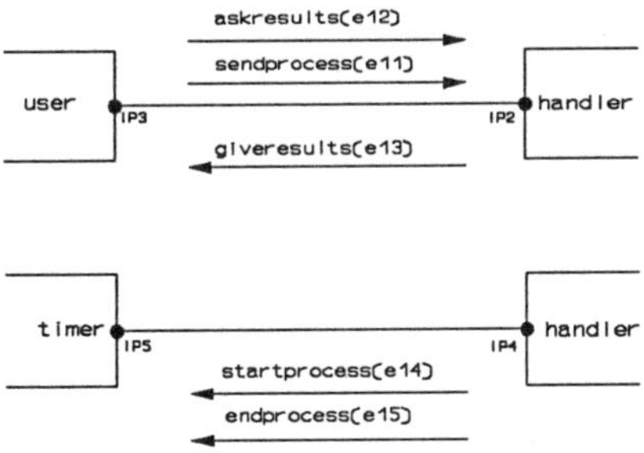

Figure 9 - Channels PC1,PC2

In this paper, we will not define the first two submodules (HANDLER, TIMER) shortly described above. We will do it for the submodule USER only, whose behaviour will be expressed by an extended finite automaton.

In a simple example of the behaviour model for the module STATION, in which each user of the computer can be in one of these five different states:

Disconnected: the module USER is disconnected and it is not working in his STATION.

Active: the module USER is in his STATION sending processes for his execution to the system.

Pickup: It asks for the results of the processes completed sent to the system.

Waiting: It waits until requested results are sent.

Studying: It studies the results that are sent to it. Then it can send more processes (active), ask for more results (pickup) or leave the system (disconnected).

Table 1 describes in an informal way the 8 posible transitions between the five states described above for this simple example.

T	FROM	TO	Informal description
1	Disconnected	Active	We suppose that when the user gets disconnected he can get connected again whenever he wants, and if a time limit is excedeed (t1) he will get connected.
2	Active	Active	A process is sent by the user, after which it stays in the same state, being able to send another process.
3	Active	Pickup	The user is going to pick up the results of the processes previously sent and already executed.
4	Pickup	Waiting	After asking for the results, it waits for their reception.
5	Waiting	Studying	When requested results arrive, it is going to study them.
6	Studying	Active	After studying the results it wants to send more processes to be executed.
7	Studying	Pickup	After studying results it wants more results.
8	Studying	Disconnected	After studying results it leaves the system.

Table 1 - Transitions of the USER automaton.

3.2 ESTELLE Specification of STATION

```
BODY stationbody FOR stationcab ;
CONST ... ;
TYPE  parameter11= ...;...
VAR   ... ;

(* CHANNELS OF STATION *)
CHANNEL PC1 (user,handler);
 BY  user:      sendprocess (e11:parameter11);
                askresults (e12:parameter12);
 BY  handler:   giveresults (e13:parameter13)
CHANNEL PC2 (handler,timer);
 BY  handler:   startprocess (e14:parameter14);
                endprocess (e15:parameter15);

(* MODULES OF STATION *)
MODULE handlercab ACTIVITY;
IP   IP4: PC2 (handler) INDIVIDUAL QUEUE ;
     IP2: PC1 (handler) INDIVIDUAL QUEUE ;
     IP3: A (station) INDIVIDUAL QUEUE ;
END;
BODY handlerbody for handlercab ;
...
END;

MODULE timercab ACTIVITY ;
IP IP5: PC2 (timer) INDIVIDUAL QUEUE ;
END:
BODY timerbody for timercab ;
...
END;
```

```
MODULE usercab ACTIVITY ;
IP IP3: PC1 (user) INDIVIDUAL QUEUE ;
END;
BODY userbody for usercab ;
CONST  t1,t2,t3,t4,t5,t6,t7 = ANY INTEGER;
TYPE  ... ;
VAR  ... ;

(* STATES OF THE USER *)
STATE disconnected, active, pickup, waiting, studying;
(* INITIAL STATE *)
INITIALIZE TO disconnected
(* INICIALIZATION VARIABLES *)
BEGIN
...
END ;
(* BEHAVIOUR OF AUTOMATON USER - 8 TRANSTIONS (table1) *)
(* TRANS1 *)
TRANS
DELAY (0, t1)
FROM disconnected
TO active
BEGIN
  (* BODY TRANS1 IN PASCAL *)
END ;
(* TRANS2 *)
TRANS
DELAY (t2)
FROM active
TO same
BEGIN
  (* BODY TRANS2 IN PASCAL *)
  ...
  OUTPUT IP3.sendprocess (...)
END;
(* TRANS3 *)
TRANS
DELAY (t3,t4)
FROM active
TO pickup
BEGIN
  (* BODY TRANS3 IN PASCAL *)
END ;

(* TRANS4 *)
TRANS
FROM pickup
TO waiting
BEGIN
  (* BODY TRANS4 IN PASCAL *)
  ...
  OUTPUT IP3.askresults (...)
END;
(* TRANS5 *)
TRANS
WHEN IP3.giveresults (...)
FROM waiting
TO studying
BEGIN
  (* BODY TRANS5 IN PASCAL *)
END;
(* TRANS6 *)
TRANS
DELAY (t5)
```

```
FROM studying
TO active
BEGIN
  (* BODY TRANS6 IN PASCAL *)
END;
(* TRANS7 *)
TRANS
DELAY (t5)
FROM studying
TO pickup
BEGIN
  (* BODY TRANS7 IN PASCAL *)
END;
(* TRANS8 *)
TRANS
DELAY (t6,t7)
FROM studying
TO disconnected
BEGIN
  (* BODY TRANS8 IN PASCAL *)
END;
END ;
(* END MODULE USER *)

(* INITIALIZATION OF MODULE STATION *)
MODVAR
   timer: timercab ;
   handler: handlercab ;
   user: usercab;
INITIALIZE
   BEGIN
   INIT timer WITH timerbody ;
   INIT handler WITH handlerbody ;
   INIT user WITH userbody ;
   ATTACH V1 TO handler.IP1 ;
   CONNECT user.IP3 TO handler.IP2 ;
   CONNECT handler.IP4 TO timer.IP5 ;
   END ;
END ;
(* END MODULE STATION *)
```

3.3 Module CPU

It is the processor of the system working in a multiprogramming mode; we will introduce a simple model that has five states:

Idle: The processor has no process to execute; when it is in this state it must ask the READY module a process to be executed.

Waiting: The processor is waiting to receive a process previously requested.

Running: The processor is executing a process; when it is in this state several changes of state can happen, that ghanges represent differents events than can happen when the CPU is executing a process.

Needmemory: if during the execution of a process more main memory is needed (for instance, a dynamic petition or the need of a page or segment not resident in main memory), the CPU will go to this state, where it stays until the process has enough main memory.

Intermediate: It is a state in which the CPU is in a transitory mode. When this state is left the CPU return to running state, with the only goal that the delay timers get initial value of zero.

Table 2 describes in an informal way 10 possible transitions between the 5 states described.

T	FROM	TO	Informal description
1	Idle	Waiting	If the processor is in the idle state it means that it is not executing any process, so with this transition it asks for a process and waits for its reception.
2	Waiting	Running	A process arrives and then it is executed.
3	Running	Idle	When the processor is executing a process, this process may be finished o interrupted. In the former case, the state will change to idle after sending the results of the finished process.
4	Running	Idle	When the processor is executing a process, it is interrupted because its quantum expires, then processor will change into the idle state after sending the process to the module READY.
5	Running	Idle	When the processor is executing a process, it is interrupted because an I/O operation is needed or it waits in a semaphore until an event occurs. This situation is simulated by this transition, where the processor changes to the idle state after sending the process to the WAITING module.
6	Running	Intermediate	If the running process wants to create a son process, it will be sent to the module ADMITTED through this transition.
7	Running	Intermediate	If the running process wants to kill a son, it will be sent to the module ADMITTED through this transition.
8	Intermediate	Running	The processor arrives to Intermediate state when the running process creates or kills another process, leaving immediatly this state through this transition.
9	Running	Needmemory	If the running process asks dynamically for more main memory, the processor changes the state Needmemory through this transition.
10	Needmemory	Running	When the demand memory is satisfied the processor continues executing the process, that is, it changes to the state running.

Table 2 - Transitions of the CPU automaton

3.4 ESTELLE Specification of CPU

```
BODY cpubody FOR cpucab ;
CONST t1,t2,t3,t4,t5,t6,t7,t8,t9,t10,t11,t12= ANY INTEGER;
TYPE  ... ;
VAR  ... ;
(* STATES OF CPU MODULE *)
STATE idle,waiting,running,needmemory,intermediate ;
(* INITIAL STATE *)
INITIALIZE TO idle
(* INITIALIZATION VARIABLES *)
BEGIN
...
END ;

(* BEHAVIOUR OF CPU MODULE - 10 TRANSITIONS OF TABLE2 *)
(* TRANS1*)
TRANS
FROM idle
TO waiting
```

```
BEGIN
  (* BODY TRANS1 IN PASCAL *)
  ...
  OUTPUT V6.askprocess
END ;
(* TRANS2 *)
TRANS
WHEN V6.processtorun (...)
FROM waiting
TO running
BEGIN
  (* BODY TRANS2 IN PASCAL *)
  ...
END ;
(* TRANS3 *)
TRANS
DELAY (t1,t2)
FROM running
TO idle
BEGIN
  (* BODY TRANS3 IN PASCAL *)
  ...
  OUTPUT V12.results(...);
  OUTPUT V6.endprocess(...)
END ;
(* TRANS4 *)
TRANS
DELAY (t3,t4)
FROM running
TO idle
BEGIN
  (* BODY TRANS4 IN PASCAL *)
  ...
  OUTPUT V6.outprocess(...)
END ;
(* TRANS5 *)
TRANS
DELAY (t5,t6)
FROM running
TO idle
BEGIN
  (* BODY TRANS5 IN PASCAL *)
  ...
  OUTPUT V7.processwait(...)
END ;
(* TRANS6 *)
TRANS
DELAY (t7,t8)
FROM running
TO intermediate
BEGIN
  (* BODY TRANS6 IN PASCAL *)
  ...
  OUTPUT V12.newprocess(...)
END ;(* TRANS7 *)
TRANS
DELAY (t9,t10)
FROM running
TO intermediate
BEGIN
  (* BODY TRANS7 IN PASCAL *)
  ...
  OUTPUT V12.killprocess(...)
END ;
(* TRANS8 *)
TRANS
FROM intermediate
TO running
BEGIN
  (* BODY TRANS8 IN PASCAL *)
  ...
END;
(* TRANS9 *)
TRANS
DELAY (t11,t12)
FROM running
TO needmemory
BEGIN
  (* BODY TRANS9 IN PASCAL *)
  ...
  OUTPUT V6.askmm(...)
END ;
(* TRANS10 *)
TRANS
WHEN V6.givemp (...)
FROM needmemory
TO running
BEGIN
  (* BODY TRANS10 IN PASCAL *)
  ...
END ;
END ;
(* END OF CPU MODULE *)
```

4. DEVELOPMENT WITH ESTELLE WORKSTATION

The whole ESTELLE specification for the multiprograming system is done in [10]. Its data are:

DATA OF THE SYSTEM SPECIFICATION

```
Number of code lines:.............  3823
Number of procedures:.............    83
Number of functions:..............    14
Number of modules:................    51
```

```
Number of modules bodys:..........     51
Number of channels:...............     36
Number of interaction points:.....    108
Number of transitions:............    190
Number of output sentences:.......    154
```

Once the system specification is specified, we can continue with the simulation of its behaviour and its later implementation using the Estelle WorkStation (EWS). EWS is the result of the ESPRIT european Project 1265 (Sedos ESTELLE Demonstrator), developed by some universities and firms from several European countries.

All software tools integrating that project run under Unix in Apollo and Sun workstations (for us, a SUN 4.0), and it is intended that the specification of any system could be validated, translated and then generated into C code, so simulations and implementations of the designed system can be performed.

First, to write the example specification of this paper we have used the editor SOE/ESTELLE [11.1] (A Syntax Oriented Editor for ESTELLE) and we store it with the name example.stl. Then, we use a ESTELLE TRANSLATOR [11.2], that validates syntacticaly the specification and gives us a listing with the results of the compilation: cross-references, information about modules and a summary of the specification; all these things are stored in the file example.stl.l. Moreover, in this step a file in an intermediate code example.if, that will be used in following phases is created.

After this, if the specification has defined values for the constants (CONST), types (TYPE), procedures (PROCEDURE) and functions (FUNCTION) we can go to following phases: to use the ESTELLE C GENERATOR [11.3] to obtain a file example.c from example.if. This is the system specification in C language (the number of generated lines is about 10 times greater than the original ESTELLE specification). Finally, the C program can be compiled and the result is used by the ESTELLE SIMULATOR [11.4] to perform the simulation, and by the ESTELLE Implementation Motor [11.5] to make the implementation.

Recently, we have a VEDA tool [12] (its name is VESAR in its later version). This tool, developed by Verilog, is a subset of ESTELLE, but has some advantages over EWS ESTELLE to simulate the behaviour of the modeled system, being able in this way to improve the study of the modeled or specified systems with the ESTELLE FDT, as it is done in this paper.

5. CONCLUSIONS

The design shown in this paper is based on a hierarchical structure for a multiprogramming operating system (fully specified in [10]); to do that we have divided modules into submodules and so on in a recursive scheme. In [10] this is also done for other kinds of operating systems: monoprogramming, multiprocess and distributed network.

This view of the system has a significant advantages: for example, if we want to study processor management in detail, there is in our design one module for the processor management (CPU), so we are able to deepen in its design, proposing different scheduling strategies, analysing and comparing the performance of the different policies.

We have already said that we have chosen the FDT ESTELLE because it fits very well with the modules and channels in which we have modeled the operating system and because we have access to some ESTELLE based tools. It would be interesting for further works to model the system using other formal description techniques: LOTOS, SDL, and so be able to compare advantages and disadvantages of each of them.

Finally, the main goal with the design shown in this paper was to prove it is possible to apply formal description techniques (in this study, FDT ESTELLE) to the field of operating systems; thus we can get the well known advantages of the traditional FDTs application, that is, communication protocols. As an operating system is a very complex concurrent system, the use of FDTs facilitates the abstraction and structuring of the operating systems definition in a concrete, clear and unambiguous way. Beside, to the advantages of the modelling of the operating system or part of it, we have to add the facilies of the Estelle WorkStation (and also VESAR), so we can perform the simulation, and also, its implementation, together with the formalitation of different policies of resource management. In this way we can study the behaviour of alternate solutions in order to choose the best.

REFERENCES

[1] H.M. Deitel. *Operating Systems, Second Edition.* Addison-Wesley, Massachusetts, 1990.

[2] G.W. Brams. *Réseaux de Petri: Théorie and Práctique, T. 1 y 2.* Ed. Masson, Paris, 1986.

[3] ISO 9074. *Information Processing Systems - Open Systems Interconnection - ESTELLE - A Formal Description Technique Based on a Extended State Transition Model.*

[4] ISO 8807. *Information Processing Systems - Open Systems Interconnection - LOTOS - A Formal Description Technique Based on a Temporal Ordering of Observational Behaviour.*

[5] CCITT. *Functional Specification an Description Language, Recommendation Z.100.*

[6] K.J. Turner. *Using Formal Description Techniques.* J.Wiley, England, 1993.

[7] M. Díaz, J. Dufau and R. Groz.. Experiences Using ESTELLE Within SEDOS ESTELLE Demonstrator. *2nd International Conference on Formal Description Techniques: FORTE89*, Vancouver, December,1989.

[8] A. Silverschatz, J. Peterson and P. Galvin. *Operating Systems Concepts, Third Edition.* Addison-Wesley, Massachusets, 1991.

[9] A.S. Tanembaum. *Modern Operating Systems.* Prentice-Hall, 1992.

[10] O. González. *Modelling of Operating Systems: Formal Specification in ESTELLE.* Doctoral Thesis. University of Oviedo (Spain), 1992. In Spanish.

[11] *SOE/ESTELLE (Syntax Oriented Editor for Estelle) User's Manual.*
ESTELLE Translator User's Manual.
ESTELLE C Generator User's Manual.
ESTELLE Simulator Interface User's Guide.
ESTELLE Implementation Motor User Guideline.
SEDOS ESTELLE Demonstrator Project 1265 Esprit, 1988 y 1989.

[12] Verilog. *VEDA- Protocol Design Verification. Administrador guide V.2.0.* Oct.1991
Verilog. *VESAR- Protocol Design Verification. V.2.1.1.* Dec. 1993.

5

The Formalization and Analysis of CCR Protocol and Service

Bairong Zhang Alistair Munro Michael Barton

Centre for Communications Research, University of Bristol,
Bristol BS8 1TR, United Kingdom

Commitment, Concurrency and Recovery protocol and service defined by ISO are studied by representing them in the notation of CCS. The behaviour of the CCR protocol and service is investigated on the Concurrency Workbench, and desirable properties of CCR are expressed and analyzed in the modal mu-calculus. The formulae are checked on the Concurrency Workbench.

Keywords: B.4.5 Reliability, Testing and Fault-Tolerance C.2.2 Network Protocols C.2.4 Distributed Systems

1. INTRODUCTION

Commitment, Concurrency and Recovery (CCR) is an application-service-element standardised by ISO as an Open System Interconnection (OSI) application layer service. It is intended to be used by other ASEs for providing the functionality of two-phase commitment to application processes. It can be referenced whenever the processing of two or more application-entity invocations (or processes) in a distributed application needs to be organized as an atomic action, or when recovery is required after failure. CCR services are used intensively by ISO OSI standards such as Transaction Processing, and Job Transfer and Manipulation which supports the creation, distribution, and management of tasks for processing by computer systems in an open network. In either case, an activity can be performed as a single, multi-branch, multi-association atomic action, or as a sequence of atomic actions. However, the different modes of operation allow a wide range of configurations to be used by a distributed application and raise questions about the security and the complexity of the recovery process after failures. Hence, proving the integrity of CCR is considered a major challenge for formal description techniques[12].

The Calculus of Communicating Systems (CCS) [9] provides a semantic basis for reasoning about transition systems [8, 14, 1, 10]. Modal logic can be interpreted on labelled transition systems and it naturally describes the transitional behaviour of CCS processes. The modal mu-calculus is a modal logic extended by fixpoint operators; this is appropriate for describing temporal properties of transition systems. The Concurrency WorkBench (CWB) is a prototype automated system which caters for the manipulation and analysis of transition systems. In this paper, the CCR protocol and service are formalized in the notation of CCS, and *atomicity* and *liveness* properties of CCR are analyzed and expressed in the modal mu-calculus, and checked on the Concurrency Workbench.

2. CCR PROTOCOL AND SERVICE

2.1. Introduction

In an Open Systems Interconnection (OSI) environment, application processes may wish to communicate with each other for a wide variety of reasons. However, any communication requires certain common services independent of the specific application. CCR provides such a "common application service element" for a single association. It consists of a set of primitive services that can be used for starting and ending a specific sequence of distributed application operations despite application or communication failure. Thus the use of CCR allows a referencing specification to define its activity as an atomic action.

An atomic action is a specific set of related distributed application operations that may be characterized by the following properties: Atomicity – a set of related operations are either all performed or none of them are performed (exactly-once semantics); Consistency – the "real effects" of a set of related operations are performed accurately, correctly and with validity, with respect to application semantics; Isolation – the partial results of a set of related operations are not accessible, except by operations of the set; Durability – all the effects of the operations are not altered by any sort of failure.

Although the CCR discussed here is specifically the ISO version, it is linked very closely to distributed two-phase commitment protocols that have been used for many years by the distributed database and transaction processing communities.

2.2. CCR Service and Procedures

There are nine CCR services defined in [4, 5], a C-CANCEL service is added in [2, 3]. Table 1 lists the CCR services, the type of service, and the requestor of the service. Those CCR services form the following procedures: Initialization; Begin branch; Prepare; Offer commitment; Order commitment; Rollback; One-phase commitment; Read-only; Branch recovery; Order commitment and begin new branch; Rollback and begin new branch.

Table 1
CCR service

Service	Type	Requestor
C-INITIALIZE	Confirmed	Association-initiator
C-BEGIN	Optionally confirmed	Either CCR service-user
C-PREPARE	Non-confirmed	Branch-initiator
C-READY	Non-confirmed	Either CCR service-user
C-COMMIT	Confirmed	Either CCR service-user
C-ROLLBACK	Confirmed	Either CCR service-user
C-ONE-PHASE	Confirmed	Either CCR service-user
C-READ-ONLY	Non-confirmed	Either CCR service-user
C-RECOVER	Confirmed or optionally confirmed	Either CCR service-user
C-CANCEL	Non-confirmed	Either CCR service-user

The combined CCR procedures have a number of possible paths in which each path is composed of some of the above procedures starting with *Initialization* and then *begin branch.* All the operations making up an atomic action are either completed successfully (leading to *commitment*) such that the *bound data* associated with an atomic action go to its *final state*; or they are *rolled back* such that the *bound data* is returned to its *initial state*; or when an atomic action is interrupted by an application or communication failure the system is left in a consistent state and *recovery* may take place at some later time.

3. CCS CONCEPT AND TOOLS

3.1. The Calculus of Communicating Systems

CCS [9] introduces five basic ways of constructing expressions for agents as the basic language. Agents are given a structured operational semantic as a labelled transition system, with labels drawn from a set of $Act = \mathcal{L} \cup \{\tau\}$ which range over by $\alpha, \beta \cdots$. Where labels $\mathcal{L} = \mathcal{A} \cup \bar{\mathcal{A}}$. $\bar{\mathcal{A}}$ is a set of complements of labels in $\mathcal{A}$, where $\bar{\mathcal{A}} = \{\bar{l} | l \in \mathcal{A}\}$. τ is a distinguished silent action which is invisible to the observer; all other actions are visible. The set of agent expressions $\mathcal{E}$ range over by $E, F, \cdots$ has the following basic forms [14]:

• $\alpha.E$, a *Prefix* ($\alpha \in Act$). $\alpha.E$ is an agent which performs an action α and then become an agent E.

• $\Sigma_{i \in I} E_i$, a *Summation* (I an indexing set). $\Sigma_{i \in I} E_i$ behaves like one of E_i; as soon as one performs its first action the others are discarded.

• $E_1 | E_2$, a *Composition.* $E_1 | E_2$ is an agent in which E_1 and E_2 may proceed independently but may also communicate by performing a pair of complementary actions.

• $E \backslash \alpha$, a *Restriction* ($\alpha \subseteq \mathcal{L}$). $E \backslash \alpha$ is an agent which behaves like E, except that action α and its complementary action $\bar{\alpha}$ are forbidden.

• $E[f]$, a *Relabelling* (f a relabelling function). $E[f]$ is an agent derived from agent E by relabelling its action using the relabelling function f, where if $f(l) = l'$ then $f(\bar{l}) = \bar{l}'$.

The inference rules for those operations give the transition semantics. They are:

$$Act \; \frac{}{\alpha.E \xrightarrow{\alpha} E} \qquad\qquad Sum_j \; \frac{E_j \xrightarrow{\alpha} E'_j}{\Sigma_{i \in I} E_i \xrightarrow{\alpha} E'_j} \; (j \in I)$$

$$Com_1 \; \frac{E \xrightarrow{\alpha} E'}{E|F \xrightarrow{\alpha} E'|F} \qquad\qquad Com_2 \; \frac{F \xrightarrow{\alpha} F'}{E|F \xrightarrow{\alpha} E|F'}$$

$$Com_3 \; \frac{E \xrightarrow{l} E' \;\; F \xrightarrow{\bar{l}} F'}{E|F \xrightarrow{\tau} E'|F'} \qquad\qquad Res \; \frac{E \xrightarrow{\alpha} E'}{E \backslash L \xrightarrow{\alpha} E' \backslash L} \; (\alpha, \bar{\alpha} \notin L)$$

$$Rel \; \frac{E \xrightarrow{\alpha} E'}{E[f] \xrightarrow{f(\alpha)} E'[f]} \qquad\qquad Con \; \frac{P \xrightarrow{\alpha} P'}{A \xrightarrow{\alpha} P'} \; (A \stackrel{\text{def}}{=} P)$$

The basic transition system is $(\mathcal{E}, \{\xrightarrow{\alpha} | \alpha \in Act\})$. The interpretation of $P \xrightarrow{\alpha} Q$ is that P may evolve into Q by performing the observed action α, whereas $P \xrightarrow{\tau} Q$ means that P may evolve autonomously into Q.

Distinguishing between two agents by any external observation of their behaviour is achieved by a notion of bisimulation which is a standard device for defining behavioural equivalence for process algebras.

P and Q are strongly bisimilar or strongly equivalent, written $P \sim Q$. In strong bisimulation, every α action of one agent must be matched by an α action of the other – even for τ actions. If we merely require that each τ action be matched by zero or more τ actions; this yields a weaker notion of bisimulation. P and Q are (weakly) bisimilar or observation-equivalent, written $P \approx Q$ (refer to [9]).

3.2. Modal and Temporal Logics

Modal and temporal logics provide a repository of useful properties [13]. Modal logics are interpreted on labelled transition systems $(\mathcal{E}, \{\stackrel{\alpha}{\longrightarrow} | \alpha \in Act\})$, where $\mathcal{E}$ is the set of agents; *Act* the action set; and $\stackrel{\alpha}{\longrightarrow}$ the α transition relation. Its formulae are:

$$A ::= true \mid \neg A \mid A \wedge B \mid [\alpha]A \tag{1}$$

A formula is either the constant true formula; or a negated formula $\neg A$; or a conjunction of formulas $A \wedge B$; or a modalized formula $[\alpha]A$ whose intended meaning is: A holds after every action α. For each formula A the agents having the property A are defined as $\|A\|$:

$$\|true\| = \mathcal{E}$$

$$\|\neg A\| = \mathcal{E} - \|A\|$$

$$\|A \wedge B\| = \|A\| \cap \|B\|$$

$$\|[\alpha]A\| = \{E \in \mathcal{E} \mid \forall E' \ if \ E \stackrel{\alpha}{\longrightarrow} E' \ then \ E' \in \|A\|\}$$

So agent E has the property $[\alpha]A$ if every E' such that $E \stackrel{\alpha}{\longrightarrow} E'$ for any α in *Act* has the property A. The dual of $[\alpha]A$ is $\langle\alpha\rangle A$ which expresses that for some E' and some $\alpha \in Act$ such that $E \stackrel{\alpha}{\longrightarrow} E'$.

Consider a recursive modal equation: $Z = \langle\alpha\rangle Z$. It can be viewed as expressing various properties of transition systems. Each of these properties is determined by a solution to the equation. On the CCS $\stackrel{\alpha}{\longrightarrow}$ transition system there are infinitely many solutions to the above equation. Especially important are the maximal solution, which is the union of all solutions, and the minimal solution, the intersection of all solutions.

To express temporal properties it is useful to add recursively defined formulae. This leads to the *modal mu-calculus* whose formulae are obtained by adding the the description to the formulae (1).

$$A ::= \cdots \mid Z \mid \nu Z.A \tag{2}$$

Where Z range over a family of propositional variables, and νZ is a fixpoint operator. The formula $\nu Z.A$ expresses the property given by the maximal solution to $Z = A$. An important derived operator is $\mu Z = \neg \nu Z. \neg A[Z := \neg Z]$, where $A[Z := \neg Z]$ is the result of substituting $\neg Z$ for each free occurrence of Z in A. The formula $\mu Z.A$ expresses the property given by the minimal solution to $Z = A$ (refer to [13]).

Maximal fixed point formulae express invariance properties which can be described as safety properties – that nothing bad ever happens. Minimal fixed point formulae express eventuality properties which can be described as liveness properties – that something good does eventually happen.

3.3. The Concurrency Workbench

The Concurrency Workbench (CWB)[11] is an automated tool designed to provide machine assistance in analysis of concurrent systems described in CCS. The CWB is useful both in understanding and in reasoning about transition systems. The main functionalities we used in CWB are:

- Environment commands for binding agent identifier; printing environment; input file; output environment or agent into a file.
- Agent Commands for constructing state space of an agent; counting state size or checking the possible deadlocks in an agent.
- Equivalences and preorders commands for exploring the behaviour of an agent; determining whether or not two agents are related by bisimulation and testing equivalence and preorders; minimizing an agent observation-equivalence.
- Model checking command for checking whether a given process satisfies a specification formulated in the modal mu-calculus.

4. CCR REPRESENTATIONS IN CCS

The CCS descriptions of CCR protocol and service are based on [4, 5, 2, 3] CCR Protocol Machine (CCRPM) state table with the constraints of CCR service-user rules, and on the dynamic behaviour part of LOTOS description of CCR [6, 7].

4.1. Introduction to the Formal Description

A CCR is modeled in Figure 1 which consists of a CCR user and a CCRPM. The CCR *request* and *response* service primitives are sent by the CCR user via CCRsu, while the CCR *indication* and *confirm* service primitives are sent by the CCRPM via CCRsp. The CCR Application Protocol Data Units (APDUs) are sent or received through CCRpeer to another CCRPM. A CCR atomic action branch is composed of two CCR users: a superior (branch requestor) and a subordinate (branch acceptor) showed in Figure 2 and Figure 3.

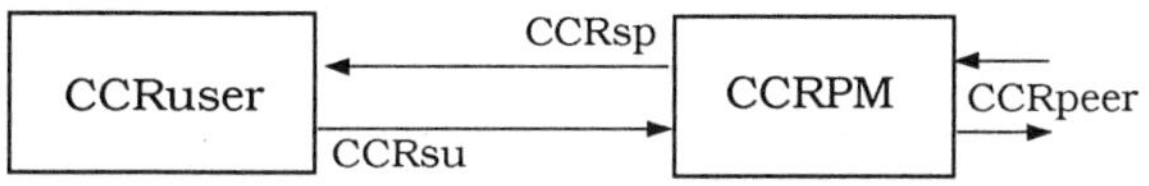

Figure 1. A CCR model

4.1.1. CCR Service

The CCR service [4, 2] CCRPM state table specifies the interrelationship between the current state of a CCRPM, the incoming service primitives and the resultant states of the CCRPM. An atomic action branch of CCR service is modeled as in Figure 2, where *sup* CCRuser represents a superior, and *sub* CCRuser represents a subordinate. Issuing a *request* primitive from one CCRuser will result in receiving an *indication* primitive in another CCRuser, while issuing a *response* primitive from one CCRuser will result in receiving a *confirm* primitive in another CCRuser.

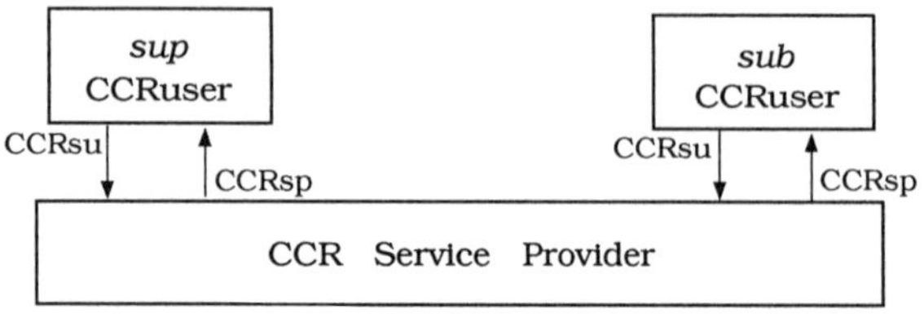

Figure 2. CCR service model

4.1.2. CCR Protocol

The CCR protocol [5, 3] CCRPM state table specifies the interrelationship between the current state of a CCRPM, the incoming events that occur, predicates and preconditions that determine what to do next, enablements that permit changes to data related to an atomic action, the action taken, outgoing events, and finally, the resultant state of the CCRPM. An atomic action branch of CCR protocol is modeled as in Figure 3. Receiving a *request* or *response* primitive from CCRsu in one CCRPM will result sending a RI (related to the Request and Indication primitives) or RC (related to the Response and Confirm primitives) CCR APDU to another CCRPM via CCRpeer, while receiving a RI or RC CCR APDU in one CCRPM from CCRpeer will result sending an *indication* or *confirm* primitive to its own CCRuser via CCRsp.

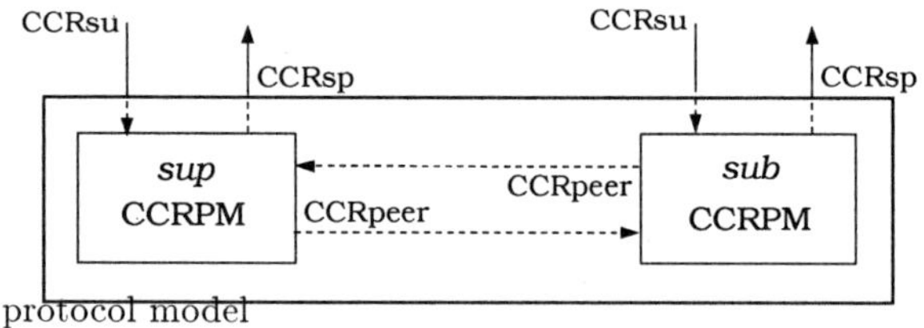

Figure 3. CCR protocol model

4.2. CCS Specification

In this paper we are concerned with a CCR atomic action branch which is composed of two CCR users: a CCR superior and a CCR subordinate. A complementary action of

l representation in CWB is $'l$.

4.2.1. Service Primitives Representation

The CCR service primitives C-INITIALIZE, C-BEGIN, C-PREPARE, C-READY, C-COMMIT, C-ROLLBACK, C-READ-ONLY, C-CANCEL, C-RECOVERY and C-ONE-PHASE are represented by **i**, **b**, **p**, **d**,**m**, **l**, **r c**, **rc** and **o** respectively. Four service primitives: request, indication, response and confirm are represented by **req**, **'req**, **rsp**, **'rsp** respectively. The recovery service primitive has five values which can be looked upon as five actions. They are **ready**, **commit**, **unknown**, **retry** and **done**.

4.2.2. CCR APDU Representation

There are sixteen CCR APDUs related to ten CCR service elements. Receiving and issuing those CCR APDUs via CCRpeer form the incoming and outgoing events in the CCRPM state table. Issuing CCR APDUs are represented as **tbga**, **tbgn**, **tcan**, **tcma**, **tcmt**, **tcmtbg**, **tina**, **tini**, **topa**, **toph**, **tprp**, **tro**, **trba**, **trbk**, **trbkbg**, **trca**, **trcv** and **trdy**. Receiving CCR APDUs are represented by adding $'$ in front of each issuing CCR APDU representation.

4.2.3. Global Constraints of the CCR

In the CCR service specification, issuing and receiving the service primitives from *sup* ccr_{user} and *sub* ccr_{user} are presented as a restricted composition in CCS. In the CCR protocol specification, issuing and receiving the CCR APDUs from *sup* ccr_{peer} and *sub* ccr_{peer} are presented as a restricted composition in CCS. The state change is caused by those incoming and outgoing events which form the dynamic behaviour of CCR.

An atomic action *Branch* is composed of a superior and a subordinate.

$$Branch = (SupCCEP|SubCCEP)\backslash\{outgoing\ event,\ incoming\ event\}$$

where CCEP means CCR Connection End Point.

4.2.4. Constraints of the Superior and the Subordinate

The CCR procedures in the superior side and subordinate side can be divided into a normal sequence and a recovery sequence. The normal sequence is made up of all the service primitives and related CCR APDUs, based on the CCRPM state table. The recovery sequence is made up of the C-RECOVERY service primitive and five actions, based on the Branch Recovery Machine (BRM) state table. The normal sequence and the recovery sequence are represented as a summation in CCS.

The CCS specification of CCR presented and studied here has five versions. The first one is based on the dynamic behaviour part of the LOTOS description of CCR [6] (see Appendix A, B, C). Only six basic CCR service C-BEGIN, C-PREPARE, C-READY, C-COMMIT, C-ROLLBACK and C-RECOVERY are concerned. This specification represented as $Branch_l$. Others are constructed from [4, 5, 2, 3] the CCRPM state table and the BRM state table. The constraints of CCR service-user rules are applied in forming a restricted composition on CCS. They are represented as $Branch_{s.91}$, $Branch_{p.91}$, $Branch_{s.94}$, $Branch_{p.94}$, BRM_{91} and BRM_{94}.

In an atomic action branch, after *committed* or *rolled back*, both sides will exit from the branch. If C-BEGIN request primitive is issued with C-COMMIT or C-ROLLBACK request primitive, the superior and the subordinate will start a new atomic action branch.

5. REASONING ABOUT CCR

A property which should be satisfied by the procedures of CCR service and protocol is to be deadlock free. Deadlock free means being always capable of executing an action. It can be expressed as a maximal fixed point formula:

$$s \models \nu Z.(\langle . \rangle true \wedge [.]Z)$$

This states that at every point on every path from s, executing an action is always possible. This formula was checked on the Concurrency Workbench. A deadlock can be found by the command '**fd**' on the Concurrency Workbench. The results from the Concurrency Workbench show that: Agents $Branch_l$, $Branch_{p.91}$, $Branch_{s.94}$ and $Branch_{p.94}$ are deadlock free. There is a deadlock state $(L3|I)$ in $Branch_{s.91}$. This deadlock can be avoided by adding an incoming event $'$**breq** from the state $L3$, then change to the state $A2$. In the amendment [2, 3] the procedures of CCR service and protocol are deadlock free. This property guarantees that CCR will always respond in some way.

The states can be counted by the '**states**' command. The results show that state D1 and state H4 are unreachable in all the *Branch* specification. D1 is a state when a C-READY request primitive has been issued, and then a C-READY indication primitive is received. H4 is a state when a C-ROLLBACK-BEGIN request primitive has been issued, and then a C-ROLLBACK indication primitive is received. The C-ONE-PHASE service has no effect on the CCRPM state table in [2]. The states K1 and J in [2] are unreachable as they are both related to the C-ONE-PHASE service. The states checked from BRM_{91} show that states from BRM5 to BRM9 are all unreachable, but they are reachable states in BRM_{94}.

5.1. Bisimulation Agent

To analyze the behaviour of a *Branch* agent, we can construct and analyze a simpler agent Br by using the relation commands. Where

$$\begin{aligned} Br &= (A|B)\backslash\{a,\ b\} \\ A &= a.A + b.P \\ B &= 'a.B + 'b.P \\ P &= e.P \end{aligned}$$

The Br agent is constructed from two agents A and B by restricted composition. The agents A and B are relevant to the superior and the subordinate in *Branch* agent. After a communication or synchronization via a pair of complementary actions a or b, A and B can either go back to the branch or exit from the branch. The results from the Concurrency Workbench show that *Branch* and *Br* are observation equivalent and observation congruent. This implies that in an atomic action branch, after a communication, both sides may exit from the branch or begin a new atomic action branch.

5.2. Liveness

How should we interpret the term *live* for the procedures of CCR? In an atomic action branch, there are a number of choices for the superior and subordinate. In a normal sequence one choice is that after committed the superior and subordinate will both exit from the branch and release the bound data in the final state. Another choice is to force the completion of an atomic action, in this case the superior and subordinate rollback and release the bound data in the initial state. A further choice is a C-BEGIN request primitive issued with C-COMMIT or C-ROLLBACK request primitive: in this case the superior and subordinate will go back and start a new atomic action. While in a recovery sequence one choice is that after a **recovery committed** request, if **recovery done** is confirmed, the superior and subordinate will both exit from the branch and release the bound data in the final state. A second choice is that after a **recovery ready** request, if **recovery unknow** is confirmed, the superior and subordinate will both exit from the branch and release the bound data in the initial state. The final choice is that if **recovery retry** is confirmed, the superior and subordinate will both go to the states where the recovery sequence started.

As CCS is a rigorous semantics language, both the interpretations and assumptions may be formulated precisely. To interpret the term *live*, we say the procedure of CCR is live if whenever at some point in a computation, all those choices described above are possible choices, and will eventually happen after some action. Since all actions in the *Branch* are complementary actions and cannot be observed outside of the *Branch*, we add an extra action at the end of each possible choice. For those paths which will exit from the *Branch* we add $exitp_i$ and $exitb_i$ in the superior and subordinate respectively, and for those paths which will go back to *Branch* we add $backp_i$ and $backb_i$ in the superior and subordinate respectively, where i range over a finite integer set. This interpretation of *live* can be expressed in the following minimal fixed point formulae:

$$Branch \models \bigwedge_i \mu Z.((\langle exitp_i\rangle true \wedge \langle exitb_i\rangle true) \vee \langle.\rangle Z)$$

$$Branch \models \bigwedge_i \mu Z.((\langle backp_i\rangle true \wedge \langle backb_i\rangle true) \vee \langle.\rangle Z)$$

This means that there is some point along some path in *Branch* consisting of some action at which both $exitp_i$ and $exitb_i$ may be performed or both $backp_i$ and $backb_i$ may be performed. The formula:

$$Branch \models$$

$$\bigwedge_i \mu Z.(((\langle backp_i\rangle true \wedge \langle backb_i\rangle true) \vee (\langle exitp_i\rangle true \wedge \langle exitb_i\rangle true)) \vee [.]Z)$$

means that along every path, containing infinitely many visible actions, one of the possible choices described above will happen eventually.

5.3. Atomicity

To say that the procedure of CCR preserves atomicity means that a set of related operations are either all performed or none of them are performed. As we have described above, superior and subordinate have several executing path choices and related bound data will be released in different states accordingly. Thus to establish that the procedure of CCR preserves atomicity it suffices to show that the superior and the subordinate cannot evolve into a state in which the bound data is in *initial state* in one side while it is in *final state* in the other side; or one exits from the branch while the other one stays; or one is in a normal sequence while the other one is in a recovery sequence. That is, it is enough to show that $exitp_i$ and $exitb_j$ where i is different from j can not be performed simultaneously; and both $backp_i$ and $backb_j$ where i is different from j can not be performed simultaneously; and both $exitp_i$ and $backb_j$ can not be performed simultaneously; and both $backp_i$ and $exitb_j$ can not be performed simultaneously. If none of above actions can be performed, then superior and subordinate will go through the same sequence and the bound data will be released in a required state.

This property can be expressed in the following maximum fixed point formulae:

$$Branch \models \bigwedge_{i} \bigwedge_{j \neq i} \nu Z.(\neg(\langle exitp_i \rangle true \wedge \langle exitb_j \rangle true) \wedge [.]Z)$$

$$Branch \models \bigwedge_{i} \bigwedge_{j \neq i} \nu Z.(\neg(\langle backp_i \rangle true \wedge \langle backb_j \rangle true) \wedge [.]Z)$$

$$Branch \models \bigwedge_{i} \bigwedge_{j \neq i} \nu Z.(\neg(\langle exitp_i \rangle true \wedge \langle backb_j \rangle true) \wedge [.]Z)$$

$$Branch \models \bigwedge_{i} \bigwedge_{j \neq i} \nu Z.(\neg(\langle backp_i \rangle true \wedge \langle exitb_j \rangle true) \wedge [.]Z)$$

This means that at every point on every path from *Branch*, consisting of any of the actions at which both $exitp_i$ and $exitb_j$; $backp_i$ and $backb_j$; $exitp_i$ and $backb_j$; $backp_i$ and $exitb_j$ cannot be performed simultaneously is always true. In this case, that superior and subordinate release the bound data in the required state is always true; or superior and subordinate go to the normal sequence or recovery sequence simultaneously is always true; or superior and subordinate exit from the *Branch* simultaneously or restart a new *Branch* simultaneously is always true, hence the atomicity property is preserved.

6. CONCLUSION

In this paper the procedures of CCR service and protocol have been formalized in the notation of CCS, and desirable properties of CCR have been interpreted and expressed by the modal mu-calculus and checked on the Concurrency Workbench.

A complex, dynamic system can be viewed as being composed of several parts called agents. Although the decomposition of a system into agents depends upon the user's intent, not upon the entities, there are often common services required to coordinate individual activities within the distributed application. CCR, which provides guarantees of atomicity, is one such service. Although the protocol analysed in this paper is described in terms related directly to the ISO OSI version, the analysis has concentrated on key properties that apply to distributed two-phase commitment in general.

From the implementation viewpoint the CCR protocol and service can be constructed as a finite state machine. Realising this directly in CCS, based on the automata given in the specification, requires a lot of states to be specified. This results in a system that is too large to be analyzed by the Concurrency Workbench. From the formal analysis viewpoint, we wish to know whether the procedures of CCR are workable in general, hence we can concentrate on those important parts which construct the main procedures of CCR and contribute to the main properties of CCR, and ignore some less important details, such as some actions related to each service primitive. We assume that actions which are related to each incoming event are performed correctly, and under the correct execution of the service primitives we are concerned whether the procedures of CCR service and protocol are workable. A more strict formal analysis of CCR, consisting of data structures in more detail is needed but this is restricted by the current tools available.

Future work will consider compositional issues: both where the protocol is used by a client application, thus capturing the semantics of an entire system; and, also, where the essence of the protocol is retained but optimised by incorporating its behaviour directly in the application protocol. In either case, the obligation is incurred to demonstrate that the system retains the properties demonstrated here.

7. Acknowledgments

The authors are grateful for the support of Prof.J.P.McGeehan of the Centre for Communications Research in this work. It was funded in part by the Science and Engineering Research Council of the UK under grant GR54399 and by the Commission of the European Union under Esprit Project 6441.

REFERENCES

1. Glenn Bruns, Stuart Anderson *The Formalization and Analysis of a Communications Protocol*, LFCS Report ECS-LFCS-91-137, University of Edinburgh, 1991
2. *Commitment, Concurrency and Recovery Service - Amendment 1: enhancements* February 1994
3. *Commitment, Concurrency and Recovery Protocol - Amendment 1: enhancements* February 1994
4. Jim Quigley *ISO 9804/AM1 – Working draft # 4*, ACSE Project, 1991
5. *Amendment to CCR protocol specification (ISO/IEC 9805) covering enhancements–Working draft 4*, September 1991
6. ISO/IEC JTC 1/SC 21 N 8502 *Revised Text of PDTR 11589, LOTOS Description of CCR Service* February 1994
7. ISO/IEC JTC 1/SC 21 N 8503 *Revised Text of PDTR 11590, LOTOS Description of CCR Protocol* February 1994
8. Joachim Parrow, *Verifying a CSMA/CD-protocol with CCS* Protocol Specification, Testing, and Verification VIII IFIP pp373-384, 1988
9. R. Milner, **Communication and Concurrency**, Prentice-Hall international, 1989

10. M. J. Morley *Modelling British Rail's Interlocking Logic: Geographic Data Correctness*, LFCS Report ECS-LFCS-91-186, University of Edinburgh, 1991
11. Faron Moller, *The Edinburgh Concurrency Workbench(Version 6.1)*, Department of Computer Science, University of Edinburgh, October, 1992
12. Alistair Munro, *Job Transfer and Manipulation(JTM)*, Centre for Communications Research, University of Bristol, UKUUG and UKnet Winter Technical Meeting, Dec. 1989
13. C. Stirling, *Temporal Logics for CCS*, Lecture Notes in Computer Science 354, pp 660-672, 1989
14. D. J. Walker *Automated Analysis of Mutual Exclusion Algorithms using CCS*, LFCS Report ECS-LFCS-89-91, University of Edinburgh, 1989

APPENDIX

A. Global constraints of the CCR service

$$\begin{aligned} Branch &= (SupCCEP|SubCCEP)\backslash\{breq, brsp, preq, dreq, mreq, mrsp, lreq,\\ &\quad lrsp, rcreq, rcrsp, mbreq, lbreq, ready, commit, unknown, retry, done\} \end{aligned}$$

B. Constraints of the superior and subordinate

B.1. Normal sequence of the superior

$$SupCCEP = SupAction + SupRecovery$$

$$SupAction = breq.ContSupAction$$

$$\begin{aligned} ContSupAction = {} & 'brsp.(preq.ContSupAction1 + ContSupAction1)+\\ & preq.('brsp.ContSupAction1 + ContSupAction1)+\\ & ContSupAction1 \end{aligned}$$

$$\begin{aligned} ContSupAction1 = {} & 'dreq.SupDecision +' lreq.(lrsp.Process+\\ & lbreq.'lrsp.ContSupAction + lreq.'lrsp.Process)+\\ & lreq.SupRollSend1 + lbreq.SupRollSend2 \end{aligned}$$

$$\begin{aligned} SupDecision = {} & mreq.'mrsp.Process + lreq.'lrsp.Process+\\ & mbreq.'mrsp.ContSupAction+\\ & lbreq.'lrsp.ContSupAction \end{aligned}$$

$$SupRollSend1 = 'lrsp.Process +' lreq.lrsp.Process$$

$$SupRollSend2 = 'lrsp.ContSupAction +' lreq.lrsp.ContSupAction$$

B.2. Recovery sequence of the superior

$$SupRecovery = \text{'}rcreq.\text{'}ready.SupRecResponse + rcreq.commit.SupRecActions$$

$$SupRecResponse = rcreq.commit.SupRecActions + rcrsp.(unknown.Process + retry.SupRecovery)$$

$$SupRecActions = \text{'}rcrsp.(\text{'}done.Process + \text{'}retry.SupRecovery)$$

B.3. Normal sequence of the subordinate

$$SubCCEP = SubAction + SubRecovery$$

$$SubAction = \text{'}breq.ContSubAction$$

$$ContSubAction = brsp.(\text{'}preq.ContSubAction1 + dreq.ContSubAction2 + Rollback) + \text{'}preq.(brsp.ContSubAction1 + dreq.ContSubAction3 + Rollback) + dreq.ContSubAction2 + Rollback$$

$$ContSubAction1 = dreq.ContSubAction3 + Rollback$$

$$ContSubAction2 = \text{'}preq.ContSubAction3 + ContSubAction3$$

$$ContSubAction3 = \text{'}mreq.mrsp.Process + \text{'}lreq.lrsp.Process + \text{'}mbreq.mrsp.ContSubAction + \text{'}lbreq.lrsp.ContSubAction$$

$$Rollback = \text{'}lreq.(lrsp.Process + lreq.\text{'}lrsp.Process) + lreq.(\text{'}lrsp.Process + \text{'}lreq.lrsp.Process + \text{'}lbreq.lrsp.ContSubAction) + \text{'}lbreq.(lrsp.ContSubAction + lreq.\text{'}lrsp.ContSubAction)$$

B.4. Recovery sequence of the subordinate

$$SubRecovery = rcreq.ready.SubRecResponse + \text{'}rcreq.\text{'}commit.SubRecAction$$

$$SubRecResponse = \text{'}rcrsp.(\text{'}unknown.Process + \text{'}retry.SubRecovery) + \text{'}rcreq.\text{'}commit.SubRecAction$$

$$SubRecAction = rcrsp.(done.Process + retry.SubRecovery)$$

6

On the Formal Specification and Verification of Network Routing Algorithms

Jim Davies
Department of Computer Science, University of Reading, Reading RG6 2AH, UK

Matt Wallis
BNR Europe Limited, Concorde Road, Norreys Drive, Maidenhead SL6 4AG, UK

Abstract This paper describes an application of two formal methods—the Z notation and the action systems formalism—to the problem of network routing. A distributed algorithm for generating routing tables is specified and proved correct. The paper shows also how parallel composition of action systems can be used to model the network as a single abstract data type; this may be implemented to provide a simulation of network behaviour.

Keywords I.2, I.3, I.8, II.4, II.12: Action Systems, IV.5, IV.12.

1. INTRODUCTION

This paper is a case study in the application of formal methods. The subject is the generation of routing tables in communication networks; the methods chosen are the Z notation [4] and the action systems formalism [1, 7, 2, 3]. The study proves a useful test of formal methods: the network routing problem is non-trivial, and the correctness of a solution is not easily established by informal argument.

The Z notation is a technique for model-based specification. It has been used with some success to specify systems in which states, and state changes, provide an adequate and concise description of system behaviour. However, Z has its limitations: there is still no accepted way to describe concurrent behaviour.

An action system consists of a state variable, an initialisation, and a set of actions. In the original approach of [1], two or more systems may interact through shared variables. An alternative approach, taken in [7], allows interaction only through shared actions. This work was extended in [2] to provide a theory of value-passing action systems.

These two formal methods are complementary. Used together, they are powerful indeed. We may use action systems to describe those aspects of a system which involve communication and concurrent execution, and the Z notation to describe state information, whether distributed or localised. Such a synthesis was proposed in [3] and is well-suited to the problem of network routing: the aim is to provide consistent state information, and the algorithms used are distributed across nodes.

The paper begins with a brief introduction to communication networks and routing algorithms. We continue with an overview of Z and action systems notation: this is necessarily brief, and the interested reader is referred to [2] for a full explanation. We then present a formal description of a simple routing algorithm, using Z to describe the data structures and operations associated with a network node, and action systems to describe the way in which the output of one node is related to the input of another.

The formal description allows us to establish that the sequence of routing information messages sent between nodes terminates: the algorithm is convergent. The remainder of our formal analysis is concerned with the state of the network after the algorithm has terminated. We are able to establish that the routing tables generated are correct, and that they correspond to minimum cost routes. The proof involves a consideration of network topology and a series of inductive arguments.

Having verified the algorithm, we show how the parallel composition of node action systems can be used to produce a sequential representation of network behaviour. This may be refined to produce a reliable software simulation of the distributed routing algorithm. The paper ends with a short discussion of the case study, including an explanation of how the work can be extended to more elaborate routing algorithms.

2. NETWORK ROUTING

A communication network consists of a set of nodes and a set of connecting links. The purpose of the network is to transmit data from one node to another, via a series of intermediate nodes if necessary. Given any pair of nodes, there will often be a choice of routes from one to the other; some of these routes will be better than others. This is quantified by associating a cost with each link in the network; the cost of a route is the sum of the costs of the links involved.

Each node maintains a routing table; these are used to determine the best route between pairs of nodes. It is not necessary to store information about complete routes; each node need only know the first step towards any given destination. The subsequent steps will be determined by successive nodes as the data reaches them. The routing table is thus a list of first steps, together with a list of route costs. By storing cost information in the routing tables, we aim to ensure that each piece of data is passed via the best route available: the route with the lowest cost.

In some networks, there may be two costs associated with each link—one measuring delay, another measuring capacity—and hence two classes of service. In these networks, the two costs are used independently, in that each has its own set of routing tables. If the generation of routing tables is independent of the utilisation of nodes and links—as it is here—then we may approach the routing problem as if we were dealing with two separate, single-cost networks.

Figure 1 depicts a small network with five nodes and seven links; each link is labelled with a cost value. There is more than one route between any pair of nodes; in some cases there is more than one minimum cost route: e.g., both *ABC* and *ADC* are minimum cost routes between *A* and *C*, each with a cost of *6*. Further, the minimum cost route may not

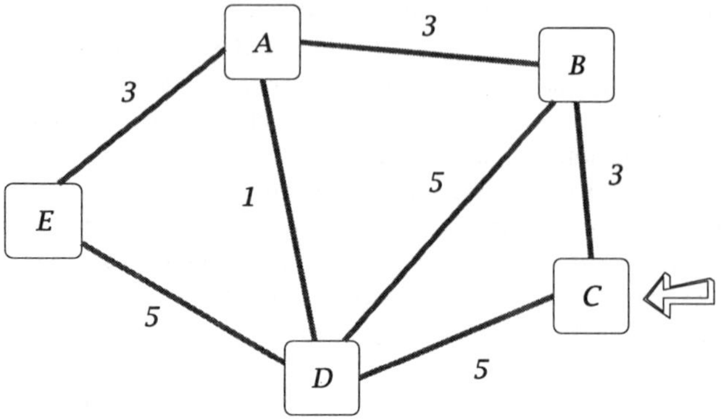

dest	cost	hop
A	6	D
B	3	B
C	0	-
D	5	D
E	9	B

Figure 1. A simple network

be the one with the fewest links—the route *CDE* has two links and a cost of *10*, while the route *CBAE* has three links and a cost of *9*.

The routing table at *C*—also shown in Figure 1—has an entry for each node in the network. The entry for node *A* tells us that the cost of the route to *A* is *6*, and the link to be used for data bound for *A* from *C* is the link to *D*. If we specify only that routing tables should contain minimum cost routes, then the entry 'A 6 B' is equally acceptable. This non-determinism in the specification is resolved in our algorithm by taking the minimum cost route that comes into service first.

Each node has link level software responsible for reporting the cost of each link. It is a function of the link level to ensure that the cost reported to the nodes at either end of the link is the same. It is also a function of the link level to report the identity of the node at the other end of a link.

Assuming that the link level software is functioning correctly, we can use the following algorithm to generate and maintain routing tables: as links appear, each node sends out offer messages, describing destinations and the costs associated with them. These correspond to the contents of the sending node's routing table. When an offer arrives, the receiving node consults its routing table; if the offer represents a cheaper route, then the table is adjusted and the neighbours are informed.

In the following sections, we will show that this algorithm is *convergent*, in that the configuration process terminates after a finite number of messages, and that the routing tables it generates are *adequate:* there is an entry for every reachable node which is added to the network; *accurate:* these entries reflect the true cost of getting there; *optimal:* these entries give the first step and the true cost of a minimum cost route.

3. FORMAL NOTATION

We use Z schemas to describe abstract data types and to encapsulate pieces of mathematical logic. Each Z schema is a named pattern:

```
┌─ SchemaName ──────────────
│ declarations
├──────────
│ predicate
└───────────────────────────
```

To model system state, we use schemas of the form

```
┌─ State ───────────────────
│ variables
├──────────
│ invariant
└───────────────────────────
```

where *invariant* is expressed in terms of *variables*, and describes some invariant property of the state.

We define operations upon the state by relating the state before the operation to the state after. We apply a special decoration $'$ to the components of a schema to indicate the values taken by the various variables after the operation has been performed. We employ the usual Z conventions concerning Δ and Ξ.

We apply other decorations to schemas to abbreviate renaming or labelling: e.g., if *State* describes a node state, we use $State_A$ to describe the state of node A. In case of multiple decorations, we do not distinguish between $State_A{}'$ and $State'_A$. We use an explicit binding notation: the pair $\langle\!| s \leadsto x, t \leadsto y |\!\rangle$ denotes a binding in which component s is bound to x and component t is bound to y.

A value-passing action system [2] consists of a state variable, an initialisation, and a set of actions. Each action consists of a guard and a command: the guard is a predicate which must be satisfied for the action to occur; the command describes the change in state associated with the action.

We use state schemas to represent the state variable and its initialisation, predicate schemas to represent the guards, and operation schemas to represent the changes in state. Every action is either an input or an output. For example,

$$\textbf{action}\ c\ \textbf{in}\ m : M :-\ G \rightarrow Op$$

corresponds to an input from communication channel c of value m with type M. The guard is schema G, while schema Op describes the resulting change in state. Some care must be taken concerning the choice of input guards if the description is to reflect our intentions. In our use of action systems, any attempt at communication will satisfy the corresponding input guards.

We use a replication construct **for** to represent indexed sets of channels. A typical action system would be:

var *State*
initially *StateInit'*
for $n \in N$ **action** *from.n* **in** $m : MessageType$:–
    $InGuard \rightarrow Input$
for $n \in N$ **action** *to.n* **out** $m : MessageType$:–
    $OutGuard \rightarrow Output$

where the guard and operation schemas may include instances of variable n.

A value-passing action system may be given a CSP-style failures-divergences semantics: e.g., the above system may engage in a synchronous communication of the form *to.n.m* whenever the *OutGuard* is satisfied. For the purposes of this paper, we are interested only in the trace component of this semantics: the possible sequences of communication actions, starting from the point of initialisation.

Action systems may be treated as CSP processes, and placed in indexed parallel compositions of the form

$$\|_{i \in I}\ \alpha Node_i : Node_i$$

where node process $Node_i$ has interface $\alpha Node_i$ in the resulting network. Two action systems operating in parallel must synchronise upon any action which appears in both interfaces. For example, if $\alpha Node_A$ includes the output action *a.b*, and $\alpha Node_B$ includes the input action *a.b*, then the two nodes may synchronise upon events of the form *a.b.m*. Such an event represents the transmission of a message m from $Node_A$ to $Node_B$. If the parallel composition is written as a single action system, this communication will appear as an output.

4. FORMAL DESCRIPTION

4.1. Node state

Each node in the network will be given a name from the set *Node*, and each link cost value will be a natural number:

$$Cost == \mathbb{N}$$

An offer message will include a destination and a price and will be passed in an envelope marked with sender and receiver:

Offer
$dest : Node$
$price : Cost$

Envelope
Offer
$sender : Node$
$receiver : Node$

Each node has a routing table indexed by node names. Each entry contains the name of a neighbouring node and a cost value; these represent the first step towards the given node, and the total cost of getting there. We declare a *null* neighbour and a suitable schema of type *RTEntry*.

RTEntry
$hop : Node$
$cost : Cost$

NullEntry
$RTEntry$

$hop = null$
$cost = 0$

The routing table has an entry for any destination for which the node has received an offer:

RTState
$nodeid : Node$
$routemap : Node \rightarrow\!\!\!\!\!\rightarrow RTEntry$

$\forall\, n : Node \bullet routemap\ n = \theta NullEntry \Leftrightarrow n = nodeid$

We represent the routing table using a partial function *routemap* and a node name *nodeid*, which records the identity of the current node. The destination cost of the current node is zero. The *NullEntry* may appear only as a node's routing table entry for itself.

Each node maintains a link table, giving the destination cost for each of its neighbours: the metric value of the corresponding link. We model this as a partial function, and refer to the domain of this function as the set *neighbours*. Each node maintains a queue of offer messages awaiting transmission to neighbouring nodes:

LState
$neighbours : \mathbb{P}\, Node$
$linkmap : Node \rightarrow\!\!\!\!\!\rightarrow Cost$

$neighbours = \mathrm{dom}\, linkmap$

QState
$queue : \mathbb{P}\, Envelope$

The correctness of the algorithm does not require a FIFO queue, so we have chosen to model the queue as a set of envelopes.

The complete state of a node includes its links, its routing table and its message queue:

NState
$LState$
$RTState$
$QState$

This abstract data type is initialised with an empty set of neighbours, an empty linkmap, and an empty queue.

4.2. Node operations

We define three operations upon this datatype. The operation *LinkUp* describes what happens when a link comes into service: the node must change its *linkmap* and *neighbours* to reflect the new link, and send its entire routing table to the new neighbour. The operation *EnvelopeOut* describes the effect of passing a message to a neighbour. This operation will fail if the queue is empty.

The third operation describes the effect of receiving an offer message from a neighbouring node. If the offered destination is unknown, or the offered cost is an improvement, then the routing table must be modified:

$$
\begin{array}{l}
\textit{OfferNewRoute} \\
\hline
\Delta QState \\
\Delta RTState \\
\Xi LState \\
e? : Envelope \\
newcost : Cost \\
\hline
e?.receiver = nodeid \\
e?.sender \in neighbours \\
newcost = linkmap\ e?.sender + e?.price \\
e?.dest \notin \operatorname{dom} routemap \ \vee \ newcost < (routemap\ e?.dest).cost \\
nodeid' = nodeid \\
routemap' = routemap \oplus \{e?.dest \mapsto \langle\!|\, cost \leadsto newcost, hop \leadsto e?.sender \,|\!\rangle\} \\
queue' = queue \cup \{Envelope \mid receiver \in neighbours \wedge sender = nodeid \wedge \\
\qquad\qquad\qquad\qquad price = newcost \wedge dest = e?.dest\} \\
\hline
\end{array}
$$

This operation has no effect upon the link table, but it changes both the routing table and the output queue.

The *routemap* component is updated to include a new entry for the node *e?.dest*—the destination component of the envelope. The *cost* component of the new entry is *newcost*; this is equal to the offer price plus the cost of getting to the first hop. The *hop* component is *e?.sender*: the name of the node which sent the envelope.

The output queue is also updated: it now includes an envelope for each of the neighbouring nodes, containing an improved price for the destination node. The precondition for this operation is quite simple: the envelope has arrived at the current node from a known neighbour, and contains a better offer for some destination.

If the alternative possibility—that the offered cost is no improvement—is covered by schema *OfferOldRoute*, then the *EnvelopeIn* operation may be defined by

$$EnvelopeIn == (OfferNewRoute \vee OfferOldRoute) \setminus newcost$$

where *newcost*, used only as a local variable, is hidden.

4.3. Node communications

We may model the communicating behaviour of a node using a value-passing action system:

var *NState*
initially *NStateInit*
for $n \in Node$ **action** *n.nodeid* **in** $e? : Envelope$:–
  $InGuard \rightarrow EnvelopeIn$
for $n \in Node$ **action** *nodeid.n* **out** $e! : Envelope$:–
  $OutGuard \rightarrow EnvelopeOut$
for $n \in Node$ **action** *link.nodeid.n* **in** $c? : Cost$:–
  $LinkGuard \rightarrow LinkUp$

This description includes the state variables (*NState*) and their initialisation (*NStateInit*). For each action, there is an input or output event, a guard, and a schema name.

The guards describe conditions that must be satisfied for that action: e.g., we allow an input from node *n* only if the *sender* component matches *n*, and an output to *n* only if there is an envelope for *n* in the queue.

InGuard
$n : Node$
$e? : Envelope$

$e?.sender = n$

OutGuard
$n : Node$
$e! : Envelope$
$queue : \mathbb{P}\, Envelope$

$e! \in queue$
$e!.receiver = n$

A similar guard is defined for the *LinkUp* operation.

The input guard is a consistency property upon our communication structure; in practice, it will be satisfied by every attempt at transmission. Similarly, the match of *receiver* and *n* is actually a state invariant of the action system; we include it here as an aid to comprehension. In each case, we have a guard which is stronger than the corresponding precondition. This is enough to guarantee that we will never call an operation in a circumstance in which it will abort.

4.4. Network communications

The communicating behaviour of the network may be modelled by a parallel combination of node action systems:

$$Network\_Process \;=\; \|_{i \in Node}\, Interface_i : Node\_Process_i$$

Each node is represented by an indexed action system of the form $Node\_Process_i$. This is obtained from the action system above by decorating all state variables with subscript *i*.

The interface to node process i is given by

$$Interface_i \;=\; \{j : Node \bullet i.j\} \cup \{j : Node \bullet j.i\} \cup \{j : Node \bullet link.i.j\}$$

We will insist later that $nodeid_i$ is identical to i.

Our proof of consistency of routing information will require results about adjacent nodes. We may use the rule for parallel composition of action systems given in [2] to rewrite the parallel composition $Node\_Process_A \parallel Node\_Process_B$ as a single action system:

var $NState_A \wedge NState_B$
initially $NStateInit_A \wedge NStateInit_B$
for $n \in Node \setminus \{B\}$ **action** $n.A$ **in** $e? : Envelope :- InGuard_A \rightarrow EnvelopeIn_A \wedge Relax_B$
for $n \in Node \setminus \{A\}$ **action** $n.B$ **in** $e? : Envelope :- InGuard_B \rightarrow EnvelopeIn_B \wedge Relax_A$
for $n \in Node \setminus \{B\}$ **action** $A.n$ **out** $e! : Envelope :- OutGuard_A \rightarrow$
$EnvelopeOut_A \wedge Relax_B$
for $n \in Node \setminus \{A\}$ **action** $B.n$ **out** $e! : Envelope :- OutGuard_B \rightarrow$
$EnvelopeOut_B \wedge Relax_A$
for $n \in Node$ **action** $link.A.n$ **in** $c? : Cost :- LinkGuard_A \rightarrow LinkUp_A \wedge Relax_B$
for $n \in Node$ **action** $link.A.n$ **in** $c? : Cost :- LinkGuard_B \rightarrow LinkUp_B \wedge Relax_A$
action $A.B$ **out** $e! : Envelope :- OutGuard_A[B/n] \wedge InGuard_B[A/n, e!/e?] \rightarrow$
$EnvelopeOut_A \wedge EnvelopeIn_B[e!/e?]$
action $B.A$ **out** $e! : Envelope :- OutGuard_B[A/n] \wedge InGuard_A[B/n, e!/e?] \rightarrow$
$EnvelopeOut_B \wedge EnvelopeIn_A[e!/e?]$

where *Relax* states that the operation has no effect upon the respective node:

$$Relax_A \;==\; [NState_A ; NState'_A \mid \theta NState_A = \theta NState'_A]$$

5. FORMAL ANALYSIS

5.1. A history variable

To show that there is a bound upon the number of output communications that a single node may perform during configuration, we add a history variable to the node state to record envelopes added to the output queue. We obtain a new data type:

NStateH
NState
history : seq *Envelope*

The initial value of *history* is $\langle\rangle$.

The history may change with *EnvelopeIn*: if the incoming envelope contains an offer which results in a change in the routing table, then a set of envelopes will be added to the queue. We don't care in which order these envelopes are added to our history sequence, as long as they are added contiguously. Accordingly, we insist only that the sequence of envelopes added contains the same envelopes as the added set.

EnvelopeInH

EnvelopeIn

$$queue' \neq queue \Rightarrow \exists\, new : \mathrm{seq}\, Envelope \bullet history' = history \frown new \land queue' \setminus queue = \mathrm{ran}\, new \land \#(queue' \setminus queue) = \#new$$

A similar modification applies to the *LinkUp* operation. The *EnvelopeOut* operation has no effect upon the history: it adds nothing to the queue.

This variable does not affect the evolution of the data type, but merely records certain steps as they are made. If we construct the corresponding action system, *Node_Process_H*, we find that $Node\_Process \sqsubseteq Node\_Process\_H$ with the abstraction invariant between them being simply *true*. The events possible in the new process (with history), are precisely those that are possible with the old process (without history) we have an equivalence for communication refinement: $traces[\![Node\_Process]\!] = traces[\![Node\_Process\_H]\!]$. Any result about traces of the enhanced data description will also apply for to the original data type.

We assume that *offers n d h* denotes the sequence of offers made to node *n* concerning destination *d* during history *h*, and that *firstprice n d h* and *lastprice n d h* denote the prices given in the first and last of these offers, respectively. With this, the following is an invariant property of the enhanced state:

Invariant

NStateH

$$\forall\, n, d : Node \bullet \#offers\ n\ d\ history \leq (firstprice\ n\ d\ history - lastprice\ n\ d\ history) + 1$$

5.2. A proof of convergence

The convergence criterion mentioned in Section 2 can be formulated in terms of the traces of the network:

$$\exists\, K : \mathbb{N} \bullet \forall\, s : traces[\![Network\_Process]\!] \bullet \#s < K$$

That is, there is a natural number bound on the length of all traces of our process. This marks the point at which there can be no further communication events pertaining to the configuration.

To demonstrate that this holds of our network action system, we show first that each component action system in our parallel combination satisfies a similar property, that for each node process there is some bound upon the number of output events. We note that

$$\#history = \Sigma_{d,n:Node}\#offers\ n\ d\ history$$

If $offers\ n\ d\ history = \langle\rangle$, then $\#offers\ n\ d\ history$ is 0. Otherwise, we know from the *Invariant* property of the previous section that

$$\#offers\ n\ d\ history \leq firstprice\ n\ d\ history - lastprice\ n\ d\ history + 1$$

in which case $firstprice\ n\ d\ history + 1$ will serve as a suitable bound. In either case, the length of the envelope history at each node is certainly bounded.

Now consider the action system: the number of output events in each local trace is equal to the length of the envelope history, and is thus bounded. The definition of parallel composition means that every input event is also an output event; the bound upon output events is also a bound upon input events. We conclude that the number of configuration events possible for the network is bounded, and hence that the algorithm is convergent.

5.3. Matching domains

Once all queues in the network are empty, if we consider any pair of nodes A and B, then the domains of their routemaps should match:

MatchingDomains

$NState_A$
$NState_B$

$$(queue_A = \{\} \wedge queue_B = \{\} \wedge A \in neighbours_B \wedge B \in neighbours_A) \Rightarrow \mathrm{dom}\ routemap_A = \mathrm{dom}\ routemap_B$$

which says that, when the configuration process is over, neighbouring nodes will know about the same set of possible destinations.

The way in which we prove this property is instructive: we search for an invariant for the action system or data type which proves the above. The property itself is true of every state, but it is of the form $Condition \Rightarrow Property$ where $Condition$ may be false in general. We are interested only in those states for which the condition is true, but to get to the next state for which that is the case, a whole sequence of operations or actions must take place. Thus we cannot hope to show that

$$Condition \Rightarrow Property \quad \vdash \quad Condition' \Rightarrow Property'$$

for an arbitrary pair of successive states.

The chosen invariant must be enough to demonstrate that

$$(queue_A = \{\} \wedge queue_B = \{\} \wedge A \in neighbours_B \wedge B \in neighbours_A) \Rightarrow \mathrm{dom}\ routemap_A = \mathrm{dom}\ routemap_B$$

so it will need to refer to the queues and the routemaps in both A and B. There is a symmetry to both the antecedent and the consequent here, so we will attempt to weaken the equality to an inequality and prove the result in each direction.

$$queue_A = \{\} \land B \in neighbours_A \Rightarrow \mathrm{dom}\, routemap_A \subseteq \mathrm{dom}\, routemap_B$$

However, this is still only vacuously true—and hence does not provide enough information to show invariance—for certain states. So we look for another property which provides at least this information. This requires the intelligence that any discrepancy between the routemaps should be compensated for by update envelopes pending in the queue. If we address this, then we should have a provable invariant:

MatchingDomainsInvA

$NState_A$
$NState_B$

$$B \in neighbours_A \Rightarrow$$
$$\mathrm{dom}\, routemap_A \subseteq (\mathrm{dom}\, routemap_B \cup \{e : queue_A \mid e.receiver = B \bullet e.dest\})$$

If there is a discrepancy, then A is trying to tell B about it: there is an offer concerning the destination in question sitting in the output queue of A, in an envelope addressed to B.

Although the invariant property is still an implication, the operation that takes us from a state in which the antecedent is false to one in which it is true provides enough information in one step to guarantee that the consequent also becomes true. We may demonstrate that a symmetric invariant (*MatchingDomainsInvB*) holds for the data type, and thus that

$$queue_B = \{\} \Rightarrow (\mathrm{dom}\, routemap_B \subseteq \mathrm{dom}\, routemap_A)$$

From the combination of these properties, we can deduce *MatchingDomains* as required.

Another useful property involves the consistency of cost information held in neighbouring nodes. If there is nothing waiting in output queues at the nodes, then taking a hop *en route* to any destination reduces the cost by precisely the cost of the link used for that hop. A proof that this property holds of any pair of nodes may be produced in exactly the same way as for *MatchingDomains*, although only one invariant is required this time.

A third property concerns the optimality of the routing table at a node, relative to all of its neighbouring nodes. For any node A, neighbour B and destination C, the cost of C via B must be greater than or equal to the route table cost of C from A. This local optimality between neighbours is used to prove global optimality of the routing tables across the network.

5.4. Network topology

To show that the nodes in our network have a consistent view of the routing information we must consider the shape of the network. This is best seen as a graph, in which the edges are links between nodes and are labelled with the metric, or cost, assigned to that link. A network can be represented as a pair $(\mathcal{N}, \mathcal{E})$, where $\mathcal{N}$ is a set of Nodes, and $\mathcal{E}$ is an edge relation (ordered pairs of nodes, each representing a unidirectional link). We assume the existence of a cost function *metric* upon the graph.

It is our intention that the routing algorithm should *discover* the shape of the network, as represented by $\mathcal{N}$, $\mathcal{E}$, and *metric*. For this to be possible, the algorithm must be provided with correct information by the link level. We call this requirement *topological consistency*:

$$
\begin{array}{l}
\textit{TopologicalConsistency} \\
\hline
\textit{NState}_A \\
\hline
\forall B : \textit{Node} \bullet B \in \operatorname{dom} \textit{linkmap}_A \Leftrightarrow (A, B) \in \mathcal{E} \wedge \textit{linkmap}_A\ B = \textit{metric}(A, B) \\
\hline
\end{array}
$$

That is, if the link level is functioning correctly, and the configuration algorithm has stabilised, then the information in each *linkmap* is consistent with the underlying metric.

A route from node A to node B is a sequence of nodes $\langle A_0, A_1, ..., A_n \rangle$ with $A_0 = A$, $A_n = B$, and $(A_{i-1}, A_i) \in \mathcal{E}$ (a link in the network), for $i = 1, \ldots, n$. Let $\mathcal{R}$ be the set of routes in a network, let *hops* r denote the number of links in route r, and let $\Sigma\, r$ denote the total cost of route r. We may define the set of all minimum cost routes as follows:

$$
\begin{array}{l}
\mathcal{M} : \mathbf{P}\operatorname{seq} \textit{Node} \\
\hline
\mathcal{M} \subseteq \mathcal{R} \\
\forall m, r : \operatorname{seq} \textit{Node} \bullet \\
\quad (m \in \mathcal{M} \wedge r \in \mathcal{R} \wedge \operatorname{first}(m) = \operatorname{first}(r) \wedge \operatorname{last}(m) = \operatorname{last}(r)) \Rightarrow \Sigma\, m \leq \Sigma\, r
\end{array}
$$

where $\operatorname{last}(m)$ is the last node in the sequence m. It is possible to prove that this set is closed under subsequence.

5.5. A proof of correctness

Adequacy If there is a route from A to B, then A has a routing table entry for B.

Proof by induction upon n—the number of hops in this route from A to B—with

$$P(n) \;==\; \forall r : \operatorname{seq} \textit{Node} \mid r \in \mathcal{R} \wedge \textit{hops}\ r = n \bullet \operatorname{last}(r) \in \operatorname{dom} \textit{routemap}_{\operatorname{first}(r)}$$

Accuracy Each routing table entry corresponds to a route that exists in the network graph: i.e., if A has an entry for B, then there is a route r between A and B such that

$$(\textit{routemap}_A\ B).\textit{cost} = \Sigma\, r$$

and if $(\textit{routemap}_A\ B).\textit{hop} \neq \textit{null}$, there is a route r from A to B with

$$\langle A, (\textit{routemap}_A\ B).\textit{hop} \rangle \leq r$$

where $r \leq s$ means that the sequence r is a prefix of the sequence s.

Proof by induction upon n—the value of $routemap_A\ B.cost$—using *TopologicalConsistency*.

Optimality If the routing table at A contains the entry $\{C \mapsto \langle\!| hop \leadsto B, cost \leadsto k |\!\rangle\}$ then any minimum route from A to C has a metric cost k, and there exists a minimum route from A to C such that the first step on this route is from A to B.

Proof by induction upon n—the number of hops in a minimum cost route—with

$$P(n) \;==\; (m \in \mathcal{M} \wedge hops\ m = n) \;\Rightarrow\; (routemap_{\mathrm{first}(m)}\ \mathrm{last}(m)).cost = \Sigma\ m$$

Note that there may be more than one minimum cost route, and that one minimum cost route need not have the same number of hops as another.

6. A NETWORK SIMULATOR

We may construct a new representation of the network state space, isomorphic to the parallel composition of action systems, using a labelling function from the set of node names to the schema type *NState*; we insist that this labelling is consistent with the value of *nodeid* at each node in the network.

We add also a component *net*, which records the set of nodes currently regarded as 'in' the network. This will be left unchanged by any of the network operations described in the previous chapters; however, we may wish to record the effect of adding nodes to the network:

```
┌─ NetworkState ─────────────────────
│ node : Node ⇸ NState
│ net : F Node
├──────────
│ ∀ n : net • (node n).nodeid = n
└────────────────────────────────────
```

The function *node* associates each node name currently in use with a node state. We insist that the name associated with the node is the same as its internal value of *nodeid*. We may initialise the network simply by initialising each node.

We relate operations upon the network state to operations upon individual node states; we are then able to describe their effect using existing operation schemas. The bindings before and after these operations are associated with the node states as returned by function *node*: e.g., consider the network operation which corresponds to the communication of a message between two nodes; this is described by the following schema:

Communication

$\Delta NetworkState$
$e! : Envelope$

$\forall n : net \mid n \notin \{e!.sender, e!.receiver\} \bullet node\ n = node'\ n$
$net = net'$
$\exists EnvelopeIn_1 \bullet \exists EnvelopeOut_2 \bullet$
$\quad node\ e!.receiver = \theta NState_1 \wedge node'\ e!.receiver = \theta NState'_1$
$\quad node\ e!.sender = \theta NState_2 \wedge node'\ e!.sender = \theta NState'_2$
$\quad e! = e?_1 = e!_2$

The change in state of the receiving node is described by the operation *EnvelopeIn*; that of the sending node is described by the operation *EnvelopeOut*. We decorate the schemas describing these operations, and associate their bindings.

The above operation has an output: the envelope passed between the nodes. We use schemas to encapsulate outputs from our network simulator:

Output

$from!, to! : Node$
$e! : Envelope$

$from! = e!.sender \ \wedge\ to! = e!.receiver$

Similarly, we may use a schema *Input* to describe the type of input values to an operation *Link*: this would model the addition of a new link to the network. With an appropriate choice of channel names, we obtain an action system description for our simulator:

var *NetworkState*
initially *NetworkStateInit*
for $m, n \in Node$ **action** *comm* **out** *out! : Output :– Communication*
for $m, n \in Node$ **action** *link* **in** *in? : Input :– Link*

In [6], the author shows how such a system may be implemented using C. The result is a reliable software simulation of the distributed routing algorithm.

7. DISCUSSION

In this paper, we have shown that formal methods may be usefully applied to the study of network routing algorithms. We have described a widely-used method of generating routing tables, and demonstrated that it satisfies certain requirements. Although only one aspect of network design has been considered, the methods used can be applied to others, such as message storage and forwarding, performance modelling, and fault-tolerance.

In [6], the author shows how the algorithm described here can be extended to support *adaptive routing*, in which link costs may be revised upwards or downwards during the operation of the network. This requires an enhanced node state: it is necessary to maintain a record of the best offers from each of the neighbouring nodes, and the order in which offers are made is important.

An argument for the correctness of this method for generating routing tables was presented in [5]. Although described as formal, the proof included in the paper lacks the clarity and discipline of a literate Z specification. The argument is hard to follow, and leaves the reader with little understanding of the operation of the algorithm at each node. A comparison of the two proofs clearly demonstrates the advantages of formal specification techniques.

The role of the routing algorithm is complementary to that of the message-passing system considered in [2]. Here, the author shows that the action systems notation may be used to describe the behaviour of nodes while storing and forwarding data packets. A fixed set of routes is assumed, and each message follows a route which is determined at its point of origin. A method of adaptive routing is also considered, but is based upon node congestion, and the algorithm is not distributed.

Apart from the specification and verification of the routing algorithm, we were also able to exhibit a specification of a simulator program. This is most useful, in that the resulting simulation is guaranteed to correspond to the network implementation. In both cases, the formal description acts as the first stage of a formal—or rigorous—development process, each action or operation can be refined towards code using standard techniques.

ACKNOWLEDGEMENTS

The authors would like to thank Carroll Morgan and Jim Woodcock for their advice and helpful suggestions. We are also grateful to the anonymous referees for their comments, and to Fleur and Gill for their proof-reading and support.

REFERENCES

1. R. J. R. Back and R. Kurki-Suonio. Decentralisation of process nets with centralised control. In *2nd ACM SIGACT-SIGOPS Symp. on Principles of Distributed Computing*, 1983.
2. M. J. Butler. *A CSP Approach to Action Systems*. Oxford University D.Phil thesis, 1992.
3. M. J. Butler. Refinement and Decomposition of Value-Passing Action Systems. In E. Best, editor, *CONCUR93*, volume LNCS 718. Springer-Verlag, 1993.
4. J. M. Spivey. *The Z Notation: A Reference Manual*. Prentice-Hall International 1989.
5. W. D. Tajibnapis. A correctness proof of a topology information maintenance protocol for a distributed computer network. *Communications of the ACM 20 7*, July 1977.
6. M. Wallis. *Formal Analysis and Software Simulation of Network Routing Algorithms*. Oxford University M.Sc thesis 1993.
7. J. C. P. Woodcock and C. C. Morgan. Refinement of state-based concurrent systems. In D. Bjørner, C. A. R. Hoare, and H. Langmaack, editors, *VDM '90*, volume LNCS 428. Springer-Verlag, 1990.

PART THREE

FDTs go Commercial

7

VALIDATION OF SUPPLEMENTARY SERVICES IN THE CONTEXT OF INTEROPERABILITY

S. Le Bricquir
CAP SESA OUEST
Rue de La Rigourdière, 35510 Cesson Sevigné, France
Tel.: 99 83 85 85; Fax: 99 83 85 81

A. Lauzanne
SETICS
194, rue de Tolbiac, 75013 Paris, France
Tel.: 45 89 74 93; Fax: 45 89 84 94; email: ss315@calvacom.fr

ABSTRACT: The purpose of this paper is to demonstrate, using a concrete experiment, how Formal Description Techniques (FDT) can help provide a high level of interoperability in multiservice communication systems.

Keywords: ISDN, Supplementary Services, SDL (Specification and Description Language) Specification, AVALON/GEODE 2, automatic simulation.

1. INTRODUCTION

The new generation of French military communication networks that will be put in service between 1995 and 2000 are of the ISDN (Integrated Services Digital Network) or broad-band ISDN type. Each military community (Army, Air Force, Navy and Gendarmerie) uses PABXs structured like a virtual private network (VPN) using the services of a common transit network with national geographic coverage to optimize the use of long distance switching and transmission resources. This transit network is designed to operate in Asynchronous Transfer Mode (ATM). The PABXs of each community can also use the services of the public switched network of France Telecom, the national operator.

This paper first describes the "System DM", and then discusses the experiment, its results and perspectives.

The "System DM" groups all specifications related to services that are offered by switched subsystems. The application of these specifications in various military programs provides a guarantee of needed interoperability.

The "System DM" is based on civilian standards specified by the relevant standardization bodies (ITU-TS, ETS, ECMA, etc.). SDL was used for the "System DM" specification because of its widespread use, even in a partial form, in ISDN standards.

In the "System DM", a formal validation (640 simulation scenarios were run to heighten confidence in the protocol's interoperability.), or verification procedure is applied to supplementary services that are specific to the military.
These services provide a typical example of end-to-end services that require a high level of interoperability and reliable specification.

The formal validation experiment described below and, more generally, the use of FDTs in the framework of specifying a "System DM" is the result of an ambitious program run by the Services Techniques de l'Electronique et de l'Informatique du Ministère de la Défense (STEI, the Electronics and Computer Science Department of the French DoD). This program supports the development of dedicated specification and validation tools based on FDTs, such as the Verilog tool used for the experiment described below. It also supports the definition of generic methods, to the extent these developments exhibit sufficient synergy for civilian-based industrial needs.

This program, which is part of an industrial policy related to FDTs, is designed to enable the French DoD engineering departments concerned to master new developments in telecommunications. For this reason, the use of FDTs that are as close as possible to those used for industrial processes (and integrated with them as much as possible) is systematically sought. This use can be extended to areas other than ISDN signaling protocols, as mentioned below.

SETICS and CAP SESA have been selected by the STEI to define and validate the "System DM", because they are both experienced in the domain of civilian and military telecommunications systems, and because CAP SESA has contributed to various projects in the FDT domain (SDL and Estelle), in particular for CNET (France Telecom), in the context of the European RACE program's PROVE project.

2. NATURE AND CONTENTS OF "SYSTEM DM"

The "System DM" comprises a set of specifications that:

- defines a reference model specific to the architecture of military models,
- identifies the support services relevant to French military networks as defined above,
- specifies the call setup procedures designed for interoperability,
- defines the supplementary services specific to the military to be provided, as well as the expanded or modified civilian supplementary services needed,
- makes a choice of basic protocols for layers 2 and 3.

The main military supplementary services specified are as follows:

- Roving User service (RU): A subscriber benefiting from this service can rove and request affiliation with any PABX in the area where he is authorized to move around. This includes all PABXs serving his own community or even PABXs serving other communities. For outgoing and incoming calls on his hosting terminal or site, the subscriber benefits from all prerogatives defined in his subscription, subject to technical compatibility with the locally available equipment.

- Implicit Partitioning supplementary service (IMP-PRT): This service allows users to form functional or application-related entities with limited incoming and outgoing access. It consists of macroscopic partitioning of large subscriber communities. This partitioning is implicit, since it does not require the calling party to specify a partition class. Each subscriber is characterized by "an outgoing partition class" and "an incoming partition class" used by the PABX during the call setup phase.

- Priority: This service allows a priority, or precedence level to be assigned to a call. In case of congestion, both transit and incoming calls are released to set up calls with higher priority. (This service is based on the CCITT MLPP multilevel precedence and preemption service.)

- Quality of service constraints: A subscriber with this service can specify the quality of service required when setting up a call in terms of BER (Bit Error Rate), end-to-end transfer time or security routing.

- Other services: These include supplementary services designed to support the security system, such as authentication, or the issuance of encryption/decryption elements, for instance.

The protocols included in the "System DM" are specified for user/network and network/network interfaces.

To create the reference model, the supplementary services specified in the "System DM" for the special needs of military communities were described according to the CCITT three-step technique:

- Step 1: Service offered.
- Step 2: Functional specifications and associated data flows.
- Step 3: Implementing functions through the physical components and protocols used.

This description mode creates a clear reference model that characterizes, on various levels, the system's compatibility as regards architecture and protocol. This mode also allows the specification work carried out in Steps 2 and 3[1] to remain valid, even if there is a change in the reference model, such as a modification of the military networks' architecture or functional implementation. It also remains valid should there be a change in the lower layers of the protocol (Levels 2 and 3), or in the generic function of supplementary service support.

In the current version of "System DM", the lower layers include an extended version of:

- various ETSI standards (ETS) derived from CCITT Rec. Q.900 series (DSS.1) for the user/network interface,
- ETS 300 172 (private network signaling protocol, called Q-SIG) for the network/network interface,
- an extended version of CCITT Rec. Q.932 for basic call setup to support the supplementary services provided. It should be noted that the equivalent Q-SIG standard called Q-SIG-GF was not yet available when these choices were made.

---

[1] i.e., the major part of the specification task.

The next figure presents a global overview of the specification and validation activities.

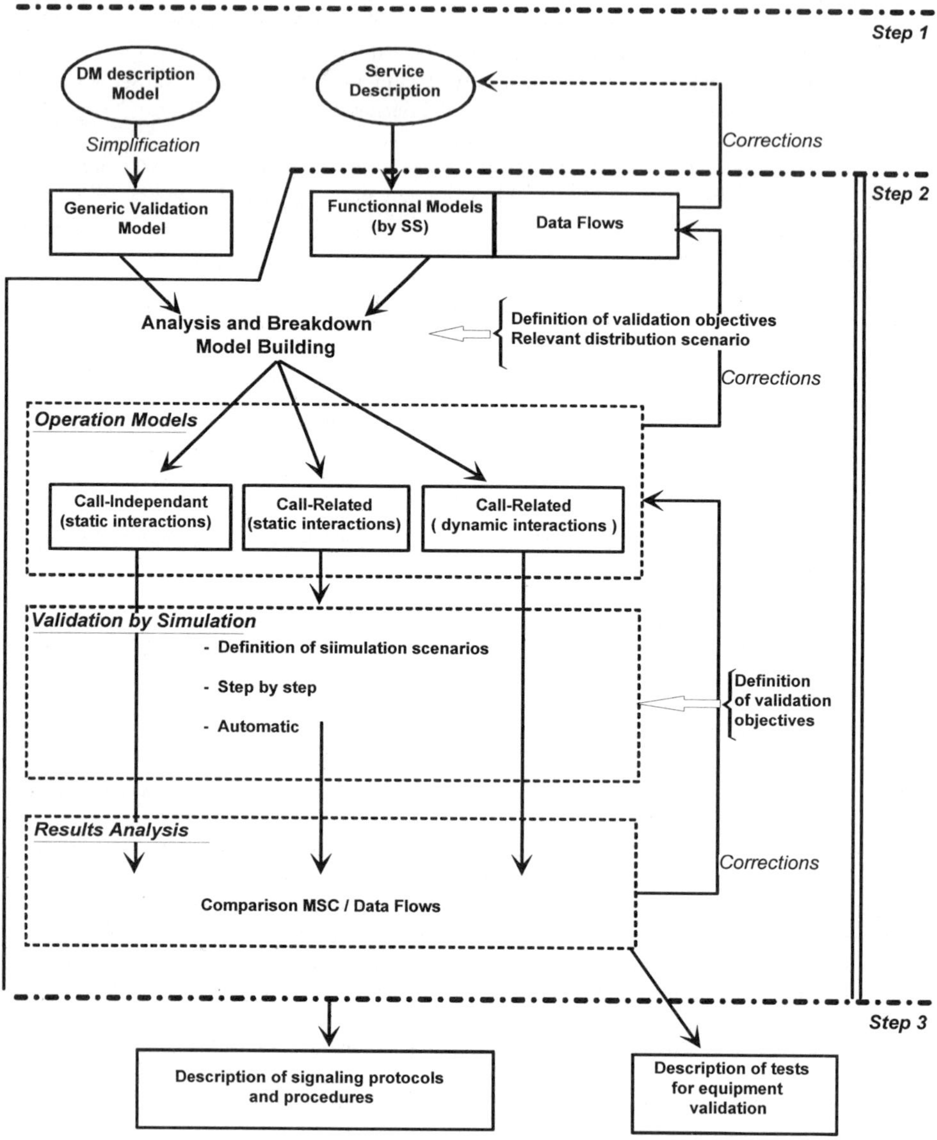

Figure 2-1: Specification and Validation Activities

3. "SYSTEM DM" DESCRIPTION MODEL

When the method described in the Appendix is used to specify supplementary services, one or more functional models are available at the outcome of Step 2, the description. The functional service entities (FSE) are then integrated in the DM description model.

The DM description model is illustrated by the layered diagram below.

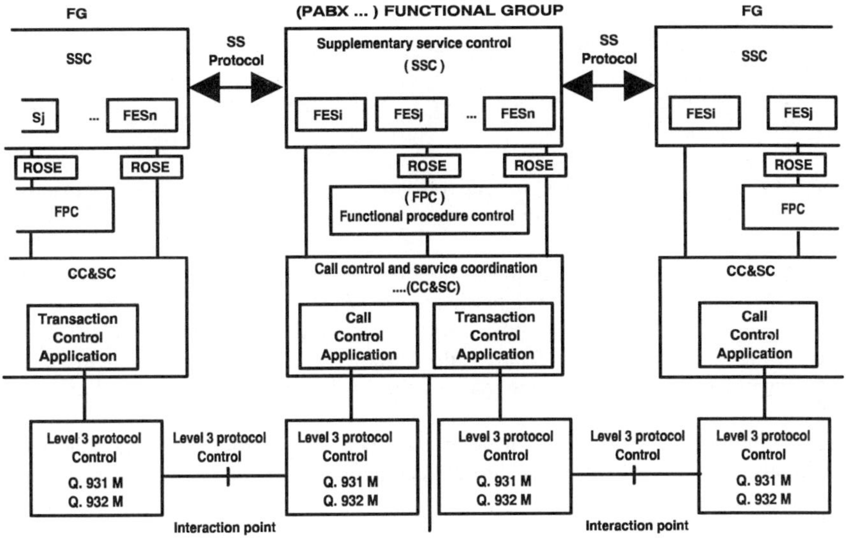

Figure 3-1: The "System DM" Description Model

For each functional group, the model includes at the application level:

- a supplementary service (SS) control function,
- a functional procedure control function,
- a call control and service coordination function.

Supplementary Service Control: This function includes the functional entities described in Step 2. A functional group can include one or more functional entities for a given service. When several functional entities of the same service are located in the same functional group, the exchanges between these entities are direct.

Functional Procedure Control: This module provides ROSE (Remote Operations Service Element, CCITT Recommendations X.219 and X.229) services. It is used in functional mode to code/decode the ROSE components transmitted to Level 3.

Call Control and Service Coordination: This function includes two types of modules, depending on the type of call used:

- a call control application that controls calls at the applications level, and controls interactions between the call and the services provided,

- a transaction control application that controls transactions at the applications level, and controls interactions with the service provided.

These modules directly communicate with Level 3 and integrate the communication and transaction services previously described.

Each call application process is associated with a particular call or transaction. These processes can communicate with one another:

- to provide a link between a transaction and a call associated with the implementation of a particular SS,
- to provide a link between several calls associated with the implementation of a particular SS.

The functional service entities are activated by call control applications, and may use the functional procedures control module.

Coordination between the call or transaction and the services is provided by procedures called within the automatic call control mechanisms. Whenever a Level 3 elementary request is received, a procedure of this type, the "message handling procedure", is called. This procedure manages interactions between SSs. It calls specific SS procedures that provide the interface with functional service entities, and may use the functional procedures control module.

This operating mode has the advantage of being able to use generic automatic mechanisms for all control applications, whatever the services implemented. Only the processing procedures have to be adapted.

4. VALIDATION OBJECTIVES

The main validation objective is to verify the consistency of the functional specification (Step 2) of the supplementary services, including error cases.

Consequently, the aim is to validate all functions implemented by the network related to the services, i.e., to analyze the behavior of the functional service entities formally specified in the services description step.

The method applied differs from the one used to specify and validate a protocol level. Indeed it involves several functional service entities, a network function, and a management entity. Each entity is specified by an SDL process.

The validation objectives for all DM supplementary services, at least for those specified in detail, have been taken into account while defining a general method.

To implement this method, a powerful description language and a high performance validation toolset have been used. They are, respectively, the SDL* language (extension of the SDL language) defined by the Verilog company, and the AVALON (GEODE 2) toolset, developed by the same company. The SDL* language compiler can handle most standard SDL code.
However, some functions are not yet supported. These include channel substructures and signal refinement, services, generic systems, inheritance types, etc...

SDL* supports three supplementary communication modes : multi-queue communications, synchronous communications (also called rendezvous), and broadcasting. AVALON supports, in addition to C, two means of describing abstract date type operators : restricted axiomatic definitions and an enhancement of SDL 92 operational definitions.
For more details about SDL* and AVALON, see B. Algayres, et. al., as referenced below.

This method has been applied to three supplementary services[2] previously specified according to the technique introduced in Chapter 2 and described in the Appendix.

For a given SS, the interesting points to study during validation are related to:

- call-independent aspects (e.g., recording a subscriber's movements),
- call-related aspects (e.g., routing a call toward a subscriber who has moved).

An interaction between SSs is also considered as a point to be validated. Two kinds of interactions should be taken into account:

- static interactions, dealing with management information shared by the various SSs,
- dynamic interactions, supposing that all concerned SSs are activated.

5. VALIDATION MODELS

Since the purpose of validation is to analyze the behavior of the functional service entities (validation of service description, Step 2), the description models can be simplified for validation purposes.

The functional procedures control module is not modeled, since coding/decoding of the ROSE components is irrelevant to validation.

The Level 3 protocol control modules are not present in Step 2 and are therefore not modeled. A network service is modeled for communication between functional entities. It provides the following functions:

- message routing,
- management of the relations between applications,
- situating the applications in the functional groups (representation of the physical entities),
- management of application status.

Whatever the implementation procedure chosen for the services (e.g., functional procedure Q.932 M), the functional service entities interface directly with the call control and service coordination module. The signals exchanged on this interface include Q.71 signals and ROSE signals.

The interface between the call control and service coordination module and Level 3 is replaced by the interface with the network module.

The figure below shows the model used for validation.

---

[2] Roving user service, implicit partitioning and call forwarding unconditional.

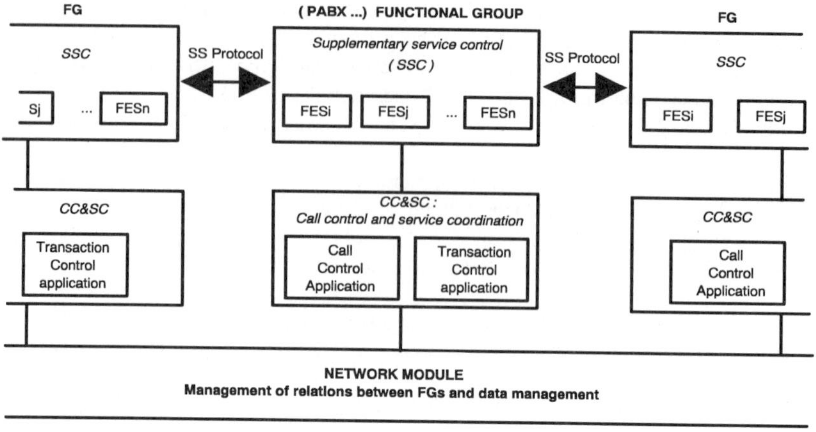

Figure 5-1: The Validation Model

Each entity (FES, Call Control Application, Transaction Control Application, Network Module) is modeled using one or several SDL processes. The Call Control Application, Transaction Control Application and Network Module entities constitute the required environment for the proper FES operations.

During the present study, the environment has been modeled in the same way, whatever the supplementary service or aspect to be validated. Indeed, as SS specification Step 2 and the validation models concept were developed concurrently, it wasn't possible to determine *a priori* the specific environmental needs of each kind of model.

The choices made take into account all needs, but generate very large models that consequently consume much development time. In most cases, however, the environment could be simplified, without any impact on validation quality.

The size of the models also prevents the use of exhaustive, or comprehensive simulation.

At the end of the study, three kinds of generic models have been identified[3]:

- the models that deal with call-independent aspects,
- the models that deal with call-related aspects,
- the models that focuses on dynamic interactions between SSs.

A generic environment that includes a subset of the above-mentioned modules can be associated with each kind of model.

[3] It is also possible to consider reduced models that allow exhaustive simulation, when the identified models are too large to use this kind of verification. Reduced model construction should be studied case by case and is not described in the present paper.

To build a validation model, the following actions must be taken, in all cases:

- describe the model's architecture, based on the SS functional model described in specification step 2,

- build the model from the FES and modules that constitute the required environment.

6. APPLICATION TO ROVING USER SERVICE (RU)

The roving user service allows any subscriber with this SS to move to a host site and have all his prerogatives available on this site.

The Step 2 description of this SS produced two functional models: one for non-call operations (recording and/or modifying the host site) and the other for operations performed during a call. Only the latter is described below.

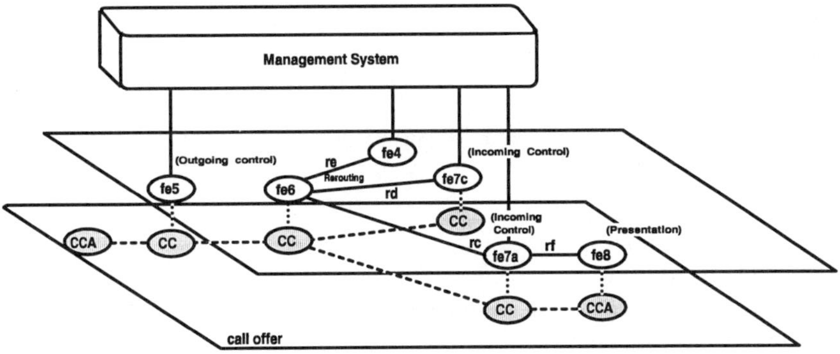

Figure 6-1: Functional Model of the Roving User SS

This service includes the following functional entities:

fe5: Checks the calling user's rights for outgoing calls. For this purpose, it queries the management system that holds the subscription files.

fe6: Critical function of the service. This function reroutes unsuccessful calls in the following cases:

- The release message specifies a host address at which the called subscriber cannot be reached.
- The release message specifies an unreachable subscriber (loss of connection, for instance).

In this case, the entity must first find a server that can indicate the host address of the called subscriber, if the subscriber is roving. It must then attempt to reroute the call to the host address, according to the answer received.

fe4: Server access function used to locate roving subscribers.

fe7: Incoming call check function. It checks that the called subscriber is affiliated to the current site. It releases the call, specifying the host address, if the subscriber has moved. fe7c allows a distinction to be made between the home site and the host site (fe7a).

fe8: Subscriber terminal interface. Option to process the call only if the called party is actually reached at home or host site. Selective call processing.

MS: Represents the management system containing all the useful system information, in particular, the subscription files.
No particular assumptions are made on whether the MS is centralized or distributed. From the FSEs' viewpoint, the services rendered are atomic (i.e., instantaneous and uninterrupted).

CCA: Call Control Agent.
CC: Call Control.

An operational model is built from the functional model for validation. The FSEs are allocated to the operational model according to a relevant physical distribution scenario.

The figure below shows such a scenario with distribution to three sites.

The physical entities include:

- TE: Terminal
- NT2: PABX
- CRU-T: Connection to the transit network
- SINAR: Data server

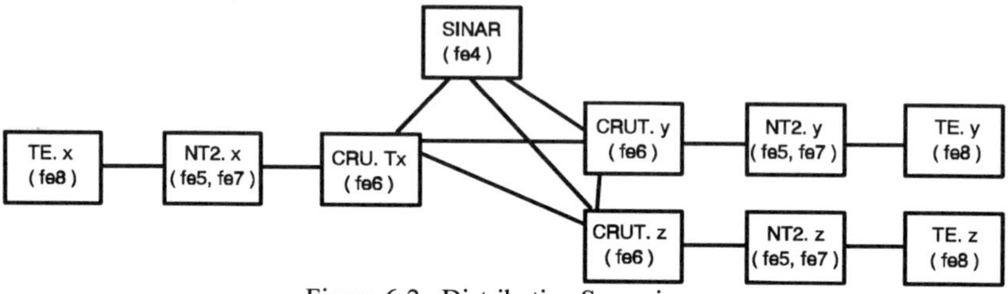

Figure 6-2: Distribution Scenario

The corresponding operational model is presented below.

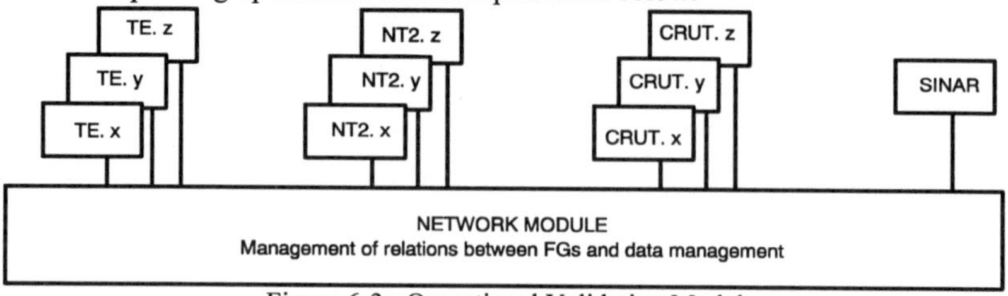

Figure 6-3: Operational Validation Model

When the operational model has been built, validation scenarios can be executed. For instance:

- Set up a call to subscriber y whose home terminal is TE.y. Subscriber y has moved to TE.z. Check that the call is offered on terminal TE.z, specifying that it is for subscriber y.

The simulation is executed:

- in single step mode; controlled manually by the operator, or
- automatically; this assumes that the model is complete so that the decisions can be made without requiring operator action, or
- exhaustively; this case corresponds to attempting all possible combinations.

Execution of the test is checked by specifying acceptable and illegal message sequences.

During execution, the simulator continuously (i.e., at each transition) checks the specified sequences and returns a positive or negative acknowledgement according to the sequences observed.

7. VALIDATION OF THE INTERACTIONS BETWEEN SS'S

The SSs are first validated separately. Then, operational models supporting the FSEs of the SSs to be validated are built to validate interactions between SSs.

The validation scenarios are then rerun and the properties specified during the separate validations are rechecked.

SSs interact with one another:

- weakly (and statically) by the modifications they make to the subscription files, involving various registration and recording operations carried out within the management system, or

- strongly and dynamically when their implementation affects the basic call service during its different phases:

  . call setup phase, in particular for quality constraints,
  . release phase: for instance, the SS rerouting process due to roving.

Static and dynamic incompatibilities can be detected through the integration, in the same operational validation model, of FSEs that are part of different SSs. To eliminate these incompatibilities and the associated inconsistencies, the FSEs of the SSs concerned must be modified.

8. THE SIMULATION PROCESS AND ITS RESULTS

Two simulation modes were used:

- interactive mode, which allows an operator to debug models, mainly the environment,

- random mode, for running validation scenarios automatically in sequence. The simulator chooses execution sequences at random.

Modifications are automatically incorporated in the models between validation runs, as needed, unless the detection of a major error requires a specification change.

750 scenarios were described for the three supplementary services and 50 for interactions between SSs.

Normal, or standard cases are executed in interactive mode. The simulator generates Message Sequence Chart (MSC) diagrams as output, which describe signal exchanges between the various SDL processes.

In random (and exhaustive) mode, a single case or several combined cases are verified dynamically, using normal MSC output to observe and track results.

Note: The exhaustive mode cannot be applied to such models, as its status diagram is too large, and the situations are heavily redundant. Specific models must be defined for this simulation mode.

The model for validating interactions between the three supplementary services contains 8500 SDL lines. The file generated for the simulations corresponds to 2.5 Mbytes of run code. 640 scenarios have been tested.

On a SUN SPARC station 2 with 16 Mbytes of RAM:

- the compilation time lasts around 15 minutes,
- the running time lasts 3 minutes for 16 scenarios.

All transitions have been executed at least once, and all states of the automatic mechanisms have been observed.

Several SS specification errors have been identified through verification, mainly during interaction between SSs.

Specification Steps 1 and 2 have been updated. Specification Step 3 (signaling) and the test objectives description have been developed on a sound basis.

With a wide range of functions and rapid execution, GEODE 2 is a high-performance tool. Its control language and interpreter are very easy and flexible to use. It provides three simulation modes and the automatic generation of simulation results in MSC form, which are displayed for consultation and modification. The GEODE 2 simulator is an efficient tool for verification of supplementary services and interactions, and meets the needs of industrial users.

9. APPLICATIONS AND PERSPECTIVES

This method can be applied to other DM SSs, with some enhancements of the models' environments, and can also be applied to every kind of ISDN SS (e.g., QSIG and QSIG GF).

Finally, the work conducted established a reliable basis for defining supplementary services. This basis can be used to analyze interoperability and to examine the relevance of solutions proposed in the context of new network connections.

ACKNOWLEDGEMENTS

We would like to thank the DGA for financing this project, particularly Cdt. Sprauel and Cpt. Plessix who launched the project and were responsible for it, as well as the other DGA participants: Mr. Caporini from STEI, and Mr. Le Pape, Mr. Ropars and Mr. Genouel from CELAR.

We would also like to thank C. Jard and B. Caillaud from IRISA who have provided valuable technical expertise. Many thanks to the Verilog team that developed and supports GEODE 2: B. Algayres, Y. Lejeune and L. Doldi.. Our strong collaboration with IRISA and Verilog considerably improved the quality of the work.

This work was carried out under the leadership of A. Lauzanne and S. Le Bricquir, with significant assistance from Ph. Le Gall from SETICS, along with Ph. Barbier and C. Conty from CAP SESA OUEST.

REFERENCES

CCITT Recommendations: I. 130, I.140, I.200 series, I.310, I.400 series, Q. 65, Q.71, Q.80, Q.500 series, Q.930, Q.931, Q.932, X.208, X.219, X.229, Z.100, Z.120.

ETSI standards: ETS 300 102, ETS 300 007, ETS 300 172.

C. Jard, B. Caillaud, J.M. Jézéquel and N. Plouzeau (IRISA, Rennes): Quelques expériences en matière de validation de protocoles industriels. Un point de vue de chercheur. CFIP'91 Ingénierie des protocoles, HERMES 91.

B. Algayres, Y. Le jeune, F. Hugonnet and F. Hantz (Verilog Toulouse): A VALidatiON Environment For SLD/MSC Descriptions. 6th SDL Forum, Darmstadt, October 1993.

APPENDIX

SERVICE DESCRIPTION METHOD

This appendix discusses the service description method used for "System DM" based on CCITT recommendations.

The method includes three steps (CCITT Recommendations I.130, I.200):

- Step 1: Service description from the standpoint of the user.
- Step 2: Functional description of the service.
- Step 3: Description of the signaling protocol(s) and associated switching functions.

These three steps correspond to the three levels that define the extent to which a system conforms to the specifications of "System DM".

A description of service use, including its operational context and services expected from other subsystems, is added to these three steps. The purpose of this description is to specify the constraints the service imposes on other network components, excluding call processing and signaling components. It is also used to discuss privileged implementation procedures.

A1. STEP 1

During Step 1, the service is described from the standpoint of the user, but without describing the user interface in detail. The description includes a specification of implementation operations. The description supplied is independent of terminal capabilities.

Step 1 includes the two following items (CCITT Recommendations I.210, I.140, Q.71 and Q.80):

1. Definition and textual description of the service offered to users: description of the semantics of the service as it may be perceived by the users. This description must be independent of the implementation. It includes procedural aspects, control, interworking and interaction with other services.

2. Static description of the service use attributes. For support services and teleservices, this concerns information transfer attributes (e.g., data rate), access attributes (e.g., signaling protocol) and general attributes (e.g., service quality). The advantage of describing such attributes for supplementary services is under study.

A2. STEP 2

Step 2 corresponds to a functional description of the service. It is therefore completely separate from its implementation. This description must be independent of the signaling which may be used.

This description includes five items:

1. Identification of the functions required to provide the service, grouping of these functions into functional entities, and identification of their interrelations. This gives a functional model of the service. In the case of supplementary services, the relations with the functional model of the basic call service is also described.

   A modular description of the services must be given, so as not to modify the description of the basic call service. The relations of other services with the basic call service are specified by defining the additional functions of the service as a separate functional entity. Its relation with a functional entity of the basic call service is identified.

   Furthermore, the functions expected of the other subsystems (data management, resource management, security, etc.) are grouped within particular, unique functional entities (one per subsystem concerned) in each service description. The functional model of the service therefore describes both the relations with the functional entities of the basic call service and with these particular functional entities.

2. Description of the data flows. The data flows show transfers between functional service entities (FSEs). They also show, where necessary, transfers with the user, with functional entities of the basic system and with functional entities of the other subsystems.
   The data flows provide a trace of actions performed by the functional entities between a signal's reception and transmission. These actions are described under Item 4.
   The data flows are described for all cases of service success or failure.

3. Description of SDL diagrams for each functional entity. These diagrams cover all the cases of service success or failure. The input and output events correspond to transfers with the user described in Step 1, and to data flows between functional entities described in Item 2 above.

   Validation of supplementary services is most significant during this step of the description. At this time, it is possible to validate all functions implemented by the network, without overloading the validation model by specifying the signaling protocol[4]. Specification by SDL diagrams is particularly advantageous if the supplementary service has been validated.

   To keep the design modular, functional service entities must be described in SDL as automatic mechanisms that are independent of the basic call's automatic mechanism.

---

[4]In effect, minor signalling protocol differences will probably always exist between the system and the DM but will have no major impact on compatibility of the services.

The interface between the automatic mechanism of the basic call and the automatic mechanisms of the services' functional entities is modeled by a procedure call within the automatic mechanism of the basic call.

The interface between the automatic mechanisms of functional service entities and the automatic mechanisms of other subsystem's functional entities is modeled by a procedure call within the automatic mechanism of the functional service entities.

The signals processed must include the parameters required for processing.

The tests results reported by the SLD automatic mechanisms are limited to signal parameter tests, or to the results of an explicitly described check.

To clarify, access by the service's functional entities to other subsystems must be named in a particular way, as must actions performed by the functional entities of the same subsystem.

4. Textual description of the actions performed for each functional entity. These actions are identified by a number appearing on the data flows, and by a comment on a task of the functional entity's SDL diagram.

5. Identification of the different physical implementation scenarios for the functional entities. The relations between two remote functional entities must be established by protocols used for communication between the two units.

A3. STEP 3

The data flows and SDL diagrams (CCITT Rec. Z.100) of Step 2 will be used as the basis for describing the signaling protocol(s) and associated switching functions.

This description must be made for each physical implementation scenario identified during Step 2 and deemed relevant.

Step 3 includes two items:

1. Description of signaling protocols and procedures. Identification of messages and data elements required to handle the data flows described in Step 2, and description of the associated procedures (CCITT Recs. I.400, Q.920 - Q.930 for the DSS.1 protocols).

2. Description of the switching functions identified in the actions performed by the functional entities (CCITT Rec. Q.500).

(End)

8

FDT based development of an interworking unit between LANs and X.25 networks

J.C. Yelmo[a], F.J. Carrasco[b], C. Sánchez[a], J.J. Gil[b], C. Ramos[c], M.I. Felipe[b]

[a]Telematic Engineering Department. Technical University of Madrid

[b]Software Technology Division, Telefónica Investigación y Desarrollo

[c]Telematic Engineering and Architectures Department. Technical University of Madrid

This paper summarizes experience and lessons learned in the framework of the MEDAS research project. The MEDAS project was devoted to the elaboration of methodological guidelines for the design of communication systems based on the use of FDTs, particularly LOTOS and SDL, and their relationship with object oriented techniques and industrial-strength software development methods.

These guidelines have been evaluated by implementing two case studies. The second of which is presented here, focusing the attention on automatic code generation from formal specifications and integration of FDTs in the software life cycle and development process used by the industrial partner.

Keyword Codes: D.2.0; D.2.2; D.2.10
Keywords: Software Engineering, General; Tools and Techniques; Design

1. INTRODUCTION

In general, FDTs provide an interesting approach to formalize the design and implementation of communication systems. They are supported by powerful reasoning frameworks that allow the assessment of system properties and validation from early stages in the development process. Particularly, both LOTOS [1] and SDL [2] nowadays provide a complete and stable notation and theoretical foundations, a wide variety of support tools and several methodological guidelines that have been proposed to guide their application [5, 4].

Nevertheless, the successful use of FDTs at industrial scale strongly depends on the integration of such techniques and associated tools with other software engineering conventional methods and development environments.

The line of research pointed out above was started in the framework of the MEDAS [1] project [9]. The main goal of the project is the elaboration of methodological guidelines for the design of communication systems based on the use of FDTs and its relationship with object oriented techniques and industrial-strength software development methods. These guidelines have been evaluated through the implementation of two case studies in

*This work has been partly sponsored by the Spanish Research Agency CDTI under contract #900120, within the MEDAS Project.

[1]MEDAS, in Spanish, stands for Advanced Methodology for Communications System Development

a controlled industrial environment, paying special attention to the integration of FDTs in the conventional software life cycle and development process used at Telefónica I+D[2]. The implementation of these case studies has been accomplished through semiautomatic code generation from LOTOS and SDL formal specifications. For this task we have used a LOTOS environment developed at DIT-UPM [3] [8, 7] and the SDT[4] development environment for SDL.

2. THE CASE STUDY: GOALS AND DESCRIPTION

A common approach to validate a new methodology from the industrial standpoint consists in developing pilot projects as similar as possible to those usually developed in the company's specific application domain. That is, controlled developments close to other products developed in the company that can be carried out without the pressure and rigid schedule of the latter.

With this purpose two case studies have been developed within the MEDAS project. The first, an interworking gateway between SS1 and SS7 ISDN signalling systems, has represented a medium complexity exploratory experience of FDTs capability for the development of real products [10]. The second one, an Interworking Unit between IEEE 802.x LANs and X.25-based networks, is a much more complex development aimed at running on a real platform that allowed us to evaluate several aspects not thoroughly covered in the first case study.

This paper summarizes the development of the second case study, because in our opinion it represents an advanced experience of industrial-strength communication software development based on FDTs, specially in the case of LOTOS, which is a more recent notation and, therefore, there is less accumulated industrial experience.

The case study consisted in the specification, design and implementation of an Interworking Unit (IWU) that allows users of a local or metropolitan area network to communicate with subscribers of a X.25 public switched data network (IBERPAC in Spain). Figure 1 depicts the general context of the IWU.

Currently, the IWU implementation runs on a computer with multilink access to IBERPAC in a 802.3 LAN environment. It allows users logged in to any host of the local network to connect to any subscriber of the IBERPAC network without logging in to the machine that actually has the X.25 links.

As figure 1 shows, the interworking problem has been solved at the data link level, in terms of the Open Systems Interconnection Architecture defined by ISO (ISO-OSI model), through the development of the following subsystems:

- A data link level LLC II entity (Logical Link Control class II), i.e. the IEEE standard for the upper data link sublayer. It is responsible for the data link connection between the IWU and LAN users that are source or destination of connections established with IBERPAC subscribers.

- A data link level HDLC entity (High-level Data Link Control), i.e. the X.25 CCITT

[2] R & D branch of the Spanish PTT

[3] Telematic Engineering Department, Technical University of Madrid

[4] SDT is a trademark of Telelogic

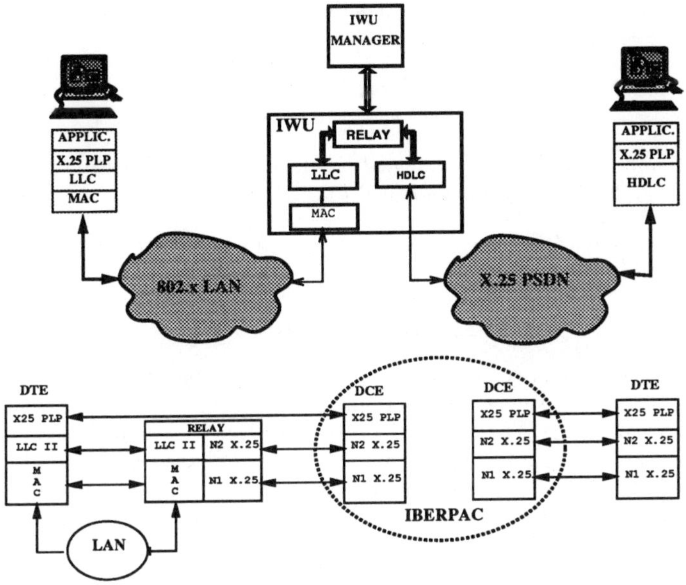

Figure 1. IWU functionality and general context

recommendation for the data link level interface to public data networks.

- A bridge entity between the two entities described above, here named RELAY, that performs service adaptation between the two concatenated data link layers [11].
- A management entity, IWU MANAGER, that may run locally or remotely with respect to the IWU on any host in the LAN. It has a X-Window graphical user interface offering a wide range of management functions like IWU configuration and enabling, entity activation, statistics, status control, etc.

The LLC and RELAY entities have been developed both using LOTOS and SDL as specification language from an object oriented analysis of the system. The RELAY entity was also developed using C++ from an object oriented design. On the other hand, the IWU MANAGER, is a conventional development based on C/X-Window. Finally, the HDLC entity comes from a commercial product.

To finish this description we would like to highlight that the case study was not aimed at implementing a complete industrial product. It was just a vehicle for exploring industry-academia technology transfer mechanisms [3], yet providing a valuable experience on industrial application of FDTs. In this sense, and in order to be specific, the main technical goals have been the following:

- Feasibility study of FDTs integration with methodologies based on the object orientation paradigm.
- Non-functional requirements analysis not well covered by FDTs: generated code optimization, performance, FDT-based maintenance, etc.
- Smooth integration of methodologies based on formal methods and the development process model defined by Telefónica I+D.
- Interoperability with conventional developments and commercial products and interchangeability of subsystems that have been developed following different paradigms.

3. PLANNING AND DEVELOPMENT PROCESS

One of the requirements expressed in the introduction is the integration of a FDT-based development process into a conventional life cycle. In our case the framework was stated by the preexistent life cycle currently in use at Telefónica I+D. This is a conventional life cycle with emphasis on the testing and implementation phases.

Telefónica I+D's life cycle does not preclude the use of any specific method in the analysis and design phases. However, interest in object oriented techniques at the company level drove our attention towards the compatibility of object oriented techniques with FDTs. Additionally, Telefónica I+D was considering the introduction of new software techniques controlling their impact in the preexistent development life cycle and management procedures. As a consequence, neither all software products should be designed with the exclusive use of a single FDT nor all components should be developed with FDT technology. This twofold perspective implied the need to evaluate the following aspects: interoperability between software components developed from different paradigms (SDL, LOTOS or C++ in our case) and between commercial software products, and interchangeability of components implemented from different techniques.

As a result, the development process used to implement this case study was defined in order to maximize concurrency for activities carried out by different working groups and, at the same time, ensure the interoperability and interchangeability goals expressed above. Figure 2 depicts IWU development planning and concurrent activities.

We would like to emphasize two different aspects: the use of incremental prototyping as a risk resolution strategy during development, and the agreement on a common architecture and interfaces before starting the work in parallel.

The analysis phase started from inputs commonly found in telecommunication products: international standards, user requirements, MEDAS documents and previous experience taken from other case studies. This phase was supported by Rumbaugh's object-oriented methodology (OMT [12]).

Architectural design was one of the most critical phases of the development. Its main purpose was to decompose the IWU into subsystems and to define their interfaces. Due to the need for common and visible interfaces between subsystems, XDR/RPC[5] were chosen for interface description and entity dialog model respectively. Figure 3 represents the high level architectural design where three main subsystems can be identified: Management

[5]eXternal Data Representation/Remote Procedure Call

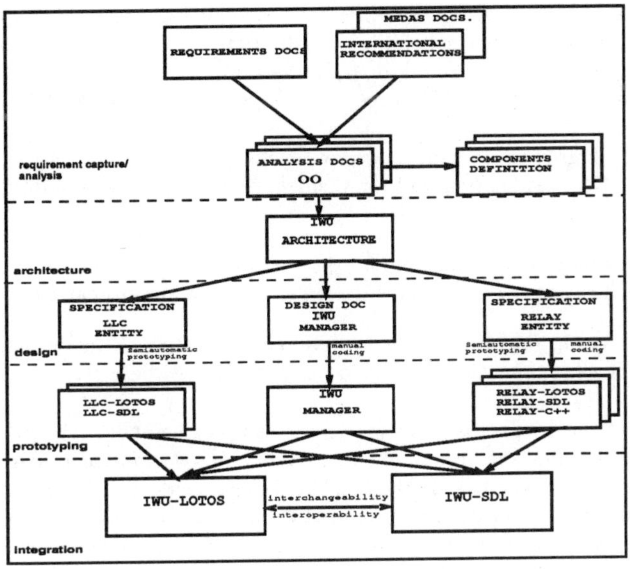

Figure 2. IWU development phases

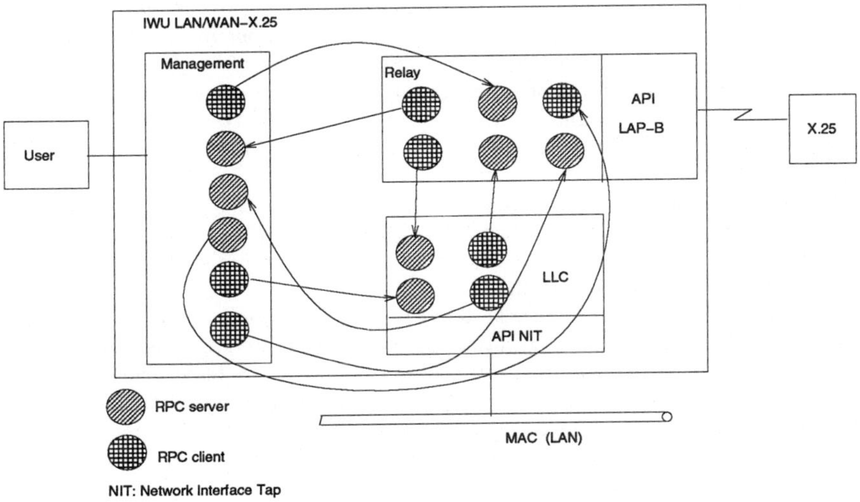

Figure 3. Architectural design

unit, RELAY and LLC. The HDLC subsystem was not implemented due to commercial products availability and the need of interoperability test with external packages; however, a specific API was developed to link it to the RELAY.

Starting from the system architecture, several groups began working in parallel:

- The LOTOS group was responsible for developing the LLC and RELAY subsystems from LOTOS specification up to implementation through semi-automatic code generation.
- The SDL group was responsible for developing LLC and RELAY subsystems from SDL specification up to implementation through semi-automatic code generation.
- The C++ group was responsible for developing the RELAY subsystem directly from the high level architecture through an object-oriented design based on OMT.

In the case of the LLC subsystem (the most complex component), SDL and LOTOS groups approached the final required functionality in two incremental steps. Initially only connectionless procedures of IEEE 802.2 were considered. Then, after validating its functionality, the subsystem was extended with connection-oriented procedures. This strategy allowed us to derive a rapid working prototype of an LLC subsystem capable to interwork with hardware resources (in our case, a LAN interface driver) and, therefore, reducing one of the major risks of our development. It is also important to notice that the effort spent in the first prototype was not lost since more than 90% of the first prototype was reused in the final LLC subsystem.

The Management subsystem was also developed incrementally. A first management unit was included in the first prototype and a more complex one in the final demo.

The integration phase was addressed under different perspectives. Due to the modular nature of the system under development and the availability of several subsystems with the same functionality, many combinations are possible. At least three integrated products were demonstrated: an all-SDL IWU, an all-LOTOS IWU and a mixed LOTOS-SDL-C++ IWU.

It is evident that we are not claiming that this multiple development approach should be used in general. Here, duplication of activities and effort (i.e. the development of the same subsystem using two or three different techniques) were aimed at demonstrating interoperability and interchangeability.

Usually, FDT experiences don't take into account management aspects of the development. Version control and configuration management are keystone techniques to cope with the complexity of managing large developments. Telefónica I+D is currently using a sophisticated software configuration management package from AT&T in their products that allowed us to automate version and release management, change control and system building for LOTOS and SDL specifications.

4. FORMAL SYSTEM SPECIFICATION

Formal specification focussed on the LLC and RELAY subsystems identified during the architecture phase. Each subsystem is described in terms of: (a) its external interfaces in

RPCL/XDR, and (b) a collection of message sequence charts showing its basic interactions with the remaining subsystems.

This section is intended to describe the structures of the LOTOS and SDL specifications. Emphasis is given to the LLC subsystem since it is the most complex and, therefore, appropriate to illustrate and compare the solution given for its formal specification in both FDTs. First, we describe the vocabulary introduced in IEEE to describe an LLC entity. Then, we give detailed descriptions of the specifications.

IEEE standard 802.2 describes the operation of the LLC entity as a dynamic and hierarchical structure of three types of components: the *Station Component* is responsible for processing the events which affect the entire LLC entity, the *Link Service Access Point (SAP) Component* is responsible for processing the events which affect a specific operating service access point, and the *Connection Component* is responsible for processing the events which affect a specific data link connection for connection-oriented procedures.

The previous components are hierarchically related; i.e., the Station Component is the *parent* of the SAP Component, which in turn is the *parent* of the Connection Component. We have extended this structure in order to include management entities for all the components but maintaining the conformance with respect to the standard.

A common problem to be solved in both FDTs is how to map XDR data types to ACT ONE, which is the abstract data type language used by SDL and LOTOS. In general, any XDR data type can be easily mapped to an equivalent ACT ONE abstract data type (ADT). However, the use of predefined ADTs in SDL, such as natural numbers, strings, structures or vectors, simplifies the data specifications in SDL systems. In the case of LOTOS, all XDR data types needed to be formally specified as complete ADTs including equations.

4.1. LOTOS specification

The LOTOS specification of the LLC subsystem is about twenty thousands LOTOS lines that have produced about ninety thousands lines of C code, counting hand coded and automatically generated code. This represents an effort close to some developments usually tackled by the software development industry, and thus presenting some of the human and technical coordination problems so common in software engineering.

Early stages of a development cycle can be characterized by a succession of tentative descriptions of the subsystems that form the architecture of the system under development and their interaction structure. These descriptions are intended to be abstract in the sense of not going into details irrelevant from the architectural viewpoint. These details can be dealt with in later stages in the development cycle, generally by successive refinements.

Graphical or visual languages are usually the most suitable for this purpose, specially when the language elements have a direct relationship with the architectural elements that are handled. This advantage is even more interesting when it is possible to automate (at least partially) the process of code derivation from graphical descriptions of the architecture.

By the time the MEDAS project started, there was no graphical notation available that could be used with this purpose, and textual LOTOS is not the most suitable way of interchanging design information and communicating solutions to project managers. In order to bridge this gap, a visual language called PTD[6] [6] was developed at DIT. This language allows to describe in a graphical form the architecture and a tool synchronization structure of a LOTOS-based development and automatically generates a textual LOTOS skeleton from the graphical design. In the remaining of the paper PTD diagrams will be used to illustrate LOTOS descriptions. PTD diagrams are quite intuitive, so it will not be necessary a more precise description of the language.

The conceptual framework provided by the 802.2 standard and the LLC entity interaction interface defined inside the MEDAS project were the starting points for the LLC LOTOS specification. This interface consists of three interaction points (gates): LLC, MAC and MGMT; where events are exchanged with upper (network) layer entities, mac sublayer and the management entity respectively. Using LOTOS notation, this general interface has the following form:

GATE ? sap: SAP ?prim: PRIMITIVE ?param: PARAMETERS;

That is, events will occur at the gate that represents the corresponding interaction point. These will be associated to a specific service access point and primitives and their corresponding parameters will be exchanged.

To finish this short description of the interface, just comment that each gate has been duplicated in order to distinguish incoming and outgoing directions.

Figure 4 shows the basic PTD structure for the LLC entity, where processes for station and SAP components can be appreciated.

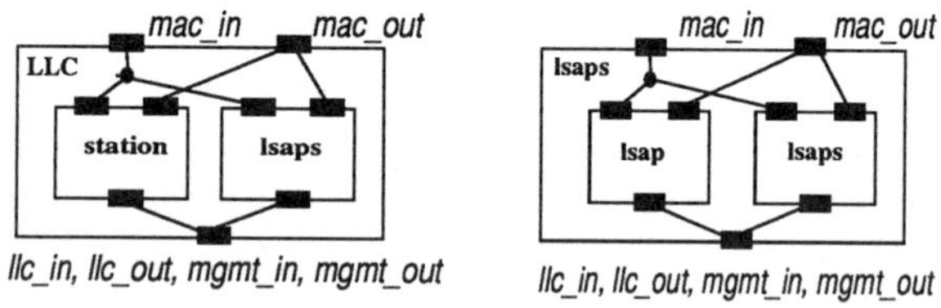

Figure 4. Basic topology of the LLC entity

Nevertheless, the diagram shown in figure 4 is not sufficient as a starting design for a complete implementation that must include management procedures such as station enabling and disabling, SAP activation and deactivation, keeping global configuration parameters, statistics, etc.

Figure 5 depicts a more complete diagram, where the process *stat-mgmt* is instantiated as manager of a disabled station and after receiving the corresponding management primitives turns into an enabled station (*station*), the registered group of SAPs (*lsaps*) and the station manager itself. This group of processes may be disabled again on request of the management entity to turn again into the original *stat-mgmt* process.

[6]PTD stands for Process Topology Diagrams

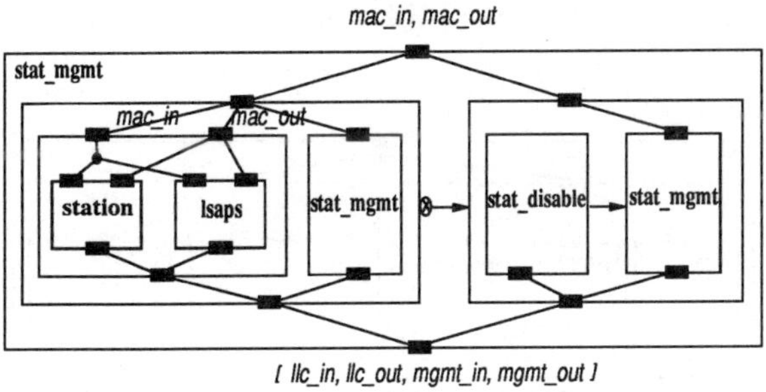

Figure 5. PTD diagram for the station component

A new refinement step on the PTD of figure 5 produces the diagram shown in figure 6. The figure details the internal structure of the process *lsaps*. This PTD follows the same structure already described for the station component, to which has been added a default SAP (*lsap-def*) to deal with messages sent to unknown or non-active SAPs.

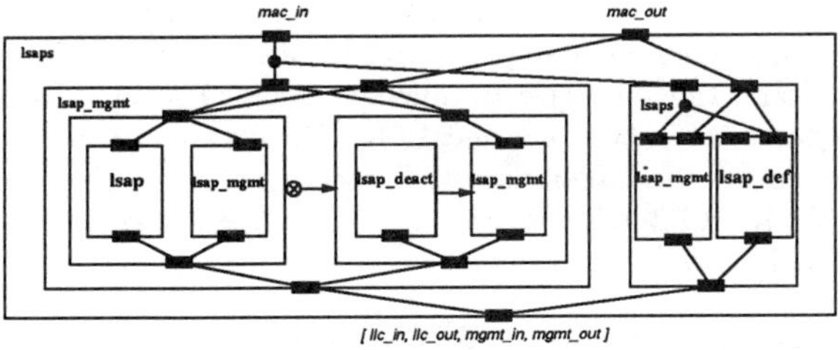

Figure 6. PTD diagram for the SAP component

Figure 7 depicts a *constraint-oriented* decomposition of the SAP component, i.e. it is specified as a conjunction of separate constraints either in the form of a *logical and* (synchronized processes) or in the form of a *logical or* (interleaved processes). In the figure, a service access point is represented as a conjunction of the constraints imposed by service (*llc-service*) and protocol (*llc-protocol*) specifications. In turn, *llc-service* is a combination of three independent services: data link service, management and MAC sublayer service. Finally, the protocol is a conjunction of connectionless procedures and connection-oriented procedures. Instances of the process *type-2-proc* correspond to the connection component described in the IEEE standard.

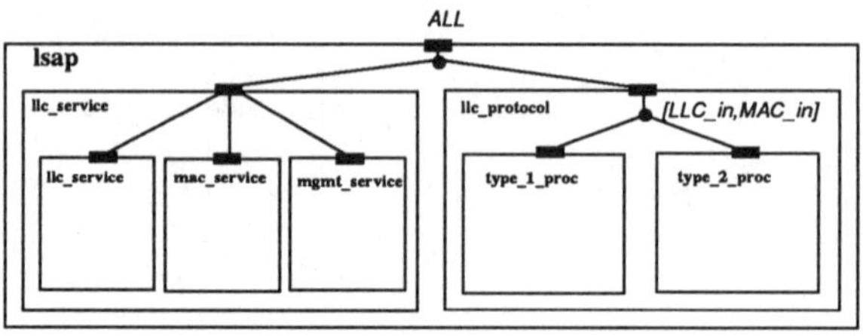

Figure 7. Constraint-oriented decomposition of the SAP component

There exists another LOTOS component not described yet in the PTDs shown in previous figures that plays an important role and links specification and implementation phases. It is the external interface, i.e. a little LOTOS process on which annotations[7] are written in order to derive the implementation with the aid of a LOTOS compiler. We will go into details on this aspect in section 5.

4.2. SDL specification

This section describes the main design principles that guided the SDL specifications of the LLC and RELAY subsystems. Even though both specifications have many aspects in common, we applied different design criteria in order to assess their impact on the formal specification and the subsequent implementation. The specifications of the LLC and RELAY subsystems are about 4K and 1.5K SDL lines and resulted in about thirty five thousands and twelve thousands lines of C code.

Firstly, the RPCL/XDR interfaces of each subsystem need to be described in SDL. This can be performed as follows:

1. adapt the operational interfaces of RPCL/XDR to the SDL asynchronous communication model.

   An operational interface is modeled in SDL as a communication channel. In order to adapt the synchronous communication model of RPC to the asynchronous communication model of SDL, each service primitive is decoupled into two signals, which correspond to the service invocation event and the service response event. These signals are communicated asynchronously between the SDL system and the environment through the communication channel associated to the interface. Therefore, a unique invocation identifier is required to both signal parameter lists in order to relate service response events to service invocation events.

2. mapping XDR data types onto ACT ONE.

[7]Annotations are additional implementation information provided to the compiler in a procedural language

Some XDR data types (i.e. byte, union and variable length vector) have not been formally specified in ACT ONE due to the difficulty to describe them in terms of predefined SDL types. Instead, these data types are directly implemented in the target programming language through annotations in the specification. This type of annotations are typically supported by commercial SDL tools.

Secondly, a high level SDL process structure needs to be defined. Both, LLC and RELAY subsystems, have in common the need to handle multiple simultaneous connections that are established and released dynamically. A single connection is, basically, a finite state machine and, thus, can be naturally modeled as a dynamic SDL process. A common problem found in SDL specifications with dynamic processes is the addressing of input signals from the environment to the SDL system.

In the case of the LLC subsystem, the SDL system uses implicit addressing mechanisms. As it is shown in figure 8 the SDL block is decomposed into two types of SDL processes: protocol processes and interface processes. The motivation behind this decomposition is to minimize the dependencies of the environment with respect to the internal process structure of the SDL system. Protocol processes model the behaviour of an LLC entity as defined in the IEEE standard. There is one protocol process for each component identified in IEEE 802.2: station component (*Station*) service access point component (*SAP*) and connection component (*Connection*). It is important to notice that processes *SAP* and *Connection* are dynamically instantiated and destroyed during the operation of the LLC subsystem. On the other hand, interface processes are responsible for addressing the SDL input signals received on a given subsystem interface (i.e. operation invocation events) to the addressed SDL process. There exists one interface process for each subsystem interface: management interface (*Mgmt*), upper layer interface (*Network*) and lower layer interface (*MAC*). For example, process *MAC* addresses the incoming data link PDUs to the process *SAP* that handles all data traffic to the addressed service access point. In turn, the latter forwards data link PDUs of the connection oriented procedures of IEEE 802.2 to the addressed data link connection. This type of process structure does not impose any special requirement on the environment but requires that the specification be responsible for addressing input signals to SDL processes, resulting in specifications which are more complex and more difficult to read due to the increased interchange of signals among SDL processes.

In the case of the RELAY subsystem, the SDL specification uses explicit addressing mechanisms. This solution imposes constrains on the environment, since it has to know the internal process structure of the SDL system, in particular, it has to keep track of the dynamic instantiation and creation of SDL processes. On the other hand, the SDL specification does not require interface processes to perform the addressing functions. The process structure of the SDL system modeling the RELAY subsystem is shown in figure 8.

The SDL block is decomposed into two protocol processes, corresponding to the station component (*RLT_Station*) and the interconnection component (*RLT_Interconnection*). Process *RLT_Interconnection* performs bridging functions between two data link connections on both sides of the IWU, and it is instantiated dynamically. There exists also an interface process with the management subsystem (*RLY_Mgmt*).

From the point of view of validating the SDL system, our experience indicates that explicit addressing of input signals from the environment facilitates the simulation of the

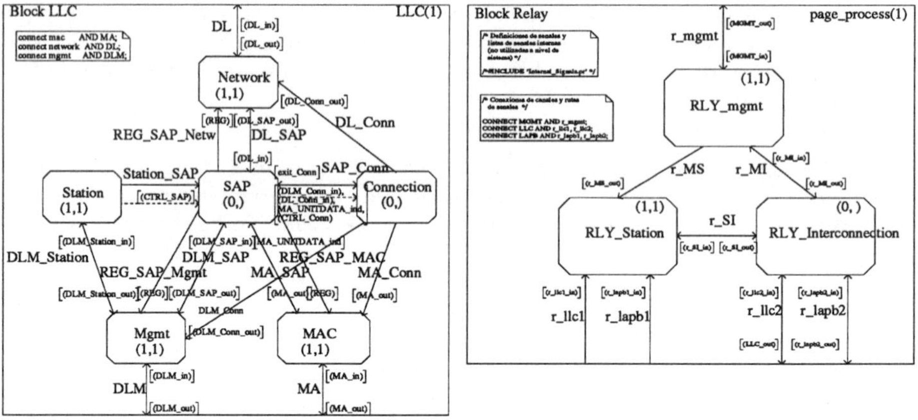

Figure 8. SDL process structure of the LLC and RELAY subsystems

SDL specification due to the increased controllability and observability of the SDL system. It also makes possible to start the simulation early since the addressing of input signals does not have to be formally specified as part of the SDL system. However, from the point of view of implementing the SDL system, the first approach reduces the requirements on the environment and, therefore, facilitates the rapid derivation of executable prototypes.

5. IMPLEMENTATION

IWU implementation was carried out using a distributed environment based on RPCs. As shown in figure 3 the IWU is composed of RPCs interconnected modules each implementing both the server and the client partners. Each module have been implemented using different FDTs approaches (basically LOTOS, SDL) and non formal approaches (C and C++).

Interfaces for modules were previously defined using the XDR/RPCL IDL (Interface Description Language) so it will be possible to pick any implementation to build a mix approach IWU as shown in figure 9.

Data types used in the interfaces were specified in XDR in the early development phases. In spite that both FDTs used ACT ONE for the abstract data types, each one defines its own syntax; so we needed to make double hand translations since automated tools was not available.

In the implementation phase, in LOTOS and SDL, we had problems with the integration of automatic generated code from the XDR/RPCL, and the code generated by the FDT compilers, for the data types. The LOTOS compiler, TOPO, and the SDL compiler, SDT, make different mapping to C, and these mappings are incompatible with the mapping from XDR to C. Since it is not possible to customize the mapping in these tools, it is necessary to write hand code in order to adapt the different automatic generated data definition.

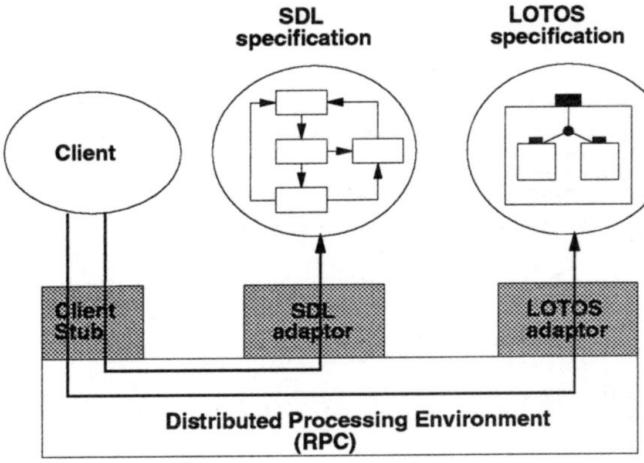

Figure 9. Distributed environment

This situation is usual in distributed systems (RPC, CORBA [13], ...). The mapping between interface languages and implementation languages is usually standard and/or will be standard, but there is not standard mapping between FDT data types and implementation languages. One solution to the problem will be the adaptation of FDT compilers to the standard mapping defined for the interface languages.

5.1. LOTOS

Introduction of automatic coding of LOTOS formal specifications in the development process requires formal specification scope to be defined based not only on application requirements, but also on the capabilities of LOTOS and its associated compilers. Features, like expressive power or completeness, desirable in the specification phase may have lower priority when efficiency is a must.

For instance, some specification styles like the *constrain-oriented* are elegant and expressive from the specification viewpoint but become forbidden, for performance reasons, when this specification serves as input to automatic implementation.

Formal specification of one of the subsystems, for instance the LLC of previous section, not only has to fulfill the service and the protocol of the standard but also has to define how to interwork with its environment. Thus it is necessary to define an *implementation oriented specification* as the *file* (output from the trade-off between formal specification and hand written code) that serves as input to the automatic code generation tool. In the case of a LOTOS based development using TOPO [8], this means a composition of the original specification with a new process, hereafter called *Environment Interface*, on which through *annotations* (syntactical and semantically defined in TOPO) the user writes the functions needed to communicate with the *environment.*

As a view of this environment specification, figure 10 shows the process *interface_in* used by the LLC subsystem to receive incoming primitives followed by a refinement of the process llc_interface_in used by the LLC to receive RPC primitives from the RELAY.

```
process interface_in [llc_in,mac_in,mgmt_in,TIMER_in] : noexit :=
    llc_interface_in[llc_in] |||
    mac_interface_in[mac_in] |||
    timer_interface_in[TIMER_in]
 where
 process llc_interface_in[LLC_in] : noexit :=

    hide LLC_extern in
      (*| wait LLC_input() |*)
    LLC_extern ? llsap (*|default llc_lsap()|*): LSAP
               ? lcei  (*|default llc_lcei()|*): CEI
               ? prim  (*|default llc_prim_type()|*): DLPT
               ? param (*|default llc_prim()|*): DLP;

    LLC_in !llsap !lcei !prim !param
      (*|C LLC_output (0,success); |*);

    llc_interface_in[LLC_in]
 endproc(* llc_interface_in *)
 ...
 endproc(* interface_in *)
```

Figure 10. Environment specification

In figure 10 the *wait* annotation is a busy waiting that does not preclude other offerings. The function called by this annotation returns TRUE if a primitive from the RELAY has arrived. This, in terms of TOPO semantics, means that the named process have been chosen. Input data conversion (from *rpcgen* [8] generated structures to data types implemented automatically by the tool from the ACT ONE specification) and buffering are functionalities accomplished before yielding control to the LTS[9] machine. Functions called by the *default* annotation pass on each *experiment-offer* the converted primitive to the implemented behavioral automaton .

Gate LLC_extern is the result of a trade-off between formal specification and hand coded implementation. As shown in figure 4 the LLC module after activation is composed of the interleaving of equal SAP components parameterized by a SAP address, each creating dynamically equal *connection* components with a distinguished *llcei* value (remote pair MAC-SAP addresses). In order to avoid the evaluation of costly guarded-expressions or selection-predicates, the specification offer to the environment *value-expressions* and exclamation symbols for the first two experiment-offers preventing (due to the semantics and syntax of TOPO annotations) direct use of the *default* annotation on the original specification LLC_in gate. So, the environment process llc_interface_in does also a sort of demultiplexing for input primitives.

The TOPO C annotation place after an *action denotation* allows the environment to pick up on the array *success* the successful synchronization values in order to carry out the inverse data types conversion done by the LLC_input function

In spite that the LTS machine created by TOPO has its own scheduler it is necessary to

[8]rpcgen is a protocol compiler from SUN Microsystems
[9]LTS stands for Labeled Transition System

integrate on this the environment. Therefore, the main loop permits the LTS to progress *step by step* interleaving between them a single call to *select* to receive *clients* requests which the RPCs server skeletons decode. Since both the RELAY and the LLC bear multiple connections, RPCs used between those entities are asynchronous by means of the *call-back* technique. As time is not yet supported in LOTOS, the specification only starts *timeout* and receives *timeout* expiration which is hand-code implemented using periodically generated signal catching.

The automatically generated and hand-written code linked together with the RPCs and TOPO libraries produce the executable version of each module.

5.2. SDL

The SDL code generator, SDT, implements an SDL machine, and offers some environment functions for environment interaction by sending and receiving signals in asynchronous mode. The compiler has system execution control in an endless loop, and the only way of interaction are the environment functions that it offers and calls. As far as data are concerned, it offers several possibilities for abstract data types signature representation, although semantic implementation has to be done by hand, following the basic operations naming rules (creation, equality, etc.).

Besides the adaptation problem between different abstract data types representations, synchronous communications offered by RPC had to be implemented on top of asynchronous interface of the generated code. In the specification, synchronous calls were unfolded in two asynchronous signals of the SDL system. When implementing, association among RPC calls and the SDL system signals had to be solved, and also had to be solved the execution control transfer problem from the RPC server to the SDL system.

6. CONCLUSIONS

The successful use of FDTs at industrial scale strongly depends on the integration of such techniques and associated tools with more conventional software development methods. The consolidation of emerging technologies such as object orientation and distributed processing will be key factors in future developments and in this paper we have described our experiences in adapting LOTOS and SDL to them.

The development of an interworking unit between LANs and X.25 networks has been our vehicle to explore this integration. The main lessons we have learned can be summarized as follows:

- satisfying non-functional requirements (i.e. performance) is only possible by means of heterogeneous implementations in which FDT based components coexist with non-FDT components. It also forces the use of implementation oriented styles in the specifications (i.e. finite state machines).

- the use of interface description languages, in particular RPCL/XDR, has proved to be quite valuable in achieving interoperability and interchangeability between heterogeneous components. We have presented a possible solution to offer this type of interfaces in LOTOS or SDL components. It consists in mapping the interfaces of a subsystem into a LOTOS or SDL skeleton of the specification, as well, as the development of specification adaptors similar to RPC stubs. We envision that tools to automate this adaptation process will be available in the near future. Also, emerging distributed processing technologies such as CORBA will have an impact on interface description languages.

- the implementation effort to adapt the automatically generated code from a LOTOS or SDL specification to the target execution environment (i.e. a distributed processing environment) cannot be neglected. The availability of libraries and tool support for the automatic generation of adaptation code would greatly reduce the effort.

- A good integration of LOTOS and SDL to object oriented methods cannot be achieved until both FDTs give better support to the object orientation paradigm. Our approach of extending the OMT method to allow describing the dynamic model of the system components in LOTOS or SDL during the design phase can be regarded as a pragmatic decision. We expect that OSDL will facilitate this integration.

REFERENCES

1. **ISO/EC**. *Information Processing Systems - Open systems Interconnection- LOTOS A formal description technique based on the temporal ordering of observational behaviour.* DIS 8807. 1988.
2. **CCITT**. *Rec. Z100. SDL: Specification and Description Language.* Blue Book. 1988.
3. **G. León**. *On the technology transfer of formal methods: An experience on LOTOS* FORTE 90. Madrid.
4. **F. Belina et al.**.*SDL with applications from protocol specification.* Prentice Hall, 1991.
5. **R.E. Booth et al.**. *Formal methods for communication systems development- the ESPRIT project LOTOSPHERE.* BT Technology Journal. Vol. 10, # 2. April 1992.
6. **J. Sánchez, G. León**. *PTD: Architectural system description support based on visual specification languages*, EUROMICRO 92, Paris, September 1992.
7. **J. Quemada et al.** *Transforming LOTOS specifications with LOLA - The Parameterized Expansion.* In Formal Description Techniques, FORTE'88, Stirling, September 1988.
8. **J. Mañas et al.**. *Tool support to implement LOTOS formal specifications.* Computer Networks and ISDN Systems 25, 1993, pp 815-839.
9. **G. León, J.C. Yelmo**. *On the intertwining of FDTs and conventional software development processes for communication systems: an incremental approach.* Software Engineering Conference proceedings pp. 555, Toulouse, December 1992.
10. **G. León et al.**. *An industrial experience on development with LOTOS and SDL.* FORTE 93. Boston, USA. October 1993.
11. **G.V. Bochmann**. *Design Principles for Communication Gateways.* IEEE Journal on Selected Areas in Communications, Vol. 8, No. 1, January 1990.
12. **J. Rumbaugh et al.** *Object Modelling and Design.* Prentice Hall, New Jersey, 1991.
13. **Object Management Group, Inc**. *Object Management Architecture Guide.* OMG TC Document 92.11.1 Rev. 2.0, Sept. 1992.

9

Using formal specification and simulation: Practical experience with distributed applications

Authors
Willem Hengeveld, Bakkenist Management Consultants
Peter van der Toorn, Bakkenist Management Consultants
Geert-Jan Houben, Eindhoven University of Technology

Bakkenist Management Consultants
P.O. Box 23 103
NL 1100 DP AMSTERDAM

Telephone : +31 20 695 6666
Fax : +31 20 698 24 26
E-mail : pim@bmc.nl

Summary
Methods for formal specification have been studied and evaluated in the academic environments already a long time. In particular the combination of formal specification and analysis with simulation, animation and code generation may change parts of the software engineering process considerable. The availability of some more user-friendly software tools, based on a sound formal basis now opens the way to application of results of fundamental academic research to important practical problems. Formal specification turns out to be realisable in times comparable to more conventional approaches, whereas they offer more possibilities to describe and analyse modern distributed software applications.

This article describes how formal software engineering techniques have been applied in a large distributed software engineering project (*RODOS*: *Ro*tterdam *Do*uane *S*ystem). We will describe the elements of the formal specification made with the tool ExSpect, which uses Petri-nets, functional languages and data-modelling. Also the way by which we come to a formal specifications is being described with the clear advantages of the approach in comparison with more traditional software engineering approaches.

1. INTRODUCTION

Quality of IT-systems is often determined by the traditional methods being used to develop the information systems. Often specifications are made with schemes, models and descriptions at different stages in a software engineering project. The role of the users is that the very voluminous documentation has to be verified, in terms of user requirements, but also often in relation to consistency and conformity with existing or presumed organisational conditions.

Any verification of the specifications is difficult, as most of the specifications are in a textual form illustrated with figures. Many software engineering tools presently available support this traditional process and offer document design and management facilities as well as dedicated drawing tools. Using structured methods with software engineering tools is now well known as CASE (Computer Aided Software Engineering).

Often it appears that after realisation many errors still remain and, even worse, the code that is written may contain functional inadequacies. If users have problems in describing requirements, one may use prototyping as a means to circumvent problems in a software development process. This however makes it necessary to use advanced tools and fourth Generation Languages (4GL). Prototypes are not always as useful in larger software engineering processes, that involve complex communication.

For large distributed applications that involve many requirements from different departments within a large organisation or that deal with communication between different organisations, the problem of describing user requirements and checking of correctness and completeness remains pertinent. It is especially in this field that formal methods can now be used, not only for technical applications (protocol implementations, etc.) but also for user applications (global banking, government applications, logistic software, etc.).

The claim of formal methods is that development of software can be checked more easily as to their correctness. Furthermore several types of analysis can be made. Often software tools that are based on formal methods allow the specification to be simulated on a computer (the specifications are executable). In this way one gets a new way for users to inspect specifications. So, formal methods, supported with a set of software tools, allow a new way to work together with end users on developing software applications.

As an example we show how to apply one way of formal specification in practice. The formal specification was made in parallel with a more traditional design. The example was made for a large distributed system for automatic customs declaration. The name of the system was: *Ro*tterdam *Do*uane *S*ystem* (RODOS). The case is presented to illustrate how a formal specification can be made, and also, what specific contributions formal specifications may add to a design phase.

By using the formal specification tool *ExSpect* (*Ex*ecutable *Spec*ification *T*ool) and its language during the design phase it turned out that in a short time we were able to come to a formal specification. It turned out that on the basis of the design all distributed aspects (in terms of EDI) could be well analysed, in a way that is not possible with traditional informal specifications.

We will start with a short description of the main ingredients of an ExSpect specification.

2. DYNAMIC MODELLING WITH EXSPECT

The way by which ExSpect specifications are made, is based on Petri-nets. In ExSpect we use Petri-nets that are an extension of the classical Petri-nets [1], for which a vast amount of scientific literature has been published. In ExSpect 'colour', 'time' and 'decomposition' are added [2]. With these additions it is possible to integrate static (binary data-model) and dynamic (process model and control flow aspects) in one formal and consistent model.
We will outline how these concepts are being combined in ExSpect models.

---

* Rotterdam Customs System

Petri-nets
Using the theory of Petri-nets means that in ExSpect a system is described in the form of a network of active components (called processors or transitions) and passive components (channels) [1]. In one ExSpect-network every processor has input and output channels. By means of a number of input channels data are offered to a processor. This is done in the form of tokens, thus following the definitions of Coloured Petri-nets [3]. In ExSpect any token is of a certain type and may therefore have any complex data structure. In the processor input tokens are being transferred into output tokens on the basis of the specification of the processor.

Functional language
For the specification of the processors a functional language is being used. This means that the value of the output tokens is determined by applying functions on values of input tokens [4]. ExSpect also allows to use 'stores'. Stores can be compared with 'buffers' in more traditional data flow diagrams. A store is in ExSpect represented as a channel with exactly one token: stores may function in this way as databases of permanent data.

Time
In ExSpect one may also include timing aspects. This is especially of importance to come to consistent distributed application. Any token receives a time stamp when it is created, and can only fire a next transition when the time of its time stamp has passed. Whenever a processor creates tokens for output channels it not only determines the values, but also the delays.

Decomposition
ExSpect allows to make hierarchic specifications. This means that different parts or functions of a system can be described independently, and finally be analysed when they are combined. Such partial Petri-networks are called systems. This approach is similar to decomposition approaches of other modelling techniques such as Data Flow Diagrams. By applying decomposition techniques it is possible to acquire libraries of standard components, thus allowing for reuse of defined components, or allowing for the design of standard-components.

The formal ExSpect models can be constructed and analysed with the tool ExSpect. We will outline in the next section its main characteristics.

3. EXSPECT TOOL

The tool ExSpect is meant to dynamically model, verify and simulate (complex) systems. It was developed by Eindhoven University of Technology and Bakkenist Management Consultants as of 1987. The main reasons for developing this CASE tool were the following needs for software development that became eminent around that time:

- software design becomes more and more system design;
- software design cannot be done without taking organisational issues into account;
- business processes may be reconsidered with development of IT;
- distributed applications become more important: networks and client-server;
- object orientation will play a more important role;
- analysis and testing has to be done at earlier stages in the development process;
- prototyping will play a more important role;
- visual techniques will get a more prominent place.

The software uses a graphical user interface. Making an ExSpect model can be done relatively easy and may include active user participation. The same interface allows to execute a specification, thus making the dynamic aspects of the systems visible to users by means of animation.

Figure 1 shows the different parts of the ExSpect environment. For more background on ExSpect we refer to the documentation (see for example [5]).

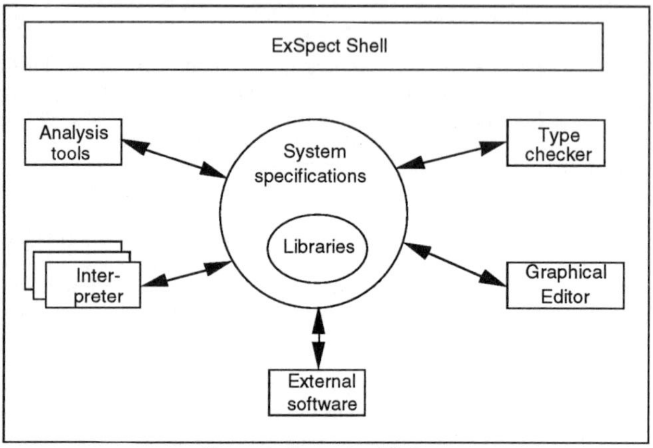

Figure 1: the tools of ExSpect

The tool ExSpect contains a graphical editor, that is able to design Petri-nets and data models in an interactive way. It allows for using several windows and is essentially mouse-driven. The user interface allows for using colour and a variety of, in principle, user defined application dependent figures. An ExSpect specification can be checked with a powerful type checker, that allows for interactive error repairs. An extensive help facility is available to help the user of ExSpect with making his specification. Several analysis tools are able to make Petri-net analyses [1, 2] like state invariants, time constraints, place invariants, etc. These can be used to check all kinds of functional properties. An interpreter for multi-user simulations is available, thus allowing not only a functional animation, but also a possibility to let different type of users work together on the same specification. An extensive overview of the advantages of simulation for modelling of systems can be found in [6].
Important is the high performance characteristics of the ExSpect simulator which allows to analyse real world problems. ExSpect is open to other software like spreadsheets, planning software, databases, etc. The ExSpect shell allows for a very user friendly way of handling the different capabilities of the tools when working on the ExSpect models. The tool ExSpect is available for UNIX and windows environments.

In the following section we will show how the tool was used for a real-world specification: the RODOS system that was developed for the Rotterdam harbour.

4. APPLICATION OF EXSPECT TO RODOS (ROTTERDAM DOUANE SYSTEM)

In the Rotterdam Harbour customs (douane) formalities are handled mainly with paper documents. This means an intensive administrative burden for the international trade and transport businesses, but also makes an effective control by the customs authorities themselves difficult. The distributed system RODOS, which is presently under development will automate the handling of all documents for incoming goods.

Distributed Characteristics
In harbours different actors are involved in the handling of incoming goods: liner agents, consignee's and consignors, expedition companies, etc. The custom authorities receive from all these organisations documents and have to check whether all formalities have been followed and necessary payments have been or will be made. Presently almost all handling is done by hand.

Communication Characteristics
Within RODOS the communication between the actors is as much as possible done with EDI (Electronic Data Interchange). Figure 2 shows the documents that have to be communicated for a simple situation. The time sequence diagrams shows with grey arrows the messages that are transferred by means of electronic communication. The other arrows show paper documents or verbal communication.

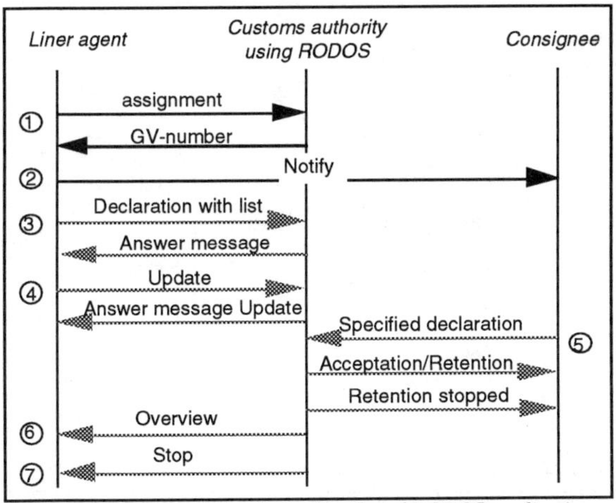

Figure 2: Communication in the distributed RODOS-application

The details of the defined protocol are as follows:
1. The liner agent informs by telephone the customs authorities.
2. The customs authorities give a GV-number.
3. The liner agent notifies the consignee or the forwarder of the arrival of goods.
4. The liner agent sends a list with all goods with EDI.
5. Optionally updates may follow.
6. Specified declarations by consignee or forwarder are made with EDI.
7. Optionally the custom authorities may need to check the incoming goods.
8. Liner agents receive message that the declaration has been handled.
9. Goods are being allowed to enter or pass through the European Union.

The above way of describing dynamic behaviour of systems in time-sequence diagrams is widely used, and not only for protocol descriptions. This way of describing is however highly informal. By using ExSpect one may both describe the time-behaviour and the systems that process the information that is exchanged between the different actors.

5. SPECIFICATION OF RODOS IN EXSPECT

A model of RODOS was made with respect to its environment: the systems that RODOS has to interact with. These systems are modelled as liner agent and consignee/forwarder. In addition to the electronic data interchanges also other types of communication (telephone calls, paper forms, etc.) between organisations in relation to the customs procedures are modelled.

Petri-net model of RODOS and its environment
Figure 3 shows as an example part of the ExSpect specification. It shows the different actors and the communication between the parties. The circles indicate the channels of the Petri-nets. These are passive components that may or may not contain tokens. The squares represent the systems, that at a different abstraction level (one hierarchical level lower) consist themselves also out of channels, systems and processors.

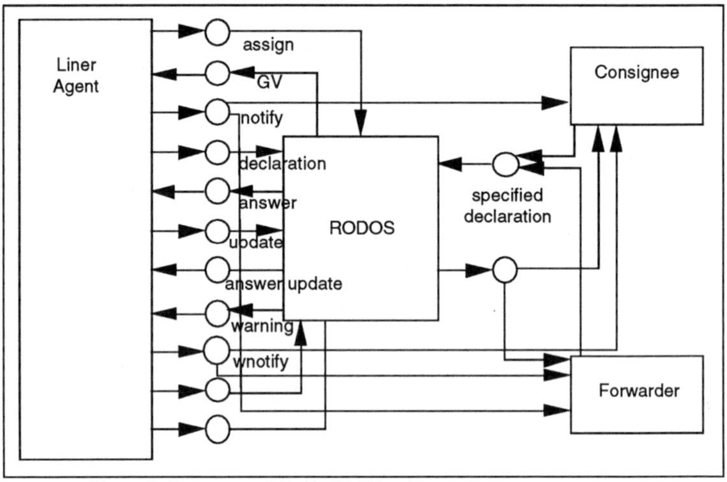

Figure 3: Process model of RODOS and its environment

Specification of the 'lowest' level
The different (sub)systems can be decomposed until one arrives at a level that consists of processors and channels only. A typical example of a specification of the lowest level is given in figure 4. The triangles represent the processors and the circles with a cross the 'stores'. In this case one single store represents a database. The specification of this database is done by means of a binary data model.

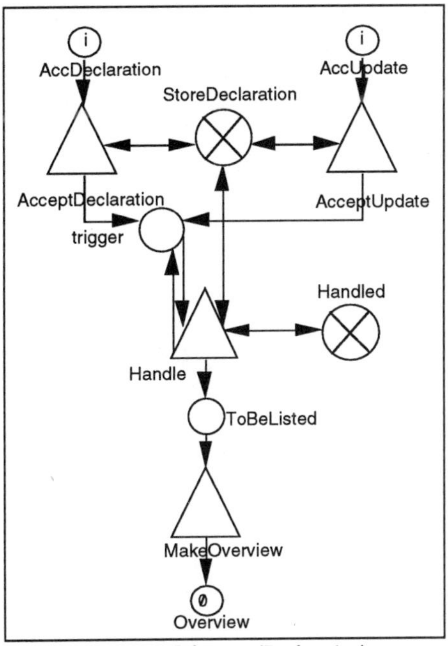

Figure 4: Subsystem 'Declaration'

Functional language
The processors are being specified by means of text, but more important with a functional language. This means that the processor contains a prescription how, given a set of input tokens, output tokens are generated. Such a prescription may include time characteristics, to model delays that processes may incur. The transformation of the process MakeOverview looks as given in figure 5.

```
proc MakeOverview [
                in ToBeListed: Listtype,
                out Overview: Overviewtype]
    :=
        Overview <- [gv: ToBeListed@gv]
    ;
```

Figure 5: example of the description of a processor: MakeOverview

The functional language used has been defined in a recursive way, thus stressing the formal aspects of the language. For performance reasons the functions have been implemented in normal code ("C"). The simple example shows how such a transformation is specified. In this case the value of the token that is being produced for the channel Overview is being determined by the value of the token in the channel ToBeListed.

In this example one has to include timing aspects. This is done by adding a simple delay function:

```
Overview <- [gv: ToBeListed@gv] delay 1. * day
```

In this case the 'delay 1. * day' means that the output tokens become available for other processes at times that occur later than 1 day after the time that the processor performs its transformation. (For more information on the functional language see [7].)

Characteristics of the RODOS specification
The overall RODOS specification and its environment consists of 3 systems and 13 channels. The overall specification contains furthermore 27 types and 18 functions. 11 Functions deal with the complex timing aspects of the specifications. In many non-formal specification techniques timing aspects are only dealt with superficially.

The RODOS system itself consists of 8 subsystems, 8 processors, 2 channels and 11 stores. The stores are necessary to deal with the data that have to be stored and checked.
The RODOS system, which is of functional nature, does include both automated functions as well as functions to be performed by the custom authority or its clients. One of the systems and 2 stores were used for a measurement system and do not belong to the actual specification. However, it may well be that for application management purposes such a measurement system may be considered useful to implement.

The environment that was modelled consisted of 4 subsystems, 22 channels, 10 stores and 24 processors. It is well known that specifying a system and its environment may well lead to errors in specifications if the environment is not modelled correctly. Fortunately ExSpect allows to let the interaction of the specification of RODOS and its environment be inspected carefully and, furthermore, it gives the possibility to let users take their place in the environment and work with the RODOS specification. ExSpect thus provides with the executable RODOS specification a working functional prototype.
The time needed to make the specification was about 15 days. The specification was made on the basis of the results of a feasibility study, that included a nice overview of the work processes of the Dutch Customs and an overview of the relevant paper and electronic forms. The time needed includes the time to communicate

The formal specification does not include a specific data model. This however was developed separately. The type definitions, that were used, have been derived by the makers of the functional design from the defined Entity Relationship diagram. ExSpect allows however to do so in an automatic way. Thus making the specification could be done even more efficiently.

Of these systems one was introduced for specific measurements on the actual specifications and thus should not be considered to be part of the specification.

Making the complete RODOS specifications was done by two functional designers. The Petri-net nature of the ExSpect models makes collaboration relatively simple. Both worked independently on parts of the RODOS system and generated the main library elements. After this the library elements were made available for the overall system specification and coupled with Petri-net channels.

6. ANALYSIS OF RODOS WITH EXSPECT

The possibilities for analysis of the RODOS specification include all different possibilities that are well known from protocol verification. Determination of deadlocks, livelocks, infinite waiting queues, consistency of protocol description with service requirements, etc. have been described in the relevant literature. ExSpect includes the possibility to perform such analysis for distributed applications. We list them in the following paragraphs.

6.1 Simulation as an important verification tool

By applying ExSpect to a real world problem it turned out, however, that the most appreciated feature of ExSpect, that is the executability of the specifications, was for the development team a highly powerful tool. The RODOS-system and the relevant interacting processes that are described with ExSpect can be simulated. In this case the different decisions that the system has to take and that are normally made by users on the basis of the contents of the incoming messages, have been imitated by means of a-select random numbers.

6.2 Animation for user requirements verification

This simulation provides a very good insight in the working of the system. This is especially supported by means of a graphical animation of the information flows in relation to processes. Users can take this animation as a possibility to verify the system, but also the designers were surprised by the actual insight that an animation could give, already at a very early stage of the development of a system. In an animation one may inspect the working of the system, the order of the different messages, the different decisions that are taken by the system, etc. On every level of abstraction these animations take place, which means that the different organisations that are involved in the development of such a distributed system may all inspect their part of the complete system and also get a clear insight in the effects that their actions have on the other organisations involved.

6.3 Inspection and verification of communication interfaces

Also the specific contents of messages may be inspected, which allows for an understanding of the reasons for inclusion of the different information streams, which in the case of the custom procedures is not always even clear to all organisations involved.

6.4 Dealing with time aspects in specifications

For the custom authorities aspects as delay times for handling documents and the times that one has to wait before goods are allowed into the European Union are highly relevant. Such figures can be determined because ExSpect allows for the specification of timing issues.

6.5 Preparing for system implementation

One step further is to analyse the impact of IT-systems on the relevant procedures and the necessary workload of personnel. Such issues are highly relevant for decision making on the basis of system specifications, which in general is not done taking into account technical aspects only.

6.6 Specification of error handling and "what if" analysis

One of the most important additions of formal executable specification came from an analysis of a large number of random declarations that was used as an input for the RODOS specification. This learned that a number of declarations were handled only partially. The main reason for this is that the exception handling was not specified adequately. As an example, it turned out that after a number of updates it was not possible to handle the following incoming messages adequately. In general such aspects of distributed systems are not described very accurately, which may lead to high correction costs, either at a later stage in the development of the relevant software or even worse, during operation of a definite system.

7. FINAL RESULTS AND LESSONS LEARNED

The quality of the formal specification was considered considerably higher than that of a traditional specification. We were able to determine this, as the two approaches, the traditional and the formal approach, were performed in parallel.
By making a formal model with ExSpect the results as described in the following sections have been reached.

7.1 Validation of EDI-protocols

With the ExSpect model the EDI-protocols that are relevant for RODOS have been validated. Therefore, at the start of the project it is clear that the defined protocols are correct and will work adequately. This is of utmost importance, as any change in the protocols or interfaces between the different systems after a test period will lead to far-reaching changes in not only the central RODOS-system, but also in the systems of the other organisations involved.
Errors like deadlocks and infinite loops can be determined. In the case of RODOS an important point was that the error handling when updates are sent by the consignee's could be analysed. This led to determining a number of problems when such updates occur too frequently. The protocol did not allow for too many updates to be sent to the RODOS-system. Furthermore the behaviour of RODOS after receiving every possible defined message could be determined. This is of interest as this allows to see whether the RODOS-system is complete with respect to the defined inputs and outputs.

7.2 Nature of the RODOS specification

When making the ExSpect model it turned out that several design-decisions have to be made that are not made during an informal specification procedure. It means that a formal specification is more complete than an informal specification, which in turn means that a functional specification that is made in a formal way will not need as much extensions during the remainder of the development process (design, coding and test). This guarantees that the functional design is complete and that during the remainder of the project less problems will occur with unexpected additional functionality.
ExSpect gives the possibility also to include timing aspect of a specification, as it is based on timed Petri-nets. This gives a good framework to specify the time behaviour of a system, which is of particular importance with distributed systems of the nature of RODOS. This may lead to determine already at an early stage how to deal with messages that are improperly dated. In informal specifications most aspects of time are not dealt with adequately.

7.3 Completeness of the RODOS specification

To test the protocol aspect of RODOS it is necessary to define also how the systems of the liner agent and the consignee are specified. It turned out that these systems were only described very globally in an informal specification, on the level of the defined protocol and interfaces. To make a specific analysis the handling of the generation and receipt of the messages by a liner agent and a consignee have to be modelled.
This shows once again that for distributed systems a formal approach provides less opportunities to forget necessary parts of a complete working distributed system. Furthermore the functional specifications of the complete systems of the consignee and the liner agent could be easily made in a consistent way, based on the formal specifications of the RODOS system.

7.4 Further options of ExSpect, not applied to RODOS

Apart form the above results, the ExSpect specification was able to help answering questions that normally can only be answered after the realisation of a complete system.

Although we did not work them out completely for this case, we mention a few of them that are of considerable interest:

- Because the ExSpect model is executable it provides a test environment for RODOS and also for the systems of the liner agent and consignee. After realisation of the software-modules the functionality can be tested. ExSpect can be applied to define a set of tests that cover the functionality of the system. During the specification phase thought arose that led to see whether the formal ExSpect specification could be used to form the basis of a real time test environment. This however has not been pursued (see also [8]).
- As the ExSpect model provides a number of input facilities in the form of forms, the screen lay-outs that were foreseen could also be defined consistently with the functional behaviour. This led to a working prototype of RODOS, that not only performed the functional behaviour, but also gave a good view of the actual working of the system to the end users. This enables us to start education and training of users already at an early stage, thereby avoiding possible time loss at a later stage.
  As far as personnel is concerned, ExSpect provides the possibility to investigate the work pressure for different departments and functions. Such an extension has however deliberately not been made in this case. In this case the need to explicitly show implications of the new system for personnel were not considered appropriate.
- The performance of the specified system can be determined beforehand. Especially with distributed systems aspects of the distribution of processing power and the capacity of data-communication connections are very important, also from a cost point of view. Usually such an analysis is made *after* a functional specification is made. Considerations that deal with performance at the design and coding phase however sometimes lead to necessary changes in functionality.
  With ExSpect the performance of the system can be calculated beforehand and if the performance analysis shows the need for changes, these can be determined right a way in the functional specification. This needs in addition to the functional specification quantitative data as the relay-time of an EDI-message, the average time for a user to check a customs declaration and the estimated needed resources for the handling of a customs declaration, the amount of declarations and the frequency by which the declarations are being offered.

8. MAIN SUMMARY AND CONCLUSIONS

A formal way of system specification has led, in our view, to a considerable improvement in quality of the functional design. The most important reason was the fact that an interactive validation of the specification by both the designers and the users is possible.

Executable specifications offer the possibility to simulate the dynamic behaviour of (complex and distributed) IT-systems. They offer the possibility to analyse all kinds of aspects that are necessary during a design, test and implementation phase.

An important condition is the availability of an adequate software support environment. ExSpect turns out to have the right qualities. The ExSpect-models can serve as adequate means to describe the functional parts of the specifications, whereas the executability offers a number of advantages.

Not only the animation possibilities help in the communication with the end user, but also the analyses that are being applied to check on consistency and correctness and that can be used to get qualitative and quantitative data on issues that are important for the remainder of the development process.

However, the most important advantage of formal specifications is that they offer the basis for automatic code generation. As far as ExSpect is concerned this issue is being worked on by Eindhoven University of Technology and Bakkenist. It is expected that in the near future such features will also be incorporated in the ExSpect-tool. This automatic generation of code will for ExSpect apply in first instance to the generation of API's and Work Flow Management software and EDI-communication systems.

Acknowledgements
The authors wish to thank prof. ir. Piet van der Vlist for his valuable suggestions on earlier versions of this paper and dr. W. van der Aalst for his contributions to the RODOS specification, that forms part of this paper.

REFERENCES

[1] Reisig, W., *Petri-nets: an introduction*, Prentice Hall, 1985

[2] Aalst, W.M.P. van der, *Timed coloured Petri-nets and their application to logistics*, Eindhoven University of Technology, 1992.

[3] Jensen, K. Coloured Petri-nets: A high level language for system design and analysis, in G.Rozenberg, editor, Advances in Petri-nets 1990, volume 483 of Lecture Notes in Computer Science, pages 342-416, 1990

[4] Hee, K.M. van, *Systems engineering: a Formal Approach*, 1994, Academic press

[5] Hee, K.M. van, Somers, L.J. and Voorhoeve, M., *Executable Specifications for distributed information systems*, in: Proceedings of the IFIP TC8/WG8.1, Elsevier, 1989, pages 139-156.

[6] Zeigler, B., Theory of modelling and simulation, Wiley Interscience, 1976

[7] Woodcock, J. and M. Loomes, Software Engineering Mathematics, Pitman, 1988

[8] W. Hengeveld and J. Kroon, *Using Checking Sequences for OSI layer conformance testing*, Protocol Specification, Testing and Verification VII, edited by Harry Rudin and C.H. West, 1987

Dr. ir. G.J. Houben is assistant professor at Eindhoven University of Technology and is involved in research on formal specification techniques and ExSpect. He is part-time employed with Bakkenist Management Consultants

Ir. P.G.A. van der Toorn is a consultant with Bakkenist and technical project manager with the Customs authorities in the Netherlands at the Rotterdam harbour

Dr. W. Hengeveld is senior consultant with Bakkenist and is involved in several projects related to advanced information and telecommunication applications.

PART FOUR

Verification

10

Representation of Process-Gate Nets in LOTOS and Verification of LOTOS Laws: the Boolean Algebra Approach

Pim Kars

TIOS Group, Department of Computer Science, University of Twente,
P.O. Box 217, 7500 AE Enschede, The Netherlands
kars@cs.utwente.nl

Process-Gate Nets (PGNs) are graphical expressions for defining synchronisation requirements between a set of processes. As a graphical language they are easier to comprehend than the linear expressions of LOTOS. On the other hand LOTOS is supported by a convenient set of tools. Therefore it may be useful in the initial phase of formal specification to first define the high-level interaction pattern of a set of LOTOS processes as a PGN, and then transform the PGN to an equivalent LOTOS expression. This raises the question whether such an equivalent expression can be found at all and, if so, to determine such an expression. It is known that some PGNs can not be represented in LOTOS, and that the class of (what we call) alpha-implicit, globally unique PGNs can always be represented using only parallel operators
We show that for a large subclass of all PGNs (including the globally unique ones) representability in LOTOS using only parallel operators can be decided by transforming the synchronisation requirements into a set of equations in a Boolean algebra and then solving the equations. Moreover this Boolean algebra method can also be used to decide the equivalence of LOTOS expressions that use only parallel operators and process variables. The method has been implemented and we give some examples of applications.

1. Introduction

A Process-Gate Net (PGN) [Bol90] is a graph of nodes denoting processes, where each node is labelled by a process name and a set of actions (the sort or alphabet of the process). Sets of nodes may be linked by edges (turning it into a hypergraph [Ber76]) that are labelled by actions (also called gates). The graph represents the synchronisation requirements on the set of processes: an edge labelled a linking a set of process nodes S prescribes that all processes in S should participate in doing a. If a label occurs on more than one edge linked to a process node this prescribes *or*-synchronisation, i.e. it offers independent possibilities for synchronisation.

Example 1

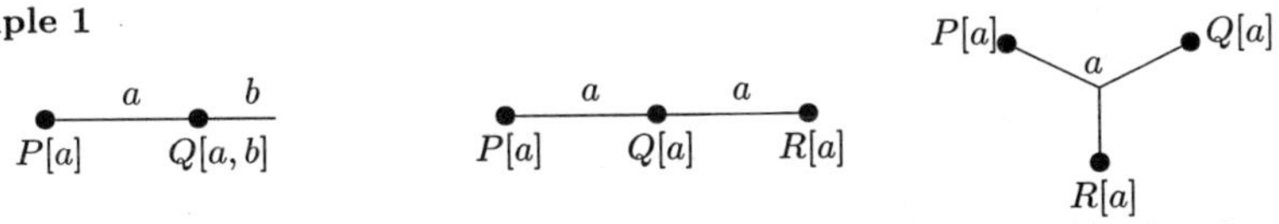

(P, Q and R are process names; a and b are actions). The first PGN prescribes that P and Q should synchronise on a actions, and that Q can do b actions on its own. No other actions are possible. The second PGN prescribes that either P and Q or Q and R should synchronise on a. The last PGN prescribes that all three of P, Q and R should synchronise on a.

PGNs can be used in the initial stage of a specification to give the top-level interconnection and synchronisation pattern of a set of processes, the behaviour of which is "filled in" later.

They form a high-level way of defining "who synchronises with whom on what". As a graphical language they are easier to comprehend than the linear expressions of LOTOS. On the other hand LOTOS is supported by a convenient set of tools. Therefore it may be useful in the initial phase of formal specification to first define the high-level interaction pattern of a set of LOTOS processes as a PGN, and then transform the PGN to an equivalent LOTOS expression. Note that PGNs and LOTOS have a common semantical domain: labelled transition systems, providing the basis for relating the two.

The problem we discuss here is to give an equivalent LOTOS expression for a PGN. "Equivalent" basically means that when behaviour expressions are substituted for the process names in the PGN and in the LOTOS expression, the resulting behaviours should be bisimular. For the second example above, an equivalent LOTOS expression is $Q[a]\,|[a]|\,(P[a]\,|||\,R[a])$.

For some PGNs is it known (see [HNS91]) that there is no LOTOS representation at all, so they extend LOTOS in capability for specifying synchronisation requirements. For others (the alpha-implicit, globally unique ones) it is known they can be represented in LOTOS using only parallel operators ([Bol90, BdFE92]).

Our main results are:

- For a large class of PGNs (the subset-unique ones) the problem whether there is an equivalent LOTOS expression using only parallel operators (and process variables) can be decided by transforming the synchronisation requirements into a set of equations in a Boolean algebra and then solving the equations (see section 4).
- This "Boolean algebra method" can equally be applied to decide the equivalence of any two LOTOS expressions consisting only of process variables and parallel operators (see section 7).

We refer to [Kar94] for most proofs and further results.

2. Definitions and notation

2.1. Boolean algebra

A Boolean algebra (see e.g. [Bro90, Rud74]) is a structure $\langle B, +, \cdot, ', 0, 1\rangle$ where B is a set, $+$ and $\cdot$ are commutative, associative binary operators on B, $'$ is a unary operator on B, and $0, 1 \in B$ $(0 \neq 1)$, satisfying the following set of axioms: for all $a, b, c \in B$:

$$\begin{array}{llcl@{\qquad}lrcl} \text{distributive} & a\cdot(b+c) & = & a\cdot b + a\cdot c & \text{complement} & a+a' & = & 1 \\ & a+(b\cdot c) & = & (a+b)\cdot(a+c) & & a\cdot a' & = & 0 \\ \text{unit} & a+0 & = & a \;=\; a\cdot 1 & & & & \end{array}$$

Operator precedence is: "$'$", "$\cdot$", "$+$". Usually "$\cdot$" is suppressed, i.e. denoted by juxtaposition. Some useful laws that hold in any Boolean algebra:

$$\begin{array}{lrcl@{\qquad}rcl} \text{absorption} & a+ab & = & a & a(a+b) & = & a \\ & a+a'b & = & a+b & a(a'+b) & = & ab \end{array}$$

A partial order ("subsumption") is defined by $a \leq b =_{\text{def}} ab' = 0$.

A *Boolean function* is a function $f : B^n \to B$ generated syntactically by

$$F \;::=\; c \;|\; x \;|\; F\cdot F \;|\; F+F \;|\; F'.$$

where $c \in B$ is a constant, $x \in \{x_1, \ldots, x_n\}$ is a variable. The semantics should be obvious.

We use Boolean algebra to calculate with sets of actions from a set Act. This means that we have the following interpretation:

$$\begin{array}{lllllllll} B & \mapsto & \mathbb{P}(Act) & 0 & \mapsto & \emptyset & a+b & \mapsto & a \cup b & \leq \ \mapsto \ \subseteq \\ a' & \mapsto & Act \setminus a & 1 & \mapsto & Act & a \cdot b & \mapsto & a \cap b & \end{array}$$

Set difference can be expressed as $a \setminus b =_{\text{def}} ab'$.
We use the notation of Boolean algebra, instead of the usual set operations, in particular whenever we want to calculate with an expression, i.e. stress its symbolic nature.

2.2. Labelled transition systems and LOTOS

We assume the reader to be already familiar with labelled transition systems and LOTOS [Bri89] with its semantics defined by transition rules and the notion of strong bisimulation equivalence, denoted by $\sim$ (cf. [Mil89]).
We assume a fixed universe of observable actions Act. The special action $\tau \notin Act$ represents an unobservable, internal action. The class of labelled transition systems over Act is denoted by $\mathcal{LTS}$. The menu of possible actions in a state B of an LTS is $out(B) =_{\text{def}} \{ a \in Act \mid \exists B' : B \xrightarrow{a} B' \}$.
The only LOTOS operators used are prefixing, (general) choice, process instantiation, parallel operators, relabelling and **stop**. Some of these operators have a slightly different syntax:

1. The binary parallel operator $|[A]|$, prescribing synchronisation on all actions in $A \subseteq Act$, is represented as $\|_A$. A will be called the *par-set* of $\|_A$.
   The interleaving parallel operator is defined by $||| =_{\text{def}} \|_\emptyset$. The CSP [Hoa85] parallel operator, denoted $|\cap|$, is defined by $P \,|\cap|\, Q =_{\text{def}} P \|_{L(P)\cap L(Q)} Q$.
2. Relabelling of actions is represented by a post-fix operator $[f]$, where $f : Act \rightarrow Act$ is a relabelling function (satisfying $f(\tau) = \tau$). f is represented by a list of substitutions a'/a, meaning that a is relabelled into a' (cf. CCS [Mil89]). E.g. where we use $P[a'/a, b'/b]$ for a process P with sort $\{a, b, c\}$, this would be represented in standard LOTOS as:

   $P[a', b', c]$ **where process** $P[a, b, c] := \ldots$
3. General choice is represented as $\sum\{ P_i \mid i \in I \}$ denoting a choice of behaviours from P_i for $i \in I$. For example, $\sum\{ P_i \mid i \in \{1, 2, 3\} \}$ corresponds to standard LOTOS $P_1 \,[]\, P_2 \,[]\, P_3$.

Definition 2 (Process names, ParExpr, Pattern, Proc)

1. A *sorted set of process names* (or process variables) is a finite set PN, together with a function $alpha : PN \rightarrow \mathbb{P}(Act)$ that assigns an alphabet to each process name.
2. A ParExpr or *LOTOS parallel expression* over a sorted set of process names PN is generated by the abstract syntax

   $$E ::= P[A] \mid E_1 \|_C E_2.$$

   where $P \in PN$, $A, C \subseteq Act$, and each process name occurs exactly once in E. The set $A = alpha(P)$ denotes the gate list of P, also known as its syntactical sort $L(P)$.
3. A *pattern* is a ParExpr without par-sets. It only shows the structure of a ParExpr. The parallel operators in a pattern are often denoted "|".
4. $proc(E)$ denotes the set of process names occurring in a ParExpr E.

The sync-set of a ParExpr E and a subset S of the process names is defined to be the set of actions on which precisely all processes in S will synchronise:

Definition 3 (Sync-set of a ParExpr)
Let E be a LOTOS parallel expression over PN. For $\emptyset \neq S \subseteq PN$ with $proc(E) \cap S \neq \emptyset$ define recursively over the structure of E

$$sync\text{-}set_{P[A]}(S) =_{\text{def}} A$$

$$sync\text{-}set_{E\|_C E'}(S) =_{\text{def}} \begin{cases} sync\text{-}set_E(S) \setminus C & \text{, if } S \cap proc(E') = \emptyset \\ sync\text{-}set_{E'}(S) \setminus C & \text{, if } S \cap proc(E) = \emptyset \\ sync\text{-}set_E(S) \cap C \cap sync\text{-}set_{E'}(S) & \text{, if } S \cap proc(E) \neq \emptyset \\ & \quad \text{and } S \cap proc(E') \neq \emptyset \end{cases}$$

$sync\text{-}set_E(S) =_{\text{def}} \emptyset$ when $S \cap proc(E) = \emptyset$.
For example, if $E = P[A] \|_X (Q[B] \|_Y R[C])$ then $sync\text{-}set_E(\{P,Q\}) = A \cap X \cap (B \backslash Y)$.
A ParExpr E may be conceived of as a binary tree *tree*(E) based on its syntactical term structure. Intermediate nodes represent parallel operators and are labelled with their par-set; leaves are labelled by process names. The sync-set of a ParExpr can be characterised in terms of its tree representation:

Lemma 4
For a LOTOS parallel expression E over PN and $\emptyset \neq S \subseteq PN$: Let E_S be the set of nodes in *tree*(E) such that both subtrees have processes in common with S and let E'_S be the set of nodes such that only one subtree has processes in common with S, then (as a Boolean expression)

$$sync\text{-}set_E(S) = \prod_{P \in S} alpha(P) \cdot \prod_{x \in E_S} par\text{-}set(x) \cdot \prod_{x \in E'_S} par\text{-}set(x)'$$

(Recall that Boolean expressions are to be interpreted as in set algebra, i.e. "$\cdot$" and "$\prod$" denote intersection, "+" and "$\sum$" denote union, and "$'$" denotes complementation with respect to *Act*.) For example, if $E = P[A] \|_X (Q[B] \|_Y R[C])$ then $sync\text{-}set_E(\{P,Q\}) = A \cdot B \cdot X \cdot Y'$, cf. the example after definition 3.

2.3. Process-Gate Nets

Definition 5 (Process-Gate Net)
A PGN over a sorted set of process names PN is a finite hypergraph $\langle PN, edges, label, nodes \rangle$ where

- PN is the set of nodes,
- *edges* is a finite set of edges,
- *label* : *edges* $\rightarrow$ *Act* gives the action label of each edge,
- *nodes* : *edges* $\rightarrow \mathbb{P}(PN) \setminus \{\emptyset\}$ gives the nonempty set of nodes linked to each edge.

subject to the requirement that each edge label is contained in the alphabet of all its nodes: for all $e \in edges$ and $P \in nodes(e)$: $label(e) \in alpha(P)$. This requirement is called the "positive cooperation condition" in [Bol93].
Note that an action (unlike a process name) may occur more than once in a PGN. To ensure a unique minimal representation, we require that no unnecessary duplicate edges occur: for all $e_1, e_2 \in edges$: $(\ label(e_1) = label(e_2)$ and $nodes(e_1) = nodes(e_2)\ ) \Longrightarrow e_1 = e_2$.
These definitions are relative to one specific PGN, which should be clear from the context. When several PGNs are used, subscripts will be used to indicate the specific PGN, e.g. $edges_N(P)$ denotes the edges of node P in PGN N.
Usually a PGN is defined as a bipartite graph, i.e. a graph with two kinds of nodes: one for process names and one for gate names (actions). Edges only occur between nodes of different

kinds, and they are unlabelled. We find our definition more natural, since it automatically prevents the occurrence of gate nodes that are not connected to any process node.

Next we define some auxiliary notions/functions for a PGN N.

Definition 6 (Sync-set, implicit alphabet)

- For $S \subseteq PN$, $\textit{sync-set}_N(S)$ is the set of edge labels on which precisely the processes in S synchronise:

  $$\textit{sync-set}_N(S) \;=_{\text{def}}\; \{\, \textit{label}(e) \mid e \in \textit{edges}, \textit{nodes}(e) = S \,\}$$

  Note that $\textit{sync-set}_N(\emptyset) = \emptyset$, and that *sync-set* is overloaded, the subscript indicates whether it refers to a PGN or a LOTOS parallel expression.

- The *implicit alphabet* of a node P is the set of action labels of edges linked to a node P:

  $$\textit{impl-alpha}_N(P) \;=_{\text{def}}\; \{\, \textit{label}(e) \in \textit{Act} \mid e \in \textit{edges}, P \in \textit{nodes}(e) \,\}$$

  Or in terms of sync-sets: $\textit{impl-alpha}_N(P) = \bigcup_{P \in S} \textit{sync-set}_N(S)$. The positive cooperation condition can now be stated as: for all $P \in PN$: $\textit{impl-alpha}_N(P) \subseteq \textit{alpha}(P)$.

Semantics of a PGN

The semantics of a PGN on a set of process names PN depends on an assignment of labelled transition systems to the process names. Given such an assignment, the PGN produces a combined transition system which may be defined operationally by a set of transition rules.

Definition 7 (Compatible assignment)

An *(alphabet-)compatible assignment* of PN is an assignment $B : PN \to \mathcal{LTS}$ such that for all $P \in PN$: $L(B(P)) \subseteq \textit{alpha}(P)$. $\textit{Ass}(PN)$ is the set of compatible assignments of PN.

The reason for only looking at compatible assignments is that a process name along with its alphabet is considered as a LOTOS process heading $P[A]$, and in LOTOS the sort (alphabet) of the body of a process should be a subset of its gatelist.

The following notation will be useful: for $B, B' \in \textit{Ass}(PN), a \in \textit{Act}, S \subseteq PN$

$$B \xrightarrow{a/S} B' \;=_{\text{def}}\; \text{for all } P \in S : B(P) \xrightarrow{a} B'(P) \text{ and for all } P \notin S : B(P) = B'(P)$$

i.e. $B \xrightarrow{a/S} B'$ when precisely the processes P in S perform an a-transition $B(P) \xrightarrow{a} B'(P)$.

Definition 8 (Semantics of a PGN)

Let N be a PGN over PN and let $B, B' \in \textit{Ass}(PN)$. The behaviour of $N(B)$ —which denotes N after substituting $P \mapsto B(P)$ for each $P \in PN$— is given by the set of transition rules:

$$\frac{B \xrightarrow{\textit{label}(e)/\textit{nodes}(e)} B'}{N(B) \xrightarrow{\textit{label}(e)} N(B')} \qquad\qquad \frac{B \xrightarrow{\tau/\{P\}} B'}{N(B) \xrightarrow{\tau} N(B')}$$

For each $e \in \textit{edges}_N$ there is a rule on the left expressing that all processes linked to the edge should synchronise on its label. For each $P \in PN$ there is a rule on the right expressing that private τ actions are allowed.

2.4. Intermezzo: Constrained-alphabet synchronisation contexts

Both PGNs and LOTOS parallel expressions can be regarded as ways of defining special *contexts*: functions that take as arguments a list of LTSs and produce an LTS the behaviour of which depends only on the behaviours of the arguments (definition by means of transition rules). They are special in two ways:

1. The alphabet of each argument is prescribed: only LTSs are meant to be substituted for which the alphabet is contained in the prescribed alphabet. We coin the term *constrained-alphabet contexts* to contexts with prescribed alphabets.
2. The transition rules have a specific format: they only allow synchronisation between arguments. Any transition rule for a context C over the set of arguments PN has the format:
$$\frac{B \xrightarrow{a/S} B'}{C(B) \xrightarrow{a} C(B')}$$
for some $a \in Act, S \subseteq PN$. We coin the term *synchronisation context* to this kind of context. For readers familiar with Larsen's "contexts as action transducers" semantics (see [Lar86]): a PGN/ParExpr over PN with $|PN| = n$ defines a context system with one n-place context C and transductions of the form: $C \xrightarrow[(a_1,\ldots,a_n)]{a} C$, where $a \in Act$ and each a_i is either a or 0, and at least one of the a_i is a. For such contexts the sync-sets may be defined by $\textit{sync-set}_C(S) =_{\text{def}} \{a \in Act \mid C \xrightarrow[a_S]{a} C\}$, where $a_S : PN \rightarrow Act \cup \{0\}$ is defined by $a_S(P) = \begin{cases} a & \text{, if } P \in S \\ 0 & \text{, if } P \notin S \end{cases}$. The previous definitions for $\textit{sync-set}_N$ and $\textit{sync-set}_E$ comply with this definition.

Summarising, both PGNs and LOTOS parallel expressions are means of defining constrained-alphabet synchronisation contexts.

2.5. LOTOS Representation of a PGN

A representation is an equivalent expression. The equivalence used is defined next and it is the one found in all the literature on PGNs.

Definition 9 (Equivalence between a PGN and a ParExpr)
Let E be a ParExpr and N a PGN over PN.

$$N \sim_\alpha E \;=_{\text{def}}\; \text{for all } B \in Ass(PN) : N(B) \sim E(B)$$

In fact, this definition applies to the equivalence of constrained-alphabet contexts in general. [BdFE92] uses $\sim\!\sim$ instead of $\sim_\alpha$ and calls it "universally strong equivalence".

We distinguish between two types of representation problems:

- *General representation problem*: Given a PGN N is there a LOTOS parallel expression E such that $N \sim_\alpha E$?
- *Pattern representation problem*: Given a PGN N and moreover a pattern E (i.e. a LOTOS parallel expression without par-sets), can the par-sets in E be chosen such that $N \sim_\alpha E$?

In both cases we are not only interested in the existence question ("is there a solution"), but also in the solution itself (preferably obtained in a constructive fashion).

2.6. Duplicate gate occurrences: a PGN hierarchy

It will be clear that the more duplicate edge labels occur in a PGN, the more difficult it will be to represent it in LOTOS. We define a hierarchy of PGNs by restricting duplicate label occurrences.

Definition 10 (PGN Hierarchy)

- A *globally unique* PGN is a PGN without duplicate edge labels. In terms of sync-sets:

$$\text{for all } S \neq T \subseteq PN : \textit{sync-set}(S) \cap \textit{sync-set}(T) = \emptyset$$

- A *locally unique* PGN is a PGN such that at each node the edge labels are unique. In terms of sync-sets:

$$\text{for all } S \neq T \subseteq PN : S \cap T \neq \emptyset \Longrightarrow \textit{sync-set}(S) \cap \textit{sync-set}(T) = \emptyset$$

- A *subset-unique* PGN is a PGN such that an edge label occurring on an edge linking a set of nodes T may not occur on an edge linking a strict subset of T. In terms of sync-sets:

$$\text{for all } S \neq T \subseteq PN : \emptyset \neq S \subset T \Longrightarrow \textit{sync-set}(S) \cap \textit{sync-set}(T) = \emptyset$$

This definition can be applied to any synchronisation context, so also to LOTOS parallel expressions. We denote the class of globally unique, locally unique, subset-unique PGNs by PGN_{global}, PGN_{local}, PGN_{subset}, resp. It will be clear from the definition in terms of sync-sets that

$$\text{PGN}_{global} \subseteq \text{PGN}_{local} \subseteq \text{PGN}_{subset} \subseteq \text{PGN}$$

Referring to example 1: the middle PGN is subset-unique but not locally unique, the other ones are globally unique. Adding an edge labelled b to P in the left PGN would turn it into a locally unique PGN.

2.7. Alphabet issues

Some PGNs are special in that for each node every action in its alphabet occurs on one of the edges linked to the node. (Note that the opposite always holds: an edge label should be present in the alphabet of the nodes linked to it.)

Definition 11 (Alpha-implicit PGN)
A PGN is *alpha-implicit* if for each node P: $\textit{alpha}(P) \;=\; \textit{impl-alpha}(P)$.

All PGNs in example 1 are alpha-implicit. A picture of an alpha-implicit PGN need not show the alphabets since they can be derived from the edge labels. By contrast, a PGN that is not alpha-implicit will be called *alpha-explicit* (an example follows). This division in alphabet-implicit and explicit PGNs is "orthogonal" to the PGN hierarchy above.

One might think (as in [dFE93]) that an alpha-explicit PGN can be turned into an "equivalent" alpha-implicit PGN by just adding to each node P edges labelled by actions from $\textit{alpha}(P) \backslash \textit{impl-alpha}(P)$. This is not the case. As a simple example consider the following PGNs:

The alpha-implicit N_1 specifies that P can do a actions, whereas the alpha-explicit N_2 specifies that a actions cannot occur! Equivalent LOTOS expressions for the two PGNs are: $N_1 \sim_\alpha P[a]$ and $N_2 \sim_\alpha P[a] \; ||_{\{a\}}$ **stop**. Summarising, alpha-explicit PGNs prescribe *blocking* on actions outside the implicit alphabet (cf. the restriction operator in CCS [Mil89]).

We will be mostly concerned with alpha-implicit PGNs, since we feel that it is their capability to express synchronisation that is of prime importance, not their capability to forbid actions.

3. Related work

[Bol90] and [BdFE92] first used alpha-implicit, globally unique PGNs, just called PGNs there. This work concerns the generalisation of results like the law of associativity of parallel composition (see section 7). Among other things they show that an alpha-implicit, globally unique PGN over PN can be represented by any LOTOS parallel expression over PN if the parallel operators between the process variables are taken to be the CSP $|\cap|$-operator, i.e. with synchronisation over all common actions. This is called the *maximal cooperation condition/principle.*

The fact that the way in which the process variables are "nested" is irrelevant, is equivalent to associativity of $|\cap|$ under bisimulation equivalence.

[BdFEOM91] generalises this work to include hiding. Globally unique PGNs are extended by a function which classifies each gate as being either an internal gate, a hidden communication gate or a visible communication gate.

Process Topology Diagrams (PTDs) ([HNS91]) extend PGNs by allowing hierarchical structures. An example is given of a PTD that has no LOTOS representation. In PGN terms it looks like

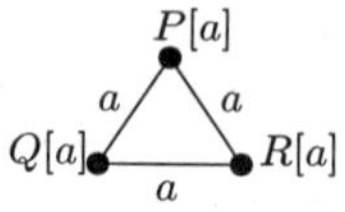

[dFE93] is the first work to consider representability of *general* PGNs as LOTOS parallel expressions. He claims to have an algorithm for deciding the general representation problem using an extension of the algorithm of [Bol90, BdFE92]. However the page limit afforded by AMAST for contributions is too small to fully appreciate the claims. He remarks that it is sufficient to consider alpha-implicit PGNs since alpha-explicit ones can be easily handled by the trick mentioned in section 2.7, which we showed to be incorrect.

[Bol93] considers the reverse problem: given a LOTOS parallel expression, derive an equivalent PGN. Specifically he shows that for each ParExpr E there is a subset unique PGN N_E such that $N_E \sim_\alpha E$, and if $E_1 \sim_\alpha E_2$ then $N_{E_1} = N_{E_2}$. A PGN is called a General Process Interaction Net, a subset unique PGN is called a Process Interaction Net and the term PGN is reserved for our alpha-implicit, globally unique PGN.

Based on the work in [Bol90] and [BdFE92], a PGN-like construct is adopted in G-LOTOS [Naj92], the graphical syntax for LOTOS that is currently being standardised.

4. The Boolean Algebra Approach

4.1. The Sync-set Principle

The problem we consider is the pattern representation problem (in section 6 we look into the general representation problem): given a PGN N and a ParExpr E over a set of processes PN, where the parallel operators in E contain uninstantiated variables: can these variables be instantiated such that $N \sim_\alpha E$?

The basic idea is that $N \sim_\alpha E$ if and only if for each subset S of PN the set of actions on which precisely all $P \in S$ synchronise in N coincides with the set of actions on which precisely all $P \in S$ synchronise in E. Formally this "Sync-set Principle" can be stated as:

$$N \sim_\alpha E \iff \text{for all } S \subseteq PN : \textit{sync-set}_N(S) = \textit{sync-set}_E(S)$$

(note that $\textit{sync-set}_N(\emptyset) = \textit{sync-set}_E(\emptyset)$ always holds, because both are $\emptyset$). Unfortunately it is not clear whether this principle holds in general, we have not been able to find a proof or a counter-example. However, for subset-unique PGNs this principle *does* hold. An important fact used in the proof is that the context defined by a LOTOS parallel expression is itself subset-unique.

Theorem 12
LOTOS parallel expressions are subset-unique.

Proof
Let E be a LOTOS parallel expression. We have to prove: $\textit{sync-set}_E(S) \cap \textit{sync-set}_E(T) = \emptyset$ for

all $S, T \subseteq PN$ such that $\emptyset \neq S \subset T$.
First we consider the case $T = S \cup \{P\}$ where $P \notin S$. Consider the path in *tree*(E) from the leaf labelled P to the root. Since $P \notin S$ there must be a node labelled x on this path such that $x \in E'_S$ (it is the lowest node such that $P \in proc(x)$ and $proc(x) \cap S \neq \emptyset$). Since $P \in T$ we will have $x \in E_T$, therefore by lemma 4 the Boolean expression $\textit{sync-set}_E(S) \cdot \textit{sync-set}_E(T)$ contains both $\textit{par-set}(x)$ and $\textit{par-set}(x)'$ hence it is 0.
Next apply induction, using that an enlargement of T affects a node different from x. □

The main result is

Theorem 13
For every PGN N and LOTOS parallel expression E over PN:

$$N \sim_\alpha E \quad \text{and} \quad N \text{ is subset-unique}$$
$$\text{if and only if}$$
$$\text{for all } S \subseteq PN : \textit{sync-set}_N(S) = \textit{sync-set}_E(S)$$

Proof (sketch)

$\Rightarrow$: Consider the "one-shot" assignment B_S, for $S \subseteq PN$, defined by

$$B_S(P) =_{\text{def}} \begin{cases} \sum\{\, a; \textbf{stop} \mid a \in alpha(P) \,\} & \text{, if } P \in S \\ \textbf{stop} & \text{, if } P \notin S \end{cases}$$

First, since $N(B_S) \xrightarrow{a} N' \iff \exists S' \subseteq S : a \in \textit{sync-set}_N(S')$ and $N' \sim N(B_{S \setminus S'})$, it follows that $out(N(B_S)) = \bigcup_{S' \subseteq S} \textit{sync-set}_N(S')$. The same goes for E.
Next, $N \sim_\alpha E$ implies (since B_S is alphabet-compatible with PN) that $N(B_S) \sim E(B_S)$ for all S. In particular $out(N(B_S)) = out(E(B_S))$, i.e. $\bigcup_{S' \subseteq S} \textit{sync-set}_N(S') = \bigcup_{S' \subseteq S} \textit{sync-set}_E(S')$.
Now use induction on the size of S to prove that subset-uniqueness of both N and E implies that for all S: $\textit{sync-set}_N(S) = \textit{sync-set}_E(S)$.

$\Leftarrow$: The first conjunct follows by showing that $\{\, \langle N(B), E(B) \rangle \mid B \in Ass(PN) \,\}$ is a strong bisimulation relation using the definition of *sync-set* for N and E.
The second conjunct follows from subset-uniqueness of E and equality of the sync-sets. □

Example 14
As a running example we take the "most general" alpha-implicit, subset-unique PGN N on $PN = \{P_1, P_2, P_3\}$:

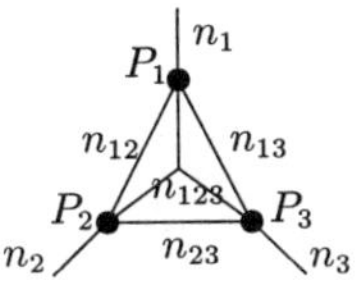

Here the edge labels do not denote actions but sets of actions: n_J denotes $\textit{sync-set}_N(\{\, P_j \mid j \in J \,\})$, e.g. $n_{12} = \textit{sync-set}_N(\{\, P_1, P_2 \,\})$. A PGN of this form (having all possible sync-sets in symbolic form) will be termed the *full* PGN on PN.
Let's try to find out whether there exists a LOTOS representation of the form:

$$N \sim_\alpha P_1 \parallel_x (P_2 \parallel_y P_3)$$

for some action sets x and y. The sync-set principle leads to the following set of equations

$$\begin{array}{llllllll} 1 & n_1 = a_1x' & 4 & n_{12} = a_1a_2xy' & 7 & n_{123} = a_1a_2a_3xy \\ 2 & n_2 = a_2x'y' & 5 & n_{13} = a_1a_3x'y & & \\ 3 & n_3 = a_3x'y' & 6 & n_{23} = a_2a_3x'y & & \end{array}$$

Here a_i denotes $alpha(P_i)$. Due to alpha-implicitness a_1, a_2, a_3 may be removed by substituting

$$\begin{array}{lll} a_1 & \mapsto & n_1 + n_{12} + n_{13} + n_{123} \\ a_2 & \mapsto & n_2 + n_{12} + n_{23} + n_{123} \\ a_3 & \mapsto & n_3 + n_{13} + n_{23} + n_{123} \end{array}$$

yielding a system of equations over the "constants" $n_1, n_2, \ldots$ and the variables x, y. The requirements for subset-uniqueness are implicitly present in these equations due to theorem 13.

4.2. Boolean equation solving

Once we have obtained a set of equations, it can be rewritten as one equation of the form $f(x, y, \ldots) = 0$ using the following laws:

$$a = b \iff a'b + ab' = 0$$
$$a = 0 \text{ and } b = 0 \iff a + b = 0$$

The first yields a set of equations of the form $f(x, y, \ldots) = 0$, and the second yields one equation of the form $f(x, y, \ldots) = 0$. The resulting expression for f may be simplified using the axioms and laws of Boolean algebra, in particular the absorption laws.

Next we use the method for solving such an equation as described in [Bro90, Chapter 6] to which we refer for a fuller exposition and proofs. The method involves elimination of variables.

Theorem 15

For a Boolean function $f : B^n \to B$, define the *eliminants* $f_0, f_1(x_1), \ldots, f_n(x_1, \ldots, x_n)$ by:

1. $f_n =_{\text{def}} f$
2. $f_{i-1} =_{\text{def}} f_i[0/x_i] \cdot f_i[1/x_i] \quad (i = n, \ldots, 1)$

where $f[0/x_i]$ denotes substitution of 0 for x_i in f. Then $f(x_1, \ldots, x_n) = 0$ has a solution if and only if the *consistency condition* (CC) $f_0 = 0$ holds, and the *subsumptive general solution* is then given by:

$$\begin{array}{rcccl} f_1(0) & \le & x_1 & \le & f_1'(1) \\ f_2(x_1, 0) & \le & x_2 & \le & f_2'(x_1, 1) \\ & & \vdots & & \\ f_n(x_1, \ldots, x_{n-1}, 0) & \le & x_n & \le & f_n'(x_1, \ldots, x_{n-1}, 1) \end{array}$$

(The CC expresses that each lower bound is contained in the corresponding upper bound.)

Simplification of the solution

The resulting system of expressions for the lower and upper bounds may be simplified using the CC. This entails that the CC is no longer reproducible from the lower and upper bounds: the simplified expressions are only valid under the hypothesis of the CC.

Furthermore, if the bounds contain variables, these variables may often be eliminated by using the bounds on the variables themselves. For example the lower bound for x_2 in theorem 15, $f_2(x_1, 0)$, may be simplified using that $f_1(0) \le x_1 \le f_1'(1)$.

For the 2-variable case: let $f(x_1, x_2)$ be a Boolean function; if $f(a_1, a_2) = f(a_1, b_2) = f(b_1, a_2) = f(b_1, b_2)$, then f is constant for $a_1 \le x_1 \le b_1$, $a_2 \le x_2 \le b_2$ (cf. [Rud74, theorem 2.4]). Since this condition is rather costly (for n variables, 2^n expressions have to be generated and $2^n - 1$ equalities checked), one can first do a quick check as follows (again only stated for $n = 2$).

Theorem 16
If $f : B^2 \to B$ is a Boolean function over x_1, x_2 and

$$f[a_1/x_1, b_1'/x_1'][a_2/x_2, b_2'/x_2'] = f[b_1/x_1, a_1'/x_1'][b_2/x_2, a_2'/x_2'] \tag{1}$$

then f is constant over the domain $a_1 \leq x_1 \leq b_1$, $a_2 \leq x_2 \leq b_2$.

($[a/x, b'/x']$ means: substitute a for x when x occurs under an even number of complementations, otherwise substitute b for x.) The main reason is that both the sum and product operators are monotonous in both operands, and complementation is anti-monotonous. So the left hand side of (1) is a lower bound and the right hand side an upper bound for f on the domain.

Example 17
Continuing our example, using the variable ordering $x_1 = y, x_2 = x$, we get:

$$\begin{aligned} f_2(x,y) &= x\,(n_1+n_2+n_3+n_{23}) + x'\,(n_{12}+n_{13}+n_{123}) + \\ &\quad\; y\,(n_2+n_3+n_{12}+n_{13}) + y'\,(n_{23}+n_{123}) \\ f_1(y) &= y\,(n_{12}+n_{13}+n_2+n_3) + y'(n_{23}+n_{123}) + f_0 \\ f_0 &= n_{12}(n_{23}+n_3) + n_{13}(n_{23}+n_2) + SU \end{aligned}$$

with $SU = n_1(n_{12}+n_{13}+n_{123})+n_2(n_{12}+n_{23}+n_{123})+n_3(n_{13}+n_{23}+n_{123})+(n_{12}+n_{13}+n_{23})\,n_{123}$. Hence, the consistency condition is

$$\mathrm{CC}: \quad n_{12}(n_{23}+n_3) + n_{13}(n_{23}+n_2) + SU = 0 \tag{2}$$

Provided this holds, by theorem 15

$$\begin{aligned} n_{23}+n_{123}+n_{12}n_3+n_{13}n_2 \;&\leq\; y \;\leq\; (n_{12}+n_{13}+n_2+n_3)' \\ y(n_2+n_3)+y'n_{23}+n_{12}+n_{13}+n_{123} \;&\leq\; x \;\leq\; (y(n_{12}+n_{13})+y'n_{123}+n_{23}+n_1+n_2+n_3)' \end{aligned}$$

This may be simplified using the CC and the simplification method of theorem 16 yielding:

> The full PGN of example 14 is equivalent to $P_1 \parallel_x (P_2 \parallel_y P_3)$ iff CC (2) holds and
>
> $$\begin{aligned} n_{23}+n_{123} \;&\leq\; y \;\leq\; (n_2+n_3+n_{12}+n_{13})' \\ n_{12}+n_{13}+n_{123} \;&\leq\; x \;\leq\; (n_1+n_2+n_3+n_{23})' \end{aligned}$$

Note that the sync-sets are not "concrete" sets of actions, they are "symbolic", hence the result will hold for *any* specific concrete sets of sync-sets $n_1, n_2, \ldots$. Consider the following concrete examples of alpha-implicit, subset-unique PGNs (now $a, b, c, d \in Act$)

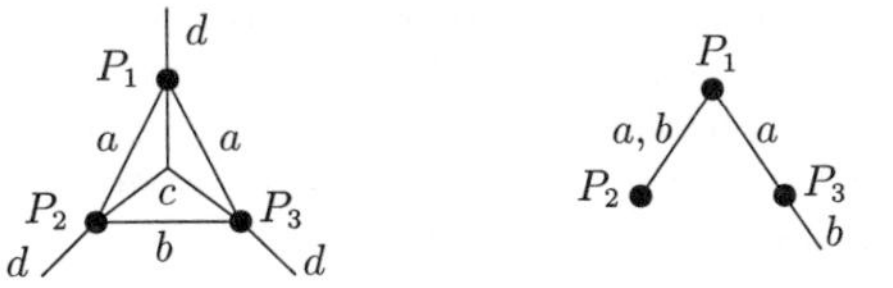

The left PGN is easily seen to satisfy the CC (2) for pattern $P_1 \parallel_x (P_2 \parallel_y P_3)$ and the LOTOS representation follows from the bounds on x and y: $P_1[a,c,d] \parallel_{\{a,c\}} (P_2[a,b,c,d] \parallel_{\{b,c\}} P_3[a,b,c,d])$. The right PGN violates the CC for the pattern $P_1 \mid (P_2 \mid P_3)$ since *sync-set*$(\{P_1,P_2\}) \cap$ *sync-set*$(\{P_3\}) = \{b\} \neq \emptyset$. Also for all other patterns ($P_2 \mid (P_1 \mid P_3)$ and $P_3 \mid (P_1 \mid P_2)$) it violates the corresponding CC, hence it has no equivalent parallel representation. It *can* be represented using also relabelling: $(P_1[a,b] \parallel_{\{a,b\}} (P_2[a,b] \,|||\, P_3[a,b][b'/b])\,)[b/b']$.

Size	*Pattern*	SOP (PYTHON)				BDD (KIMWITU/C)		
n		*gen.*	*solve*	*simplify*	*total*	*gen.*	*solve*	*total*
2	$P_1 \mid P_2$				2			< 1
3	$P_1 \mid (P_2 \mid P_3)$	2	6	2	10			< 1
4	$P_1 \mid (P_2 \mid (P_3 \mid P_4))$	41	182	42	265	2	1	3
4	$(P_1 \mid P_2) \mid (P_3 \mid P_4)$	40	154	50	244	2	1	3
5	$P_1 \mid (P_2 \mid (P_3 \mid (P_4 \mid P_5)))$	6392	13365	2047	21804	19	7	26
5	$((P_1 \mid P_2) \mid P_3) \mid (P_4 \mid P_5)$	6473	12644	2923	22040	15	6	21
6	$((P_1 \mid P_2) \mid P_3) \mid ((P_4 \mid P_5) \mid P_6)$	—	—	—	—	193	40	233

Figure 1: Times in seconds of sample runs on full PGNs.

5. Implementation

The Boolean algebra method for subset-unique PGNs was first implemented in a language suitable for prototyping algorithms called PYTHON [Ros93]. Boolean expressions are maintained in so-called "sum of products" form, using a list of lists of (complemented) letters. Using this implementation on a 24 MB SPARC station IPC the representation problem for some full PGNs and various patterns was solved. The time to generate and solve the system, and to simplify the solution is given in figure 1 under the heading "SOP". These figures clearly show an exponential time complexity with respect to the number of processes n. One of the reasons is that $2^n - 1$ equations are generated using $n - 1$ variables and $2^n - 1$ constants. But note that the size of the PGN is already exponential in n itself. Moreover we are not solving the pattern representation problem for one particular instance but for all (subset-unique) PGNs on a set of processes and a specific pattern at once. One should also take into account that PYTHON is an interpreted language.

Improved implementation

To get a more efficient implementation, the system was rewritten into C, this time using Reduced Ordered Binary Decision Diagrams [Bry92], BDDs for short, to represent Boolean expressions. The implementation was alleviated by the use of the "term processor" KIMWITU [EB92]. Basically KIMWITU produces C code from some user-provided definitions of recursive data types (including uniquely stored ones using hash tables) and functions over these data types (including memofunctions); these are precisely the ingredients needed for the implementation of BDDs following the techniques described in [BRB90]. Instead of the simplification idea of theorem 16, the BDD implementation uses the `constrain` operation of [CBM89]. The procedure for solving and simplifying is shown in figure 2. It recursively eliminates a variable until no more variables are left. It returns to its caller the combined constraint up to that point. Note that the consistency condition is a constraint of the form $c = 0$, but `constrain` uses a constraint of the form $c = 1$, hence "`return NOT(f)`" in case "`vs == empty`". In the other case we know that $c = 1$ and $low \leq v \leq upb$ which is the same as $c(v \cdot upb + v' \cdot low') = 1$.

Figure 1 gives the times to generate and solve various representation problems with the improved implementation under the heading "BDD". Note the drastic decrease in execution time. These times were obtained using the following variable ordering: the variables labelling the par-sets were ordered "top down", according to their place in the pattern, and were also eliminated in this order. For example for the pattern $P_1 \parallel_x (P_2 \parallel_y P_3)$ the order is $x < y$. The most logical order for the constants $n_1, n_2, \ldots$ would seem *after* the par-set variables to ensure fast elimination

```
boolexpr solve (f,vs)
boolexpr f; varlist vs;
{
   if ( vs == empty ) then
   {
        print("CC: 0 = ",f);
        return NOT(f);
   }
   else {
        var v = head(vs);
        boolexpr f0 = eval(f,v,0);
        boolexpr f1 = eval(f,v,1);
        boolexpr c = solve( AND(f0,f1), tail(vs) );
        boolexpr low = constrain(     f0,  c );    print(v," >= ", low);
        boolexpr upb = constrain( NOT(f1), c );    print(v," <= ", upb);
        return IfThenElse( v, AND(c,upb), AND(c,NOT(low)) );
   }
}
```

Figure 2: Solving and simplification in the BDD implementation.

of variables. Indeed with this order the performance was about twice as fast compared to the times in figure 1. However we were not able to get rid of all the variables in the upper and lower bounds. Therefore we settled on an order where the constants preceed the variables.

Results from the implementation

Analysing the results obtained by running the implementation on all patterns with up to five processes, a pattern evolved in the solutions and consistency conditions. We have not completed a proof of this result yet, so we state it as a conjecture:

Conjecture 18

Let N be an alpha-implicit, subset-unique PGN on $PN = \{P_1, \ldots, P_n\}$, and E a LOTOS pattern on PN where each parallel operator is labelled by a distinct variable. Provided the CC (see below) holds, for each subexpression $E_1 \|_x E_2$ of E the solution is given by

$$\underline{x} \leq x \leq \overline{x}$$

where
$$\underline{x} = \sum\{\, \textit{sync-set}_N(S) \mid S \subseteq PN : \textit{proc}(E_1) \cap S \neq \emptyset \text{ and } \textit{proc}(E_2) \cap S \neq \emptyset \,\}$$
$$\overline{x} = \prod\{\, \textit{sync-set}_N(S)' \mid S \subseteq PN : \textit{proc}(E_1) \cap S \neq \emptyset \text{ xor } \textit{proc}(E_2) \cap S \neq \emptyset \,\}$$

Note the fact that the upper and lower bounds for each variable are independent of the other variables (for alpha-explicit PGNs this is not the case). The simplification method of theorem 16 proves to be sufficient to accomplish this (for the SOP-implementation).

As to the consistency condition: for patterns where for each parallel operator one of the subpatterns is a leaf (like $P_1 \mid (P_2 \mid P_3)$ and $P_1 \mid (P_2 \mid (P_3 \mid P_4))$) the CC is simple: it is obtained by setting $\underline{x} \cdot \overline{x}' = 0$ for all $x \in E$. For other patterns like $(P_1 \mid P_2) \mid (P_3 \mid P_4)$ extra terms have to be added (see [Kar94]).

6. The General Representation Problem

The general parallel representation problem may be solved by considering the pattern representation problem for each possible pattern. For example, for PGNs on 3 processes we have to look at the patterns (modulo commutativity of the parallel operator): $P_1 \parallel_x (P_2 \parallel_y P_3), P_3 \parallel_x (P_1 \parallel_y P_2)$ and $P_2 \parallel_x (P_1 \parallel_y P_3)$.
On a parallel machine this could even be done in parallel. Since all these patterns have the same structure and only the position of the process names is permuted, as soon as we know the representability condition for one of them the condition for another pattern can be simply obtained by applying the appropriate permutation to the known condition. Using our previous results the condition for parallel representability then becomes (apart from subset-uniqueness)

$$\begin{aligned}&n_{12}(n_{23}+n_3)+n_{13}(n_{23}+n_2)=0 \text{ or}\\&n_{12}(n_{13}+n_3)+n_{23}(n_{13}+n_1)=0 \text{ or}\\&n_{13}(n_{12}+n_2)+n_{23}(n_{12}+n_1)=0\end{aligned}$$

It would be nice if *one* set of equations could be derived that corresponds to general representability. Unfortunately there is no Boolean operator O such that for all Boolean expressions f, g: $f = 0$ or $g = 0 \iff (f\ O\ g) = 0$ (this would allow to combine the equations for the various patterns into one equation). [Kar94] gives estimates on the number of (essentially different) patterns and permutations that have to be considered for the general representation problem.

7. Application to LOTOS Laws

The application of this method is not limited to relating PGNs and LOTOS expressions, but it may also be used to relate LOTOS parallel expressions to each other. In fact, the sync-set principle is now fully applicable since both sides are subset-unique (see theorem 13)!

Example 19
As an example we will derive the *law of associativity of parallel composition* (from [VSSB91]):

$$\text{if } L(P) \cap B \subseteq A \text{ and } L(R) \cap A \subseteq B \text{ then } (P \parallel_A Q) \parallel_B R \;\sim\; P \parallel_A (Q \parallel_B R)$$

A proof of this law can be found in [VSSB91]. We show that the requirements are not only sufficient but also necessary using the sync-set principle. The question can be stated as: for what x and y do we have that for all $P, Q, R \in BExpr$:

$$P \parallel_x (Q \parallel_y R) \;\sim\; (P \parallel_x Q) \parallel_y R$$

(we may use $\sim$ instead of $\sim_\alpha$ here since we work with *symbolic* sorts). Deriving the sync-sets on each side leads to the following equations (p denotes $L(P)$, etc.):

$$\begin{array}{llllllll}1 & px' = px'y' & 4 & pqxy' = pqxy' & 7 & pqrxy = pqrxy\\2 & qx'y' = qx'y' & 5 & prxy' = prx'y & & \\3 & rx'y' = ry' & 6 & qrx'y = qrx'y & & \end{array}$$

Equations 2, 4, 6 and 7 are trivially satisfied for any x, y. Using the method to obtain one equation from the remaining ones yields

$$px'y + rxy' = 0$$

which translates into (using $a + b = 0 \iff a = 0$ and $b = 0$, and $ab' = 0 \iff a \le b$):

$$L(P) \cap y \subseteq x \quad\text{and}\quad L(R) \cap x \subseteq y$$

The method may easily be used to get conditions for the associativity of more than two parallel operators.

Example 20
As a second example we consider a law mentioned in [Bol90]:

$$P \parallel_{PQ \cup PQR \cup PR} (Q \parallel_{PQR \cup QR} R) \sim Q \parallel_{PQ \cup PQR \cup QR} (P \parallel_{PQR \cup PR} R)$$

where

$$\begin{array}{llll} PQ =_{\text{def}} & (L(P) \cap L(Q)) \backslash L(R) & QR =_{\text{def}} & (L(Q) \cap L(R)) \backslash L(P) \\ PR =_{\text{def}} & (L(P) \cap L(R)) \backslash L(Q) & PQR =_{\text{def}} & L(P) \cap L(Q) \cap L(R) \end{array}$$

The validity of this law —forbidding as it may look— should not come as a surprise. Note that the first par-set may be rewritten as $L(P) \cap (L(Q) \cup L(R))$, the second as $L(Q) \cap L(R)$, the third as $L(Q) \cap (L(P) \cup L(R))$ and the fourth as $L(P) \cap L(R)$. Hence this law becomes $P \,|\cap|\, (Q \,|\cap|\, R) \sim Q \,|\cap|\, (P \,|\cap|\, R)$, expressing the associativity of $|\cap|$.

In order to check this law, we can encode it in several ways.

1. Encode it as stated using constant Boolean expressions for the par-sets. A sample run showed within a second that the consistency condition of the generated set of equations is $0 = 0$, meaning that the law is true.
2. Encode it using variables for some of the sets. As an example we fixed the sets on the left hand side and determined the most general solution for the par-sets on the right-hand side, i.e. the question is for which values of x, y do we have

   $$P \parallel_{L(P)\cdot(L(Q)+L(R))} (Q \parallel_{L(Q)\cdot L(R)} R) \sim Q \parallel_x (P \parallel_y R)$$

   The result showed that equivalence holds provided

   $$\begin{array}{rcl} L(Q)\cdot(L(P)+L(R)) \leq & x & \leq L(Q)\cdot(L(P)+L(R)) + L(P)'\cdot L(Q)'\cdot L(R)' \\ L(P)\cdot L(R) \leq & y & \leq L(P)\cdot L(R) + L(P)'\cdot L(R)' \end{array}$$

These examples show that a variety of laws may be derived/checked by the method.

8. Concluding remarks

The Boolean algebra method discussed here allows to decide whether a subset-unique PGN can be represented as a LOTOS parallel expression and to derive conditions for bisimilarity of LOTOS parallel expressions. A unique feature of the method is that it may be used to decide the representation problem for all PGNs on a given set of process names at once.

[Kar94] gives further results like an algorithm that shows that locally unique PGNs are always representable in LOTOS using parallel operators, relabelling and a process called *Run* (defined by $Run(A) =_{\text{def}} \sum\{\, a; Run(A) \mid a \in A \,\}$ with $A \subseteq Act$).

Some things are not settled yet, e.g. does the sync-set principle also hold for nonsubset-unique PGNs? Even though we want to stress the Boolean algebra method and not our particular implementation nor its results when applied to particular problems, a direct proof of conjecture 18 would be welcome.

Extending the method to allow *Run* in the LOTOS representation of PGNs is easy. The inclusion of hiding should not be too difficult either. However including relabelling looks problematic: the method works because the sync-sets can be expressed in Boolean algebra; for relabelling, the sync-sets will include functions which cannot be expressed in Boolean algebra.

Finally, using the results of this paper one might consider extending G-LOTOS [Naj92] with

PGN-constructs that are more permissive than the ones adopted now.

REFERENCES

[BdFE92] T. Bolognesi and D. de Frutos-Escrig. Regrouping parallel processes. Chapter 4 in [Bol92].

[BdFEOM91] T Bolognesi, D. de Frutos-Escrig, and Y. Ortega-Mallén. Graphical composition theorems for parallel and hiding operators. In J. Quemada, J. Mañas, and E. Vázquez, editors, *Formal Description Techniques, III*, pp. 459–470. North-Holland, 1991.

[Ber76] C. Berge. *Graphs and hypergraphs.* North-Holland, Second revised edition, 1976.

[Bol90] T. Bolognesi. A graphical composition theorem for networks of LOTOS processes. In *Proc. 10th Int. Conf. on Distributed Computing Systems*, pp. 88–95. IEEE, 1990.

[Bol92] T. Bolognesi, editor. *Catalogue of LOTOS Correctness Preserving Transformations.* LotoSphere Task 1.2 Final Deliverable. The LOTOSPHERE Consortium, 1992.

[Bol93] T. Bolognesi. Deriving graphical representations of process networks from algebraic expressions. *Information Processing Letters*, 46:289–294, 1993.

[BRB90] K. S. Brace, R. L. Rudell, and R. E. Bryant. Efficient implementation of a BDD package. In *27th ACM/IEEE Design Automation Conf.*, pp. 40–45. IEEE, 1990.

[Bri89] E. Brinksma, editor. *Information Processing Systems, Open Systems Interconnection, LOTOS - A Formal Description Technique Based on the Temporal Ordering of Observational Behaviour.* International Standard IS-8807. ISO, 1989.

[Bro90] F. M. Brown. *Boolean Reasoning — The Logic of Boolean Equations.* Kluwer, 1990.

[Bry92] R. E. Bryant. Symbolic boolean manipulation with ordered binary-decision diagrams. *ACM Computing Surveys*, 24:293–318, 1992.

[CBM89] O. Coudert, C. Berthet, and J. C. Madre. Verification of synchronous sequential machines based on symbolic execution. In *Proc. Workshop on Automatic Verification Methods for Finite State Systems.* Grenoble, 1989.

[dFE93] D. de Frutos-Escrig. A characterization of LOTOS representable networks of parallel processes. In P. Scollo, editor, *AMAST'93*, 1993.

[EB92] P. van Eijk and A. Belinfante. The term processor *kimwitu* — manual and cookbook. Memorandum INF-92-67, University of Twente, 1992.

[HNS91] J. Hinterplattner, H. Nirschl, and H. Saria. Process topology diagrams. In J. Quemada, J. Mañas, and E. Vázquez, editors, *Formal Description Techniques, III*, pp. 459–470. North-Holland, 1991.

[Hoa85] C. A. R. Hoare. *Communicating Sequential Processes.* Prentice-Hall, 1985.

[Kar94] P. Kars. Representation of process-gate nets in LOTOS and verification of LOTOS laws: the Boolean algebra approach. University of Twente, 1994. In preparation.

[Lar86] K. G. Larsen. *Context-Dependent Bisimulation Between Processes.* PhD thesis, University of Edinburgh, 1986. Technical report CST-37-86.

[Mil89] R. Milner. *Communication and Concurrency.* Prentice Hall, 1989.

[Naj92] E. Najm, editor. *Information Processing Systems, Open Systems Interconnection, LOTOS - A Formal Description Technique Based on the Temporal Ordering of Observational Behaviour, Amendment 1: G-LOTOS.* ISO 8807/DAM1. ISO, 1992.

[Ros93] G. van Rossum. Python tutorial. Available by anon. FTP from `ftp.cwi.nl`, 1993.

[Rud74] S. Rudeanu. *Boolean Functions and Equations.* North-Holland, 1974.

[VSSB91] C. A. Vissers, G. Scollo, M. van Sinderen, and E. Brinksma. Specification styles in distributed systems design and verification. *Theor. Comp. Science*, 89:179–206, 1991.

11

Reachable state space analysis of LOTOS specifications* †

A. Kerbrat
VERIMAG‡

We present a symbolic analysis technique for LOTOS programs with integer variables on which only linear expressions are allowed. The technique is applicable to models generated by the LOTOS compiler of the CÆSAR-ALDÉBARAN toolbox which are Petri nets extended with guarded commands. It allows to compute a predicate on variables characterizing the set of the reachable states or an upper approximation of it. Predicates are represented as systems of linear inequalities on program variables. We implemented a tool for performing the operations necessary for the analysis such as conjunction, disjunction, widening operation as well as comparison of predicates. The method is applied to two examples showing that non trivial relations between program variables can be discovered

1. Motivations

Most formal verification methods are based on *model checking*: starting from a specification written in any Formal Description Technique with well defined operational semantics, a model is build and properties are checked on this model. The majority of the model checking verification techniques are based on a more or less exhaustive exploration of the graph, either for deadlock searching, checking of *logical properties* or checking of *equivalence* between the model and a property. The performances and applicability of these methods depend greatly on the size of the model to be checked. Unfortunately, real life examples tend to produce huge models: this problem is known as the *state explosion problem*.

Various improvements have been introduced on the classical enumerative exploration of the fully generated model in order to cope with this problem. The first class of improvement try to optimize the exploration of the model, either by avoiding its complete generation (*"on the fly" techniques* [11, 24]) or its complete exploration (*Partial order techniques* [15, 24]), or by exploration of an *abstraction* of the model [16] . A second class of improvement concerns the *representation* of the model: the aim of *symbolic techniques* is to avoid the explosive enumerative representation of a model by the representation and manipulation of *sets* of states instead of individual states. One particular technique, Binary Decision Diagrams [5] (BDD) has already a story of success in hardware verification [7, 23] and has been applied successfully in other concurrent programming domains [3, 12]. However, the effectiveness of BDDs depends greatly on the structure

*This work has been partly supported ESPRIT-BRA action "REACT"

†Author address: VERIMAG, Miniparc ZIRST, 38330 Montbonnot, France.
e-mail: Alain.Kerbrat@imag.fr

‡Verimag is a joint laboratory of CNRS, Institut National Polytechnique de Grenoble, Université Joseph Fourier and Verilog SA associated with IMAG.

of the model to represent. In particular, one major cause of the state explosion is the presence of variables in the program such as counters, which cannot be handled directly by BDDs.

Another idea is to use symbolic techniques based on an *implicit representation* of *numerical* values. Such techniques are widely applied in the context of semantic analysis of imperative languages (e.g. [9, 17, 4]) and are now introduced for the analysis of real-time systems (e.g. [1, 21]).

The aim of this paper is to present an analysis technique based on a partly symbolic representation of the model. The idea is to keep the control structure of the program explicit and to represent its data part symbolically by systems of linear inequalities known as *polyhedra*. This approach offers several advantages, such as the possibility to deal with infinite models and the checking of properties expressed in terms of variables of the program instead of actions.

Together with this model, we present a *approximate forward analysis* method successfully applied in other contexts [9, 19]. Applied with polyhedra, this method allows the analysis of *linear relations* among variables of the program and the computation of an approximation of the *reachable states space*. We will show the various applications of these methods to the analysis of LOTOS specifications with integer variables and the results given by our implementation in the CÆSAR-ALDÉBARAN toolbox [13] context.

2. Definitions

We propose to work on a model which keeps a distinction between the *control part* and the *data part*, presented below as an *extended automaton*.

An extended automaton is a tuple $M = (C, V, T, q_0)$ where

- C is a finite set of *locations*
- V is a finite set $\{x_1, x_2, \ldots, x_n\}$ of *variables* of $\mathbb{R}$. We will call a *valuation* the function v which associates a real value $v(x)$ to each variable $x \in V$. A *state* of the underlying model is defined as a couple (l, v) with $l \in C$ and v is a valuation. We will note $Q \subseteq C \times \mathbb{R}^n$ the (possibly infinite) set of all possible states.
- T is a finite set of transitions. Each transition is a tuple $(l_1, l_2, \gamma, \alpha)$ where :
  - $l_1, l_2 \in C$ are respectively the *source* location and the *target* location
  - γ is the *guard* of the transition. It is a conjunction of m *linear constraints*, each constraint being defined as a triple (a, b, r) where a is a vector of $\mathbb{Z}^n$, $b \in \mathbb{Z}$ and r is a comparison operator belonging to $\{<, \leq, =, \geq, >\}$. The guard defines a *convex* set of points of $\mathbb{R}^n$ and is called a *linear system*. Given a set Φ of valuations, the value of $\gamma(\Phi)$ is given by the intersection of the set of points defined by γ and the set Φ. If the set Φ is convex, then the intersection $\Phi \cap \gamma$ is convex too.
  - α is an action, defined as a *linear assignment* from valuations to valuations. It is given as a couple (A, B) such that $A \in \mathbb{Z}^{n \times n}$ is a square matrix of integer coefficients and $B \in \mathbb{Z}^n$ is a vector of integer coefficients. Given a

linear assignment $\alpha = (A, B)$ and a set Φ of valuations, we note $\alpha(\Phi)$ the transformation by α of Φ:

$$\alpha(\Phi) = \{Av + B \mid v \in \Phi\}$$

- $q_0 = (l_0, v_0)$ is the *initial* state.

We will note $q_1 \xrightarrow{\gamma,\alpha} q_2$ the transition between $q_1 = (l_1, v_1)$ and $q_2 = (l_2, v_2)$ such that $\exists t = (l_1, l_2, \gamma, \alpha) \in T, v_1 \in \gamma \wedge v_2 = \alpha(v_1)$.

A *run* of this automaton is a finite or infinite sequence $q_0 \xrightarrow{\gamma_0,\alpha_0} q_1 \xrightarrow{\gamma_1,\alpha_1} \ldots$. We will say that a state $q = (l, v)$ is *reachable* if and only if there exists a run $q_0 \xrightarrow{\gamma_0,\alpha_0} q_1 \xrightarrow{\gamma_1,\alpha_1} \ldots$ such that $\exists i \geq 0, q = q_i$.

We will note *Reach* the set of all reachable states.

More precisely, we are interested in the computation of the *set of reachable valuations* for any location $l \in$ L, that we will note ϕ_l:

$$\phi_l = \{v \mid (l, v) \in \phi_l\}$$

In the following, we will call ϕ_l the *context* of the location l.

2.1. From Lotos specification to extended automaton

Given the type of variables used in an extended automaton, we will first restrict ourselves to LOTOS specifications with integer or enumerated variables, and linear expressions on these variables. The translation of a LOTOS program into an extended automaton suitable for analysis is done in two steps, via the *interpreted Petri Net* provided by the CÆSAR [14] compiler. This interpreted Petri Net is a *safe* Petri Net decorated with guards and operations on variables. The Petri Net is then simulated without taking account of the variables, to provide the control part of the extended automaton. This simulation can in some cases generate a extended automaton with unreachable locations. However, this automaton is generally much smaller than the explicit model, due to the implicit representation of the numerical variables. The whole process is best resumed by the figure 1.

During the translation, all the correspondences between locations of the automaton and places of the Petri Net are kept. In particular, all transitions in the extended automaton refer to a transition in the Net. As the inverse relation is also kept, we can associate to a given transition in the Net the set of corresponding transitions in the extended automaton. These informations give the possibility to interpret *automatically* the properties deduced on the automaton as properties on the Net.

The current version of the CÆSAR compiler dœs not allow us to extend this backward interpretation up to the LOTOS program, but still the Petri Net generated matches closely the structure of the source program, so any information obtained on the Net is easy to understand on the LOTOS specification.

3. Forward analysis

The computation of a set of reachable states from the initial state is usually performed by the resolution of a *least fixpoint* equation $x = F(x)$, where x represents sets of states

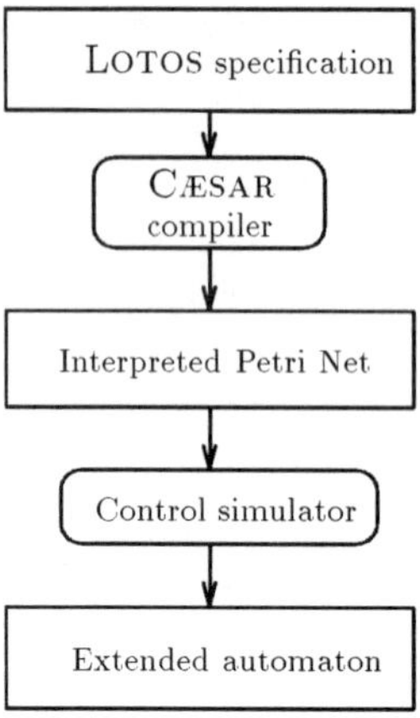

Figure 1. LOTOS to extended automaton

and F is defined as a transformation on sets of states. When dealing with programs using numerical variables, we face two problems:

1. How do we represent sets of states ? We need an *efficient* representation which can be *normalized* (for comparison of elements) and allows the definition of an *ordering relation* (for fixpoint computation).

2. As the set of all states can be infinite, how do we perform (possibly) infinite fixpoint computations ?

Both problems can find an answer in the context of *abstract interpretation*.

3.1. Abstract interpretation

Abstract Interpretation is a useful method proposed by [8] for computing *formally* and *automatically* an approximation of the dynamic behaviour of a program. It gives a way to establish a correspondence between the *concrete* domain and an *abstract* domain on which representation of elements becomes simpler and more efficient. An application of this framework proposed in [9] gives us an answer to our first question: they propose to represent a concrete set ϕ of valuations of variables by an *abstract* representation given by the *convex hull* Φ of the set ϕ. This convex hull can be defined as the set of solutions of a system of linear constraints, i.e. a *convex polyhedron*.

We first recall the definition of a convex polyhedron, then we present how the usual operators on sets such as inclusion, union, ... translate in terms of operators on polyhedra.

Convex polyhedron

A convex polyhedron P can be expressed as one of the following dual representations:

- a *system of linear inequalities* also called its *constraint system*: $P = \{\vec{x} \mid A\vec{x} \geq B\}$ where $\vec{x} = \{x_i \mid i \in [1 \ldots n] \wedge x_i \in I\!R\}$, $A \in Z\!\!Z^{m \times n}$ and $B \in Z\!\!Z^m$.

- a *system of generators*, which is a couple (V, R) where V and R are finite sets of vectors of $I\!R^n$, called respectively *vertices* and *rays*:

$$P = \{\sum_{v_i \in V} \lambda_i.v_i + \sum_{r_j \in R} \mu_j.r_j \mid \lambda_i \geq 0, \mu_j \geq 0, \sum_i \lambda_i = 1\}$$

We use an efficient algorithm initially proposed by [6] then improved and implemented by [25] to compute one representation from the other. This algorithm also allows to minimize each representation, by the detection of redundant constraints in the constraint system and conversely redundant generators.

Most of the operators needed for the computations of fixpoints are directly defined as operators on polyhedra. Given two convex polyhedra P and P', their *minimal* system of inequalities (A, B) and (A', B') and their *minimal* system of generators (V, R) and (V', R'), we define the following operators:

Test operators:

- **Emptyness:** P is empty if and only if $V = \emptyset$

- **Inclusion** $\sqsubseteq$: $P \sqsubseteq P' \Leftrightarrow \forall v \in V, A'v \geq B'$ and $\forall r \in R, A'r \geq 0$ (each vertex and ray of P satisfies the system of inequalities of P')

- **Equality** $=$: $P = P' \iff P \sqsubseteq P'$ and $P' \sqsubseteq P$

Set operators:

- **Intersection** $\sqcap$: the polyhedron $P \sqcap Q$ is defined by the system of inequalities given by the conjunction of (A, B) and (A', B')

- **Union:** the union of two convex polyhedra is not (in general) a convex polyhedron. We will approximate this operator by the *convex hull* operator

- **Convex hull** $\sqcup$: The convex hull of P and P' is the convex polyhedron defined by the system of generators $(V \cup V', R \cup R')$

- **Linear assignment:** given $\alpha = (A'', B'')$ a linear assignment, then $\alpha(P)$ is the convex polyhedron defined by the system of generators (V'', R'') where $V'' = \{A''v + B'' \mid v \in V\}$ and $R'' = \{A''r \mid r \in R\}$

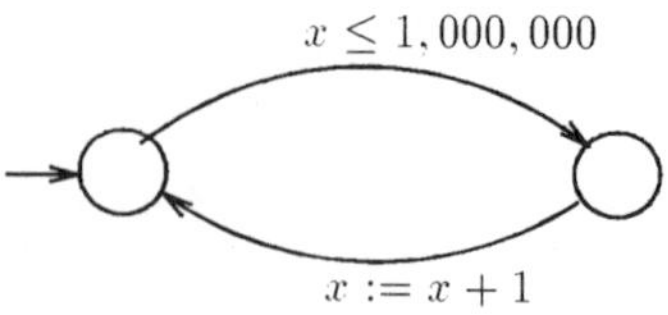

Figure 2.

3.2. Fixpoint computation acceleration

Choosing convex polyhedra as abstract domain dœs not provide an answer to our second problem: the lattice of convex polyhedra is of infinite depth, so we can still be faced with infinite fixpoint computations. It would be possible to choose an abstract domain either finite or of finite depth: in that case, all fixpoint computations converge in a finite number of steps. However, abstract domains of this kind are usually uninteresting, giving too inaccurate approximations. Furthermore, even if a fixpoint terminates, it may do so in too many steps to be of any practical interest: if we consider the example of figure 2, a classical iteration starting with $x := 0$ would take 1,000,000 steps before giving the obvious result $\{0 \leq x \leq 1,000,000\}$.

To complete the framework of abstract interpretation, [8] propose another kind of approximation, which enforces the convergence of an infinite fixpoint computation by extrapolation of its limit. This technique is based on the use of a *widening operator* and allows to compute an *upper approximation* of a least fixpoint, giving a safe approximation of the desired result.

If we consider the fixpoint equation $\Phi = F(\Phi)$, it can be written as the limit of the sequence

$$\begin{cases} \Phi^{(0)} = \bot \\ \Phi^{(n+1)} = F(\Phi^{(n)}) \end{cases} \tag{1}$$

We define a widening operator as follows:

Definition 3.1 (Widening operator)
Given a complete lattice L, a widening operator $\nabla : L \times L \rightarrow L$ satisfies the following properties:

- $\forall x, y \in L, x \cup y \sqsubseteq x \nabla y$
- *for all increasing chains $x_0 \sqsubseteq x_1 \sqsubseteq \cdots$ the chain $y_0 = x_0, y_{i+1} = y_i \nabla x_{i+1}$ is not strictly increasing, i.e. the limit of the sequence can be computed in a finite number of steps.*

The sequence (1) then becomes:

$$\begin{cases} \phi^{(0)} = \bot \\ \phi^{(n+1)} = \phi^{(n)} \nabla F(\phi^{(n)}) \text{ if } F(\phi^{(n)}) \not\sqsubseteq \phi^{(n)} \\ \phi^{(n+1)} = \phi^{(n)} \text{ if } F(\phi^{(n)}) \sqsubseteq \phi^{(n)} \end{cases}$$

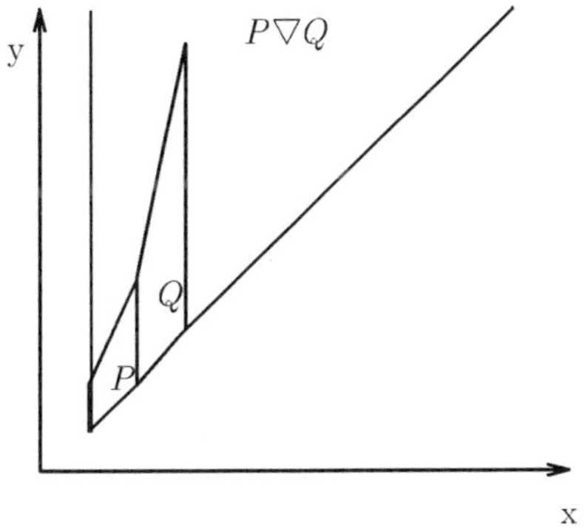

$$P = \left\{ (x,y) \mid \begin{pmatrix} 1 & \leq & x & \leq & 2 \\ x & \leq & y & \leq & 2x \end{pmatrix} \right\}$$

$$Q = \left\{ (x,y) \mid \begin{pmatrix} 2 & \leq & x & \leq & 3 \\ x & \leq & y & \leq & 3x \end{pmatrix} \right\}$$

$$P \nabla Q = \{ (x,y) \mid (\; 1 \; \leq \; x \; \leq \; y \;) \}$$

Figure 3. Widening operator

3.3. Widening operator for polyhedra

A widening operator for polyhedra has been first proposed in [18]. Given two polyhedra P and Q, the basic idea for computing $P \nabla Q$ is to keep only the constraints of P satisfied by Q:

Given $P = \{(x,y) \mid 1 \leq x \leq 2, x \leq y \leq 2x\}$ and $Q = \{(x,y) \mid 2 \leq x \leq 3, x \leq y \leq 3x\}$ then (see figure 3)

$$P \nabla Q = \{(x,y) \mid 1 \leq x, x \leq y\}$$

This widening operator can be improved when P is not full dimensional. In that case, its system of inequalities can be rewritten in order to maximize the number of constraints satisfied by Q:
Given $P = \{(x,y) \mid 1 \leq x \leq 2, y = 1\}$ and $Q = \{(x,y) \mid 1 \leq y \leq x \leq 2\}$, if we apply the basic idea of the widening operator, we obtain

$$P \nabla Q = \{(x,y) \mid 1 \leq x \leq 2, 1 \leq y\}$$

If we first rewrite P into $P = \{(x,y) \mid 1 \leq y \leq x \leq 2, y \leq 1\}$, the result of the widening becomes

$P \nabla Q = \{(x,y) \mid 1 \leq y \leq x \leq 2\}$ This operator satisfies the properties of a widening operator: the result of the widening contains its operands and it cannot be applied infinitely without convergence, as if $P \neq Q$, the system of inequalities of $P \nabla Q$ will be a strict subset of the system of inequalities of P.

3.4. Analysis of extended automaton with polyhedra

We can now gather the results of abstract interpretation with convex polyhedra to apply them on extended automata.

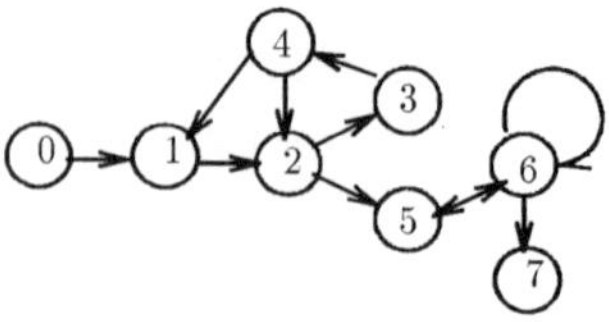

Figure 4. SCCs computation example

As the set of states Q of an extended automaton is partitioned with respect to the set of locations C, the computation of the reachable states is given by a *system* of fixpoint equations. Each equation is associated with a given location l and defines the context of this location as follows:

$$\phi_l = F_l(\{\phi_{l_1} \mid l_1 \in C\}) \tag{2}$$

In particular, the context of a location l depends only on the contexts of locations l_1 such that $\exists(l_1, l, \gamma, \alpha) \in T$. So the equations (2) become:

$$\phi_l = \bigcup_{(l_1,l,\gamma,\alpha)\in T} \{\alpha(\phi_{l_1} \cap \gamma)\}$$

We can write this system of fixpoint equations directly in terms of polyhedra and operators on polyhedra. In particular, the guard γ of each transition is itself a polyhedron (conjunction of linear inequalities) and the action α can be directly defined in terms of linear assignments on polyhedra:

$$\Phi_l = \bigsqcup_{(l_1,l,\gamma,\alpha)\in T} \{\alpha(\Phi_{l_1} \sqcap \gamma)\} \tag{3}$$

Due to the introduction of the convex hull operator, the forward analysis using polyhedra computes an upper approximation of the reachable states.

3.5. Choice of widening locations

The basic idea to ensure the convergence of the system of fixpoint equation (3) is to introduce a widening operation in every equation. This system then becomes:

$$\Phi_l = \Phi_l \nabla \bigsqcup_{(l_1,l,\gamma,\alpha)\in T} \{\alpha(\Phi_{l_1} \sqcap \gamma)\}$$

As the widening operator is an important source of loss of accuracy, it is important to apply it only when necessary. A good intuition is that every cycle of the extended automaton must contain a widening location, in order to ensure convergence. This condition is not always necessary, depending on the nature of the automaton: it is not necessary to set a widening location for a loop doing only an assignment of the form $X := K$.

We need to choose a *minimal* (in number of elements) set of widening locations, i.e. a minimal set of locations v such that every cycle of the automaton contains at least one state of v. This problem being NP-complete, we use an algorithm of graph decomposition into Strongly Connected Components (SCC), this decomposition being applied recursively until no further decomposition is possible. The algorithm gives also for each SCC its *entry point* for a depth-first exploration, we will choose this entry point as the widening location for the SCC. The result of this decomposition is an *admissible* set of widening locations, possibly not minimal. The decomposition into SCCs of the example of Figure 4 gives the following result:

$$0\ (\underline{1}\ (\underline{2}\ 3\ 4))(\underline{5}\ (\underline{6})\ )\ (7)$$

where the $(\ldots)$ operator indicates a SCC. The *admissible* set of widening locations computed is $\{1\ 2\ 5\ 6\}$ and a *minimal* set of widening locations computed would be $\{2\ 6\}$.

3.6. Iteration strategy

Given this system of equations and an admissible set of widening locations, we still have to choose an iteration order for the computation and stabilization of each fixpoint equation. All orders are not equivalent: it is better to delay the computation of a context Φ_l until a maximum of the contexts Φ_{l_1} it depends on are themselves updated. Hopefully the decomposition in SCCs of the extended automaton will help us to choose a consistent iteration strategy, as it allows to sort the locations in a *topological* order.

The iteration strategy itself consists in a *recursive* iteration as presented in [4]. The whole stabilization algorithm (Figure 5) is based on a depth first exploration of the hierarchy of SCCs. During this exploration, we will stabilize first the innermost SCCs of the automaton.

The main call to this algorithm is $Stabilize(C)$ (C is the set of all locations). The context of the initial location is defined by the initial value of the variables. The context of all other locations is empty. The successors of the initial location are marked not stable.

The application of this strategy on the example of figure 4 could be given by the expression

$$0\ (\underline{1}\ (\underline{2}\ 3\ 4)^*)^*(\underline{5}\ (\underline{6})^*\ )^*\ (7)$$

where the * operator's meaning is " iterate until stabilization".

4. Analysis example

We have developed a prototype implementation of the approximate analysis described in the previous sections. We present now the application of this tool on two different LOTOS programs.

4.1. Read-write protocol

The first example is a simple readers-writers protocol. A common resource is shared by a pool of potential writers and readers. Reading and writing are mutually exclusive, only one writer is allowed at a time, but all readers can read together if they want. All writing demands are counted, and a reader can start to read only when there is no more writers waiting to write, thus giving priority to the writers over the readers.

The process controlling the access to the resource is given by the following LOTOS process:

```
procedure Stabilize(Loc : set of locations)

while there exists a non stable location in Loc do
    let l ∈ Loc be the least (wrt iteration order) non stable location
    Φ'_l = ⊔_{(l1,l,γ,α)∈T} {α(Φ_{l1} ⊓ γ)}
    Mark l as stable
    if Φ_l ⊏ Φ'_l then
        if l is a widening location then
            Φ_l = Φ_l ∇ Φ'_l
        else
            Φ_l = Φ_l ⊔ Φ'_l
        fi
        foreach (l, l2, γ, α) ∈ T do
            Mark l2 as not stable
        od
    fi
    if l is an entry point of a sub SCC then
        Stabilize(SCC(l))
    fi
od
```

Figure 5. Stabilization algorithm

```
process RESOURCE[S_READ,E_READ,WRITE_DEMAND,S_WRITE,E_WRITE](N_READERS, N_W_DEMANDS, WRITING:NAT) : noexit :=
(* N_READERS gives the number of readers currently reading *)
(* N_W_DEMANDS gives the number of writers asking to write *)
(* WRITING indicates that a writer is currently writing *)
[(WRITING eq 0)]->
(
  [N_W_DEMANDS eq 0] ->
     S_READ ; RESOURCE [S_READ,E_READ,DEMAND_WRITE,S_WRITE,E_WRITE](N_READERS+1, N_W_DEMANDS, WRITING)
  []
  [(N_READERS eq 0) and (N_W_DEMANDS gt 0)] ->
     S_WRITE ; RESOURCE [S_READ,E_READ,DEMAND_WRITE,S_WRITE,E_WRITE](N_READERS, N_W_DEMANDS-1,1)
)
[]
WRITE_DEMAND ; RESOURCE [S_READ,E_READ,DEMAND_WRITE,S_WRITE,E_WRITE](N_READERS, N_W_DEMANDS+1,0)
[]
E_WRITE ; RESOURCE [S_READ,E_READ,DEMAND_WRITE,S_WRITE,E_WRITE](N_READERS, N_W_DEMANDS,0)
[]
E_READ ; RESOURCE [S_READ,E_READ,DEMAND_WRITE,S_WRITE,E_WRITE](N_READERS-1, N_W_DEMANDS,WRITING)
endproc
```

The global parallel expression for 3 readers and 2 writers is

```
(READER[S_READ,E_READ] ||| READER[S_READ,E_READ])
     |[S_READ,E_READ]|
RESOURCE[S_READ,E_READ,S_WRITE,E_WRITE](0, 0)
     |[S_WRITE,E_WRITE]|
(WRITER[S_WRITE,E_WRITE] )
```

The following results are given for M readers and N writers if we compute the union of the non empty contexts of the extended automaton after forward analysis, we obtain the following convex polyhedron, which represents an approximation of the smallest invariant of the program:

$$P = \left\{ \begin{array}{ll} \text{N\_W\_DEMANDS} & \geq 0 \\ \text{N\_READERS} & \geq 0 \\ \text{WRITING} & \geq 0 \\ \text{N\_W\_DEMANDS} + \text{WRITING} & \leq N \\ \text{N\_READERS} + M * \text{WRITING} & \leq M \end{array} \right\}$$

As WRITING can only take the value 0 and 1, this invariant shows that the number N_READERS can vary from 0 to M *if and only if* WRITING is equal to 0, thus ensuring the mutual exclusion. Note that the constraint $N\_READERS \leq M$ dœs not appear anywhere in the LOTOS program, this information has been deduced by the analysis process and is derived from the control part, not only from the data part of the program.

Another interesting result of this analysis is the detection of unreachable locations (empty context) in the extended automaton (see the size of the *explored* part of the automaton in table 1). This unreachability detection could be later used by the simulation phase of CÆSAR in order to remove some transitions of its Petri Net and reduce its state vector size .

4.2. Scheduler

This example is a task scheduler which handles several classes of tasks, each class is assigned a specific job whose completion takes a certain number of time units (TASK#LENGTH). All these tasks are constrained to run on a single processor. Each class is activated by a specific interrupt, these interrupts being generated at constant time intervals (INT#LENGTH). We describe further the version of the protocol running with two classes of tasks.

The LOTOS process generating the interrupts for two classes of tasks is the following:

```
process Interrupts[TICK,INT1,INT2](t1,t2:nat):noexit :=
[t1 ge Int1Length] -> INT1 ?X:nat;Interrupts[TICK,INT1,INT2](0,t2)
[]
[t2 ge Int2Length] -> INT2 ?X:nat;Interrupts[TICK,INT1,INT2](t1,0)
[]
TICK ;Interrupts[TICK,INT1,INT2](t1+1,t2+1)
endproc
```

When no tasks are running, the control is taken by an *Idle* process which reacts to the first interruption coming:

```
process Idle[TICK,INT1,INT2]:noexit :=
   INT1 ;Task1[TICK,INT1,INT2](0,1)
   []
   INT2 ;Task2[TICK,INT1,INT2](0,0,0,1)
   []
   TICK;Idle[TICK,INT1,INT2]
endproc
```

For each class of task, the number of incomplete tasks (either running or waiting) is accumulated in a variable (O#). Furthermore, we want to give priority to interrupts of the second class other the first class. The first class has just to keep count of its own interrupts and of the time it has already spent running a job. When interrupted by a second class interrupts, it saves its context by giving it to task2.

```
process Task1[TICK,INT1,INT2](L1,01:nat): noexit :=
(
   TICK;(
     [L1 eq Task1Length]-> (* Current task completed *)
       ([01 le 1]->Idle[TICK,INT1,INT2] (* No more tasks *)
        []
        [01 gt 1]->Task1[TICK,INT1,INT2](0,01-1) (* begin another task *)
       )
    []
    [L1 lt Task1Length]->Task1[TICK,INT1,INT2](L1+1,01) (* Running the current task*)
   )
   []
   INT1 ;Task1[TICK,INT1,INT2](L1,01+1) (* Keeping count of interrupts of same class*)
   []
   INT2 ;Task2[TICK,INT1,INT2](L1,0,01,1) (* Prioritary interrupt *)
)
endproc
```

As the second class has maximum priority, it must keep a trace of all interrupts counters and length of incomplete tasks of lower priority.

```
process Task2[TICK,INT1,INT2](L1:nat,L2:nat,01:nat,02:nat): noexit :=
(
   TICK;(
     [L2 eq Task2Length]-> (* Current task completed *)
       ([02 le 1]-> (* No more task of this class *)
        (
        [01 ge 1]-> Task1[TICK,INT1,INT2](L1,01) (* Continue or begin a task of class 1*)
         []
         [01 eq 0]-> Idle[TICK,INT1,INT2] (* No task of any class *)
        )
        []
        [02 ge 2]->Task2[TICK,INT1,INT2](L1,0,01,02-1) (* Begin another task of this class *)
       )
    []
    [L2 lt Task2Length]->Task2[TICK,INT1,INT2](L1,L2+1,01,02) (* Running current task *)
   )
   []
   INT1 ;Task2[TICK,INT1,INT2](L1,L2,01+1,02) (* Keeping count of lower priority interrupts *)
   []
   INT2 ;Task2[TICK,INT1,INT2](L1,L2,01,02+1) (* Keeping count of same interrupts *)
)
endproc
```

Given that priority is given to tasks of the second class over tasks of the first class, we want to verify that tasks of the second class never wait if the time interval between the generation of interrupts of class 2 is greater than the duration of tasks of class 2 (INT2LENGTH > TASK2LENGTH). It is equivalent to verify that the number of waiting *or* running tasks of class 2 never exceed 1 ($0 \leq \text{O2} \leq 1$) under the previous condition.

An important remark is that exact forward analysis on this example will not terminate, as the variables of the interrupts generator are not bounded. It is therefore necessary to approximate the computations in order to terminate. The results obtained for this example need a little more interpretation than the previous example: if we compute only the union of all non empty contexts, we obtain $\mathbb{R}^n$ (n being the number of variables) as a result. To obtain more accurate results, it is necessary to investigate at the transitions level of the Petri Net. In particular, we want to know the possible values of the variable O2 in the process task2. As the only line of this process able to change the value of O2 is the line

```
INT2 ;Task2[TICK,INT1,INT2](L1,L2,01,02+1)
```

we have to check the corresponding transition of the Petri Net, which expands to a set of transitions in the extended automaton. This is done automatically by the tool, by the insertion of a question mark {O2?} at this transition in the Petri Net. The analysis gives the following results: the union of the contexts of *source* and *target* locations of these transitions is computed, giving the following polyhedra (for all examples such that INT2LENGTH > TASK2LENGTH):

$$\textbf{Source locations } P = \left\{ \begin{array}{rcl} 0 & \leq & \mathrm{T1} \\ 0 & \leq & \mathrm{T2} \leq \mathrm{TASK2LENGTH} \\ 0 & \leq & \mathrm{L1} \leq \mathrm{TASK1LENGTH} \\ \mathrm{T2} & = & \mathrm{L2} \\ \mathrm{O2} & = & 1 \\ 0 & \leq & \mathrm{TASK1LENGTH} * \mathrm{O1} - \mathrm{L1} \\ 0 & \leq & \mathrm{T1} + \mathrm{T2} + \mathrm{TASK2LENGTH} * \mathrm{T1} \end{array} \right\}$$

$$\textbf{Target locations } P' = \emptyset$$

and the answer to the question mark is {O2 = 1}.
This result shows that:

1. Whenever the control arrives at this transition, O2 = 1

2. The transition is never enabled (target contexts are all empty), so O2 cannot be incremented.

So O2 can never be greater than 1; tasks of the second class can never wait.

Note that the analysis performances in this case do not depend on the size of the variables: the example has been run with values for (TASK1LENGTH,INT1LENGTH) and (TASK2LENGTH, INT2LENGTH) going from {(2,4) (3,5)} or {(4,8) (10,20)} up to {(30,40) (40,50)}, with the same performances, thanks to the widening operator. The limitation here comes from the internal representation of numbers in our package, leading to some overflows during the minimization of the polyhedra representations.

Another important remark is that the scheduler example is difficult to verify with enumerative tools like CÆSAR: the 2 tasks version remains workable, as the graphs generated by CÆSAR are in the order of 100,000 states, but if we consider the 3 tasks version of the scheduler, even if we bound the variables $T1$, $T2$ and $T3$ to at most max(INT1LENGTH, INT2LENGTH, INT3LENGTH) (otherwise, the resulting graph is infinite), and we choose small values ($\leq$ 10) for the durations of the different tasks and the time intervals between interrupts, then the resulting graph generated by CÆSAR is more than two millions states big (when we stopped the generation), well beyond the capacity of classical verification tools.

Program name	Petri Net		Extended automaton		Explored automaton		Variables number	Analysis time
	places	transitions	states	transitions	states	transitions		
3 readers 2 writers	20	19	320	608	166	250	3	1.3s
4 readers 3 writers	25	24	2156	4748	754	1257	3	15s
5 readers 3 writers	27	26	4364	10412	1454	2677	3	42s
Scheduler 2 tasks	17	27	27	51	24	48	8	4.4s
Scheduler 3 tasks	26	46	46	91	43	88	15	25.9s

Table 1
Analysis performances on a SPARC station 10

5. Conclusion

We have presented a symbolic representation of models well-suited for the approximate analysis of reachable states space. Although the analysis technique as well as the symbolic representation presented are not new [18] , their application to concurrent programming languages like LOTOS is as far as we know an innovation. Similar works for protocols can be found in [22] and [2], but based on a less powerful extrapolation operator than the widening operator used in our work.

One of the applications of this analysis technique is of course the verification of some invariant properties on the variables of the program, as shown on the examples presented. But we also use our implementation as an *abstract debugging* tool: it is often useful to know what are the bounds on variables or relations between them to understand some behaviours (and often unwanted behaviours) of the program. As our tool allows finely tuned analysis, at the level of places and transitions of the Petri Net, we can see as a *symbolic* debugger of LOTOS programs.

The analysis technique proposed here is based on a semi-symbolic model, an automaton being used to describe the control part and a symbolic representation being used for the data part. The control part of this automaton is derived from a simulation of a Petri Net. Yet, this automaton is only used for a partitioning of the analysis, in order to obtain more accurate results. An improvement would be to explore the automaton "on the fly" during the analysis, to avoid storing it completely and generating unreachable states and transitions, in the same spirit as those developed in [11, 24], and also partial-orders methods such as [15, 24].

A last direction of interest is the application of these techniques to timed automata, already under investigation in [19] for delay analysis and [20] for linear hybrid systems.

An adaptation of our tool to this kind of systems could complement the tool presented in [10] for the verification of an extension of LOTOS with timing requirements .

Acknowledgments

Thanks are due to Jean Claude Fernandez and Laurent Mounier for their helpful comments and careful readings of this paper.

REFERENCES

1. R. Alur, C. Courcoubetis, N. Halbwachs, D. Dill, and H. Wong-Toi. Minimization of timed transition systems (extended abstract). In *CONCUR'92, Stony Brook*. LNCS 630, Springer Verlag, august 1992.
2. Bernard Boigelot and Pierre Wolper. Symbolic verification with periodic sets. *6th Workshop on Computer-Aided Verification,Stanford*, 1994.
3. A. Bouali and R. de Simone. Symbolic bisimulation minimisation. In G. Bochmann, editor, *Proceedings of the fourth workshop on Computer-Aided Verification (Montreal,Canada)*, june 1992.
4. Francois Bourdoncle. *Semantique des languages imperatif d'ordre superieur et interpretation abstraite*. PhD thesis, Ecole Polytechnique, 1992.
5. R. E. Bryant. Graph-based algorithms for boolean function manipulation. *IEEE Transactions on Computers*, C-35(8), 1986.
6. N. V. Chernikova. Algorithm for discovering the set of all solutions of a linear programming problem. *U.S.S.R. Computational Mathematics and Mathematical Physics*, 8(6):282–293, 1968.
7. O. Coudert, J.C. Madre, and C. Berthet. Verifying temporal properties of sequential machine without building their state diagram. In R.P. Kurshan and E.M. Clarke, editors, *Proceedings of the Workshop on Computer-Aided Verification (Rutgers, USA)*. DIMACS, June 1990.
8. P. Cousot and R. Cousot. Abstract interpretation: a unified lattice model for static analysis of programs by construction or approximation of fixpoints. In *4th POPL*, January 1977.
9. P. Cousot and N. Halbwachs. Automatic discovery of linear restraints among variables of a program. In *5th. Annual Symp. on Principles of Programming Languages*, pages 84–87, 1978.
10. C. Daws, A. Olivero, and S. Yovine. Verifying et-lotos programs with the tool kronos. In *7th international Conference on Formal Description Techniques for Distributed Systems and Communication Protocols*, 1994.
11. J.C. Fernandez, C. Jard, T. Jéron, and L. Mounier. "on the fly" verification of finite transition systems. *Formal Methods in System Design, Kluwer Academic Publishers*, 1992.
12. J.C. Fernandez, A. Kerbrat, and L. Mounier. Symbolic equivalence checking. In C. Courcoubetis, editor, *Proceedings of the 5th Workshop on Computer-Aided Verification (Heraklion, Greece)*, 1993.
13. Jean-Claude Fernandez, Hubert Garavel, Laurent Mounier, Anne Rasse, Carlos Rodríguez, and Joseph Sifakis. A toolbox for the verification of lotos programs. In

Lori A. Clarke, editor, *Proceedings of the 14th International Conference on Software Engineering ICSE'14 (Melbourne, Australia)*, pages 246–259, New-York, May 1992. ACM.

14. Hubert Garavel and Joseph Sifakis. Compilation and verification of lotos specifications. In L. Logrippo, R. L. Probert, and H. Ural, editors, *Proceedings of the 10th International Symposium on Protocol Specification, Testing and Verification (Ottawa, Canada)*, Amsterdam, June 1990. IFIP.
15. P. Godefroid and P. Wolper. Using partial orders for the efficient verification of deadlock freedom and safety properties. In Kluwer Academic Publishers, editor, *Formal Methods in System Design*, April 1993.
16. S. Graf and C. Loiseaux. A tool for symbolic program verification and abstraction. In *5th Conference on Computer-Aided Verification (Elounda, Greece)*. LNCS 697, Springer Verlag, jul 1993.
17. Philippe Granger. Static analysis of linear congruences. In *International Journal of Computer Mathematics*, 1989.
18. N. Halbwachs. *Détermination automatique de relations linéaires vérifiées par les variables d'un programme*. Thèse de 3e cycle, Université de Grenoble, March 1979.
19. N. Halbwachs. Delay analysis in synchronous programs. In *5th Conference on Computer-Aided Verification (Elounda, Greece)*. LNCS 697, Springer Verlag, jul 1993.
20. N. Halbwachs, Y.E. Proy, and P. Raymond. Verification of linear hybrid systems by means of convex approximations. In *First International Static Analysis Symposium*. Springer Verlag, September 1994.
21. T. Henzinger, X. Nicollin, J. Sifakis, and S. Yovine. Symbolic model-checking for real-time systems. In *LICS'92*, June 1992.
22. A.S. Krishnakumar. Reachability and recurrence in finite state machines: Modular vector addition systems. In C. Courcoubetis, editor, *Proceedings of the 5th Workshop on Computer-Aided Verification (Heraklion, Greece)*, 1993.
23. H. Touati, H. Savoj, B. Lin, R. K. Brayton, and A. Sangiovanni-Vincentelli. Implicit state enumeration of finite state machines using BDDs. In *International Conference on Computer Aided Design (ICCAD)*, November 1990.
24. A. Valmari. On-the-fly verification with stubborn sets. In C. Courcoubetis, editor, *Proceedings of the fifth workshop on Computer-Aided Verificatio n (Elounda, Crete)*. DIMACS, june 1993.
25. H. Le Verge. *Un environnement de transformations de programmes pour la synthèse d'architectures régulières*. PhD thesis, Université de Rennes, 1992.

12

An Improvement in Formal Verification

Gerard J. Holzmann and Doron Peled

AT&T Bell Laboratories, Computer Science Research, Murray Hill, NJ 07974, USA.

Critical safety and liveness properties of a concurrent system can often be proven with the help of a reachability analysis of a finite state model. This type of analysis is usually implemented as a depth-first search of the product state-space of all components in the system, with each (finite state) component modeling the behavior of one asynchronously executing process. Formal verification is achieved by coupling the depth-first search with a method for identifying those states or sequences of states that violate the correctness requirements.

It is well known, however, that an exhaustive depth-first search of this type performs redundant work. The redundancy is caused by the many possible interleavings of independent actions in a concurrent system. Few of these interleavings can alter the truth or falsity of the correctness properties being studied.

The standard depth-first search algorithm can be modified to track additional information about the interleavings that have already been inspected, and use this information to avoid the exploration of redundant interleavings. Care must be taken to perform the reductions in such a way that the capability to prove both safety and liveness properties is fully preserved. Not all known methods have this property. Another potential drawback of the existing methods is that the additional computations required to enforce a reduction during the search can introduce overhead that diminishes the benefits. In this paper we discuss a new reduction method that solves some of these problems.

Index Terms: D.2, F.1,2,3.

1. *INTRODUCTION*

The depth-first search algorithm that is used for on-the-fly reachability analyses can explore many execution sequences that are not strictly required to prove the safety and liveness properties of a concurrent system. In the last few years, several proposals have been made for revised search algorithms that can avoid some or all of this redundancy, e.g., [V90], [G90], [GW91], [HGP92], [V93], [P93].

The methods that have been studied so far can be classified as 'dynamic reduction methods.' They attempt to compute mostly at runtime (i.e., during the search) which parts of the reachability analysis are redundant and can be skipped. Unavoidably, the additional computations also consume resources: they require memory to store additional data structures, and they require CPU time to discover the redundancies. This overhead reduces the amount of improvement that can be achieved. In some cases, the costs of improvement outweigh the gains, which means that the unoptimized full search can sometimes outperform the 'optimized' reduced search.

This paper considers the feasibility of performing the computations before the search, instead of during the search. We prove for one such reduction strategy that it preserves both the safety and the liveness properties of a concurrent system. The new reduction strategy is therefore generally usable for linear time temporal logic (LTL) model checking.

Section 2 discusses the main concepts needed for a definition of the static reduction method. Section 3 introduces the algorithm itself, and Section 4 contains its proof of correctness. Section 5 discusses an implementation of the static reduction method, as an experimental addition to the verification tool SPIN [H92]. Section 6 contains an evaluation of the performance of this implementation, and a comparison against both the classic search method and an existing dynamic reduction method, all implemented as part of the same verification system. Section 7 summarizes the results.

2. *DEFINITIONS*

We consider any verification problem that can be formalized as a reachability analysis problem in a finite labeled transition system (LTS). This specifically includes the problems of proving safety, liveness, and linear time temporal logic properties for any finite state concurrent system.

An **LTS** is defined as a triple $\{S, s_0, T\}$, where S is a finite set of states, s_0 is a distinguished initial state in S, and T is a finite set of transitions, with $T \subseteq (S \times S)$. In a simple form, an LTS can be used to formalize the behavior of a single sequential process. It can also formalize the combined behavior of a finite number of interacting and asynchronously executing sequential processes. Each transition of the LTS then corresponds to the execution of a specific atomic statement within one of the processes, in accordance with a standard interleaving semantics of concurrency. The LTS can be represented by a graph with nodes corresponding to the states in S and directed edges corresponding to the transitions in T. A connected path through this graph then defines the effects of a possible execution in the underlying concurrent system. There will be at least one path through the graph for every possible way in which the execution of atomic process statements could be interleaved in time.

Given a transition $t \in T$ in an LTS, we will use the notation $Label(t)$ to refer to the process statement that is represented by transition t, and we will use $Pid(t)$ to refer to the sequential process that contains the statement $Label(t)$. Without loss of generality, we assume that the mapping from transitions to process statements is unique. The reverse mapping will in general not be unique. Note that, in general, a compound process statement (such as a selection or a repetition structure) could correspond to .a series of transitions in the LTS. In the remainder of this section, therefore, the term 'statement' will always refer to a simple statement (i.e., not a compound), and can therefore be used interchangeably with the term 'transition.'

The semantics of a statement $a = Label(t)$ are defined by two functions *Cond* and *Act*, where

$Cond(a)$ is the subset of S where a is enabled (or 'executable' [H92]), and
$Act(a,s)$ is that state of S that is reached when a is executed in a given $s \in Cond(a)$.

Normally, a statement in a sequential process is 'enabled' or 'executable' only if it is pointed to by the current program counter of the sequential process that contains that statement. In a concurrent system, however, we can define additional constraints on the enabledness or executability of statements. A message send operation, for instance, can be defined to be enabled only if also the destination message buffer is non-full, and a message receive operation can be defined to be enabled when also the source message buffer is non-empty.

Two statements a and b are defined to be **independent** at state $s \in S$, written as $\{a,b\} \in Ind(s)$, if and only if the following five conditions are met:

(1) $s \in Cond(a)$, i.e., statement a is enabled in s,
(2) $s \in Cond(b)$, i.e., statement b is enabled in s,
(3) $Act(a,s) \in Cond(b)$, i.e., the execution of a cannot disable b,
(4) $Act(b,s) \in Cond(a)$, i.e., the execution of b cannot disable a,
(5) $Act(b, Act(a,s)) = Act(a, Act(b,s))$, i.e., the effect of executing a followed by b is indistinguishable from that of executing b followed by a.

Note that two statements from the same sequential process, i.e., with $Pid(a) = Pid(b)$, can not be independent. If the two statements are executed sequentially, they cannot be simultaneously enabled. If they appear together in a single selection, the execution of either one statement will disable the other. Two statements from distinct sequential processes can be independent under certain conditions. Two send operations on distinct message queues will in general be independent, but two send operations on the same message queue will not. The send operation that executes first may disable the second if its message fills the queue to capacity, which violates requirements (3) and (4). In addition, the order in which the two statements are executed can be distinguished by the order in which the messages appear in the destination queue, which violates requirement (5).

Statements a and b are defined to be **globally independent** if and only if they are independent in every possible state where they are simultaneously enabled:

(6) $s \in (Cond(a) \cap Cond(b)) \rightarrow \{a,b\} \in Ind(s).$

Note that a and b are trivially globally independent when $Cond(a) \cap Cond(b) = \varnothing$. Two assignment statements from two distinct sequential processes, i.e., $Pid(a) \neq Pid(b)$, that access only local variables within each process, will in general also be globally independent.

Because it is known that both safety and liveness properties can be expressed by next-time-free linear-time temporal logic (LTL) formulae [W83], we will focus on a method for proving the satisfiability of LTL formulae (cf. [V90], [V93], [P93]). The **LTL formulae** we consider may contain boolean propositions on system-states, the boolean operators $\wedge$, $\vee$, ! (not), and the temporal operators $\Box$ (always), $\Diamond$ (eventually), and U (until), but not the temporal operator $\circ$ (next-time).

Wolper [W83] showed that any next-time-free LTL formula can be formalized as a nondeterministic Büchi Automaton with a predefined initial state, and a finite set of acceptance states. The *transitions* in the Büchi Automaton carry predicate labels, each of which represents a boolean proposition. In our case, the boolean propositions can refer only to the (global) system-state of the labeled transition system for which the LTL formula formalizes a property. The Büchi Automaton itself can be represented by an LTS with predefined acceptance states. The satisfaction of an LTL formula can now be proven by detecting acceptance cycles in the synchronous product of two labeled transition systems: one representing the concurrent system and one representing the Büchi Automaton (e.g., [CVWY92]). The absence of acceptance cycles can similarly prove that the LTL formula cannot be satisfied [H92].

The **synchronous product** $F \times G$ of a labeled transition system F, representing a concurrent system, and a Büchi Automaton G, derived from a next-time-free LTL formula, is defined as follows. Let $F = (S_F, f_0, T_F)$ and $G = (S_G, g_0, T_G)$. Each state of the synchronous product $F \times G$ is a pair (f,g), with $f \in S_F$ and $g \in S_G$. Each transition, similarly, is a pair (v,w), with $v \in T_F$ and $w \in T_G$. We define the LTS for the synchronous product $F \times G$ recursively as follows. The initial state of $F \times G$ is (f_0, g_0). For each state (f,g) there is a successor state (h,k), reachable via transition (v,w), if and only if:

(1) $v = (f,h) \in T_F$, i.e., h is a successor of f via v in F,
(2) $w = (g,k) \in T_G$, i.e., k is a successor of g via w in G, and
(3) The boolean proposition defined by $Label(w)$ is *true* in state $f \in S_F$.

A statement a in F is said to be **observable** by Büchi Automaton G if there exists a label in G for which the corresponding proposition can have a different truth-value in at least one system-state $s \in Cond(a)$ and in $Act(a,s)$. The statement a can now be said to be

- **Safe** if a is non-observable to G and globally independent from every statement b with $Pid(a) \neq Pid(b)$, and
- **Conditionally Safe** for condition $P(s)$, if a is safe in every state s where $P(s)$ holds.

The reduction algorithm that we will describe in the next section relies on the fact that the safety or conditional safety of statements can in many cases be determined statically.

3. *REDUCTION ALGORITHM*

We first consider the standard depth-first search algorithm that implements the generation of the labeled transition system F from a specification of a concurrent system. We then consider how this search can be extended to generate the synchronous product $F \times G$, where G is a Büchi Automaton that encodes an LTL formula, and to detect the existence of acceptance cycles in that product.

The initialization of the search is illustrated in Figure 1a. First, the basic transition structure of the concurrent system is obtained and optimized. The optimization step, can, as we shall argue, also include a precomputation of independence relations, with a static identification of all safe and conditionally safe process-statements. Two sets of states are then initialized with the predefined initial system state s_0: the *Statespace* and the *Stack*. The search begins with a call of the depth-first search routine, *Dfs*( ), with parameter 1. The relevance of the parameter will become clear shortly.

```
1 start_search(s0)
2 { derive and optimize transition structures
3   enter s0 into Statespace;
4   push  s0 onto Stack;
5   Dfs(1);   /* see Figure 1c */
6 }
```

Figure 1a – Initialization

Figure 1b first shows the expansion step for process statements, in a routine called *dfs*( ). In the absence of a Büchi Automaton, the calls on line 5 in Figure 1a and on line 16 in Figure 1b could be implemented as calls on *dfs*(N), instead of the capitalized version *Dfs*(1).

```
 7 dfs(N)
 8 { s = top(Stack);
 9   for each sequential process i
10   {     nxt = all transitions in F enabled in s with Pid(t)=i
11         for all t in nxt
12         {     s' = successor of s after t;
13               if {s',N} NOT in Statespace
14               {     enter {s',N} into Statespace;
15                     push s' onto Stack;
16                     Dfs(N);
17   }     }     }
18   pop s from Stack
19 }
```

Figure 1b – Expansion Step for the Sequential Processes

In the presence of a Büchi Automaton the calls on lines 5 and 16 invoke the routine *Dfs*(N) shown in Figure 1c, which implements the expansion step for the transitions in the Büchi Automaton. The state of the Büchi Automaton is part of compound system state s. Because the transitions in the Büchi Automaton represent boolean propositions from an underlying LTL formula, a transition $t \in T_G$ in Büchi Automaton G will only be enabled (cf. line 22 in Figures 1c-e) if and only if proposition *Label*(t) holds. The synchronous coupling of system F and Büchi Automaton G is achieved by alternating the calls to `Dfs(N)`, on line 16, and `dfs(N)` on line 28. Each pair of subsequent calls, explores one synchronous transition of $F \times G$.

Figure 1c shows only the basic expansion step without the extra hooks that are required to detect

```
20 Dfs(N)
21 { s = top(Stack);
22   nxt = all transitions in G enabled in s; /* Büchi Autmt */
23   for all t in nxt
24   {     s' = successor of s after t;
25         if {s',N} NOT in Statespace
26         {     enter {s',N} into Statespace;
27               push s' onto Stack;
28               dfs(N);
29   }     }
30   pop s from Stack
31 }
```

Figure 1c – Interleaved Transitions of Büchi Automaton

the presence of acceptance cycles in the synchronous product of concurrent system and Büchi Automaton. To enable also the detection of acceptance cycles, we can check for every reachable acceptance state in *G* if that state is also reachable from itself. We do so with a second depth-first search, in post-order, in a separate state space. Two separate values for parameter N serve to indicate in which part of the search the algorithm operates. To initiate the second search, we include four extra lines between lines 28 and 29 of Figure 1c, as illustrated in Figure 1d. If the seed state is reachable from itself when *N* equals 2 this can be detected and reported at line 24, as illustrated by lines 24a-d in Figure 1d.

```
20 Dfs(N)
21 { s = top(Stack);
22   nxt = all transitions in G enabled in s; /* Büchi Autmt */
23   for all t in nxt
24   {     s' = successor of s after t;
24a        if N == 2 and s' == seed
24b        {     report acceptance cycle
24c              return
24d        }
25         if {s',N} NOT in Statespace
26         {     enter {s',N} into Statespace;
27               push s' onto Stack;
28               dfs(N);
28a              if N == 1 and s is an accepting state in G
28b              {     seed = s
28c                    dfs(2)
28d              }
29   }     }
30   pop s from Stack
31 }
```

Figure 1d – Extension for Cycle Detection

A description, and correctness proof, for this method of cycle detection was given in [CVWY92]. The algorithm generates at least one example of an acceptance cycle, if one or more such cycles exist. It is not guaranteed to generate *all* such cycles. If, however, the Büchi Automaton is used to formalize an undesirable behavior, i.e., the violation of a correctness requirement, a proof of either the existence or the absence of acceptance cycles that satisfy the claim is always sufficient for a conclusive verification result.

Note that when the existence of an acceptance cycle is discovered, its complete traversal is contained in the *Stack*, and can be generated as a counter-example to the correctness claim.

3.1 *Static Reduction*

To implement a static reduction technique, it suffices to modify only the algorithm from Figure 1b, since the 'safety' of transitions, as defined at the end of section 2, applies only to the transitions in the sequential processes, not to those of the Büchi Aütomaton. The change is illustrated in Figure 1e. The aim of the reduction method is to find a smaller set of transitions that suffices to perform the expansion (given that we want to preserve both safety and liveness properties). Clearly, the expansion cannot be complete unless for every transition selected, we also select all those simultaneously enabled transitions that are dependent with it (i.e., not independent from it). This means that if we select a, we must minimally also select all simultaneously enabled transitions b with $Pid(b) = Pid(a)$ (cf. line 10 in Figure 1e).

```
 7 dfs(N)
 8 { s = top(Stack);
 8a  order processes; /* (cond.) safest process first - see text */
 9   for each sequential process i
 9a  {   boolean NotInStack = true
 9b      boolean AtLeastOneSuccessor = false
10       nxt = all transitions t in F enabled in s with Pid(t)=i
11       for all t in nxt
12       {    s' = successor of s after t;
13            if {s',N} NOT in Statespace
14            {    enter {s',N} into Statespace;
15                 push s' onto Stack;
16                 Dfs(N);
16a           } else if s' in Stack  /* reduction proviso */
16b                NotInStack = false
16c           AtLeastOneSuccessor = true
16d      }
16e      if AtLeastOneSuccessor ∧ NotInStack
16f           break /* from the loop over processes */
17   }
18   pop s from Stack
19 }
```

Figure 1e – Reduced Expansion Step

In the static reduction method we try to identify at least one process that can execute only safe, or conditionally safe, transitions. Such a process can be found by a prescan of the processes. In Figure 1e this critical step is performed on line 8a and is used to re-order the processes in such a way that processes that perform only (conditionally) safe transitions can be selected first for the expansion step on line 9. If the expansion succeeded (more about this below) we can ignore the (independent) transitions from all other processes by breaking out of the loop over processes on line 16f. The ordering step itself introduces virtually no runtime overhead. In the implementation discussed in Section 5, for instance, it is implemented by a table-lookup for unconditionally safe transitions, and by the evaluation of a precomputed boolean condition for conditionally safe transitions.

A check is added on lines 16a-b, to see if the last transition explored returned the search to a state s' that is already contained in the search stack or not. If there is at least one such transition, the value of a local boolean variable *NotInStack* is set to *false*. Once all transitions of the process have been explored, the values of *NotInStack* and *AtLeastOneSuccessor* are inspected. The reduction attempt fails unless all transitions explored for the current process have produced successor states that are currently *not* contained in *Stack*. If this requirement is not met, the algorithm will try to make another selection of transitions, by moving to the next process in the outer for-loop. In the worst case, this will mean that the reduced expansion step will explore all enabled transitions, just as it did in Figure 1b.

The condition on line 16e that has to be fulfilled for the reduction attempt to be considered successful is known as the **reduction proviso**. The need for such a proviso was first recognized by Valmari in [V90]. The version of the proviso used here was first proposed in [P93]. A weaker version of the same test, for the preservation of safety properties only, was discussed in [HGP92]. In the next section we will show that the stronger proviso from [P93] guarantees the preservation of both safety and liveness properties.

The tests on line 16a and 16e introduce virtually no overhead to the algorithm.

4. *PROOF OF CORRECTNESS [skip on first reading]*

We will give the main proof argument that supports the correctness of the reduction algorithm. The remaining steps that are required for a rigorous proof are briefly indicated.

An *execution sequence* σ of an LTS can be defined either as a sequence of transitions or as the sequence of states that is traversed by these transitions. Let $Eq(\sigma)$ be the set of all execution sequences that can be obtained from σ by zero or more permutations of adjacent, globally independent, transitions. A finite prefix of at least one sequence in $Eq(\sigma)$ is called a **permuted prefix** of σ. A **generalized permuted prefix** of σ is any finite sequence that can be transformed into a permuted prefix of σ by omitting zero or more non-observable transitions.

To prove the correctness of the reduced search algorithm, we first prove the following Lemma. The Lemma states that each transition that could be generated from a given reachable state in an unreduced search, is necessarily also generated by the reduced search, though possibly after some intervening non-observable transitions.

Lemma – At each reachable state s, for every execution sequence σ that begins at s with a transition a, when placed in s the reduced search algorithm generates at least one generalized permuted prefix ρ of σ that begins at s and that ends with transition a. □

Proof – The proof is by induction on the order in which states are removed from the depth-first stack in the reduced search algorithm.

[1.] For the induction basis, consider the first state that is removed from the stack in the reduced search algorithm. There are two cases to consider, depending on the number of enabled transitions in that state.

[1.1.] The state has no enabled transitions, and thus no successor states. In this case there exist no further executions from this state, and the Lemma holds.

[1.2.] The state has enabled transitions. All these transitions must have returned the search to previously visited states: they cannot be new states because such states would have been removed from the stack before the current one (a property of the depth-first search). Since no states were previously removed from the stack, all previously visited states are still contained in the stack. The reduction proviso from the reduced search algorithm will in this case force a complete exploration of all enabled transitions from this state (line 16a, Figure 1e). This set includes the first transition a from σ. This transition a is a generalized permuted prefix ρ of length one. The Lemma therefore holds for this case.

[2.] Next, we must show that if the Lemma holds for the first N states that are removed from the stack, it necessarily also holds for the $(N+1)$-th state. Let s be that state. There are again two cases to consider.

[2.1.] The set of enabled transitions in s does *not* contain a true subset of (conditionally) safe transitions that includes all the enabled transitions for one sequential process, none of which leads to a successor state on the stack. In this case, the reduced search algorithm explores all enabled transitions from s and the Lemma holds by the same construction as was used in the proof of step **[1.2.]**.

[2.2.] The set of enabled transitions in s *does* contain a true subset of (conditionally) safe transitions that includes all enabled transitions for one sequential process, and none of which leads to a successor state on the stack. Call that subset x, and call the (non-empty) set of all remaining enabled transitions y. The reduced search algorithm explores only the sequences that start with a transition from x. Note that the transitions in x are necessarily non-observable.

There are two cases to consider.

[2.2.1.] If σ starts with a transition from x, the Lemma again holds.

[2.2.2.] Next, consider the case where σ starts from state s with a transition a not from x, but from y. We distinguish two further sub-cases.

[2.2.2.1.] Consider first the case where at least one transition from x appears somewhere after a in σ. Let b be the first such transition in σ. The sequence of transitions that appear before b must then be independent of b, since b is itself either safe or conditionally safe (or else it could not have been in set x). This means that there must exist another sequence σ' in $Eq(\sigma)$ that starts with the two transitions $b.a$. Call s' the state that is reached after the execution of b. Let σ'' be the sequence obtained from σ' after removing the initial b. σ'' is an execution sequence that starts from state s'. By the induction hypothesis we now know that the reduced search algorithm will generate a generalized permuted prefix ρ of σ'', from state s'. But then, $b.\rho$ must also be a generalized permuted prefix of σ that starts at state s. Thus the Lemma also holds for this case.

[2.2.2.2.] Next, consider the case where no transition in x appears in σ, which means that all transitions in x are globally independent of all transitions in σ. If b is a transition from x, then $b.\sigma$ corresponds to a possible execution sequence from s. Call s' the state that is reached after the execution of b in this last sequence. Again, by the induction hypothesis we know that the reduced search algorithm will generate a generalized permuted prefix ρ of σ from state s'. But then, since b is non-observable, $b.\rho$ is trivially also a generalized permuted prefix for σ that starts at state s. This completes the proof of the Lemma. □

The Lemma can be shown to imply that for every execution sequence σ, the reduced search algorithm explores at least one execution sequence that becomes equivalent to σ when a finite number of non-observable transitions are deleted from it. (This proof step is not detailed here.) Next we must show that this property is sufficient for the completeness of the search itself. To do this, we must take a closer look at the synchronous product of a concurrent system and a Büchi Automaton.

Given a concurrent system C and a Büchi Automaton M, we can construct an ordered set of predicates $P(M)$ with one predicate for each boolean proposition on the states of C that appears in M. For each reachable system state of C, each predicate in $P(M)$ then uniquely defines a boolean value, and the set $P(M)$ similarly defines a unique vector of boolean values. For given $P(M)$, an execution sequence of C corresponds to a sequence of boolean value vectors. Call that sequence 'the vector-sequence induced by M.'

We define two execution sequences to be M-**equivalent**, for given Büchi Automaton M, if and only if the corresponding vector-sequences induced by M are equal up to stuttering, i.e., if the two sequences are equal when each series of two or more consecutive occurrences of the same value vector v is replaced by a single occurrence of v.

The Lemma implies that the reduced search algorithm generates at least one M-equivalent sequence for each execution sequence of the concurrent system. The intuition for this is that all non-observable transitions correspond to stuttering steps. (This proof step is not further detailed here.) The correctness of the reduced search algorithm can now be formalized in the following theorem.

Theorem – If there exist acceptance cycles in the synchronous product of a Büchi Automaton and a concurrent system, the reduced search algorithm will detect at least one of these cycles. □

Proof – by the Lemma and the fact that the set of sequences satisfying a next-time-free LTL formula is closed under stuttering [L83]. The reduced search generates at least one M-equivalent sequence for each complete sequence that satisfies the LTL formula. All sequences that satisfy the LTL formula are detected in the non-reduced depth-first search as acceptance cycles in the synchronous product of the corresponding Büchi Automaton and the concurrent system (e.g., [W83][CVWY92][H92]). Therefore, if at least one M-equivalent sequence for such a satisfying sequence is generated in the reduced search, at least one acceptance cycle is detected. □

A more general version of the proof of correctness for this reduction method can be found in [P94]. Ideally, we believe it should be possible to produce proofs such as these in a more mechanical fashion, using automated theorem provers.

5. *IMPLEMENTATION*

For the implementation of the static reduction technique in the verification system SPIN and its specification language PROMELA [H92], we identified five types of statements that can be marked statically as unconditionally safe when they appear separately, and conditionally safe when they appear as guards in selection structures.

(1) Any access to exclusively local variables. Any atomic process-statement that reads or writes exclusively objects that are non-observable to other processes, is also non-observable to the PROMELA *never* claim (which formalizes the Büchi Automaton).

(2) Any receive operation on a message queue q, provided that no more than one process can either receive messages from q or test the contents or length of q. We mark such a queue with a special status: *exclusive receive-access.* Exclusive receive-access implies that a *never* claim contains no propositions on the contents of q.

(3) Any send operation on a message queue q, provided that no more than one process can send messages to q, or test the contents or length of q. We say that such a queue has *exclusive send-access.* Exclusive send-access implies that a *never* claim contains no propositions on the contents of q.

(4) The boolean test *nfull*(q), that returns *true* when message queue q is currently non-full, and false otherwise, provided that the statement is performed by a process that has exclusive send-access to that queue.

(5) The boolean test *nempty*(q), that returns *true* when message queue q is currently non-empty, and false otherwise, provided that the statement is performed by a process that has exclusive receive-access from that queue.

The statements of types (1)-(5) are *conditionally safe* if they do appear as guards in selection structures. The condition for the conditionally safe statements is defined as the logical *and* combination of the following clauses for each type of guard: (1) *true* (i.e., these statements contribute no additional constraints), (2) and (5) *nempty*(q), and (3) and (4) *nfull*(q). Note that statements of type (2-5) can only contribute constraints of two statically determined types.

We extended the PROMELA grammar with the two new primitives *nfull*( ) and *nempty*( ), referred to in (4) and (5). A simple grammar rule in the parser prevents attempts to include negations of these two tests.

The observability of the effect of statements to the propositions of the Büchi Automaton (i.e., the PROMELA *never* claim) is already guaranteed by the scope rules of PROMELA: in the absence of remote referencing, the *never* claim can only refer to global objects in the specification. All safe and conditionally safe operations are therefore necessarily non-observable to the claim. Any reference to a queue, for instance, breaks the exclusive access status of that queue, and automatically marks the send or receive operations as observable, and therefore non-safe.

Because PROMELA allows the dynamic creation of a finite number of processes, it is not always possible to determine a priori which processes will be able to access which queues. Exclusive send and receive access, in our implementation, is therefore entered into the PROMELA specification as a logical assertion, which can be checked at runtime. The Appendix shows an example of a complete PROMELA specification for a leader election protocol from [DKR82], with the exclusive send and receive assertions added. It can easily be shown that the validity of an assertion of this type can be proven by both the non-reduced and the reduced search, even when the reduction is based on an invalid assertion of this type. The intuition behind this is that the reduced search can only *permute* globally independent statements, it cannot prevent their execution alltogether. Therefore, at least one send or receive operation that violates an exclusive access assertion will eventually be executed in the reduced search, though perhaps at a different place then in the non-reduced search. There is, of course, also the possibility that the reduced search is stopped on the detection of an acceptance cycle *before* the violation of an exclusive access assertion can be demonstrated. In that case, however, the search has already reached its goal: it has detected the existence of at least one error (i.e., an acceptance cycle). If the violation of an exclusive access assertion can be demonstrated first, our implementation also reports an error (i.e., an assertion violation), which in that case means that the reduction itself was invalid.

For the correctness of the reduction algorithm itself it must be demonstrated that if there exist one or more acceptance cycles in $F \times G$, the reduced search algorithm will always report at least one of them. The proof of this property is given in [P94]. Note that it is not guaranteed, neither for the reduced nor for the standard algorithm, that *all* acceptance cycles will be reported.

6. *PERFORMANCE*

We have measured the performance of the new reduction algorithm on five sample protocols, including a best-case example, a worst-case example, and two average protocol applications produced independently by users of SPIN [H92]. The performance is compared both to the non-reduced 'classic' verification algorithm, and to an existing dynamic reduction method based on Godefroid's sleep-set method [G90], [HGP92], which is publicly available (in binary form) via anonymous FTP from the University of Liege as an extension to SPIN. In the comparisons, it should be observed that the dynamic reduction method preserves only safety properties, while the other two methods preserve both safety and liveness properties, and thus provide a stronger and more general model checking system. There exist other reduction methods that preserve both safety and liveness properties (e.g., [V90], [V93]), but at the time of writing no implementation of these methods was available for these comparisons.

Not all types of reduction can be computed with a static algorithm. For instance, if multiple sequential processes share access to the elements of a data array, the precise values of the array indices may only be known dynamically, which makes it impossible for a static algorithm to

Table I – Measurements

Protocol	Algorithm	States	Transitions	Time(sec.)	Memory (Mb)
Best-Case	Non-Reduced	100,001	450,002	13.2	4.3
	Static Reduction	47	47	(<0.1)	1.0
	Dynamic Reduction	47	47	0.1	1.4
Worst-Case	Non-Reduced	100,001	450,002	14.5	5.0
	Static Reduction	100,001	450,002	16.7	5.1
	Dynamic Reduction	100,001	450,002	84.5	5.3
Tpc	Non-Reduced	3,918,286	11,762,426	630.6	268.4
	Static Reduction	391,534	466,753	30.6	26.2
	Dynamic Reduction	267,204	295,395	131.4	18.9
Snoopy	Non-Reduced	91,920	305,460	14.4	11.5
	Static Reduction	16,279	23,532	1.7	3.2
	Dynamic Reduction	7,158	8,459	6.8	2.6
Pftp	Non-Reduced	417,321	1,244,865	73.2	62.3
	Static Reduction	53,244	67,901	6.8	9.3
	Dynamic Reduction	125,718	163,459	105.5	20.6
Leader	Non-Reduced	45,885	185,032	8.1	9.6
	Static Reduction	79	79	0.1	1.1
	Dynamic Reduction	79	79	0.2	1.4

determine if the access of any given element is safe. Because every reduction that can be made with a static algorithm can also be made with a dynamic algorithm, we should expect that dynamic algorithms can achieve greater reductions than static ones in terms of the numbers of states and transitions explored. If the bottom-line commodities of run-time and memory used are measured, however, the results are less predictable, and it should be possible for a static algorithm, with lower overhead, to defeat a dynamic one.

The best and the worst case examples are artificial cyclic models, containing only local and only global statements, respectively. *Tpc* is a model of a telephone switch, specified in 279 lines of PROMELA. *Snoopy* is a model of a cache coherence protocol specified in 255 lines of PROMELA. *Pftp* is a version of the file transfer protocol of 204 lines from [H92]. *Leader* is the leader election protocol with 5 processes, as shown in the Appendix. All measurements were made on a Sparc-10 workstation with 128Mbyte of RAM. The runtimes are the sum of system-time and user-time.

The static reduction method gave the shortest run-time in all cases tested, despite the fact that it sometimes searched a larger statespace compared to the dynamic method. In one case (Pftp) the static reduction method even completes its search in a smaller search space than the dynamic method, but presumably, this is a deficiency in the dynamic method that could be remedied. The amount of memory used by the static and the dynamic reduction method is comparable, and not decisively in favor of either method. For both the dynamic and the static reduction method, the amount of memory used is significantly lower than in a non-reduced search.

In the worst case, the performance of the reduced search is not significantly different from that of a non-reduced search, which is not true for the dynamic method. The best-case application, as expected, shows a (literally) exponential reduction of both runtime and memory requirements.

6.2 *Exploiting Structure*

The reduction rules for the static reduction algorithm also have an unexpected, and sometimes quite dramatic, positive effect on our capability to prove essential properties of large protocol specifications. Because it can lower the complexity of the search if we can mark queues as having exclusive send-access or exclusive receive-access, the user can consciously avoid using queues that do not have this property. It is often possible to rewrite a specification in this way, without changing its functionality. As an example, consider the system from Figure 2a, which corresponds to a validation model that we built in 1988 for the validation of the IEEE 802.2 Logical Link Control Protocol [IEEE84].

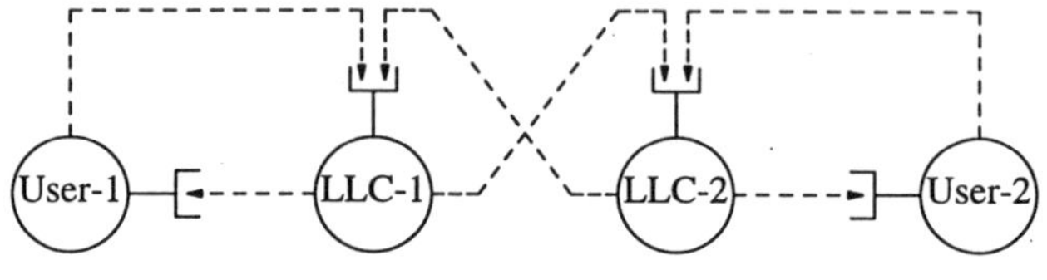

Figure 2a – Model of IEEE LLC 802.2

An exhaustive search for this protocol generates 1,851,049 reachable states. The reduced search brings this down to 111,159 reachable states. Notice, however, that the input queues of the LLC processes are non-exclusive. By splitting these two queues, as illustrated in Figure 2b, using a separate queue for each source of information, all queues in the revised model obtain both exclusive receive- and exclusive send-access.

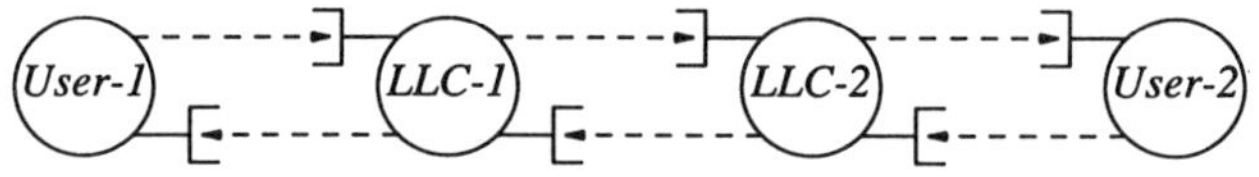

Figure 2b – Revised Model

The reduction algorithm will be able to do a better job in this case, but the revision also has an effect on the complexity of the *non-reduced* search, by avoiding the generation of all non-deterministic interleavings of messages from different sources in the shared input queues of the LLC processes. The search space reduces from 1,851,049 reachable states to 19,407 reachable states, for the non-reduced algorithm. The reduction algorithm reduces this further to 2,000 reachable states. The reduction method in this case successfully guides us to a solution that requires three orders of magnitude less resources than before.

7. *CONCLUSION*

We have described a new static reduction algorithm that preserves the capability of a depth-first search to prove both safety and liveness properties of concurrent systems. We have shown that this static reduction method performs considerably better than a non-reduced search. It also approaches the reduction achieved by a dynamic method closely, while providing a greater correctness proving power. The static reduction method has no significant worst-case behavior.

Because it is known which types of statements can be exploited by the algorithm, with a static reduction method the designer of a protocol can reduce the complexity of a validation *a priori*, be choosing a specific structure for the model. In certain cases, it may thus become possible to perform true 'structured design validation' for concurrent systems.

Our intention is to release the extended version of SPIN as Version 2.0, in source form, within the next few months.

8. *REFERENCES*

[CVWY92] C. Courcoubetis, M. Vardi, P. Wolper, M. Yannakakis, ''Memory efficient algorithms for the verification of temporal properties,'' Formal Methods in Systems Design I, 1992, pp. 275-288.

[DKR82] D. Dolev, M. Klawe, and M. Rodeh, ''An O(n log n) unidirectional distributed algorithm for extrema finding in a circle,'' Journal of Algorithms, Vol 3. (1982), pp. 245-260.

[G90], P. Godefroid, ''Using partial orders to improve automatic verification methods,'' Proc. 2nd Workshop on Computer Aided Verification, LNCS 531, Eds. R. Kurshan and E. Clarke, New Brunswick, New Jersey, June 18-21, 1990.

[GW91], P. Godefroid, P. Wolper, ''A partial approach to model checking,'' 6th LICS, 1991, Amsterdam, pp. 406-415.

[H92] G.J. Holzmann, ''Design and Validation of Computer Protocols,'' Prentice Hall, 1992.

[HGP92] G.J. Holzmann, P. Godefroid, and D. Pirottin, ''Coverage preserving reduction strategies for reachability analysis,'' Proc. IFIP, Symp. on Protocols Specification, Testing, and Verification, June 1992, Orlando, Fl. pp. 349-364.

[IEEE84] ANSI/IEEE Standard 8802/2, Logical Link Control, ISBN 0-471-82748-7, New York, 1984.

[L83] L. Lamport, ''What good is temporal logic?,'' Information Processing 83: Proc. of the 9th IFIP World Computer Congress. Ed. R.E.A. Mason, Elsevier Publ., pp. 657-668.

[P93] D. Peled, ''All from one, one for all – on model checking using representatives,'' 5th Int. Conf. on Computer Aided Verification, Greece, 1993, LNCS 697, Springer Verlag, pp. 409-423.

[P94] D. Peled, ''Combining Partial Order Reductions with On-the-fly Model Checking,'' 6th Int. Conf. on Computer Aided Verification, Stanford, Ca., June 1994. LNCS 818, Springer Verlag, pp. 377-390.

[V90] A. Valmari, ''A stubborn attack on state explosion,'' 2nd Int. Conf. on Computer Aided Verification, Rutgers University 1990, Dimacs Series, Vol 3, pp. 25-42.

[V93] A. Valmari, ''On-the-fly verification of stubborn sets,'' 5th Int. Conf. on Computer Aided Verification, 1993, LNCS 697, Springer Verlag, pp. 397-408.

[W83] P. Wolper, M.Y. Vardi, and A.P. Sistla, ''Reasoning about infinite computation paths,'' Proceedings of 24-th IEEE symposium on the foundations of computer science, Tuscan, 1983, pp. 185-194.

APPENDIX

The following PROMELA specification lists the complete model of the leader election protocol in unidirectional ring that was used in the performance study. The protocol was defined in [DKR82].

```
/*
 * PROMELA validation model - Leader Election - [DKR82]
 */

#define I3    /* The node to have smallest identifier */
#define N5    /* number of processes */
#define L10   /* size of message buffer */

mtype    = { first, second };

chan q[N] = [L] of {byte,byte};       /* global FIFO buffer */

proctype node(chan in, out; byte id)
{    byte number, maxi=id, neighbourR;    /* local */
     bit active=1;                /* local */

     xr in;   /* assert exclusive receive access to chan in */
     xs out;  /* assert exclusive send access to chan out   */

     out!first(id);
end: do             /* repetition : valid end-state */
     :: in?first(number) ->
          if        /* selection */
          :: active ->
               if   /* selection */
               :: number != maxi ->
                    out!second(number);
                              neighbourR = number
               :: number == maxi ->
                    assert maxi == N
               fi
          :: !active -> out!first(number)
          fi
     :: in?second(number) ->
          if        /* selection */
          :: active ->
               if   /* selection */
               ::   neighbourR > number && neighbourR > maxi ->
                    maxi = neighbourR;
                    out!first(neighbourR)
               :: !(neighbourR > number && neighbourR > maxi) ->
                    active = 0
               fi
          :: !active -> out!second(number)
          fi
     od
}
```

```
init {
        byte n=1;              /* a local variable */
        atomic {               /* non-interleaved */
                do             /* repetition */
                :: n <= N ->
                        run node(q[n-1], q[n%N], (N+I-n)%N+1);
                        n = n+1
                :: n > N ->
                        break  /* end repetition */
                od
}       }
```

Position Statements

13

A Modal Based Verification for LOTOS (abstract)

Hacène Fouchal and Ana R. Cavalli
Institut National des Télécommunications
9, Rue Charles Fourier, 91011 Evry Cédex France
E-mail: {fouchal, cavalli}@int-evry.fr

This paper presents a verification method based on a modal logic for the LOTOS language. The method has been inspired by the works on verification methods for the CCS language. These works are extended in this paper to the LOTOS operators (parallel, enabling and disabling). It is also shown how to decompose a general formula into smaller parts in order to be verified by parts of the program which describes a system. A full version of this paper is presented in [1].

1. INTRODUCTION

The verification of distributed systems is a very important step in the software design process. The problem of verification can be approached in two different ways: model based methods, where the specification is translated into a model expressing the semantics of the described system, and axiomatic methods, where the verification is done at the source level, specifications are considered as formulae describing the expected properties of the system.
The method presented in this paper is related to axiomatic methods. Properties are expressed in a modal logic and for each LOTOS operator, a rule is defined, which considers the composition of properties verified by sub-processes.

2. A VERIFICATION METHOD FOR LOTOS

2.1. A logic for specifications

We shall define a simple modal logic *ML*, whose formulae express conditions on the behavior of processes described with the LOTOS language. This logic defines two modal operators <a> and [a]. The semantics of these two operators is given by the following rules:
$p \models \langle\alpha\rangle A$ if $\exists\, \alpha \,/\, p \overset{\alpha}{\Rightarrow} p'$ and $p' \models A$
$p \models [\alpha] A$ if $\forall\, \alpha$, if $p \overset{\alpha}{\Rightarrow} p'$ then $p' \models A$
where $p \models A$ means that the process p verifies the property A which is a modal formula. $p \overset{\alpha}{\Rightarrow} p'$ means that a process p executes an observable action α to become p'. The definition of this relation implies also that p can execute a sequence of internal actions before α and before reaching the process p'.

2.2. Satisfaction relation between formulae

In the previous section we have defined a satisfaction relation, $\models$, between processes and formulae. In this section we shall define the semantic rules for the satisfaction relation $A \circ B \models C$, between formulae of the modal logic defined in the previous section.

This relation is defined essentially to prove properties of the parallel, enabling and disabling operators of the LOTOS language. We will show that if it is true that $A \circ B \models C$ and processes p_1 and p_2 are such that $p_1 \models A$ and $p_2 \models B$ then it can be concluded that $p_1 * p_2 \models C$ (where $\circ$ is a particular composition relation between formulae and * is one of the above mentioned LOTOS operators).

2.3. Adaptation to LOTOS language

The adaptation of the relation to *LOTOS* parallel operators is performed by the three relations $|_G$, $|_F$ et $|_I$. Their meanings are stipulated to be:

$$A \,|_G\, B \models C \;\; if \;\; \forall p, q \;\; if \;\; p \models A \;\; and \;\; q \models B \;\; then \;\; p|[g_1,..,g_n]|q \models C \;\; and \;\; G = \{g_1,..,g_n\}$$

$$A \,|_F\, B \models C \;\; if \;\; \forall p, q \;\; if \;\; p \models A \;\; and \;\; q \models B \;\; then \;\; p||q \models C$$

$$A \,|_I\, B \models C \;\; if \;\; \forall p, q \;\; if \;\; p \models A \;\; and \;\; q \models B \;\; then \;\; p|||q \models C$$

We propose two new relations between formulae $A \rhd B \models C$ and $A \div B \models C$. Their meanings are stipulated to be:

$$A \rhd B \models C \;\; if \;\; \forall p, q \;\; if \;\; p \models A \;\; and \;\; q \models B \;\; then \;\; p \gg q \models C$$

$$A \div B \models C \;\; if \;\; \forall p, q \;\; if \;\; p \models A \;\; and \;\; q \models B \;\; then \;\; p \;[> q \models C$$

The definition of these relations preserves the compositionality of the method. The semantics of these relations is defined by the following clauses:
(1) $C \,\Delta\, C' \models$ **true**
(2) $D \,\Delta\, D' \models A$ *iff* $\forall\, C \in D, \forall\, C' \in D', C \,\Delta\, C' \models A$
(3) $C \,\Delta\, C' \models A$ *iff* $C=$ **false** *or* $C'=$ **false**
(4) $C \,\Delta\, C' \models A \vee B$ *iff* $C \,\Delta\, C' \models A$ *or* $C \,\Delta\, C' \models B$
(5) $C \,\Delta\, C' \models A \wedge B$ *iff* $C \,\Delta\, C' \models A$ *and* $C \,\Delta\, C' \models B$
where Δ could be one of the following relations : $\rhd$, $\div$, $|_F$, $|_I$ or $|_G$ and D is in a normal form, i.e., D = $D_1 \vee .. \vee D_n$. C $\in$ D iff $\exists$ i such as C = C_i.
The clauses (1) to (5) are similar to those defined for the *CCS* language and they are defined in the same way for the relations $|_I$, $|_G$, $|_F$, $\rhd$ or $\div$.
In the following, we will give the semantics of the relations $|_I$, $|_G$ and $|_F$. For the relations related to the enabling and disabling operators see the full paper [1].
(6) $C \,|_I\, C' \models \langle\alpha\rangle B$ *iff* $\exists\, D, D'$
$\quad \langle\alpha\rangle D \in C$ *and* $D \,|_I\, C' \models B$ *or*
$\quad \langle\alpha\rangle D' \in C'$ *and* $C \,|_I\, D' \models B$
(7) $C \,|_I\, C' \models [\alpha]B$ *iff* $\forall\, D, D'$
$\quad$ *if* $[\alpha]D \in C$ *then* $D \,|_I\, C' \models B$ *and*
$\quad$ *if* $[\alpha]D' \in C'$ *then* $C \,|_I\, D' \models B$

The clauses (6) and (7) apply to the interleaving operator of the *LOTOS* language. For the *eventuality* operator ($\langle\alpha\rangle$), the basic action α can either be obtained from the left side or from the right side. In the case of the *necessity* operator ($[\alpha]$), the basic action α can also be obtained from both sides, but both resulting formulae must verify the formula *B*.

(8) $|_F\ C' \models \langle\alpha\rangle B$ *iff* $\exists\ D, D'$

$\langle\alpha\rangle D \in C$ *and* $\langle\alpha\rangle D' \in C'$ *and* $D\ |_F\ D' \models B$

(9) $C\ |_F\ C' \models [\alpha]B$ *iff* $\forall\ D, D'$

if $[\alpha]D \in C$ *then* $[\alpha]D' \in C'$ *and* $C\ |_F\ D' \models B$ *and* $D\ |_F\ C' \models B$

or

if $[\alpha]D' \in C'$ *then* $[\alpha]D \in C$ *and* $C\ |_F\ D' \models B$ *and* $D\ |_F\ C' \models B$

The clauses (8) and (9) further describe the semantics of the full synchronization operator. The basic action α must be obtained from both sides because this action reflects a *rendez-vous* that must be executed by both processes.

(10) $C\ |_G\ C' \models \langle\alpha\rangle B$ *iff* $\exists\ D, D'$

if $\alpha \in G$ *then*

$\langle\alpha\rangle D \in C$ *and* $D\ |_G\ C' \models B$ *and*

$\langle\alpha\rangle D' \in C'$ *and* $C\ |_G\ D' \models B$

if $\alpha \notin G$ *then*

$\langle\alpha\rangle D \in C$ *and* $D\ |_G\ C' \models B$ *or*

$\langle\alpha\rangle D' \in C'$ *and* $C\ |_G\ D' \models B$

(11) $C\ |_G\ C' \models [\alpha]B$ *iff* $\forall\ D, D'$

if $\alpha \in G$ *then*

if $[\alpha]D \in C$ *then* $[\alpha]D' \in C'$ *and* $C\ |_G\ D' \models B$ *and* $D\ |_G\ C' \models B$

or

if $[\alpha]D' \in C'$ *then* $[\alpha]D \in C$ *and* $C\ |_G\ D' \models B$ *and* $D\ |_G\ C' \models B$

if $\alpha \notin G$ *then*

if $[\alpha]D \in C$ *then* $D\ |_G\ C' \models B$

or

if $[\alpha]D' \in C'$ *then* $C\ |_G\ D' \models B$

The clauses (10) and (11) present a combination of the preceding clauses. When the action α belongs to the synchronisation set, the result obtained is similar to the result of the application of clauses (8) and (9). If it is not the case, the result is similar to the result obtained by application of clauses (6) and (7).

3. CONCLUSIONS AND FUTURE WORK

This paper presented a proof method for LOTOS based on a modal logic. We have shown how to adapt this logic to the parallel LOTOS operators. The main feature of this method is the compositionality of the proof of properties. We intend to implement this method into our development tool, which already includes a modal formula checker for LOTOS specifications. We are also working on extending the method to the LOTOS language including data.

REFERENCES

1. Hacène Fouchal and Ana R. Cavalli, A Modal Based Verification for LOTOS, Research Report 94-07-06, Institut National des Télécommunications, August 1994.

14

Tableau methods to decide Strong Bisimilarity on LOTOS processes involving pure interleaving and enabling

A. Fantechi♦♠, S. Gnesi♦ and R. Sacchelli♦

♦ I.E.I. - C.N.R.,
via S Maria 46, I-56126 Pisa, Italy

♠ Dip. di Ingegneria dell'Informazione, Univ. di Pisa,
via Diotisalvi 2, I-56126 Italy

Position statement

1. INTRODUCTION

Several methodologies and most automatic tool support for LOTOS [1] can only be used if the specifications considered describe finite-state machines [2,3,4]. For example, an equivalence verifier can only manipulate specifications in this class; an attempt to verify the equivalence of two non finite-state processes may not terminate with an answer. However, as it is known, LOTOS gives the possibility to describe also non finite-state systems. So it could be useful to provide methodologies and tools for verification on this class of systems. Indeed more powerful verification techniques have recently been exploited for non-finite state systems defined in other process algebras.

Among these verification techniques we wish to recall the bisimulation decision algorithms on context free processes of Basic Process Algebra (BPA) [5] due to Baeten, Bergstra and Klop and to Hüttel and Stirling [6,7] and the algorithm to decide the bisimilarity between Basic Parallel Processes (BPP) given in [8].

In this paper we present a decision algorithm to prove the bisimulation between non finite-state systems on a subset of LOTOS, $\text{LOTOS}_{\text{CF+|||}}$, able to describe the sequential composition of normed interleaved recursive processes. The class of processes so defined covers all context free processes and some context dependent processes [9]. This subset, called from now on $\text{LOTOS}_{\text{CF+|||}}$, can be seen as the combination of two process languages. The first, LOTOS_{CF}, includes action prefix, exit, enabling and recursion. The second, $\text{LOTOS}_{|||}$, includes action prefix, exit, pure interleaving and recursion.

2. THE TABLEAU DECISION METHOD

In [6,7,10] it was proved that the bisimulation equivalence is decidable on BPA processes. The proof of this statement is based on the regularity properties of the computation trees, starting from the system of recursive equations defining such processes. Moreover in [8] the decidability of normed BPP terms has been proved with a similar algorithm. These results can be mapped on $LOTOS_{CF}$ and on $LOTOS_{|||}$ processes making some extensions to the original algorithms, since both $LOTOS_{CF}$ and $LOTOS_{|||}$ processes can be expressed as systems of recursive equations [9]. Here, we briefly present an algorithm to prove the bisimulation equivalence between $LOTOS_{CF+|||}$ processes. The presented algorithm is obtained by a careful combination of the two algorithms defined in literature respectively for BPA [7] and for BPP [8]. So far, to make the combination possible, we have had to consider only processes having an enabling as outermost operator and not under the scope of an interleaving operator.

The termination of the composed algorithm is guaranteed by the termination conditions of the component algorithms and by the following result:

Theorem 1

Let α=P>>Q and β=R|||T, they can never be bisimulation equivalent if P, Q are non-trivial (that is, different from **exit**) normed $LOTOS_{CF}$ processes and R,T are non-trivial normed $LOTOS_{|||}$ processes.

The bisimulation decision algorithm on $LOTOS_{CF+|||}$ terms, based on the construction of semantic tableaux following the structure of the LOTOS terms, applies for each transition the definition of strong bisimulation equivalence itself: for each action performed by a process it checks if the other process is able to perform the same action, and then it checks the bisimilarity of the processes resulting from the execution of this action. The algorithm goes on in this way until an unsuccessful or successful termination condition is reached.

Let $(X|B_X)$ and $(Y|B_Y)$ be two $LOTOS_{CF+|||}$ systems of recursive equations, normed and in 2-V form[1]. To test whether X=Y, we build a tableau using a goal-directed technique as defined in [7]. The tableau rules are built around equations of the form E>>α=F>>β, where α and β can be also empty sequences of variables and E, F, α and β are $LOTOS_{CF+|||}$ behaviour expressions. In the case α or β are empty we consider equations having only the left side of the corresponding enabling operator. Each rule has the form:

---

[1]For each process variable P, we define the *norm* of P, |P|, as: $|P|=\min\{ lth(w) \mid P \xrightarrow{w} stop\}$.

We say that P is *normed* iff $|P| < \infty$.

A $(X_i|B)$ system is normed if all process variables X_i are normed

A $(X_i|B)$ system is in *2-V normal form* if the recursive equations in B are defined as:
$X_i=([]_k\ a_k;\alpha_k\ )[]([]_j\ a_j;\beta_j\ )$ where $a_k, a_j \in Act$, $\alpha_k=X_k>>Y_k \mid \alpha_k=X_k \mid$ **exit** and $\beta_j=X_j|||\ Y_j \mid \beta_j=X_j \mid$ **exit.**

$$\frac{E{>>}\alpha = F{>>}\beta}{E_1{>>}\alpha_1 = F_1{>>}\beta_1 \cdots E_n{>>}\alpha_n = F_n{>>}\beta_n}$$

The premise of the rule represents the goal (i.e. $E{>>}\alpha=F{>>}\beta$) while the consequents are the subgoals. Each of these rules is backwards sound, insofaras the consequents are true (i.e. their equations relate strongly bisimilar processes) then so is the antecedent.

A tableau for the equation $X{>>}\alpha=Y{>>}\beta$ can be seen as a finite proof tree whose root is labelled $X{>>}\alpha=Y{>>}\beta$, where the equations which label the immediate successors of a node are determined by an application of one of the rules. The algorithm goes on applying the rules unless no more rules are applicable, hence producing some leaves labelled with equations of the type $\alpha = \beta$. These leaves can represent either terminal or non-terminal nodes of the tableau. Terminal nodes are either successful or unsuccessful. If at any point we reach an unsuccessful terminal, we terminate and consider the resulting tableau as being unsuccessful. On the other hand, a successful tableau is a finite depth proof tree, all of whose leaves are successful terminals.

REFERENCES

1. Bolognesi, T., Brinskma, E.: Introduction to the ISO Specification Language LOTOS, Computer Networks & ISDN Systems, 14, 1, 25-29 (1987).

2. Bolognesi, T., Caneve, M.: Equivalence Verification: Theory, Algorithms and a Tool, in van Eijk, P.H.J., Vissers C.A., Diaz M., eds., The Formal Description Technique LOTOS, pp. 303-326, North-Holland, 1989.

3. Fernandez, J.C., Garavel, H., Mounier, L., Rasse, A., Rodriguez, C., Sifakis, J.: A Toolbox for the Verification of LOTOS Programs. 14th ICSE, pp. 246-261, 1992.

4. Madeleine, E., Vergamini, D.: AUTO: A Verification Tool for Distributed Systems Using Reduction of Finite Automata Networks, Formal Description Techniques - II, pp. 61-66, North-Holland, 1989.

5. Bergstra, J. A., Klop, J. W: The Algebra of Recursively Defined Processes and the Algebra of Regular Processes, ICALP '84, LNCS vol. 172, pp. 82-94, 1984.

6. Baeten, J. C., Bergstra, J. A., Klop, J. W: Decidability of bisimulation equivalence for processes generating context-free languages, Journal of ACM 40,3, pp. 653-682,,1993.

7. Hüttel, H., Stirling, C.: Actions Speak Louder than Words: Proving Bisimilarity for Context-Free Processes, LICS 91, Computer Society Press, pp. 376-386, 1991.

8. Christensen, S , Hirshfeld, H , Moller, F: Bisimulation Equivalence is Decidible for Basic Parallel Processes, CONCUR'93, LNCS vol. 715, pp. 143-157, 1993.

9. Fantechi, A., Gnesi, S., Mazzarini, G., "How Much Expressive are LOTOS Behaviour Expressions?" Formal Description Techniques - III, pp. 17-32, North-Holland, 1990.

10. Christensen, S., Hüttel, H., Stirling, C.: Bisimulation Equivalence is Decidable for All Context-Free Processes, CONCUR'92, LNCS vol. 630, pp. 138-147, 1992.

15

Simulator for Lotos to study the Independence and Causality of Events

B. Botma, R. Langerak [a]

[a] University of Twente, Department of Computer Science, Tele-Informatics and Open Systems group, 7500 AE Enschede, The Netherlands. Email: {botma,langerak}@cs.utwente.nl.

The standard interleaving semantics, by introducing all possible interleavings of all concurrent actions, creates choices and dependencies in the final transition system that do not really reflect a branching of the system behaviour. This is especially problematic for a simulation tool as the user of such a tool has to be bothered with making irrelevant choices. In this paper we use one particular causality based model, the *bundle event structure* model, in which the creation of unwanted choices and dependencies is avoided by moving towards a partially ordered set transition system. This forms the basics of an interactive simulation tool called *SLICE*. Simulations with this tool show that we can greatly improve upon simulation in an interleaving framework in terms of number of states.

1. Introduction

For the purpose of designing systems one does not only need a specification formalism, but also a semantical model to reason about specifications. In this paper we use the *bundle event structure* model[1, 2]. Event structures provide the opportunity to avoid the state space explosion caused by modelling all interleavings of concurrent processes by taking advantage of the intrinsic relations between events. The bundle event structure model defines two relations between events, namely bundles and asymmetric conflicts, that are used to model causal and conflict relations between events. Independence of events can then be defined as the absence of the above two relations. The approach taken by this paper is not to remove unwanted choices, i.e. by taking confluence properties into account, or to factor out unwanted choices in the transition system, i.e. by using dependency relations, but to avoid the creation of unwanted choices by moving towards a so-called partially ordered set transition system. Partially ordered sets (posets) are sets of events together with an ordering relation, often referred to as the *precedence relation*, over these events. A poset transition system is a transition system in which the transitions are labelled with partially ordered sets.

1.1 Example (Poset transition system).
Consider the behaviour $B = P[a, p_1, p_2] \ |[a]| \ Q[a, q_1, q_2]$ where

$$\begin{array}{ll} \mathit{process}\ P[a, p_1, p_2] : \mathit{noexit} := & p_1; p_2; a; \mathtt{stop}\ \mathit{endproc} \\ \mathit{process}\ Q[a, q_1, q_2] : \mathit{noexit} := & q_1; (q_2; \mathtt{stop}\ [\,]\ a; \mathtt{stop})\ \mathit{endproc} \end{array}$$

The labelled transition of B is given by figure 1. It clearly shows the interleavings of the actions p_1, p_2, q_1, q_2, and a single transition labelled with a. The bundle event structure corresponding to B is given in figure 2. It shows four bundles and a single symmetric conflict between a and q_2.

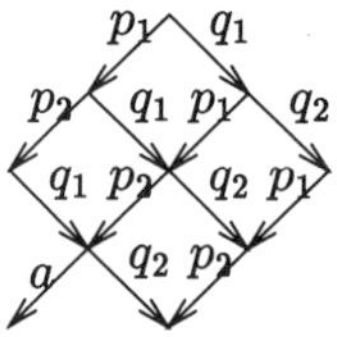

Figure 1. Transition system.

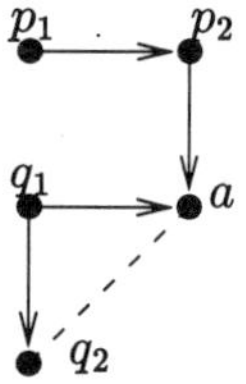

Figure 2. Bundle event structure.

We can build the transition system from the bundle event structure as follows: the events p_1 and q_1 do not depend on any other event and are not in conflict with any other event; therefore these events may happen in any order. We let these happen and obtain a new bundle event structure with only three events. Of these three events, the event p_2 does not depend on any other event, and the events a and q_2 are in mutual conflict. Again, we let p_2 happen and obtain a new bundle event structure with only two events a and q_2 which are in mutual conflict. This conflict however does reflect the branching of the system behaviour: the choice behaviour expression in process Q. Figure 3 gives one possible transition system derived this way. The next step is to combine the sequence of transitions p_1, q_1 and p_2 into one larger transition but somehow preserving the fact that p_1 must happen before p_2. This leads to a poset transition which is a transition labelled with a set of events together with a precedence relation which tells precisely which events must happen before which other events. The final transition system is depicted in figure 4. The arrow between p_1 and p_2 tells us that in this poset transition p_1 must occur before p_2. By combining transitions into larger (poset) transitions we have reduced the state space of the behaviour expression B from 6 states to 4 states. The reduced transition system contains in general less information than the original transition system; for instance, from the poset transition system it is not clear that the system has e.g. trace $q_1 q_2$.

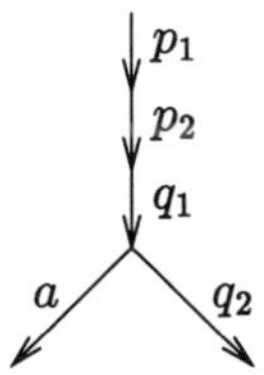

Figure 3. One of the transition systems.

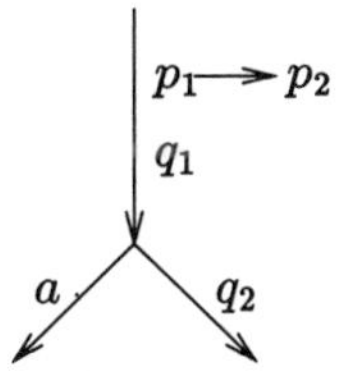

Figure 4. Poset transition system.

However, because the poset transition system contains all states corresponding to maximal traces, the poset transition does contain all deadlocking states. Furthermore, with the information that q_1 and q_2 are independent from p_1 and p_2 we can reconstruct the original transition system from the poset transition system.

The derivation of the poset transition system forms the basics of a tool called *SLICE* (Simulator for Lotos to study the Independence and Causality of Events).

Our approach bears a strong resemblance to the work of Valmeri[3] on *persistent sets*. A persistent set in a given state is the set of enabled transitions whose occurrences can not be affected by the evolution of the system by transitions outside the persistent set in that state. This strongly relates to our notion of *auto_init* set defined in the full paper. However, Valmeri uses the notion of "independent" transitions to factor out equivalence classes of sequences of transitions, whereas in our model such equivalence classes are avoided.

The full paper shows how to compute a transition system for basic LOTOS, provides some state space numbers and extends the transition system with data; it also gives some suggestions for further developments.

Bibliography

[1] R. Langerak. Event structures for the design and transformation in lotos. In *Fourth International Conference on Formal Description Techniques*, pages 271–287, 1991.

[2] R. Langerak. Bundle event structures. In M. Diaz and R. Groz, editors, *Fifth International Conference on Formal Description Techniques*. North-Holland, 1992.

[3] A. Valmari. Stubborn sets for reduced state space generation. In *Advances in Petri Nets*, volume 483 of *Lecture Notes in Computer Science*, pages 491–515. IEEE Computer Society Press, Springer-Verlag, 1991.

PART FIVE

Verification and Real-Time

16

Verifying ET-LOTOS programs with KRONOS *

C. Daws, A. Olivero and S. Yovine [a] † ‡

[a]VERIMAG, Miniparc-Zirst, Rue Lavoisier - 38330 Montbonnot St. Martin, France

ET-LOTOS is a timed extension of LOTOS proposed for modeling real-time systems. KRONOS is a tool that checks whether an automaton extended with clocks (called timed automaton) satisfies a real-time requirement expressed as a formula of the logic TCTL. This paper shows that real-time systems described in a reasonable subset of ET-LOTOS can be verified with KRONOS by compiling them into timed automata. We illustrate the practical interest of our approach with a case study: the Tick-Tock protocol.
Keywords: I.1 / II.2: ET-LOTOS, I.3, I.6 / III.1: Timed Automata / III.2: TCTL, I.7 / I.8: Tick-Tock protocol, I.10: Real-time

1. Introduction

The development of services and protocols requires timing mechanisms such as timeouts, watchdogs, delays and isochronous data transfers. In the last few years, many FDTs with constructs to model the above mechanisms have been proposed. These "timed" FDTs are mainly based on extensions of asynchronous process algebras, see [11, 7] for an overview. FDTs need supporting tools to allow automatic verification of requirements.

ET-LOTOS [8] is a timed extension of LOTOS that allows the modeling of real-time behaviors. KRONOS [12, 10] is a tool that checks whether a real-time system described by a timed automaton (automaton extended with clocks) satisfies a requirement expressed as a formula of the logic TCTL [1].

The need of verification methods and tools for ET-LOTOS has been stated in [8]. The main issue of this work is to show that real-time systems described in ET-LOTOS can be verified with the tool KRONOS by compiling them into timed automata. A similar approach has also been successfully applied to the algebra of timed processes ATP [12]. To illustrate the practical interest of our approach we apply it to a case study: the Tick-Tock protocol.

This paper is organized as follows. In section 2, we review the description of the Tick-Tock protocol. The ET-LOTOS specification of the protocol is given in section 3. We also review here the syntax and semantics of ET-LOTOS. Section 4 is devoted to timed automata. We illustrate the main ideas of the translation method of ET-LOTOS into timed automata with the example of the protocol. In section 5 we formalize the translation method. In section 6 we use the tool KRONOS to check that the ET-LOTOS description of the protocol satisfies the timing requirements.

---

*This work has been partially supported by ESPRIT-BRA REACT-P 6021.

†E-mail: {Conrado.Daws,Alfredo.Olivero,Sergio.Yovine}@imag.fr

‡VERIMAG is a joint laboratory of CNRS, INPG, Université Joseph Fourier, and Verilog SA, associated with IMAG.

2. A case study: the Tick-Tock protocol

The Tick-Tock protocol [9] is a case study for the assessment of timed FDTs. Although this protocol intends to be a realistic system, all details that are not relevant to timing requirements are simplified. However, it exhibits most of the essential timing constraints that need to be described by timed FDTs.

The Tick-Tock protocol is composed of three entities called *sender*, *receiver* and *service*. *Service* interacts with *sender* and *receiver* through their SAPs, respectively called Ss-SAP and Sr-SAP. Figure 1 shows a sketch of the system. To keep things simple, we restrict ourselves to the specification of *service*.

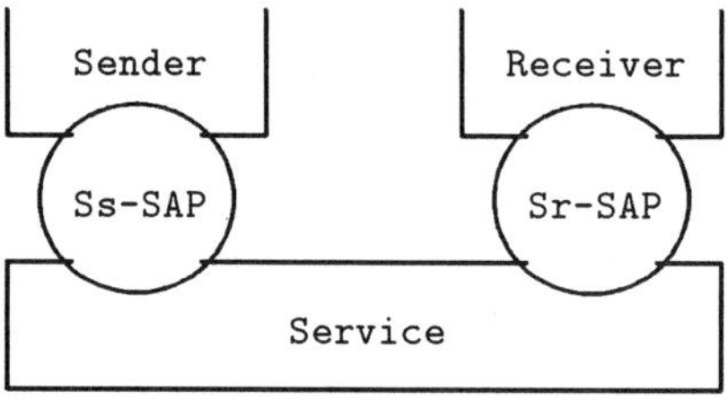

Figure 1. Sketch of the protocol.

Description of service

Service transmits data from *sender* to *receiver*. Exchanges through the corresponding SAPs are instantaneous and atomic, and carry a data *cell* as parameter. *Service* satisfies the following timing requirements.

Isochronism. The behavior of *service* is *isochronous*, that is, a cell from *sender* is only accepted at precise, punctual instants with a given period of π time units. *Sender* may neglect an opportunity of emission. Just one cell can be exchanged at any instant.

Transmission delays. *Service* always proposes a cell to *receiver* between τ_{min} and τ_{max} time units after its emission.

Spacing between deliveries. There is always a delay of at least α time units between two successive offers of cells at Sr-SAP.

Immediate acceptance. A cell offered by *service* to *receiver* must be immediately accepted by *receiver*, otherwise the *service* losses the cell immediately.

Loss-free transmission. The previous requirement describes the only way a cell received by *service* from *sender* can be lost: no cell is lost during its transit through *service*.

3. Specification in ET-LOTOS

ET-LOTOS [8] is an extension of LOTOS that allows the specification of timing requirements. In order to illustrate the meaning of the temporal operators of ET-LOTOS we first specify the behavior of *service*. We then recall the semantics of ET-LOTOS which is given in terms of labeled transition systems. The formal semantics of full ET-LOTOS, as well as its compatibility with respect to LOTOS, is given in [8].

3.1. Specification of Service

Our specification is basically the one given in [8] with slight changes. The specification of *service* is given in a constrained-oriented style, that is, each requirement is given as an ET-LOTOS process, and the overall behavior of the system corresponds to the parallel composition of all of them. Internal synchronizations are hidden using the `hide` operator.

Figure 2 shows the general structure of *service*. It is composed of four processes, namely, `Isochronous`, `TransDels`, `SpacingDeliveries` and `ImmAccept`.

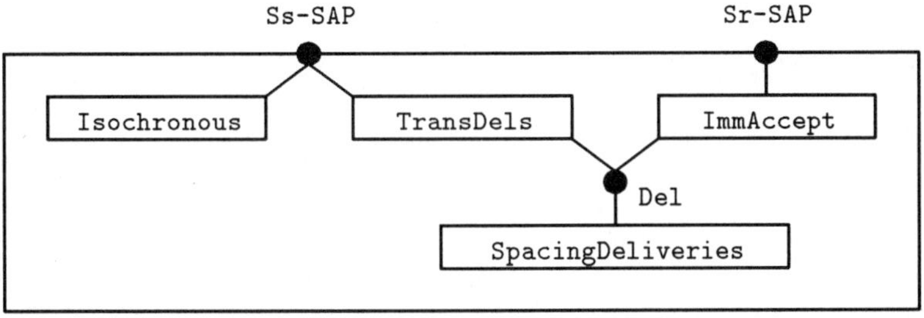

Figure 2. Structure of *service*.

Notice that `TransDels`, `SpacingDeliveries` and `ImmAccept` synchronize on the internal gate `Del`. The gate `Del` becomes enabled only when *service* is ready to deliver a cell to *receiver*. In this case, the cell must be transmitted immediately or lost.

Isochronism. The isochronous behavior of *service* is:

```
process Isochronous [Ss-SAP] : noexit :=
Ss-SAP{0}; Δ^π Isochronous
[]
Δ^π Isochronous
endproc
```

Process `Isochronous` is a choice between two possible behaviors, and this choice is resolved immediately. The action-prefix `Ss-SAP{0}` means that the offer `Ss-SAP` of *service* to *sender* has a zero duration, that is, either the *sender* is ready to interact at this punctual instant and the action occurs, or the offer is removed without executing the subsequent behavior. So, either the offer is accepted immediately, or the process behaves like `stop`. In the first case, a new occurrence of `Isochronous` is called after a delay π, which is expressed by the operator Δ^{π}. In the second case, a new communication through `Ss-SAP` will be offered after a delay of π units.

This example illustrates the two basic timed operators of ET-LOTOS, namely, the extended prefixing `g{t}`, called *life reducer*, and the *delay* operator Δ^t. The expression `g{t}` means that `g` is only offered during the interval `[0,t]`, that is, if `g` has not occurred after a delay `t`, the offer is removed and the process behaves like `stop`. It is important to note that the attribute `{t}` *does not force* `g` to be executed before `t`. A main assumption is that observable actions, i.e., actions different from `i`, may occur at any time: that is, if no time attribute `{t}` is associated with an offer `g`, the execution of `g` may be delayed without bound. The delay operator Δ^t expresses that the subsequent behavior will be delayed by `t`. We write `g{min,max}` as a shorthand notation for $\Delta^{\mathtt{min}}$ `g{max-min}`.

Transmission delays. This requirement is expressed by:

```
process TransDels [Ss-SAP,Del] :  noexit :=
Ss-SAP; i{τmin, τmax}; Del; stop
|||
TransDels [Ss-SAP, Del]
endproc
```

An important property of the action `i` is that it *must* occur within the interval associated with it. By default, when no time parameter is present, we assume that it is equal to 0, and we say that `i` is *urgent*. So, `TransDels` introduces a minimum delay τ_{min} after the reception of a cell through `Ss-SAP` and an additional and non-deterministic delay less than or equal to $\tau_{max} - \tau_{min}$. Since `i` must occur, the offer `Del` is enabled within the interval $[\tau_{min}, \tau_{max}]$. However, `Del` may not be executed immediately because of the minimal delay imposed between two consecutive deliveries. We will see later in section 6 that under some assumptions on the parameters, the `Del` will never occur after τ_{max}.

Spacing between deliveries. The following ET-LOTOS process ensures that two successive offers of `Del` are spaced out at least by α time units:

```
process SpacingDeliveries [Del] :  noexit :=
Del; Δ^α SpacingDeliveries
endproc
```

Immediate acceptance. This requirement is specified as follows:

```
process ImmAccept [Del,Sr-SAP] : noexit :=
Del;
( Sr-SAP{0}; ImmAccept[Del,Sr-SAP]
[]
ImmAccept[Del,Sr-SAP] )
endproc
```

The behavior of `ImmAccept` is as follows: either it outputs the cell immediately through `Sr-SAP`, or if the output is not possible, it looses the cell and waits for the next one.

Service. The above processes synchronize on the internal gate `Del`. As `Del` is an internal action, it must be hidden. Process `Service` is defined as follows:

```
process Service [Ss-SAP] : noexit :=
```

```
( Isochronous [Ss-SAP]
|[ Ss-SAP ]|
( hide Del in
( TransDels [Ss-SAP,Del]
|[ Del ]|
SpacingDeliveries [Del]
|[ Del ]|
ImmAccept [Del,Sr-SAP] ) ) )
endproc
```

In ET-LOTOS, hidden actions must occur as soon as they become enabled. This behavior is ensured by the semantics of the `hide` operator. However, notice that even if `Del` is forced to occur as soon as possible, it is not clear whether it will always happen within the required interval of time. In section 6 we use the tool KRONOS to prove that this requirement is fulfilled.

3.2. Semantics of ET-LOTOS

The formal meaning of ET-LOTOS is given in terms of labeled transition systems. There are two kind of transitions, namely discrete and timed. A discrete transition corresponds to the execution of an action, and is therefore labeled with the name of the action. $P \xrightarrow{a} P'$ means that P may engage in action a and then behave like P'. Timed transitions correspond to time elapses. The set of possible timed values is called the *time domain.* If D is the time domain, $+$ the addition in D, 0 the neutral element of D, then $(D, +, 0)$ is a commutative monoid, $(D, \leq)$ is a total order and ∞ is the absorbent element for $+$. D can be discrete or dense. Hereafter we assume that D is the set of rational numbers. Now, $P \xrightarrow{d} P'$ means that P may idle during a time d and then behave like P'.

Table 1 shows the formal operational semantics of a subset of ET-LOTOS. We restrict ourselves to the subset containing all the constructs used in the example modeled above. The semantics of full ET-LOTOS is given in [8]. The semantics is given as a set of axioms and inference rules, with $d \in D - \{0\}$, $d_1 \in D$, $a \in G \cup \{\delta, \mathtt{i}\}$ and $\Gamma \subseteq G$, where G is the set of observable actions. We have also $a \neq \delta$ in (AP1), (AP2), (AP3). The left and right columns correspond to discrete and timed transitions, respectively.

4. Extended automata with clocks

In this section we model timing requirements by automata extended with a set of *clocks.* Clocks are variables that count the time elapsed between events. These automata are called *timed automata.* Many different definitions of timed automata have been given, see for instance [1, 12, 13]. Here we adopt the definition given in [10].

A timed automaton has a finite set of clocks. The values of the clocks increase uniformly with time. Timing constraints are expressed by associating enabling conditions with transitions. A transition is enabled, that is, may be executed, if the condition is satisfied by the current values of the clocks. A clock can be reset with any transition. At any instant the value of a clock is equal to the time elapsed since the last time it was reset.

The semantics of timed automata and ET-LOTOS is given in terms of the same class of labeled transition systems. Besides, the parallel composition operator of ET-LOTOS can also be defined on timed automata.

Table 1
Operational semantics of ET-LOTOS.

		(S)	$\mathtt{stop} \xrightarrow{d} \mathtt{stop}$
(AP1)	$a\{d_1\}; P \xrightarrow{a} P$	(AP2)	$a\{d_1 + d\}; P \xrightarrow{d} a\{d_1\}; P$
		(AP3)	$a\{d_1\}; P \xrightarrow{d} \mathtt{stop} \quad (a \neq i, d > d_1)$
(D1)	$\dfrac{P \xrightarrow{a} P'}{\Delta^0 P \xrightarrow{a} P}$	(D2)	$\Delta^{d_1+d} P \xrightarrow{d} \Delta^{d_1} P$
		(D3)	$\dfrac{P \xrightarrow{d} P'}{\Delta^0 P \xrightarrow{d} P'}$
(CH1)	$\dfrac{P \xrightarrow{a} P'}{P[]Q \xrightarrow{a} P'}$	(CH2)	$\dfrac{P \xrightarrow{d} P', Q \xrightarrow{d} Q'}{P[]Q \xrightarrow{d} P'[]Q'}$
	(+CH1')		
(P1)	$\dfrac{P \xrightarrow{a} P'}{P\vert[\Gamma]\vert Q \xrightarrow{a} P'\vert[\Gamma]\vert Q} \quad (a \notin \Gamma \cup \{\delta\})$	(P3)	$\dfrac{P \xrightarrow{d} P', Q \xrightarrow{d} Q'}{P\vert[\Gamma]\vert Q \xrightarrow{d} P'\vert[\Gamma]\vert Q'}$
	(+P1')		
(P2)	$\dfrac{P \xrightarrow{a} P', Q \xrightarrow{a} Q'}{P\vert[\Gamma]\vert Q \xrightarrow{a} P'\vert[\Gamma]\vert Q'} \quad (a \in \Gamma \cup \{\delta\})$		
(H1)	$\dfrac{P \xrightarrow{a} P'}{\mathtt{hide}\Gamma\mathtt{in}P \xrightarrow{a} \mathtt{hide}\Gamma\mathtt{in}P'} \quad (a \notin \Gamma)$	(H3)	$\dfrac{P \xrightarrow{d} P', \forall a \in \Gamma . P \not\xrightarrow{a}}{\mathtt{hide}\Gamma\mathtt{in}P \xrightarrow{d} \mathtt{hide}\Gamma\mathtt{in}P'}$
(H2)	$\dfrac{P \xrightarrow{a} P'}{\mathtt{hide}\Gamma\mathtt{in}P \xrightarrow{i} \mathtt{hide}\Gamma\mathtt{in}P'} \quad (a \in \Gamma)$		
(In1)	$\dfrac{[g/h]P \xrightarrow{a} P', Q[h] := P}{Q[g] \xrightarrow{a} P'}$	(In2)	$\dfrac{[g/h]P \xrightarrow{d} P', Q[h] := P}{Q[g] \xrightarrow{d} P'}$

4.1. Specification of Service

We specify the behavior of *service* to illustrate the syntax and semantics of timed automata. For each ET-LOTOS process we show a timed automaton that describes the same behavior. *Service* is modeled as the parallel composition of the following timed automata.

Isochronism. The timed automaton of process `Isochronous` is shown in figure 3. This automaton has two control *locations*, namely 0 and 1, and one clock, namely x. The initial location is 0 and the initial value of x is 0, which is denoted by the small arrow. At location 0, the automaton behaves like follows: either it takes the transition `Ss-SAP` when the value of x is equal to 0, or it waits until the value of x is π to reset x to 0 in order to allow a new offer of `Ss-SAP`. The label Δ means that the transition corresponds to the delay operator of ET-LOTOS. At location 1, the automaton waits π time units and then moves to location 0.

In both locations the value of x *cannot increase beyond* π, which forces the transitions to be taken when the value of x reaches π. Consequently, the time spent in one visit at each location is less or equal than π. This behavior is ensured by associating with each location a predicate called the *time-can-progress* predicate, noted **tcp**, that determines the amount of time the automaton may stay at the location. The **tcp** predicate is formally defined in section 4.2.

Transmission delays. Recall the definition of process `TransDels` given before. Clearly, this process cannot be represented by a finite automaton since the use of recursion through the parallel operator generates an unbounded number of instances. Instead, we consider the following definition of process `TransDels`:

```
process TransDels [Ss-SAP,Del] :  noexit :=
Ss-SAP; i{τmin,τmax}; Del; TransDels [Ss-SAP, Del]
endproc
```

Remark 4.1 *Notice that in this case,* service *has no capacity to buffer cells and it can only accept one cell at a time. If buffering of cells is required, we can put in parallel as many of these processes as the size of the buffer.*

Figure 4 shows the timed automaton corresponding to process `TransDels`. Location 0 is the initial location with z equal to 0. The system may rest at locations 0 and 2 a time without bound, but it must leave location 1 at some time within the interval $[\tau_{min}, \tau_{max}]$. We assume that the **tcp** predicate ensures that the value of z cannot increase beyond τ_{max} at location 1.

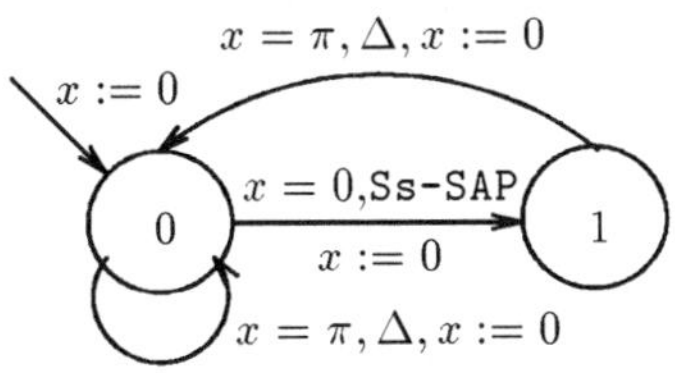

Figure 3. Isochronism.

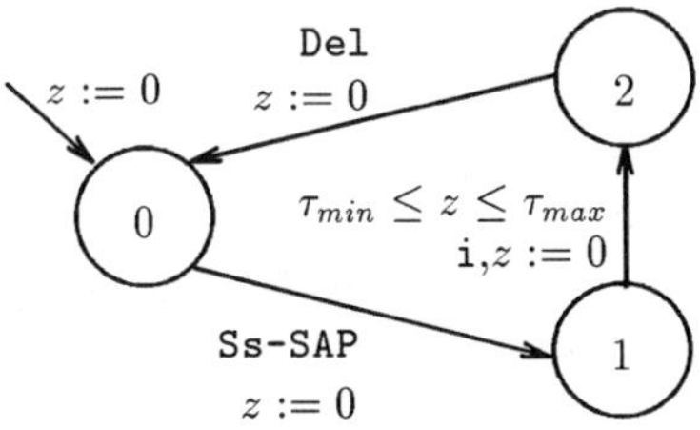

Figure 4. Transmission delays.

Spacing between deliveries. The timed automaton of figure 5 describes the same behavior than the ET-LOTOS process `SpacingDeliveries`. At location 0, the system may take the transition `Del` at any time. Location 1 models the delay of α required between two consecutive offers of `Del`. At this location, the **tcp** predicate ensures that the value of y cannot increase beyond α, so the transition must be taken when the value of y reaches α. The Δ-transition models the expiring delay.

Immediate acceptance. Figure 6 shows the timed automaton of `ImmAccept`. After a transition `Del`, the value of w is reset to 0 and a `Sr-SAP` is offered. Since the enabling condition of the transition `Sr-SAP` is $w = 0$, either it is taken immediately, or it becomes disabled and the opportunity is lost until the next `Del`. The `tcp` predicate is true in both locations.

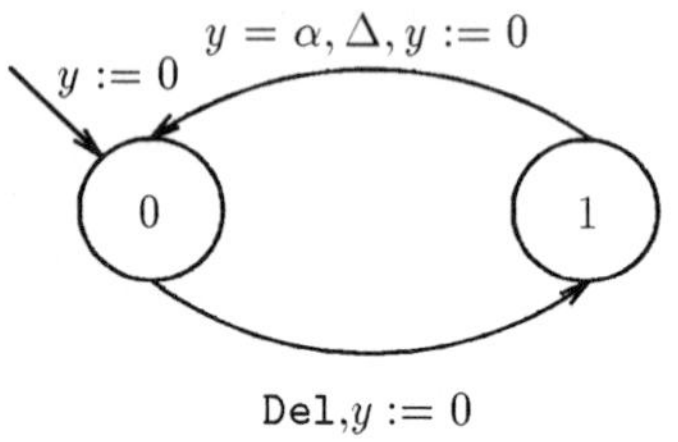

Figure 5. Spacing between deliveries.

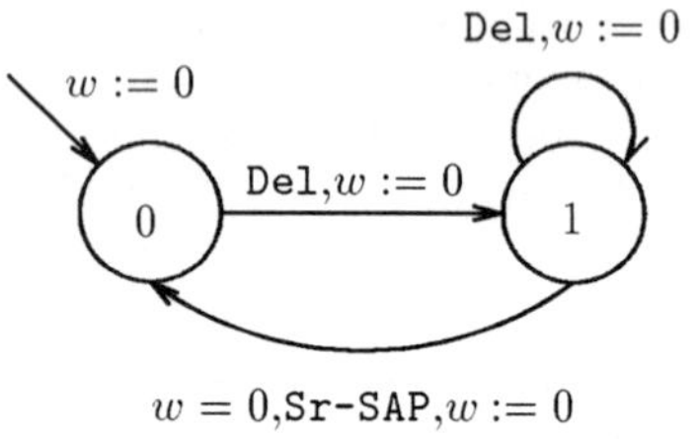

Figure 6. Immediate acceptance.

4.2. Operational semantics

Formally, a timed automaton is a tuple $T = \langle S, X, E, s^0, \mathtt{tcp}\rangle$ where:

1. S is a finite set of vertices called locations.
2. X is a finite set of variables called clocks. A valuation v of clocks is a function that assigns a value $v(x) \in D$ to each clock $x \in X$. V denotes the set of valuations.

   A *state* q is a pair (s, v) consisting of a location $s \in S$ and a valuation $v \in V$. We write Q for the set of states.
3. E is a finite set of edges called transitions. Each transition $e = (s, l, \psi, \rho, s')$ consists of a source location $s \in S$, a target location $s' \in S$, a label l, a guard ψ, and a set $\rho \subseteq X$ of clocks to be reset to zero simultaneously with the transition.

   A guard is any boolean combination of atoms of the form $x\#c$ where $x \in X$, $c \in \mathsf{N}$ and $\#$ is any binary relation in the set $\{<, \leq, >, \geq, =\}$.

   The transition e is *enabled* at a state (s, v) if v satisfies the guard ψ. We write $v[\rho := 0]$ for the valuation v' such that $v'(x) = 0$ for $x \in \rho$ and $v'(x) = v(x)$ otherwise. The state $(s', v[\rho := 0])$ is the discrete successor of state (s, v) by transition e.
4. s^0 is the initial location.
5. For each $s \in S$, `tcp` is the *time-can-progress* predicate that determines for each state (s, v) and time value $d \in D$, if the system may let time progress by an amount of d at location s. In this case, the values of the clocks increase by d.

   We write $v + d$ for the valuation v' such that $v'(x) = v(x) + d$ for all clocks $x \in X$. The state $(s, v + d)$ is the timed successor of state (s, v).

The operational semantics of a timed automaton is a labeled transition system where the set of states is $\mathcal{Q}$ and the transition relation $\rightarrow$ is defined by the rules in table 2.

Table 2
Operational semantics of timed automata.

(T1) $\dfrac{e = \langle s, l, \psi, \rho, s' \rangle \in E,\ \psi(v)}{(s, v) \xrightarrow{l} (s', v[\rho := 0])}$	(T2) $\dfrac{\mathsf{tcp}(s)(v, d), d \in D - \{0\}}{(s, v) \xrightarrow{d} (s, v + d)}$

Note For instance, the tcp predicate of location 0 of the timed automaton of figure 3 is $\mathsf{tcp}(0)(v, d) \equiv v(x) + d \leq \pi$. From this definition and rule (T2) above it is easy to see that the amount of time the automaton may stay at location 0 is less than or equal to π. For a detailed description of the tcp predicate see [10].

5. The principle of the translation method

The example above suggests a general method for translating ET-LOTOS specifications into timed automata. In this section we show the principle of this method for the subset of ET-LOTOS considered in section 3. Our translation method is based on the one developed for the timed process algebra ATP presented in [12, 13]. We use the label Δ to represent transitions corresponding to expiring events of the delay operator, l denotes an element of $G \cup \{\delta, \mathtt{i}, \Delta\}$.

Stop. The timed automaton corresponding to the process `stop` is

$$\langle \{s\}, \{x\}, \emptyset, s, \mathsf{tcp} \rangle$$

where $\mathsf{tcp}(s)$ is the predicate *true*, that is, $\mathsf{tcp}(s)(v, d)$ is true for all valuations v of x and for all time values $d \in D$. This means that `stop` is a process that only lets time pass (rule (S) in table 1).

Prefixing. Let P be the term $a\{d_1\}; P_1$ and $T_1 = \langle S_1, X_1, E_1, s_1^0, \mathsf{tcp}_1 \rangle$ be the timed automaton of P_1. The timed automaton of P is

$$\langle S_1 \cup \{s\}, X_1 \cup \{x\}, E_1 \cup \{(s, a, x \leq d_1, X_1, s_1^0)\}, s, \mathsf{tcp} \rangle$$

where the initial location $s \notin S_1$ is a new one, and $x \notin X_1$ is a new clock, and the new transition represents that the gate a is enabled for at most d_1 time units (rules (AP1) and (AP2) in table 1).

For all locations $s_1 \in S_1$, $\mathsf{tcp}(s_1)$ is like $\mathsf{tcp}_1(s_1)$. For the initial location $s \in S$, $\mathsf{tcp}(s)$ depends on whether a is an observable action or $a = \mathtt{i}$. If $a \neq \mathtt{i}$, then $\mathsf{tcp}(s)$ is *true* which means that time can progress without bound at the initial location s (rule (AP3) in table 1). If $a = \mathtt{i}$, then $\mathsf{tcp}(s)(v, d)$ is true iff $v(x) + d \leq d_1$, which means that the value of x cannot increase beyond d_1, expressing the necessity for the transition to be taken within the interval $[0, d_1]$.

Delay. Let P be the term $\Delta^{d_1} P_1$ and $T_1 = \langle S_1, X_1, E_1, s_1^0, \mathsf{tcp}_1 \rangle$ be the timed automaton of P_1. The timed automaton of P is

$$\langle S_1 \cup \{s\}, X_1 \cup \{x\}, E_1 \cup \{(s, \Delta, x = d_1, X_1, s_1^0)\}, s, \mathsf{tcp} \rangle$$

where the initial location $s \notin S_1$ is a new one, $x \notin X_1$ is a new clock, and the new transition represents that the system moves to the initial state of T_1 after idling d_1 time units at the initial location s. For all locations $s_1 \in S_1$, $\mathsf{tcp}(s_1)$ is $\mathsf{tcp}_1(s_1)$. For the initial location s, $\mathsf{tcp}(s)(v, d)$ is true iff $v(x) + d \leq d_1$, which means that the value of x cannot increase beyond d_1, forcing the transition Δ to be executed when the value of x is equal to d_1.

Nondeterministic choice. Let P be the term $P_1[]P_2$ and $T_i = \langle S_i, X_i, E_i, s_i^0, \mathsf{tcp}_i \rangle$ be the timed automaton of P_i, for $i \in \{1, 2\}$. The timed automaton of P is

$$\langle S_1 \cup S_2 \cup S_1 \times S_2, X_1 \cup X_2, E \cup E_1 \cup E_2, (s_1^0, s_2^0), \mathsf{tcp} \rangle$$

where the set of transitions E is

$$\begin{aligned} E \;=\; & \{((s_1, s_2), a, \psi, \rho, s_1') \mid (s_1, a, \psi, \rho, s_1') \in E_1\} & (1)\\ & \cup \{((s_1, s_2), a, \psi, \rho, s_2') \mid (s_2, a, \psi, \rho, s_2') \in E_2\} & (2)\\ & \cup \{((s_1, s_2), \Delta, \psi, \rho, (s_1', s_2)) \mid (s_1, \Delta, \psi, \rho, s_1') \in E_1\} & (3)\\ & \cup \{((s_1, s_2), \Delta, \psi, \rho, (s_1, s_2')) \mid (s_2, \Delta, \psi, \rho, s_2') \in E_2\} & (4) \end{aligned}$$

(1) and (2) express that a choice is resolved by the execution of an action (rules (CH1) and (CH1') in table 1), and (3) and (4) express that the expiration of a delay does not resolve a choice (rule (CH2) in table 1). For all locations $s_i \in S_i$, $i = 1, 2$, $\mathsf{tcp}(s_i)$ is $\mathsf{tcp}_i(s_i)$. For all locations $(s_1, s_2) \in S_1 \times S_2$, $\mathsf{tcp}(s_1, s_2)$ is $\mathsf{tcp}_1(s_1) \wedge \mathsf{tcp}_2(s_2)$ which means that time passes at the same pace in both processes (rule (CH2) in table 1).

Parallel composition. Let P be the term $P_1|[\Gamma]|P_2$ and $T_i = \langle S_i, X_i, E_i, s_i^0, \mathsf{tcp}_i \rangle$ be the timed automaton of P_i, for $i \in \{1, 2\}$. The timed automaton of P is

$$\langle S_1 \times S_2, X_1 \cup X_2, E, (s_1^0, s_2^0), \mathsf{tcp} \rangle$$

where the set of transitions E is

$$\begin{aligned} E \;=\; & \{((s_1, s_2), l, \psi, \rho, (s_1', s_2)) \mid (s_1, l, \psi, \rho, s_1') \in E_1, l \notin \Gamma \cup \{\delta\}\} & (5)\\ & \cup \{((s_1, s_2), l, \psi, \rho, (s_1, s_2')) \mid (s_2, l, \psi, \rho, s_2') \in E_2, l \notin \Gamma \cup \{\delta\}\} & (6)\\ & \cup \{((s_1, s_2), l, \psi_1 \wedge \psi_2, \rho_1 \cup \rho_2, (s_1', s_2')) \mid (s_i, l, \psi_i, \rho_i, s_i') \in E_i, i = 1, 2, l \in \Gamma \cup \{\delta\}\} & (7) \end{aligned}$$

(5), (6) and (7) correspond to rules (P1), (P1') and (P2) in table 1, respectively. For all locations $(s_1, s_2) \in S_1 \times S_2$, $\mathsf{tcp}(s_1, s_2)$ is $\mathsf{tcp}_1(s_1) \wedge \mathsf{tcp}_2(s_2)$ which means that time passes at the same pace in both processes (rule (P3) in table 1).

Hide. Let P be the term $\texttt{hide}\Gamma\texttt{in}P_1$ and $T_1 = \langle S_1, X_1, E_1, s_1^0, \mathsf{tcp}_1\rangle$ be the timed automaton of P_1. The timed automaton of P is

$$\langle S_1, X_1, E, s_1^0, \mathsf{tcp}\rangle$$

where the set of transitions E is

$$E = \{(s_1, l, \psi, \rho, s_1') \mid (s_1, l, \psi, \rho, s_1') \in E_1, l \notin \Gamma\} \quad (8)$$
$$\cup\{(s_1, \texttt{i}, \psi, \rho, s_1') \mid (s_1, l, \psi, \rho, s_1') \in E_1, l \in \Gamma\} \quad (9)$$

(8) and (9) correspond to rules (H1) and (H2) in table 1. For all locations $s \in S_1$, $\mathsf{tcp}(s)(v,d)$ is the predicate $\mathsf{tcp}_1(s)(v,d) \wedge \forall l \in \Gamma.\neg\mathsf{enable}(s,l)(v)$ where $\mathsf{enable}(s,l)(v)$ is true if an outgoing transition from s with label l is enabled for the valuation v. This definition of $\mathsf{tcp}(s)$ captures exactly the meaning of rule (H3) in table 1.

The syntax-driven construction presented in this section introduces Δ-transitions to represent expirations of delays. These transitions, which are also used in the translation method of ATP [12, 13], play a role similar to the ϵ-transitions used in [4] and are introduced in order to obtain a compositional method of construction. In many cases these transitions may be eliminated by propagating the corresponding timing constraints. The ET-LOTOS program and the automaton have the same behaviors. The proof of correctness is similar to the one given for ATP in [14]. It relies on the definition of an equivalence relation between transition systems that appropriately handles Δ-transitions introduced by the translation method. To keep the presentation simple we do not deal with recursion here. For a formal treatment of recursion see for instance [4, 14].

It is important to note that the complexity of the algorithm is independent of the values of the parameters of the timing constructs of ET-LOTOS.

6. Verifying timing requirements with KRONOS

We address here the problem of verifying whether a given ET-LOTOS specification satisfies a requirement that is expressed in the real-time temporal logic TCTL [1]. In order to do so, we first apply the translation method proposed in section 5 to obtain the timed automaton corresponding to the ET-LOTOS process. Then we use the model-checker KRONOS [12, 10] which checks the validity of the formula on the timed automaton.

6.1. The logic TCTL

The real-time temporal logic TCTL extends the temporal operators $\exists\mathcal{U}$ (*there exists a run*) and $\forall\mathcal{U}$ (*for all runs*) of CTL [3] with timing constraints that allow "quantitative" temporal reasoning. Let $\mathcal{P}$ be a set of basic predicates. The formulas of TCTL are defined by the following grammar:

$$\varphi \ ::= \ p \mid \neg\varphi \mid \varphi_1 \vee \varphi_2 \mid \varphi_1 \exists\mathcal{U}_I \varphi_2 \mid \varphi_1 \forall\mathcal{U}_I \varphi_2$$

where $p \in \mathcal{P}$ is a basic predicate and I is an interval with positive integer end-points (I may be open, closed, bounded or unbounded).

Intuitively, $\varphi_1 \exists\mathcal{U}_I \varphi_2$ means that there exists a run of the system with a finite prefix such that φ_2 holds in the last state at a time t within the interval I and φ_1 continuously holds in all the previous states. $\varphi_1 \forall\mathcal{U}_I \varphi_2$ means that for all runs the above property holds.

For example, we can write $true\exists\mathcal{U}_{<5}p$ to say that along some run, proposition p becomes true before 5 time units.

We will use the typical abbreviations such as $\forall\Diamond_I\varphi$ for $true\forall\mathcal{U}_I\varphi$, $\exists\Diamond_I\varphi$ for $true\exists\mathcal{U}_I\varphi$, $\exists\Box_I\varphi$ for $\neg\forall\Diamond_I\neg\varphi$ and $\forall\Box_I\varphi$ for $\neg\exists\Diamond_I\neg\varphi$. The unrestricted temporal operators of CTL correspond to TCTL operators subscripted by ≥ 0.

The set $\mathcal{P}$ of basic predicates is defined by the following grammar:

$$p \;::=\; \mathsf{init} \mid \mathsf{enable}(a) \mid \mathsf{after}(a) \mid x \in I$$

where $x \in X$ is a clock of the timed automaton we want to verify, **init** defines the initial state (i.e. the initial location with all clocks set to 0) and for any action a, **enable**(a) defines the set of states having an enabled transition labeled a and **after**(a) defines the set of states reached by a transition labeled a.

6.2. The tool KRONOS

Given a timed automaton T and a TCTL-formula φ, the model-checker KRONOS computes the set of states of the labeled transition system of T that satisfies the formula φ. This set is called the characteristic set of φ and it is denoted $[\![\varphi]\!]$. KRONOS implements the symbolic model-checking algorithm developed in [5]. This algorithm is based on fixpoint characterizations of the operators $\exists\mathcal{U}$ and $\forall\mathcal{U}$ in terms of a binary next operator $\triangleright$. In [5] it is shown that the fixpoints can be computed iteratively and the procedure always terminates. Here, we show how KRONOS iteratively computes the characteristic set of some of the TCTL-formulas.

Reachability. The set of states from where it is possible to reach a state satisfying the formula φ within the interval I is expressed in TCTL by the formula $\exists\Diamond_I\varphi$. The characteristic set of this formula is iteratively computed as $\bigcup_i Y_i[z := 0]$ with

$$\begin{aligned} Y_0 &= [\![\varphi \wedge z \in I]\!] \\ Y_{i+1} &= Y_i \cup (true \triangleright Y_i) \end{aligned}$$

The variable z is a new clock added to count the time elapsed since the beginning of the run. So, the algorithm starts with the set of states that satisfy $\varphi \wedge z \in I$ and computes the predecessors and so on. At the end, the value of z is reset to 0.

Invariance. To verify if the formula φ is an invariant of the system we check if the initial state is contained in the characteristic set of the formula $\forall\Box\varphi$. This characteristic set can be iteratively computed as $\bigcap_i Y_i$ with

$$\begin{aligned} Y_0 &= [\![\varphi]\!] \\ Y_{i+1} &= Y_i \cap \neg(true \triangleright \neg Y_i) \end{aligned}$$

Bounded-time response. The real-time response property asserting that a given event must occur within a certain time bound c is expressed in TCTL by a formula of the form $\forall\Diamond_{\leq c}\varphi$. It is shown in [5] that this formula is equivalent to $\neg(\neg\varphi\exists\mathcal{U}_{>c}true)$, that is, φ must occur within c time units if there exists no run for which the negation of φ continuously holds for a time greater than c. The characteristic set can be iteratively computed as $\neg \bigcup_i Y_i[z := 0]$ with

$$\begin{aligned} Y_0 &= [\![z > c]\!] \\ Y_{i+1} &= Y_i \cup ([\![\neg\varphi]\!] \rhd Y_i) \end{aligned}$$

where the variable z is a new clock added to count the time elapsed since the beginning of the run.

6.3. Requirements of Service

Isochronism. This requirement states that offers at Ss-SAP occur at punctual instants with a period π. Isochronism is specified by a set of formulas. The first formula

$$\mathsf{init} \Rightarrow \forall\Box(\mathsf{enable}(\texttt{Ss-SAP}) \Rightarrow \forall\Diamond_{=\pi}\mathsf{enable}(\texttt{Ss-SAP})) \tag{10}$$

states that always an offer at Ss-SAP is inevitably followed by another offer exactly π time units later. The second formula

$$\mathsf{init} \Rightarrow \forall\Box(\mathsf{enable}(\texttt{Ss-SAP}) \Rightarrow \neg\exists\Diamond_{(0,\pi)}\mathsf{enable}(\texttt{Ss-SAP})) \tag{11}$$

states that an offer at Ss-SAP will never follow the previous one before a time π. The third formula

$$\neg(\mathsf{enable}(\texttt{Ss-SAP})\exists\mathcal{U}_{>0}\mathsf{enable}(\texttt{Ss-SAP})) \tag{12}$$

says that there exists no run where enable(Ss-SAP) holds for some time greater than zero, or equivalently, offers at Ss-SAP have zero duration.

Transmission delays. This requirement is specified by the following formula:

$$\mathsf{init} \Rightarrow \forall\Box(\mathsf{after}(\texttt{Ss-SAP}) \Rightarrow \forall\Diamond_{[\tau_{min},\tau_{max}]}\mathsf{enable}(\texttt{Sr-SAP})) \tag{13}$$

that is, always an offer at Sr-SAP must occur within the interval $[\tau_{min}, \tau_{max}]$ after its reception from *sender*.

Spacing between deliveries. There is always a delay of at least α time units between two successive offers of cells at Sr-SAP. The formula

$$\mathsf{init} \Rightarrow \forall\Box(\mathsf{enable}(\texttt{Sr-SAP}) \Rightarrow \neg\exists\Diamond_{(0,\alpha)}\mathsf{enable}(\texttt{Sr-SAP})) \tag{14}$$

states that there exists no run starting at a state satisfying after(Sr-SAP) where an offer at Sr-SAP is enabled before a time α.

Immediate acceptance. A cell offered by *service* to *receiver* must be immediately accepted or lost. The formula

$$\neg(\mathsf{enable}(\texttt{Sr-SAP})\exists\mathcal{U}_{>0}\mathsf{enable}(\texttt{Sr-SAP})) \tag{15}$$

says that offers at Sr-SAP have zero duration.

Table 3
Sizes.

buffer size	trans	states	clocks
1	64	24	5
2	240	72	6
3	864	216	7

Table 4
Results for a buffer of size 1.

	A			B			C			D		
	$\tau_{min} = 50$ $\alpha = 90$			$\tau_{min} = 10500$ $\alpha = 12000$			$\tau_{min} = 13$ $\alpha = 1050$			$\tau_{min} = 75$ $\alpha = 50$		
	$\tau_{max} = 80$ $\pi = 100$			$\tau_{max} = 13666$ $\pi = 15000$			$\tau_{max} = 813$ $\pi = 1000$			$\tau_{max} = 120$ $\pi = 100$		
	iter	time	eval	iter	time	eval	iter	time	eval	iter	time	eval
10	8	3.61 s	`true`	8	3.66 s	`true`	36	5.83 s	`false`	16	6.75 s	`false`
11	7	0.53 s	`true`	7	0.51 s	`true`	6	0.33 s	`true`	7	0.63 s	`true`
12	1	0.03 s	`true`	1	0.03 s	`true`	1	0.05 s	`true`	1	0.06 s	`true`
13	12	4.46 s	`true`	8	6.05 s	`true`	19	4.96 s	`false`	9	9.86 s	`true`
14	7	1.81 s	`true`	7	1.56 s	`true`	8	2.25 s	`true`	6	1.56 s	`true`
15	1	0.15 s	`true`	1	0.16 s	`true`	1	0.13 s	`true`	1	0.16 s	`true`

We show here the results obtained with the tool KRONOS to verify whether the ET-LOTOS specification of *service* given in section 3 satisfies the TCTL-formulas given above.

Table 3 shows the number of states, transitions and clocks of the timed automata of service for different sizes of the buffer (see remark 4.1). Tables 4, 5, 6 show the results obtained with KRONOS for timed automata with buffer size equal to 1,2 or 3 respectively. Column "iter" gives the number of iterations required for computing the characteristic set of each formula using the algorithms sketched in section 6.2, column "time" gives the running times measured in seconds (on a 16 MB SUN 4 Sparc Station) and column "eval" shows the result of the evaluation.

We have evaluated the formulas for different values of $\tau_{min}, \tau_{max}, \alpha$ and π. Notice that

Table 5
Results for a buffer of size 2.

	E			F			G		
	$\tau_{min} = 75$ $\alpha = 50$			$\tau_{min} = 50$ $\alpha = 150$			$\tau_{min} = 75$ $\alpha = 50$		
	$\tau_{max} = 120$ $\pi = 100$			$\tau_{max} = 75$ $\pi = 100$			$\tau_{max} = 220$ $\pi = 100$		
	iter	time	eval	iter	time	eval	iter	time	eval
10	15	527.45 s	`true`	33	45.70 s	`false`	22	69.15 s	`false`
11	9	6.93 s	`true`	7	2.60 s	`true`	9	3.15 s	`true`
12	1	0.20 s	`true`	1	0.21 s	`true`	1	0.23 s	`true`
13	17	664.53 s	`true`	19	92.06 s	`false`	13	149.83 s	`true`
14	7	16.98 s	`true`	8	16.38 s	`true`	7	10.20 s	`true`
15	1	0.41 s	`true`	1	0.41 s	`true`	1	0.41 s	`true`

Table 6
Results for a buffer of size 3.

	H		
	$\tau_{min} = 75$	$\alpha = 50$	
	$\tau_{max} = 220$	$\pi = 100$	
	iter	time	eval
10	23	1278.71 s	`true`
11	10	30.35 s	`true`
12	1	2.40 s	`true`
13	13	2582.61 s	`true`
14	8	78.48 s	`true`
15	1	3.51 s	`true`

the values of the parameters affect the validity of the formulas:

- Let b be the size of the buffer. We can see that when $b.\pi < \tau_{max}$, like for instance in columns D and G, the formula 10 is not satisfied. For these values, *service* may not be ready to accept a new cell at `Ss-SAP` because the buffer is full. The formula is verified if we increment the size of the buffer such that $b.\pi \geq \tau_{max}$, like in columns E and H.

- When $\alpha > \pi$, like in columns C and F, the requirement *transmission delay* (formula 13) is not verified. Consider two consecutive emissions separated by π. In the worst case, the first one is delivered in τ_{max}. Then *service* must deliver the second one after $\tau_{max} + \alpha$ to satisfy the spacing between deliveries but before $\pi + \tau_{max}$ as specified by the requirement *transmission delay*. But this is impossible if $\alpha > \pi$. Notice that formula 10 is false whenever formula 13 is false, because the buffer may become full in this case.

Also notice that formulas 10 and formula 13 are harder to verify than the others. Besides, changes of the values of the parameters do not significantly change the running times, except when these changes affect the truth value of the formula.

7. Conclusion and further work

This paper shows that real-time systems described in a reasonable subset of ET-LOTOS can be verified with the tool KRONOS by compiling them into timed automata. It is important to note that the complexity of the translation method and the size of the constructed automaton is independent of the values of the time parameters of the timing constructs of the language. We have also shown that the model-checker KRONOS allows the verification of real-time systems with arbitrarily large time delays.

The practical application of our approach is illustrated with a case study: the *service* entity of the Tick-Tock protocol, specially designed for the assessment of real-time FDTs and tools. We are working on the verification of the complete Tick-Tock protocol described in [9].

In order to keep things simple, we have restricted ourselves to a subset of ET-LOTOS. Nevertheless, our method can be extended to full ET-LOTOS. An efficient compiler is envisaged by combining the techniques used by the tool CÆSAR [4] and the ATP compiler [12]. The main problem here is how to deal with the `@t` ("at t") operator. This operator declares a time variable `t` used to store the time of occurrence of an action. The variable `t` can be considered as an *integrator* [6, 2], i.e., a clock that can be stopped. The problem is that unrestricted use of integrators leads to undecidability of verification.

Finally, we are currently implementing optimizations to improve on the tool KRONOS in order to be able to cope with larger systems.

Acknowledgements We are grateful to Joseph Sifakis for the helpfull discussions.

REFERENCES

1. R. Alur, C. Courcoubetis, and D. Dill. Model-checking for real-time systems. In *Proc. 5th LICS*, pages 414–425. IEEE, 1990.
2. A. Bouajjani, R. Echahed, and J. Sifakis. On model-checking for real-time properties with durations. In *8th. Symp. on Logic in Computer Science*. IEEE, 1993.
3. E. Clarke, E. Emerson, and A. Sistla. Automatic verification of finite-state concurrent systems using temporal logic specifications. *ACM TOPLAS* 8(2):244–263, 1986.
4. H. Garavel and J. Sifakis. Compilation and verification of LOTOS specifications. In *Proc. 10th PSTV*, Canada, 1990. IFIP, North-Holland.
5. T.A. Henzinger, X. Nicollin, J. Sifakis, and S. Yovine. Symbolic model-checking for real-time systems. In *Proc. 7th LICS*, pages 394–406. IEEE, 1992.
6. Y. Kesten, A. Pnueli, J. Sifakis, and S. Yovine. Integration graphs: a class of decidable hybrid systems. In *Workshop on Theory of Hybrid Systems*, pages 179–208. LNCS 736, Springer-Verlag, 1993.
7. G. Leduc and L. Léonard. A timed LOTOS supporting dense time domain and including new timed operators. In *Formal Description Techniques V*, pages 87–102. North-Holland, 1993.
8. G. Leduc and L. Léonard. A formal definition of time in LOTOS. In J. Quemada, editor, *Revised draft on enhancements to LOTOS*. Annex G of document ISO/IEC JTC1/SC21/WG1/Q48.6, 1994.
9. G. Leduc, L. Léonard, and A. Danthine. The Tick-Tock case study for the assessment of timed FDTs. In *The OSI95 transport service with multimedia support on HSLAN's and B-ISDN*, 1994.
10. X. Nicollin, A. Olivero, J. Sifakis, and S. Yovine. An approach to the description and analysis of hybrid systems. In *Workshop on Theory of Hybrid Systems*, pages 149–178. LNCS 736, Springer-Verlag, 1993.
11. X. Nicollin and J. Sifakis. An overview and synthesis on timed process algebras. In *Proc. 3rd Workshop on Computer-Aided Verification*, pages 376–398, Denmark, 1991. LNCS 575, Springer-Verlag.
12. X. Nicollin, J. Sifakis, and S. Yovine. Compiling real-time specifications into extended automata. *IEEE TSE Special Issue on Real-Time Systems*, 18(9):794–804, 1992.
13. X. Nicollin, J. Sifakis, and S. Yovine. From ATP to timed graphs and hybrid systems. *Acta Informatica*, 30:181–202, 1993.
14. S. Yovine. Méthodes et outils pour la vérification symbolique de systèmes temporisés. Thèse, Institut National Polytechnique de Grenoble, Grenoble, France, 1993.

17

Automatic Verification of Real–Time Communicating Systems by Constraint–Solving *

Wang Yi, Paul Pettersson and Mats Daniels

Department of Computer Systems, Uppsala University, Box 325, S751 05, Uppsala, Sweden. Email: {yi,paupet,matsd}@docs.uu.se

In this paper, an algebra of timed processes with real–valued clocks is presented, which serves as a formal description language for real–time communicating systems. We show that requirements such as "a process will never reach an undesired state" can be verified by solving a simple class of constraint systems on the clock–variables. A complete method for reachability analysis associated with the language is developed, and implemented as an automatic verification tool based on constraint–solving techniques. Finally as examples, we study and verify the safety–properties of Fischer's mutual exclusion protocol and a railway crossing controller.

1. INTRODUCTION

Correct timing plays an important role in ensuring the correct operation of real–time systems. Since such systems are often embedded in safety–critical environments, it is important to formally verify that certain crucial requirements are always met by the systems. This creates a need for formalisms to describe the abstract behavior of timed systems (i.e. modeling) and to check logical properties of the abstract descriptions (i.e. verification). During the past few years, researchers have developed various formal techniques for modeling and verifying real–time systems, e.g. automaton based, [1–3, 14] and process algebra based [24, 7, 9, 15, 22, 17, 13, 20, 29]. One of the most successful approaches is *timed automata* due to Alur and Dill [3], which are the classical finite-state automata extended with variables modeling system clocks.

In this paper, we study real-time communicating systems. Such a system may consist of a number of components with their own or shared clocks. The components may communicate with each other, and the environment through channels according to the timing constraints on the values of the clocks. Naturally, we can use timed automata to describe the components. However, it is not obvious how to combine the component descriptions to achieve the whole system description. Originally, the parallel composition of timed automata is interpreted as logical conjunction, which is similar to the strong (multi–) synchronization operator from process algebras, defined by the rule:

$$\frac{P \xrightarrow{a} P' \qquad Q \xrightarrow{a} Q'}{P\&Q \xrightarrow{a} P'\&Q'}$$

*This research is supported in part by the Swedish Board for Technical Development (NUTEK), project No. 93-3244.

Intuitively, it means that the whole system described by $P\&Q$ may make a move (i.e. doing a) only if the components described by P and Q can do the same. That is, all components of a concurrent system must synchronize on every action at every time point. Otherwise, the system will be deadlocked. This seems to be a strong restriction for practical application of timed automata, as real systems are often highly distributed and in many cases, a system component may only want to communicate with the environment or a particular component, without synchronizing with the others. Therefore, we introduce a CCS–like parallel composition operator for timed automata, to describe one–to–one communication and interleaving.

As the first contribution of this paper, we present an algebra of timed automata, which provides a number of algebraic operators including the parallel composition operator to model communication and concurrency. The operators can be used to construct complex automata (i.e. complex system descriptions) in terms of simpler ones (i.e. component descriptions). Thus, the algebra may serve as a *structural* description language for real–time communicating systems.

As the second contribution of the paper, we develop a verification tool based on constraint–solving techniques, for the type of systems described above. There have been a number of verification techniques developed for timed automata, e.g. [2, 1, 14]. However, most of the existing algorithms are based on the notion of region graphs, which always construct the whole reachability graph for a given automaton first and then check properties of the graph. Though there have been efficient algorithms to construct minimal reachability graphs such as [2], the problem of state–explosion is still an obstacle for automatic verification. In particular, when the systems to be analyzed include many components, it would be impossible to construct the whole reachability graph even in the untimed setting.

It has been pointed out in [12] and elsewhere that the practical goal of verification of real–time systems is to verify simple logical properties, which does not need the whole power of model–checking (e.g. for timed CTL). We shall only consider simple safety–properties, which can be verified without constructing the whole reachability graph of a timed system. For instance, a railway control system (see section 4) should guarantee that "at most one train can cross a critical point at the same time". This is a typical safety–property meaning that bad things can never happen. However, we can also verify properties requiring that a *good* thing will eventually happen within a certain time limit. For example, "a train should be able to pass a critical point (such as a bridge), within a bounded delay". We will show that such properties can be verified by solving a simple class of linear constraint systems.

The rest of the paper will be organized as follows: In section 2, we present an algebra of timed processes, in which a syntactical term describes a timed automaton; any timed automaton can be expressed in the algebra. In section 3, we study the reachability problems associated with the algebra. An algorithm is presented, and proved to be sound (i.e. it always provides the right answer) and complete (i.e. it always terminates). It is implemented as a tool, based on an existing constraint–solving program. In section 4, as examples, we study a variant of the railway crossing problem and Fischer's mutual exclusion protocol. Finally, in section 5 we give some concluding remarks.

2. AN ALGEBRA OF PROCESSES WITH CLOCKS

Process algebras provide a clean and general paradigm for compositional specification of communicating processes. We present an algebra of timed automata, serving as a structural description language for real-time communicating systems. The idea is to use algebraic operators to construct complex system descriptions in terms of simpler ones (or component descriptions). Following the tradition of process algebra, we shall call the algebraic terms *processes* instead of timed automata.

2.1. Syntax

Traditionally, a prefix expression $\alpha.P$ in process algebras like CCS describes a process which may perform an α-transition and then continue with P. But no timing information is given on when the transition may be taken.

Following Alur and Dill [3], we assume a set of clocks to specify timing constraints on transitions. Conceptually, the clocks may be considered as the system clocks of a concurrent system, owned or shared by processes in the system. The processes may test the clocks by comparing the clock values with integer constants and reset the clocks (i.e. assigning clock values to 0). Further, assume that all clocks proceed at the same rate and measure the amount of time that has been elapsed since they were reset or started.

We extend the action prefix $\alpha.P$ to the form $(g, \alpha, \phi).P$ where g is a predicate over the clock values and ϕ is a subset of clocks to be reset. Intuitively, $(g, \alpha, \phi).P$ describes a timed process which may perform an α-transition instantaneously when g is true of the current clock values and then continue with P with the clocks in ϕ being reset (and the other clocks will proceed with their old values).

Enabling Conditions. We use C to denote the set of clocks, ranged over by x, y, z. An *enabling condition* g is a logical formula generated by the following syntax:

$$g ::= \mathsf{tt} \mid \mathsf{ff} \mid \mathsf{A} \mid \mathsf{g} \wedge \mathsf{g}$$

where A is an atomic formula of the form: $x \prec n$ for $\prec \in \{\leq, \geq, <, >\}$ and n being a natural number. We could allow a more general form of formulas such as disjunction $g \vee f$. However, it will not give more expressive power to the description language we are going to develop. In fact, logical disjunction can be modeled by the behavioural choice operator.

The language is essentially CCS extended with the timed action prefix $(g, \alpha, \phi).P$. As in CCS, we assume a set $\Lambda = \Delta \cup \bar{\Delta}$[2] with $\bar{\bar{\alpha}} = \alpha$ for all $\alpha \in \Lambda$, ranged over by α, β representing external actions, and a distinct symbol τ representing internal actions. We use $\mathcal{A}ct$ to denote the set $\Lambda \cup \{\tau\}$ ranged over by a, b, c representing both internal and external actions. Further, assume a set of process variables ranged over by X, Y (and sequences of letters).

We shall see that the algebraic structure of a process expression P represents the control–structure of a process. This will be clear when we present the operational semantics. We adopt a two-phase syntax according to two types of control–structures: *regular* and *concurrent*.

[2]The action $\bar{\alpha}$ is called the co-action of α. In our example, we shall use $\alpha!$ instead of $\bar{\alpha}$ to denote an *output* event and $\alpha?$ instead of α to denote an *input* event.

Processes with Regular Control-Structure. We start with processes whose control-structure is regular in the sense that no concurrency is involved. The regular process expressions are generated by the following grammar:

$$E ::= \mathsf{nil} \;\Big|\; X \;\Big|\; (g, a, \phi).E \;\Big|\; E + F \;\Big|\; X \stackrel{def}{=} E$$

We shall restrict expressions to be *well-guarded* in the following sense:

Definition 2.1 *X is well-guarded in E if and only if every free occurrence of X in E is within a subexpression (a guard) of the prefix form $(g, a, \phi).F$ in E. E is well-guarded if and only if every free variable in E is well-guarded in E, and for every subexpression of the form $X \stackrel{def}{=} F$ in E, X is well-guarded in F.* □

Let $\mathcal{A}$ denote the set of closed and well-guarded expressions generated by the grammar above. We call $\mathcal{A}$ *regular timed processes.* Note that if we consider the prefix (g, a, ϕ) to be a single guard (or structured action), $\mathcal{A}$ corresponds precisely to the set of CCS regular processes.

Processes with Concurrent Control-Structure. We shall study concurrent processes in the form:

$$(P_1|...|P_n)\backslash L$$

where $P_i \in \mathcal{A}$ describing the components and $L \subseteq \Lambda$ representing the set of internal channels connecting the components.

We use $\mathcal{P}$ to denote the set of timed concurrent processes, ranged over by P, Q and R. For simplicity, we have ignored the relabelling operator. The results of this paper can easily be extended to more general types of processes modeled by the combination of parallel composition, restriction and relabelling.

2.2. Semantics

We interpret $\mathcal{P}$ in terms of clock assignments. A clock assignment $\rho : C \longrightarrow \mathcal{R}_{+0}$ is a function mapping each clock x to a non-negative real $\rho(x)$. We assume that a process is always started with an initial clock assignment.

Before going further, we need to define some notation. Assume that d is a non–negative real and ϕ is a set of clocks. We use $\rho + d$ to denote the clock assignment which maps each clock x to $\rho(x) + d$, and $\phi[\rho]$ to denote the clock assignment which maps x to 0 if $x \in \phi$ or $\rho(x)$ otherwise. Furthermore, given a predicate g over C, we write $g(\rho)$ to mean the truth value of g, relative to assignment ρ.

A global state (or a configuration) of a process is a pair (P, ρ) where $P \in \mathcal{P}$ stands for the current control-state and ρ denotes the current clock values. A process may make two types of transitions from state to state:

- Timed transition: $(P, \rho) \stackrel{d}{\rightsquigarrow} (P, \rho + d)$ following the rules given in definition 2.2.
- Action transition: $(P, \rho) \stackrel{a}{\rightsquigarrow} (P', \rho')$ following the rules given in definition 2.3.

The timed transition relation describes the pure passing of time; the action transition relation describes the instantanesous occurrence of actions and, possibly, the resetting of clocks.

Definition 2.2 *(time transition relation)*

$$\frac{d \leq M(E,\rho)}{(E,\rho)\overset{d}{\rightsquigarrow}(E,\rho+d)}$$

where $M(E,\rho)$ is the maximal delay of (E,ρ), defined inductively as follows:

$$\begin{aligned}
M(\mathsf{nil},\rho) &= \infty \\
M((g,a,\phi).E,\rho) &= \sup\{t|g(\rho+t)\} \\
M(E+F,\rho) &= \max\{M(E,\rho),M(F,\rho)\} \\
M(X,\rho) &= M(E,\rho) \text{ if } X \overset{def}{=} E \\
M(E|F,\rho) &= \min\{M(E,\rho),M(F,\rho)\} \\
M(E\backslash L,\rho) &= M(E,\rho)
\end{aligned}$$

□

For example, assume $\rho(x) = 0.4$ and $g \equiv x < 1$. Then $\sup\{t|g(\rho+t)\} = 0.6$ and $M((g,a,\phi).E,\rho) = 0.6$.

Intuitively, $M(P,\rho)$ is the maximal time that (P,ρ) may stay in the same control-state i.e. P before it must switch to another control-state. This is also assumed by the *maximal progress assumption* adopted in timed process algebras [9, 29]. For example, the control-state of $X \overset{def}{=} (x<1,\tau,\{\}).Q$ will become Q by doing the τ–action before the clock value of x proceeds to 1. However, when the value of x is larger than, or equal to 1, the control-state of $Y \overset{def}{=} (x<1,\tau,\{\}).Q + (x \geq 1, a, \{x\}).R$ will remain the same, i.e. Y, but τ–action will be disabled and a–action will be enabled.

Definition 2.3 *(action transition relation)*

$$\frac{g(\rho)}{((g,\alpha,\phi).E,\rho)\overset{a}{\rightsquigarrow}(E,\phi[\rho])} \qquad \frac{(E,\rho)\overset{a}{\rightsquigarrow}(E',\phi[\rho])}{(E+F,\rho)\overset{a}{\rightsquigarrow}(E',\phi[\rho])} \qquad \frac{(F,\rho)\overset{a}{\rightsquigarrow}(F',\phi[\rho])}{(E+F,\rho)\overset{a}{\rightsquigarrow}(F',\phi[\rho])}$$

$$\frac{(E,\rho)\overset{a}{\rightsquigarrow}(E',\phi[\rho])}{(X,\rho)\overset{a}{\rightsquigarrow}(E',\phi[\rho])}\ [X \overset{def}{=} E] \qquad \frac{(E,\rho)\overset{a}{\rightsquigarrow}(E',\phi[\rho])}{(E|F,\rho)\overset{a}{\rightsquigarrow}(E'|F,\phi[\rho])} \qquad \frac{(F,\rho)\overset{a}{\rightsquigarrow}(F',\phi[\rho])}{(E|F,\rho)\overset{a}{\rightsquigarrow}(E|F',\phi[\rho])}$$

$$\frac{(E,\rho)\overset{a}{\rightsquigarrow}(E',\phi[\rho]) \quad (F,\rho)\overset{\bar{a}}{\rightsquigarrow}(F',\varphi[\rho])}{(E|F,\rho)\overset{\tau}{\rightsquigarrow}(E'|F',(\phi\cup\varphi)[\rho])} \qquad \frac{(E,\rho)\overset{a}{\rightsquigarrow}(E',\phi[\rho])}{(E\backslash L,\rho)\overset{a}{\rightsquigarrow}(E'\backslash L,\phi[\rho])}\ [a,\bar{a} \notin L]$$

□

3. VERIFYING SAFETY–PROPERTIES OF PROCESSES

The language developed in the previous section can be used as a tool to construct the abstract model of an existing system or a system to be designed. In this section, we discuss how to verify properties of such systems in terms of their abstract models.

3.1. Verification by Reachability Analysis

It has been pointed out in [12] and elsewhere that the practical goal of verification of real–time systems, in particular safety–critical systems is to verify simple safety–properties. The type of properties is usually formalized as temporal logic formulas in the form $\Box\neg F$ read as "it is impossible that F will be true in the future". Here, F describes a certain undesired situation or logical property. For example, to verify a railway control system, the first question to ask would be: is it possible that two trains are crossing a certain critical point at the same time? For finite–state systems, this kind of properties can be verified simply by checking all reachable states whether they satisfy F or not, that is, by "reachability analysis". Unfortunately, the systems concerned here are infinite–state because the clock values range over the reals.

Definition 3.1 *(Simple Reachability Problem) Assume $P_0, P_f \in \mathcal{P}$ and ρ_0, ρ_f are clock assignments. We say that (P_f, ρ_f) is reachable from (P_0, ρ_0) iff there is a natural number n and a sequence of transitions starting from (P_0, ρ_0) and ending up with (P_f, ρ_f), i.e. $(P_0, \rho_0)\overset{\sigma_1}{\leadsto}(P_1, \rho_1)...(P_{n-1}, \rho_{n-1})\overset{\sigma_n}{\leadsto}(P_f, \rho_f)$ for $\sigma_i \in \mathcal{A}ct \cup \mathcal{R}_{+0}$.* □

More generally, we will consider the reachability problem for sets of clock assignments.

Definition 3.2 *(General Reachability Problem) Assume $P_0, P_f \in \mathcal{P}$ and D_0, D_f are sets of clock assignments. We say that (P_f, D_f) is reachable from (P_0, D_0) iff there exists $\rho_0 \in D_0$, and $\rho_f \in D_f$ such that (P_f, ρ_f) is reachable from (P_0, ρ_0).* □

We shall develop an algorithm based on constraint–solving techniques, for solving the General Reachability Problem.

3.2. Reachability Analysis by Constraint–Solving

Given a process to be analyzed, we assume that its clocks are ordered as a vector $< x_1, x_2, ..., x_n >$. Then a clock assignment can be considered as a vector of reals or a point in the n-dimensional space $\mathcal{R}^n_{+0}$. We shall use linear constraint systems to describe regions of points in such a space (as their solution sets), and solve the reachability problems by manipulating a simple class of linear constraint systems.

3.2.1. A Class of Linear Constraint Systems

By a *linear constraint system*, we simply mean a set of linear inequalities over a set of variables ranging over $\mathcal{R}_{+0}$ (in our case, the clock variables). A *solution* to such a system is an assignment that maps each variable to a value, which satisfies the set of inequalities. In general, a constraint system may have more than one solution. In the rest of the paper, we shall simply call a constraint system D a *region*, which means its solution set. We shall write (1) $D = \{\}$ to mean D is not satisfiable, (i.e. its solution set is empty), (2) $D \subseteq D'$ to mean that D implies D' (i.e. the solution set of D is included in the solution set of D'), and $D \wedge D'$ to mean the intersection of the solution sets of D and D'.

As we are only allowed to compare clock variables with natural numbers in enabling conditions, the class of constraint systems we need to deal with is restricted to a simple class which we call *time regions.* We shall use $\mathcal{D}$ to denote this class of constraint systems ranged over by D, D'.

Definition 3.3 *(time region) Let $C = \{x_1 ... x_n\}$ be a set of clocks. A time region of C is a constraint system in the following form*[3]*:*

$$D = \{a_i \prec x_i | i \leq n\} \cup \{x_i \prec b_i | i \leq n\} \cup \{x_i - x_j \prec d_{ij} | i, j \leq n\}$$

where $\prec \in \{\leq, <\}$, and a_i, b_i, d_{ij} are natural numbers. In particular, b_i, d_{ij} may be ∞. □

Intuitively, a_i is the lower bound of x_i in D, b_i is the upper bound of x_i in D and d_{ij} is the maximal distance between x_i and x_j in D. We shall need a few operations on time regions in doing reachability analysis.

Definition 3.4 *Assume that D is a time region.*

1. *(Weakest Pre-Condition):* $\mathsf{wp}(D) = \{ \rho \mid \exists d \in \mathcal{R}_{+0} : \rho + d \in D\}$
2. *(Border-Line):* $\mathsf{border}(x, D) = \{ \rho \mid \rho \in D \text{ and } \rho(x) = 0 \}$
3. *(Free-Variable):* $\mathsf{free}(x, D) = \{ \rho[x := d] \mid \rho \in D \text{ and } d \in \mathcal{R}_{+0} \}$ □

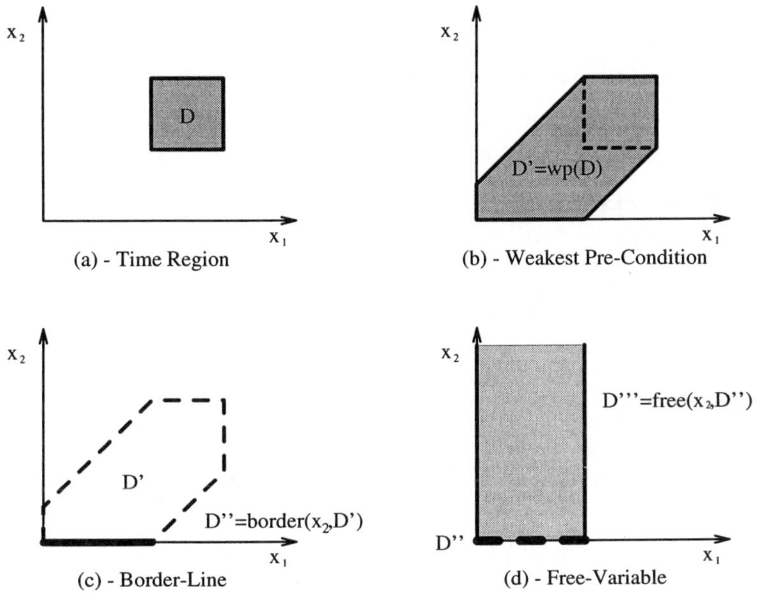

Figure 1. Operations on Time Regions

[3]Note that the symbol ∪ here means the union of the sets of constraints. It does not mean the union of the solution sets, rather the intersection of the solution sets.

The three operations on time regions are illustrated in Fig 1 for the case of two clocks. The weakest pre–condition, $\mathsf{wp}(D)$ is the largest region of points that will eventually reach D after some delay, $\mathsf{border}(x_2, D')$ is the border–line of D' on x_1–coordinate (that is the region $D' \wedge \{x_2 = 0\}$) and $\mathsf{free}(x_2, D'')$ is the largest region that has the same projection as D'' on x_1–coordinate.

We will apply the operations $\mathsf{border}(x, D)$ and $\mathsf{free}(x, D)$ on sets of variables. Assume that ϕ is a set of clock variables and $\phi = \{x\} \cup \phi'$. We define $\mathsf{border}(\phi, D) = \mathsf{border}(x, (\mathsf{border}(\phi', D)))$, $\mathsf{free}(\phi, D) = \mathsf{free}(x, (\mathsf{free}(\phi', D)))$ and in particular, $\mathsf{border}(\{\}, D) = \mathsf{free}(\{\}, D) = D$. Furthermore, we define conjunction $D \wedge D'$ of two time regions D, D' in the standard way, that is, $\{\rho | \rho \in D \text{ and } \rho \in D'\}$.

It can be established that the class of constraint systems known as time regions is closed under the operations: "Conjunction", "Weakest Pre–Condition", "Border–line" and "Free-Variable".

Proposition 1 *Assume that C is a set of clocks, D and D' are time regions of C, and $\phi \subseteq C$. Then $D \wedge D', \mathsf{wp}(D), \mathsf{free}(\phi, D)$ and $\mathsf{border}(\phi, D)$ are also time regions of C.* □

We shall use these operations for backward reachability analysis. Similar operations such as strongest post–condition can be defined in order to do forward reachability analysis.

3.2.2. An Algorithm and Its Correctness

Having introduced the notion of time regions, in the following we will simply call (P, D) a region of states, and extend the transition relation $\rightsquigarrow$ to regions.

Definition 3.5 *Assume $a \in \mathcal{A}ct$ and a new symbol ε representing delays.*

1. *$(P, D) \overset{a}{\rightsquigarrow} (P', D')$ iff for all $\rho \in D$, $(P, \rho) \overset{a}{\rightsquigarrow} (P', \rho')$ for some $\rho' \in D'$, and vice versa for all $\rho' \in D'$, $(P, \rho) \overset{a}{\rightsquigarrow} (P', \rho')$ for some $\rho \in D$.*

2. *$(P, D) \overset{\varepsilon}{\rightsquigarrow} (P, D')$ iff for all $\rho \in D$, $(P, \rho) \overset{d}{\rightsquigarrow} (P', \rho')$ for some $d \in \mathcal{R}_{+0}$ and $\rho' \in D'$, and vice versa for all $\rho' \in D'$, $(P, \rho) \overset{d}{\rightsquigarrow} (P, \rho')$ for some $d \in \mathcal{R}_{+0}$ and $\rho \in D$.* □

Now, the general reachability problem can be reformalized as follows:

Proposition 2 *Given an initial region (P_0, D_0) and a final region (P_f, D_f), (P_f, D_f) is reachable from (P_0, D_0) iff there exists a finite number n, $\sigma_i \in \mathcal{A}ct \cup \{\varepsilon\}$ and $D_i \in \mathcal{D}$ for all $i \leq n$, such that $P_n \equiv P_f$, $D_n \wedge D_f \neq \{\}$ and $(P_0, D_0) \overset{\sigma_1}{\rightsquigarrow} (P_1, D_1) ... (P_{n-1}, D_{n-1}) \overset{\sigma_n}{\rightsquigarrow} (P_n, D_n)$.* □

To achieve a decision algorithm for the problem, we shall take the approach of backward reachability analysis. Usually, to verify safety–properties, a backward analysis algorithm may terminate much faster than a forward analysis algorithm for the following reason: In case that a system does not contain an undesired state, the backward analysis needs not to check the whole reachable state–space of the system (but the forward analysis does), and the probability for a safety–critical system to contain a serious error is often very small.

The general principle of backward analysis is to start from the final and search back to the initial. If the initial is found, the algorithm terminates with answer "yes", otherwise "no". However, our backward analysis method can be easily adopted to forward analysis.

First, we need to study the control–structures more carefully. It has been said earlier that the algebraic structure of a term P describes the control–structure of a process. In fact, the set of subexpressions of P is a superset of the control–states of P, and the transitions among the control–states obey the rules in Fig. 2.

$$\frac{}{(g,a,\phi).E \xrightarrow{g,a,\phi} E} \qquad \frac{E \xrightarrow{g,a,\phi} E'}{E+F \xrightarrow{g,a,\phi} E'} \qquad \frac{F \xrightarrow{g,a,\phi} F'}{E+F \xrightarrow{g,a,\phi} F'} \qquad \frac{E \xrightarrow{g,a,\phi} E'}{X \xrightarrow{g,a,\phi} E'}\ [X \stackrel{def}{=} E]$$

$$\frac{E \xrightarrow{g,a,\phi} E'}{E|F \xrightarrow{g,a,\phi} E'|F} \qquad \frac{F \xrightarrow{g,a,\phi} F'}{E|F \xrightarrow{g,a,\phi} E|F'} \qquad \frac{E \xrightarrow{g,a,\phi} E' \quad F \xrightarrow{f,\bar{a},\varphi} F'}{E|F \xrightarrow{g\wedge f,\tau,\phi\cup\varphi} E'|F'} \qquad \frac{E \xrightarrow{g,a,\phi} E'\ [a,\bar{a} \notin L]}{E\backslash L \xrightarrow{g,a,\phi} E'\backslash L}$$

Figure 2. Transition Rules for Control–States.

It should be obvious that each term P_0 describes a timed automaton, i.e. $< C_S, P_0, \longrightarrow>$ where C_S is all control–states reachable from P_0, $\longrightarrow$ is the least transition relation defined by the transitional rules. In particular, note that C_S is finite.

The reachability analysis algorithm is based on the following idea: Assume that we want to decide whether (P', D') may reach (P, D) in one step (i.e. without passing other control–states) or not. The first thing to check is whether it is possible for P' to switch to P directly. If this is not the case, that is, $P' \xrightarrow{g,a,\phi} P$ for no P', g, a, ϕ, we can conclude immediately that (P, D) is not reachable from (P', D') in one step. Now, assume $P' \xrightarrow{g,a,\phi} P$. To reach (P, D), there should be time regions D_0, D_1, D_2 such that $D' \cap D_0 \neq \{\}$ and $(P_0, D_0) \overset{\varepsilon}{\leadsto} (P_0, D_1) \overset{a}{\leadsto} (P, D_2) \overset{\varepsilon}{\leadsto} (P, D)$. Note that D' and D are given. We need to find D_0, D_1, D_2. Clearly, we may choose

$$\begin{aligned} D_2 &= \mathsf{border}(\phi, \mathsf{wp}(D)) \\ D_1 &= g \wedge \mathsf{free}(\phi, D_2) \\ D_0 &= \mathsf{wp}(D_1) \end{aligned}$$

That is, $D_0 = \mathsf{wp}(g \wedge [\mathsf{free}(\phi, \mathsf{border}(\phi, \mathsf{wp}(D)))])$. In fact, D_0 is the largest region of points that may (1) pass the guard g and (2) be reset by ϕ, and finally (3) reach D. In general, for any given g, ϕ and D, we define $\mathsf{image}(g, \phi, D) = \mathsf{wp}(g \wedge [\mathsf{free}(\phi, \mathsf{border}(\phi, \mathsf{wp}(D)))])$.

Now, we are ready to present the algorithm, shown in Fig. 3 for backward reachability analysis. We use two buffers for saving regions (of states): passed and waiting where passed stands for the set of regions that have been examined and waiting for the set of regions that are to be examined next.

Algorithm. (Input: $P_0, P_f \in \mathcal{P}$ and $D_0, D_f \in \mathcal{D}$ and Output: answer = 'yes' or 'no'.)

1. **Initial:** passed := {} and waiting := $\{(P_f, D_f)\}$.

2. **Repeat**

   for all $(P, D) \in$ waiting, **do**
   begin
   (1) If there is no D' such that $D \subseteq D'$ and $(P, D') \in$ passed **then**
   begin
   (a) passed := passed $\cup \{(P, D)\}$ and
   (b) waiting := waiting $\cup \{(P', D')\}$ for all P', g, a, ϕ such that $P' \xrightarrow{g,a,\phi} P$, and $D' = \text{image}(g, \phi, D) \neq \{\}$.
   end;
   (2) waiting := waiting $- \{(P, D)\}$
   end

   until waiting = {} or $(P_0, D_0') \in$ waiting for some D_0' such that $D_0 \wedge D_0' \neq \{\}$.

3. **Termination:** If waiting = {} then answer := 'no'; otherwise answer := 'yes'.

Figure 3. An Algorithm for Reachability Analysis.

The algorithm is started with passed = {} and waiting = $\{(P_f, D_f)\}$, and then repeatedly examines the regions in waiting. If a region (P, D) found in waiting is smaller than a region (P, D') (with the same control–state) in passed, then (P, D) does not need to be examined further. Otherwise, put all the regions that may reach (P, D) in one step into waiting to be examined later, and put (P, D) into passed. The algorithm will terminate when waiting is empty (i.e. nothing is left to be examined, and therefore fails to find the initial region) or a region (P_0, D_0') is found, which includes a part of the initial region (P_0, D_0) (i.e. $D_0 \wedge D_0' \neq \{\}$).

It is easy to prove the partial correctness (soundness) of the algorithm: given proper inputs, it always provides the right answer.

Theorem 1 *(Partial Correctness) For all initial regions (P_0, D_0) and final regions (P_f, D_f), if the algorithm terminates with* 'yes', *then (P_f, D_f) is reachable from (P_0, D_0). Otherwise, (P_f, D_f) is not reachable from (P_0, D_0).* □

It is slightly more difficult to prove the total correctness (completeness) of the algorithm: given proper inputs, it always terminates with an answer.

Theorem 2 *(Total Correctness) For all initial regions (P_0, D_0) and final regions (P_f, D_f), the algorithm always terminates with an answer which is either* 'yes' *or* 'no'. □

3.2.3. Implementation

In describing the reachability algorithm, we did not explain how to (1) perform the four operations (*wp, free, border,* $\wedge$) defined on time regions, (2) check the emptiness of a time region (or satisfiability of a constraint system), and (3) set–inclusion (i.e. $\subseteq$) between time regions. In fact, these functions are often provided by constraint solvers or straight forward to implement using primitive functions of a constraint solver.

We have implemented the algorithm as a tool, based on a constraint solver developed at the Swedish Institute of Computer Science called Prolog Constraint Solver (PCS) [21]. Several examples have been used to test the tool (see next section), which show that the implementation is fairly efficient.

4. EXAMPLES

In this section, we present examples which have been verified by our tool. In addition to clock variables, in describing the examples, we shall also use ordinary variables. These variables do not change their values automatically as the clock variables; they can only be assigned to values from finite domains, and therefore they will not cause infinite–stateness. Fortunately, the implementation of our tool is based on a general constraint solver which can handle logical constraints and assignments including ordinary variables in the same way as timing constraints and clock assignments.

4.1. Fischer's Mutual Exclusion Protocol

The protocol was proposed originally by Fischer and described by Lamport [18]. It is to guarantee mutual exclusion in a concurrent systems consisting of several processes using a shared variable (among the processes) and properly timing the processes in changing the shared variable. Each of the processes is assumed to have a local clock. The idea behind the protocol is that the timing constraints on the local clocks are set so that only one process can change the global variable to its own process number, then read the global variable later and if the shared variable is still equal to its own number, enter the critical section.

Assume a concurrent system with n processes $\mathsf{P}_1 ... \mathsf{P}_n$. We use x_i to model the local clock for each process P_i. The formal description of P_i is given in Fig. 4, and illustrated[4] in Fig 5.

This is a simplified version of the original protocol and has been studied by researchers, e.g. [4, 25], which permits only one process to enter the critical section and never exits it. Recovery actions from failure to enter the critical section are omitted. However, the protocol can be extended to an actual mutual exclusion algorithm.

The processes, P_i, may be in either of the four local states $\mathsf{A}_i, \mathsf{B}_i, \mathsf{C}_i, \mathsf{CS}_i$. Initially, all processes are in their A–state and the shared variable v is initially 0. A process, P_i, that tries to enter the critical section changes state from A_i to B_i if it sees $\mathsf{v}=0$. In B_i, it will move to C_i before the clock x_i proceeds to **const**, and in doing so, reset the clock x_i (i.e. $\mathsf{x}_i := 0$) and assign v to its own process number (i.e. $\mathsf{v} := \mathsf{i}$). From C_i, it can move to the critical section CS_i if v is still equal to its process number (i.e. $\mathsf{v} = \mathsf{i}$) when the clock value of x_i is larger than **const**.

Intuitively, the protocol behaves as follows: The constraints on the shared variable v

[4]In figures, we adopt the convention that when a transition is not labelled with a timing constraint, it means implicitly that the constraint is tt, that is, the transition can be taken at any time. We shall also adopt the convention that when a transition is not labelled with an action, it means that the transition is an internal one, that is, labelled with τ.

$$
\begin{aligned}
P_i &\stackrel{def}{=} A_i \\
A_i &\stackrel{def}{=} (\{v = 0\}, \tau, \{x_i\}).B_i \\
B_i &\stackrel{def}{=} (\{x_i < \mathbf{const}\}, \tau, \{v := i, x_i\}).C_i \\
C_i &\stackrel{def}{=} (\{v = i, x_i > \mathbf{const}\}, \tau, \{\}).CS_i \\
CS_i &\stackrel{def}{=} \mathsf{nil}
\end{aligned}
$$

Figure 4. The Formal Description of Fischer's Protocol.

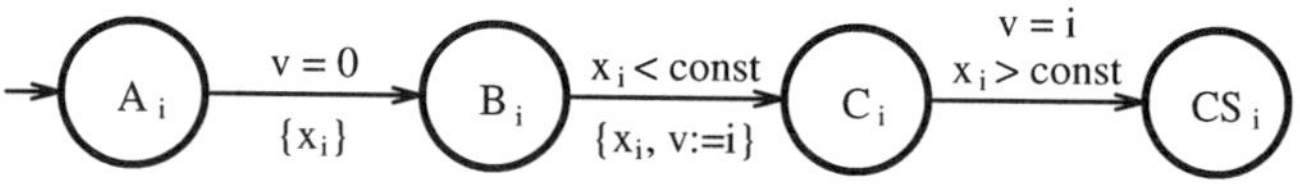

Figure 5. Fischer's Mutual Exclusion Protocol

ensure that a process must reach B–location before any process reach C–location; otherwise, it will never move from A-location to B-location. The timing constraints on the clocks ensure that all processes in C–location must wait until all processes in B–location reach C–location. The last process that reached C–location and set v to its own process number gets the right to enter its critical section. In fact, the protocol will guarantee mutual exclusion for any non–zero constant **const**.

We need to verify that the mutual exclusion property is satisfied, i.e. there will never be more than one process which may reach the critical section, $\mathsf{CS_i}$. The requirement can be formalized as follows: The concurrent system, with an initial state where the control–state is $\mathsf{A_1|\ldots|A_n}$ and arbitrary variable assignment, will never reach a state where the control–state is in the form

$$\mathsf{S_1|\ldots|CS_k|\ldots|CS_l|\ldots|S_n}$$

for some $\mathsf{k,l \le n}$ and $\mathsf{S_i \in \{A_i, B_i, C_i, CS_i\}}$.

We have used our tool and verified a system consisting of 10 processes and $\mathbf{const} = 1$, which satisfies the property. We are in progress to extend the tool to treat the number of processes as a variable and verify that the property is satisfied by systems with arbitrary number of processes.

4.2. A Simple Railway Control System

We consider a railway control system to automatically control trains passing a critical point such as a bridge. The idea is to use a computer to guide trains from several tracks crossing a single bridge instead of building many bridges. Obviously, a safety–property of such a system is to avoid the situation where more than one train are crossing the bridge at the same time.

Assume that the whole system consists of n trains and a simple controller. We model the system by the following process:

$$(\mathsf{C}|\mathsf{Train}_1|\ldots|\mathsf{Train}_n)\backslash\mathsf{A}$$

where Train_i describe the behavior of trains, C describes the behavior of the controller, and $\mathsf{A} = \{\mathsf{appr}_i, \mathsf{stop}_i, \mathsf{leave}_i, \mathsf{go}_i\}$ is the set of internal channel names (or signals) between the trains and the controller.

To describe timing constraints, we use x and x_i to model the local time of the controller and the trains respectively. The controller uses a list L for the trains waiting to cross the bridge. The formal descriptions of Train_i's and C are given in Fig. 6 and illustrated in Fig. 7.

$$\begin{array}{rcl}
Train_i & \stackrel{def}{=} & Safe_i \\
Safe_i & \stackrel{def}{=} & (\{\mathtt{tt}\}, appr_i!, \{x_i\}).Appr_i \\
Appr_i & \stackrel{def}{=} & (\{x_i \geq 0 \wedge x_i \leq 10\}, stop_i?, \{x_i\}).Slow_i \\
 & & +(\{x_i \geq 11 \wedge x_i \leq 20\}, \tau, \{x_i\}).Cross_i \\
Cross_i & \stackrel{def}{=} & (\{x_i \geq 3 \wedge x_i \leq 5\}, leave_i!, \{x_i\}).Safe_i \\
Slow_i & \stackrel{def}{=} & (\{x_i \geq 5 \wedge x_i \leq 7\}, \tau, \{x_i\}).Stop_i \\
Stop_i & \stackrel{def}{=} & (\{\mathtt{tt}\}, go_i?, \{x_i\}).Start_i \\
Start_i & \stackrel{def}{=} & (\{x_i \geq 7 \wedge x_i \leq 15\}, \tau, \{x_i\}).Cross_i \\
 & & \\
C & \stackrel{def}{=} & Occ_1 \\
Occ_1 & \stackrel{def}{=} & (\{\mathtt{tt}\}, leave_i?, \{L := L - i\}).Free + (\{\mathtt{tt}\}, appr_i?, \{n := i, x\}).Occ_2 \\
Occ_2 & \stackrel{def}{=} & (\{x < 10\}, stop_n!, \{L := L :: n\}).Occ_1 \\
Free & \stackrel{def}{=} & (\{L = empty\}, appr_i?, \{L := [i]\}).Occ_1 \\
 & & +(\{L \neq empty\}, go_i!, \{i := hd(L)\}).Occ_1
\end{array}$$

Figure 6. The Formal Description of the Railway Control System.

Intuitively, when a train, Train_i, approaches the bridge it sends a signal to the controller within a certain distance. If the bridge is occupied the controller sends a stop signal stop_i within 10 time units to prevent the train from entering the bridge. Otherwise, if the approaching train does not receive a stop signal within 10 time units, it will start to cross the bridge within 20 time units (but it will take at least 11 time units for a train to enter the bridge). The crossing train is assumed to leave the bridge within 3 to 5 time units; a stopped train will slow down and eventually stop after some delay. When the bridge is

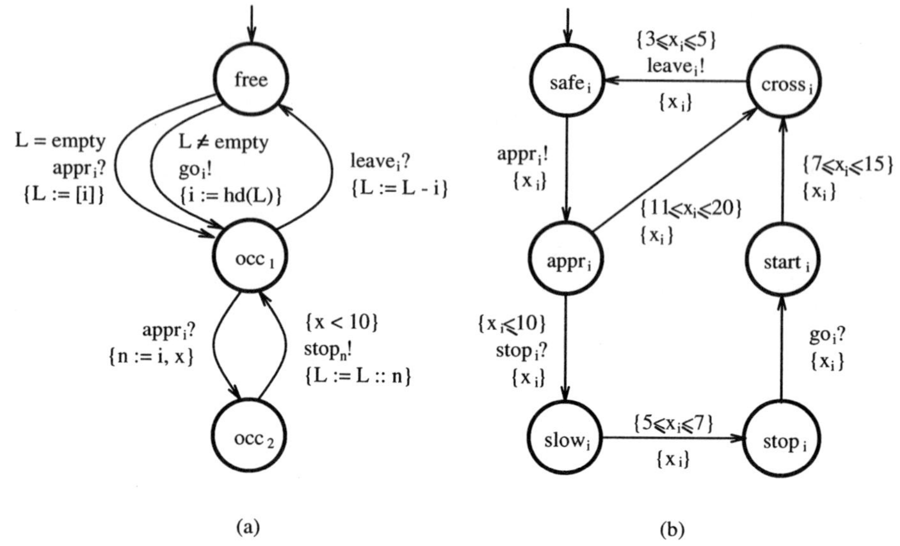

Figure 7. (a) – Controller, (b) – Train

free again and the controller signals (by sending $\mathsf{go_i}$) the first train in the waiting list to cross.

Assume that the system is started with the following control-state:

$$(\mathsf{Free}|\mathsf{Safe}_1|\ldots|\mathsf{Safe}_n)\backslash\{\mathsf{appr_i}, \mathsf{stop_i}, \mathsf{leave_i}, \mathsf{go_i}\}$$

and all clocks are initialized to 0.

We need to guarantee that the system will never reach a control-state where two trains are in location Cross (the clocks may have any values). That is, a state in the form:

$$(\mathsf{S_i}|\mathsf{T}_1|\ldots|\mathsf{Cross_k}|\ldots|\mathsf{Cross_l}|\ldots|\mathsf{T_n})\backslash\{\mathsf{appr_i}, \mathsf{stop_i}, \mathsf{leave_i}, \mathsf{go_i}\}$$

for some $\mathsf{k}, \mathsf{l} \leq \mathsf{n}$, $\mathsf{S_i} \in \{\mathsf{Free}, \mathsf{Occ}_1, \mathsf{Occ}_2\}$ and $\mathsf{T_i} \in \{\mathsf{Safe_i}, \mathsf{Appr_i}, \mathsf{Slow_i}, \mathsf{Stop_i}, \mathsf{Start_i}\}$.

We have verified a system consisting of 6 trains by our tool, which satisfies the safety–requirement. As in the previous example (Fisher's protocol), we can only check a system with a fixed number of trains. We hope to extend our system to deal with any number of trains.

5. CONCLUSION

The first contribution of this paper is an algebra of processes with clocks, which extends timed automata with algebraic operators. The algebra may serve as a formal description language for real–time communicating systems. In particular, a parallel composition operator is introduced for timed automata to model communication and concurrency, which can be used to construct complex system descriptions in terms of component descriptions.

The second contribution of this paper is a reachability analysis algorithm for the description language, based on constraint–solving techniques. The algorithm is proved to be sound (i.e. always provides the right answer) and complete (i.e. always terminates with an answer). It has been implemented as an automatic verification tool, for verifying safety–properties of real–time communicating systems, based on an existing constraint–solver. Several examples have been used to test the tool. In particular, we have studied and verified Fisher's mutual exclusion protocol and a railway controller using our tool.

There have been many proposals for verifying timed systems e.g. [2, 24, 1, 7, 14, 17]. However, most of them are intended to construct the whole reachability graph of a system or to obtain more efficient model–checking algorithms with respect to a real–time temporal logic, or to check equivalences between abstract specifications. We believe in that the goal of verifying real–time systems, in particular safety–critical systems is to check simple logical properties, which can be done without constructing the whole reachability graph or the full power of model–checking. We are of the opinion that our approach is simpler as it is based directly on constraint–solving techniques and can be fairly efficient in verifying systems consisting of many components as it avoids to explore the whole state–space.

We are in progress to extend our tool to deal with more general types of variables such as lists, in addition to clock variables. In particular, we will treat the number of components in a concurrent system as a parameter (i.e. an ordinary variable) in order to verify systems with many similar components such as the trains in the railway controller and the processes in Fisher's protocol in a more efficient way.

Acknowledgements: We would like to thank Parosh Abdula and Bengt Jonsson for discussions, and the anonymous referees for comments and suggestions on the paper. We would also like to thank Bengt Asker and Ulf Olsson from the steering group of the Swedish Board for Technical Development (NUTEK) for supervising the project.

REFERENCES

1. Rajeev Alur, Costas Courcoubetis, and David Dill. Model–checking for real–time systems. In *Proceedings of the Fifth IEEE Symposium on Logic in Computer Science*, 1990.
2. R. Alur, C Courcoubetis, N. Halbwachs, D. Dill, H. Wong–Toi. Minimization of Timed Transition Systems. CONCUR92, LNCS 630, 1992.
3. Rajeev Alur and David Dill. Automata for modelling real–time systems. In *Automata, Languages and Programming: Proceedings of the 17th ICALP*, LNCS 443. Springer-Verlag, 1990.
4. Martin Abadi and Leslie Lamport. An Old-Fashioned Recipe for Real Time. *Lecture Notes in Computer Science*, volume 600, Springer-Verlag, 1993.
5. J.C.M. Baeten and J.A. Bergstra. Real time process algebra. Technical Report P8916, University of Amsterdam, 1989.
6. B. Berhomieu and M. Diaz. Modeling and Verification of Time Dependent Systems Using Time Petri Nets. In *IEEE trans. on Software Engineering*, pages 259–273, Vol 17, March 1991.
7. Karlis Cerans, Jens Chr. Godskesen and Kim G. Larsen. Time Modal Specification – Theory and Tools. *Proceedings of the 5th Int. Conf. on Computer Aided Verification*, 1993.
8. M. Diaz. Modeling and analysis of communication and cooperation protocols using Petri net

based models. *Computer Networks*, Dec. 1982.

9. Jim Davis and Steve Schneider. An introduction to timed CSP. Technical Report PRG–75, Oxford University Computing Laboratory, 1989.
10. Willem Jan Fokkink and Steven Klusener. Real time algebra with prefixed integration. Technical report, CWI, Amsterdam, 1991.
11. Jens Chr. Godskesen and Kim G. Larsen. Real–time calculi and expansion theorems. In *Twelfth Conference on the FST and TCS*, Lecture Notes in Computer Science. Springer-Verlag, December 1992.
12. Nicolas Halbwachs. Delay Analysis in Synchronous Programs. In the *Proceedings of CAV'93*, volume 697 of *Lecture Notes in Computer Science.* Springer-Verlag, 1991.
13. Uno Holmer, Kim Larsen, and Yi Wang. Deciding properties of regular timed processes. In Kim G. Larsen and Arne Skou, editors, *Proceedings of the Third Workshop on Computer Aided Verification,*, volume 575 of *Lecture Notes in Computer Science.* Springer-Verlag, 1991.
14. T. Henzinger, X. Nicollin, J. Sifakis, and J. Yovine. Symbolic Model Checking for Real–Time Systems. Proceedings of the 7th IEEE Symposium on Logic in Computer Science, 1992.
15. Matthew Hennessy and Tim Regan. A process algebra for timed systems. Technical Report 5/91, University of Sussex, 1991.
16. P.C. Kanellakis and S.A. Smolka, CCS Expressions, finite state processes, and three problems of equivalence. Information and Control Vol 86, 1990.
17. Kim G. Larsen and Wang Yi, Time Abstracted Bisimulation: Implicit Specification and Decidability. In the proceedings of MFPS93 , New Oleans, USA, 1993. Lecture Notes in Computer Science No. 802, 1994.
18. Leslie Lamport. A fast mutual exclusion algorithm. *ACM Trans. on Computer System*, pages 1–11, volume 5(1), February 1987.
19. Robin Milner. *Communication and Concurrency.* Series in Computer Science. Prentice–Hall International, 1989.
20. Faron Moller and Chris Tofts. A temporal calculus of communicating systems. In *CONCUR'90*, volume 458 of *Lecture Notes in Computer Science.* Springer-Verlag, 1990.
21. Martin Nilsson. Piecewise Linear Constraints and Entailment Technical report, Swedish Institute of Computer Science, August 1993.
22. X. Nicollin, J.-L. Richier, Joseph Sifakis, and J. Yovine. ATP: an algebra for timed processes. In *Proceedings of the IFIP TC 2 Working Conference on Programming Concepts and Methods*, Sea of Gallilee, Israel, April 1990.
23. Xavier Nicollin, Joseph Sifakis, and Sergio Yovine. From ATP to timed graphs and hybrid systems. In *Real–Time: Theory in Practice*, volume 600 of *Lecture Notes in Computer Science.* Springer-Verlag, 1991.
24. Kārlis Čerāns. Decidability of bisimulation equivalences for processes with parallel timers. In the Proceedings of CAV'92, 1992.
25. N. Shankar. Verification of Real–Time Systems Using PVS. *Proceedings of the 5th Int. Conf. on Computer Aided Verification*, 1993.
26. Frits Vaandrager and Nancy Lynch. Action Transducers and Timed Automata. In *CONCUR '92*, volume 630 of *Lecture Notes in Computer Science.* Springer-Verlag, 1992.
27. Yi Wang. Real–time behaviour of asynchronous agents. In *CONCUR '90*, volume 458 of *Lecture Notes in Computer Science.* Springer-Verlag, 1990.
28. Yi Wang. *A Calculus of Real Time Systems.* PhD thesis, Chalmers University of Technology, Göteborg, Sweden, 1991.
29. Yi Wang. CCS + time = an interleaving model for real time systems. In *ICALP '91, LNCS 510.* Springer-Verlag, 1991.

18

Verifying Timing Properties of Concurrent Algorithms

Victor Luchangco, Ekrem Söylemez, Stephen Garland, and Nancy Lynch[a*]

[a]MIT Laboratory for Computer Science, Cambridge, MA 02139

This paper presents a method for computer-aided verification of timing properties of real-time systems. A timed automaton model, along with *invariant assertion* and *simulation* techniques for proving properties of real-time systems, is formalized within the Larch Shared Language. This framework is then used to prove time bounds for two sample algorithms—a simple counter and Fischer's mutual exclusion protocol. The proofs are checked using the Larch Prover. Keywords: I.3, I.8, I.6/II.12: Larch, III.1, IV.8

1. Introduction

Techniques based on *simulations* are widely accepted as useful for verifying the correctness of (untimed) concurrent systems. These methods involve describing both the problem specification and an implementation as state machines, establishing a correspondence known as a *simulation mapping* between their states, and proving that the mapping is preserved by all transitions of the implementation. Such methods are attractive because they provide insights into a system's behavior, appear to be scalable to systems of substantial size, and provide assistance in modifying system descriptions and proofs.

It is usually possible to describe the transitions of the specification, the transitions of the implementation, and the simulation relation, all as equations involving states. Then the proof that the simulation mapping is preserved is an exercise in equational deduction. Such deductions are natural candidates for partial automation. Proofs of this sort for untimed systems have already been automated, for example, using HOL [8], Isabelle [14], and the Larch Prover [19].

Recently, the simulation method has been extended to proofs of correctness and timing properties for timing-based systems [11, 12, 10]. The extended method is based on the *timed automaton* model of Merritt, Modugno and Tuttle [13]. Both the specification and implementation are described as timed automata, which include *timing conditions* in their states. The implementation's conditions represent timing assumptions, and the specification's conditions represent timing upper and lower bounds to be proved. As in the untimed case, a simulation mapping is defined between the states of the implementation and those of the specification; but now the mapping typically includes inequalities involving the timing conditions. The proof that the mapping is preserved by all transitions has a similar deductive flavor to the proofs in the untimed case, but now the deductions involve inequalities as well as equations.

*Research supported in part by the Advanced Research Projects Agency of the Department of Defense, monitored by the Office of Naval Research under contracts N00014-92-J-1795 and N00014-92-J-4033, by the National Science Foundation under grants 9115797-CCR and 9225124-CCR, and by the Air Force Office of Scientific Research and the Office of Naval Research under contract F49620-94-1-0199.

The simulation method for timed systems has the same attractions as for untimed systems. Furthermore, it is capable of proving performance as well as correctness properties. Examples of proofs done by hand using this method appear in [11, 10, 18, 6, 9].

Just as in the untimed case, the timed proofs are amenable to automation. Specifically, the notions of timed automata, invariant assertions, and simulation mappings are formalized using the Larch Shared Language [5], and this formal infrastructure is used to specify, verify, and analyze two sample algorithms—a simple counter [11] and Fischer's mutual exclusion protocol. Fischer's algorithm has been verified many times by many people [1, 16, 17], including some with machine assistance [17]. But in addition to the usual correctness property of mutual exclusion, we prove a more difficult timing property—an upper bound on the time from when some process requires the resource until some process acquires it.

The rest of the paper proceeds as follows. We introduce our techniques by way of a simple example in Section 2. Then we use these techniques to verify Fischer's mutual exclusion protocol in Section 3.

2. A Simple Example

In this section, we verify the correctness and timing properties of a simple timed automaton. We present both manual and machine-checked proofs. Our model of timed automata is based on work by Merritt, Modugno, and Tuttle [13] and by Lynch and Attiya [11]. We describe this model by means of an example in this section.

Consider a counting automaton C_k, which decrements a counter with initial value k and issues a single report when the counter reaches 0. We will verify that C_k implements the specification given by another automaton R, which just issues a single report. We will also establish bounds a_1 and a_2 on how long it takes the specification automaton R to issue its report based on k and the time bounds c_1 and c_2 for the actions of the implementation automaton C_k. Figure 1 defines the two automata.

The untimed part of each automaton is a simple state-transition system. Actions are said to be *enabled* in the states satisfying their preconditions. Actions are classified as *external* or *internal* so that we may compare an implementation with its specification.

To describe timing properties, the actions are partitioned into *tasks*. A task is *enabled* when any of its actions are enabled. Lower and upper bounds, $lower(C)$ and $upper(C)$, on each task C specify how much time can pass after C becomes enabled before either one of its actions occurs or the task is disabled. The upper bound can be infinite.

The timed part of each automaton contains three additional state components: a real-valued variable *now* representing the current time, and two functions *first* and *last* representing the earliest and latest times that some action from each task can occur. All times are absolute, not incremental. All tasks that are not enabled have trivial *first* and *last* components (i.e., 0 and ∞). In a start state, $now = 0$, and $first(C) = lower(C)$ and $last(C) = upper(C)$ for each enabled task C.

A timed action is a pair associating either an untimed action or a special *time-passage* action with the time it occurs. The time-passage action (ν, t) modifies only the *now* component of the state, setting it equal to t; it cannot let time pass beyond any task's upper bound, i.e., $t \leq last(C)$ for all tasks C. Other actions (π, t) are viewed as happening

Specification automaton: $R(a_1, a_2)$	Implementation automaton: $C_k(c_1, c_2)$
State	**State**
reported, initially *false*	*reported*, initially *false*
	count, initially $k \geq 0$
Actions	**Actions**
External *report*	**External** *report*
Pre: $\neg reported$	Pre: $count = 0 \wedge \neg reported$
Eff: $reported \leftarrow true$	Eff: $reported \leftarrow true$
	Internal *decrement*
	Pre: $count > 0$
	Eff: $count \leftarrow count - 1$
Tasks	**Tasks**
$\{report\}$: $[a_1, a_2]$	$\{report\}$: $[c_1, c_2]$
	$\{decrement\}$: $[c_1, c_2]$

Figure 1. A counting process and its specification

instantaneously at time t. They must not occur before the lower bound for their tasks (i.e., $first(task(\pi)) \leq now$), and they do not modify *now*. They reset the values of *first* and *last* for their task and for any other tasks that are newly enabled or disabled as a result of their effect on the untimed part of the state. We write $s \xrightarrow{(\pi,t)} s'$ to denote a transition of the timed automaton.

An execution of a timed automaton is *admissible* if time increases without bound. A state is *reachable* if it appears in some execution. Properties that are true of every reachable state are *invariants*. The visible behavior of a timed automaton is characterized by its *admissible timed traces*, which are the sequences of external timed actions in admissible executions. We say that one timed automaton *implements* another if any admissible timed trace of the first is also an admissible timed trace of the second.

2.1. Manual Proofs

We seek to show that $C_k(c_1, c_2)$ implements $R(a_1, a_2)$ when $a_1 = (k+1)c_1$ and $a_2 = (k+1)c_2$. Note that our notion of correctness for timed automata incorporates both safety properties (e.g., that C_k issues no more than one report) and liveness properties (e.g., that it issues its report in time at most $(k+1)c_2$).

The key steps in the proof are (1) proving that the states of C_k satisfy an invariant and (2) defining a *simulation mapping* between the states of C_k and those of R. Given such a mapping f, a straightforward proof by induction shows that f maps any admissible execution of C_k to some admissible execution of R. We say that a binary relation f between states of C_k and states of R is a *simulation mapping* from C_k to R if it satisfies the following conditions:

1. If $f(s, u)$, then $u.now = s.now$.
2. If s is a start state of C_k, then there is a start state u of R such that $f(s, u)$.
3. If s and u are reachable states such that $f(s, u)$ and $s \xrightarrow{(\pi,t)} s'$, then there is a state u' of R such that $f(s', u')$, and a sequence of timed actions that takes R from u to

u' and has the same visible behavior as (π, t).

For the first step, we prove that C_k preserves the invariant $count > 0 \Rightarrow \neg reported$. This invariant is trivially true in C_k's initial state. Only the *report* action can make *reported* true, and that can happen only if $count = 0$. Thus, every action preserves the invariant.

For the second step, we define $f(s, u)$, where s is a state of C_k and u is a state of R, to hold if and only if the untimed components of the two states are the same and the timing components are properly related, i.e., if and only if

- $u.now = s.now$
- $u.reported = s.reported$
- $u.first(report) \leq \begin{cases} s.first(decrement) + s.count \cdot c_1 & \text{if } s.count > 0 \\ s.first(report) & \text{otherwise} \end{cases}$
- $u.last(report) \geq \begin{cases} s.last(decrement) + s.count \cdot c_2 & \text{if } s.count > 0 \\ s.last(report) & \text{otherwise} \end{cases}$

We prove that f is a simulation mapping from $C_k(c_1, c_2)$ to $R(a_1, a_2)$ when $a_1 = (k+1)c_1$ and $a_2 = (k+1)c_2$. If $f(s, u)$, then $u.now = s.now$ by definition. It is also easy to see that $f(s_0, u_0)$, where s_0 and u_0 are the start states of C_k and R. Finally, suppose s and u are reachable states of C_k and R such that $f(s, u)$ and that $s \xrightarrow{(\pi,t)} s'$. We show that there is a sequence of timed actions with the same visible behavior as (π, t) that takes R from u to some state u' such that $f(s', u')$. There are three possibilities for π.

1. If $\pi = report$, we show that R can take a *report* step, resulting in a state u' such that $f(s', u')$. Because $f(s, u)$ and $(report, t)$ is enabled in s, we have $u.reported = s.reported = false$, $s.count = 0$, and $u.first(report) \leq s.first(report) \leq t$. Hence $(report, t)$ is enabled in u and $f(s', u')$, because $u'.now = u.now = s.now = s'.now$.

2. If $\pi = decrement$, we show that R need not take any step. Since *decrement* is internal, it suffices to show that $f(s', u)$. Because $f(s, u)$ and *decrement* occurred, we have $u.now = s.now = s'.now$, $u.reported = s.reported = s'.reported$, $s.count > 0$, and $u.first(report) \leq s.first(decrement) + s.count \cdot c_1 \leq s.now + s.count \cdot c_1$. We consider two cases. If $s.count > 1$, then $u.first(report) \leq s.now + c_1 + (s.count - 1) \cdot c_1 = s'.first(decrement) + s'.count \cdot c_1$, because the time bound for *decrement* is reset. If $s.count = 1$, then $\neg s.reported$ by the invariant for C_k and $u.first(report) \leq s.now + c_1 = s'.first(report)$, because *report* is newly enabled. Similary, $u.last(report) \geq s'.last(decrement) + s'.count \cdot c_2$ if $s.count > 1$ and $u.last(report) \geq s'.last(report)$ if $s.count = 1$.

3. If $\pi = \nu$, we show that R can take a corresponding (ν, t) step, resulting in a state u' such that $f(s', u')$. Since $t \geq s.now = u.now$, to show that (ν, t) is enabled in u', we only need to check that $t \leq u.last(report)$. If $s.count > 0$, then $t \leq s.last(decrement) < u.last(report)$. Otherwise, $t \leq s.last(report) \leq u.last(report)$. Since time-passage actions modify only the *now* components of the states, and $u'.now = t = s'.now$, we have $f(s', u')$.

```
AutomatonCount (C, k): trait
  includes Automaton(C), CommonActionsRC, Natural
  States[C] tuple of count: N, reported: Bool
  introduces
    k                   : → N
    decrement, report : → Actions[C]
  asserts
    sort Actions[C] generated freely by report, decrement
    sort Tasks[C] generated freely by task
    ∀ s, s': States[C], a, a': Actions[C]
      isExternal(report); isInternal(decrement); common(report) = report;
      start(s)                  ⇔ ¬s.reported ∧ s.count = k;
      enabled(s, report)        ⇔ s.count = 0 ∧ ¬s.reported;
      effect(s, report, s')     ⇔ s'.count = s.count ∧ s'.reported;
      enabled(s, decrement)     ⇔ s.count > 0;
      effect(s, decrement, s') ⇔ s'.count + 1 = s.count ∧ s'.reported = s.reported;
      inv(s)                    ⇔ s.count > 0 ⇒ ¬s.reported
  implies
    Invariants(C, inv)
    ∀ s: States[C], a: Actions[C]
      enabled(s, task(decrement)) ⇔ enabled(s, decrement);
      enabled(s, task(report)) ⇔ enabled(s, report);
      a = report ∨ a = decrement
```

Figure 2. LSL trait defining untimed part of automaton C_k

2.2. Machine-Checked Proofs

In order to check this simulation proof mechanically, we must first create machine-readable versions of the definitions and abstractions used in the manual proof, filling in details normally suppressed in careful, but not completely formal proofs. To this end, we use the Larch Shared Language (LSL), which provides suitable notational and parametrization facilities. Later, we use the Larch Prover (LP), which provides assistance for reasoning in first-order logic. The versions of these tools used for this paper are enhancements of the versions described in [5, 4]; the primary differences are that both tools now support full first-order logic, and that LP now has features for reasoning about linear inequalities [15] similar to those in the Boyer-Moore prover [2, 3] and in PVS [17].

2.3. Machine-Readable Definitions

Figure 2 contains an LSL definition of the untimed part of automaton C_k. This formal definition mimics the definition given in Figure 1. It builds upon a library of LSL specifications that defines general notions related to timed automata and that can be reused in simulation proofs like the ones in this paper.

The basic unit of specification in LSL is a *trait*, which introduces symbols for *sorts* (such as `Actions[C]` and `States[C]`) and *operators* (such as `decrement` and `enabled`), and which constrains their properties by axioms expressed in first-order logic. Sort symbols denote disjoint nonempty sets of values; operator symbols denote total mappings from

```
SimulationRC: trait
  includes
    TimedAutomaton(R, br, TR), AutomatonReport(R),
    TimedAutomaton(C, bc, TC), AutomatonCount(C, k)
  introduces
    a, c :                          → Bounds
    f    : States[TC], States[TR] → Bool
  asserts ∀ u: States[TR], s: States[TC], cr: Tasks[R], cc: Tasks[C]
    br(cr) = a;              % bounds [a1, a2] for tasks of R
    bc(cc) = c;              % bounds [c1, c2] for tasks of C
    c.bounded;
    a = (k+1)*c;
    f(s, u) ⇔
        u.now = s.now
      ∧ u.basic.reported = s.basic.reported
      ∧ (if s.basic.count > 0
          then s.bounds[task(decrement)] + (s.basic.count * c)
          else s.bounds[task(report)]) ⊆ u.bounds[task(report)]
  implies SimulationMap(TC, TR, f)
```

Figure 3. LSL trait defining the timed simulation of R by C_k

tuples of values to values. When a trait *includes* another, it inherits the other trait's symbols and axioms. Thus `AutomatonCount` inherits general properties of automata from the library trait `Automaton` and properties of the natural numbers from the trait `Natural` in the Larch handbook [5]. Because LSL requires sorts to represent disjoint nonempty sets, `AutomatonCount` also includes the following trait `CommonActionsRC`, and it defines a map `common` from the actions of `C` to a new sort `CommonActions` so that the traces of `C` (whose actions have sort `Actions[C]`) can be compared with those of `R` (whose actions have sort `Actions[R]`).

```
CommonActionsRC: trait
  introduces report: → CommonActions
  asserts CommonActions generated by report
```

When a trait *implies* another, its theory is claimed to include that of the other. The `implies` clause in `AutomatonCount` claims that the predicate `inv` satisfies the axioms of the library trait `Invariants`; Figure 4 contains an LP proof of this claim. The `implies` clause also lists several lemmas that are easy to verify with LP, but are not noticed automatically by the prover.

The specification of R's untimed part is similar to, but shorter than C_k's. The trait `SimulationRC` in Figure 3 uses the library trait `TimedAutomaton` to extend these two specifications to the timed parts of C_k and R. It also claims that a particular relation `f` is a simulation mapping, i.e., that `f` satisfies the properties of the library trait `SimulationMap`. Later we use LP to verify this claim.

```
execute AutomatonCount
set proof-methods ⇒, normalization
prove start(s) ⇒ inv(s)
  qed
prove inv(s) ∧ isStep(s, a, s') ⇒ inv(s') by cases on a
  qed
```

Figure 4. LP proof of invariance for automaton C_k

The most notable feature of the formalization process is that it is quite mechanical to move from definitions such as those in Figure 1 to LSL definitions. In fact, one could write a compiler to perform the translation.

2.4. Machine-Checkable Proofs

This section contains two entire LP proof scripts, one showing that automaton C_k preserves its invariant, and the other that `f` is indeed a timed forward simulation. LP's proof mechanisms include proofs by cases and induction, equational term rewriting (for simplifying hypotheses and conjectures), and decision procedures for proving linear inequalities.

The LP proof of invariance in Figure 4 is virtually identical to the manual proof. It begins with commands that load the axioms of the trait `AutomatonCount` and that set LP's proof methods. That the invariant holds in the initial state is proved without human guidance. That the invariant is preserved by all actions requires exactly the same guidance as in the manual proof: separate consideration of each action.

The proof that `f` is a simulation mapping in Figure 5 is considerably longer than the proof of invariance, but similar in length and structure to the manual proof.[1] The user guides the proof that each start state `s` of C_k corresponds to a start state `u` of R by producing an explicit description of `u` and showing LP why "it is easy to see that `f(s, u)`."[2] In the induction step of the proof, `s'c` and `sc` are fresh constants that LP generates and substitutes for the variables `s` and `s'` when it assumes the hypotheses of the implication it is trying to prove. In addition to suggesting separate consideration of each action, and to providing the simulating execution fragment for each action, the user provides guidance for the induction step of the proof using the `set immunity` and `instantiate` commands, which call LP's attention to instances of the hypotheses (and other facts) used by the decision procedure for linear arithmetic.

3. Fischer's Mutual Exclusion Algorithm

In this section, we use timed automata to model Fischer's well-known timing-based mutual exclusion algorithm, which uses a single shared read-write register [7]. We use

[1]Two periods `..` in this proof script mark the end of a multiline LP command; they do not indicate any elision of the script.

[2]While the length of this proof suggests room for improvement in LP, the need to consider the case $k = 0$ separately suggests room for clarification in the manual proof.

```
execute SimulationRC
set proof-methods ⇒, normalization
prove f(s, u) ⇒ u.now = s:States[TC].now
  qed
prove start(s:States[TC]) ⇒ ∃ u (start(u) ∧ f(s, u))
    resume by specializing u to [[false], 0, update({}, task(report), a)]
      instantiate c:Tasks[C] by task(report) in *Hyp
      instantiate c:Tasks[C] by task(decrement) in *Hyp
      resume by specializing a:Actions[R] to report
        resume by case k = 0
          resume by cases on c:Tasks[R]
          resume by cases on c:Tasks[R]
  qed
declare variables u: States[TR], alpha: StepSeq[TR]
set immunity ancestor
prove
  f(s, u) ∧ isStep(s:States[TC], a, s') ∧ inv(s:States[TC]) ∧ inv(u:States[TR])
    ⇒ ∃ alpha (execFrag(alpha) ∧ first(alpha) = u ∧ f(s', last(alpha))
                ∧ trace(alpha) = trace(a:Actions[TC]))
  by cases on a:Actions[TC]
  ..
      resume by cases a1c = report, a1c = decrement
        % Case 1: simulate report action
        resume by specializing alpha to
          ({uc}) {addTime(report, uc.now),
                  [[true], uc.now, update(uc.bounds, task(report), [false,0,0])]}
          ..
          resume by cases on c:Tasks[R]
        % Case 2: simulate decrement action
        resume by specializing alpha to {uc}
          instantiate c:Tasks[C] by task(report) in *impliesHyp
          instantiate c:Tasks[C] by task(decrement) in *impliesHyp
          resume by case s'c.basic.count = 0
            instantiate t:Time by c.first, n by s'c.basic.count in Real
            instantiate t:Time by c.last, n by s'c.basic.count in Real
      % Case 3: simulate passage of time
      resume by specializing alpha to ({uc}) {nu(lc), [uc.basic, lc, uc.bounds]}
        resume by cases on c:Tasks[R]
          instantiate c:Tasks[C] by task(report) in *Hyp
          resume by case sc.basic.count = 0
            instantiate c:Tasks[R] by reportTask in *Hyp
            instantiate n by sc.basic.count in TimedAutomaton
            instantiate c:Tasks[C] by task(decrement) in *Hyp
  qed
```

Figure 5. LP proof that f is a simulation mapping

State

$region_i \in \{remainder, trying, critical, exit\}$ for $i \in I$, initially *remainder*

Actions

External try_i
Pre: $region_i = remainder$
Eff: $region_i \leftarrow trying$

External $crit_i$
Pre: $region_i = trying$
for all j, $region_j \neq critical$
Eff: $region_i \leftarrow critical$

External $exit_i$
Pre: $region_i = critical$
Eff: $region_i \leftarrow exit$

External rem_i
Pre: $region_i = exit$
Eff: $region_i \leftarrow remainder$

Tasks

$\{try_i\}$: $[0, \infty]$
$\{crit_i : i \in I\}$: $[0, 5a + 2c]$
$\{exit_i\}$: $[0, \infty]$
$\{rem_i\}$: $[0, 2a]$

Figure 6. Automaton M: a simple specification for mutual exclusion

simulations to prove not only mutual exclusion, but also an upper bound on the time to reach the critical region, which is much harder to prove than mutual exclusion. We believe that the use of simulations both gives insight into the algorithm and yields a convincing proof that can be checked using automated provers like LP.

3.1. A Specification for Mutual Exclusion

We begin with the specification in Figure 6 of a *mutex object* M described as a timed automaton that keeps track of the regions of all processes (with indices in I) and ensures that at most one process is in its critical region at any time.

Notice that all *crit* actions belong to the same task. Intuitively, this means that if one or more processes are trying to acquire the resource when it is free, then one will succeed within time $5a + 2c$. (The parameters a and c here are derived from the bounds we will impose on the tasks of Fischer's algorithm.)

3.2. Fischer's Timed Mutual Exclusion Algorithm

In this algorithm, shown in Figure 7, there is a single shared register. Intuitively, if some process has the resource, the register contains the index of that process; and if no process has, wants, or is releasing the resource, the register contains 0.[3] Each process trying to obtain the resource tests the register until its value is 0, and then sets it to its own index. Since several processes may be competing for the resource, the process waits for the register value to stabilize, and then checks the register again. The process whose index remains in the register (the last one to set it) gets the resource, and the others return to testing until the register is 0 again. When a process exits, it resets the register to 0.

One problem with this algorithm as described so far is that a fast process might not wait

[3]We assume $0 \notin I$.

State

$pc_i \in \{remainder, test, set, check, leave\text{-}trying, critical, reset, leave\text{-}exit\}$ for $i \in I$, initially *remainder*
$x \in I \cup \{0\}$, initially 0

Actions

External try_i
Pre: $pc_i = remainder$
Eff: $pc_i \leftarrow test$

Internal $test_i$
Pre: $pc_i = test$
Eff: if $x = 0$ then $pc_i \leftarrow set$

Internal set_i
Pre: $pc_i = set$
Eff: $x \leftarrow i$
$pc_i \leftarrow check$

Internal $check_i$
Pre: $pc_i = check$
Eff: if $x = i$
then $pc_i \leftarrow leave\text{-}trying$
else $pc_i \leftarrow test$

External $crit_i$
Pre: $pc_i = leave\text{-}trying$
Eff: $pc_i \leftarrow critical$

External $exit_i$
Pre: $pc_i = critical$
Eff: $pc_i \leftarrow reset$

Internal $reset_i$
Pre: $pc_i = reset$
Eff: $x \leftarrow 0$
$pc_i \leftarrow leave\text{-}exit$

External rem_i
Pre: $pc_i = leave\text{-}exit$
Eff: $pc_i \leftarrow remainder$

Tasks
Assume $a < b \leq c$

$\{try_i\}$: $[0, \infty]$
$\{test_i\}$: $[0, a]$
$\{set_i\}$: $[0, a]$
$\{check_i\}$: $[b, c]$

$\{crit_i\}$: $[0, a]$
$\{exit_i\}$: $[0, \infty]$
$\{reset_i\}$: $[0, a]$
$\{rem_i\}$: $[0, a]$

Figure 7. Automaton F: Fischer's algorithm

long enough, check the register before a slow process has managed to set it, and so proceed to its critical region. The slow process might then overwrite the register with its own index, which would remain there until the slow process checked it and entered its critical region as well, violating mutual exclusion. This situation can be avoided by a simple time restriction that requires every process to wait long enough for any other process to see the new value in the register, or else to overwrite it. Formally, $upper(set_i) < lower(check_j)$ for all $i, j \in I$.

Notice that every action is a task by itself, corresponding to our intuition that each process acts independently of the other processes. We define timing conditions for all the tasks other than try_i and $exit_i$ in order to prove the timing conditions for the specification.[4]

Finally, we use the following invariants in our proofs of the simulations. The last, which we call *strong mutual exclusion*, clearly implies mutual exclusion.

[4]We can show tight, slightly better bounds at the cost of additional complexity. See [9].

State

$region_i \in \{remainder, trying, critical, exit\}$ for $i \in I$, initially *remainder*
$status$, an element of $\{start, seized, stabilized\}$, initially *start*

Actions

External try_i
Pre: $region_i = remainder$
Eff: $region_i \leftarrow trying$

Internal $seize$
Pre: for some i, $region_i = trying$
$status = start$
for all i, $region_i \neq critical$
Eff: $status \leftarrow seized$

Internal $stabilize$
Pre: $status = seized$
Eff: $status \leftarrow stabilized$

External $crit_i$
Pre: $region_i = trying$
$status = stabilized$
Eff: $region_i \leftarrow critical$
$status \leftarrow start$

External $exit_i$
Pre: $region_i = critical$
Eff: $region_i \leftarrow exit$

External rem_i
Pre: $region_i = exit$
Eff: $region_i \leftarrow remainder$

Tasks

$\{try_i\}$: $[0, \infty]$
$\{seize\}$: $[0, 3a + c]$
$\{stabilize\}$: $[0, a]$
$\{crit_i : i \in I\}$: $[0, a + c]$
$\{exit_i\}$: $[0, \infty]$
$\{rem_i\}$: $[0, 2a]$

Figure 8. Automaton I: an intermediate milestone automaton

1. If $x = i$, then $pc_i \in \{check, leave\text{-}trying, critical, reset\}$.
2. If $x = i \neq 0$, $pc_i = check$, and $pc_j = set$ then $first(check_i) > last(set_j)$.
3. If $pc_i \in \{leave\text{-}trying, critical, reset\}$, then $x = i$ and $pc_j \neq set$ for all j.

3.3. Milestones: An Intermediate Abstraction

While we could give a simulation mapping directly from F to M, it seems useful to introduce an intermediate level of abstraction that we believe captures the intuition behind Fischer's algorithm. We then define two intuitive simulation mappings, one from the algorithm to the intermediate automaton, and one from the intermediate automaton to the specification, thereby proving that the algorithm implements the specification.

The intermediate automaton, shown in Figure 8, expresses two *milestones* toward the goal of some process reaching its critical region. The first occurs when a process sets the register from 0 to its index; we say that the register is *seized* at this point. After this, the register will have some non-zero value until some process reaches its critical region and resets the register as it exits. Thus only processes that have already tested the register will set it. The second milestone, a *stabilize* event, occurs when the last process sets the register, i.e., when no other process has $pc = set$.

We need one easy invariant for this automaton:

If $status \neq start$, then $region_i = trying$ for some i and $region_j \neq critical$ for all j.

3.4. Simulations

3.4.1. Simulation from Intermediate to Specification

We define a relation g between the states of I and M, where $g(s,u)$ if and only if:

- $u.now = s.now$
- $u.region_i = s.region_i$
- $u.last(crit) \geq \begin{cases} s.last(seize) + 2a + c & \text{if } seize \text{ is enabled in } s \\ s.last(stabilize) + a + c & \text{if } stabilize \text{ is enabled in } s \\ s.last(crit) & \text{if } crit_j \text{ is enabled in } s \text{ for some } j \end{cases}$
- $u.last(rem_i) \geq s.last(rem_i)$ if $s.region_i = exit$

It is straightforward to show that g is a simulation mapping. This simulation corresponds to the notion that *seiz*ing and *stabiliz*ing are just steps that need to be done before a process can enter its critical region. Note, however, that *seize* and *stabilize* are not actions of individual processes, but of the entire system.

3.4.2. Simulation from Algorithm to Intermediate

We define a relation f between the states of F and I, where $f(s,u)$ if and only if:

- $u.now = s.now$
- $u.region_i = \begin{cases} trying & \text{if } s.pc_i \in \{test, set, check, leave\text{-}trying\} \\ critical & \text{if } s.pc_i = critical \\ exit & \text{if } s.pc_i \in \{reset, leave\text{-}exit\} \\ remainder & \text{if } s.pc_i = remainder \end{cases}$
- $u.status = \begin{cases} start & \text{if } s.x = 0 \text{ or } s.pc_i \in \{critical, reset\} \text{ for some } i \\ seized & \text{if } s.x \neq 0,\ s.pc_i \notin \{critical, reset\} \text{ for all } i, \text{ and} \\ & \quad s.pc_i = set \text{ for some } i \\ stabilized & \text{if } s.x \neq 0 \text{ and } s.pc_i \notin \{set, critical, reset\} \text{ for all } i \end{cases}$
- $u.last(seize) \geq \begin{cases} s.last(reset_i) + 2a + c \quad \text{if } s.pc_i = reset \\ \min_i\{w(i)\} \quad \text{if } s.x = 0 \text{ where } w(i) = \begin{cases} s.last(test_i) + a & \text{if } s.pc_i = test \\ s.last(set_i) & \text{if } s.pc_i = set \\ s.last(check_i) + 2a & \text{if } s.pc_i = check \\ \infty & \text{otherwise} \end{cases} \end{cases}$
- $u.last(stabilize) \geq s.last(set_i) \quad$ if $s.pc_i = set$
- $u.last(crit) \geq \begin{cases} s.last(check_i) + a & \text{if } s.pc_i = check \text{ and } s.x = i \\ s.last(crit_i) & \text{if } s.pc_i = leave\text{-}trying \end{cases}$
- $u.last(rem_i) \geq \begin{cases} s.last(reset_i) + a & \text{if } s.pc_i = reset \\ s.last(rem_i) & \text{if } s.pc_i = leave\text{-}exit \end{cases}$

The *now* and *region* correspondences are straightforward; that for *status* follows naturally from the intuition given earlier about the *seize* and *stabilize* milestones. The first inequality for *seize* says that if some process is about to *reset*, then the simulated state must allow the register to be seized at least up to $2a+c$ after the *reset* occurs. The second inequality for *seize* says that if $x = 0$ (so, by strong mutual exclusion, no process is about to *reset*) then the time until the register must be seized is determined by the minimum of a set of possible times, each corresponding to some candidate process that might set x. For instance, if some process i is about to set x, then the corresponding time is only the maximum time until it does so, while if i is about to test x, then the corresponding time is an additional a after the *test* occurs. The interpretations for the remaining inequalities are similar.

Most of the proof that f is a simulation mapping involves straightforward but tedious checking that each action of F preserves the mapping, since the corresponding behavior in I is easy to intuit. (It is the same action if it is external, and no action if not.) The one exception to this is the *set* action. Recall the intuition here is that, if it is the first time the register is set (i.e., it was previously 0), then there must be a corresponding *seize* action. If no other process is about to set the register (i.e., no other process has $pc = set$), then this is the last *set* before some process enters its critical region, and so there must be a corresponding *stabilize* action. We examine this case and its proof in more detail.

If s and u are reachable states of F and I such that $f(s,u)$ and $s \xrightarrow{(set_i,t)} s'$, then $s.pc_i = set$, $s'.pc_i = check$, and $s'.x = i \neq 0$. By strong mutual exclusion, $s'.pc_j \notin \{critical, reset\}$ for all j. We have the following cases:

1. If $s.x = 0$, let u' be such that $u \xrightarrow{(seize,t)} u'$. The state exists because $u.status = start$, $u.region_i = trying$, and $u.region_j \neq critical$ for all j.
   (a) If $s.pc_j \neq set$ for all $j \neq i$, then let u'' be such that $u' \xrightarrow{(stabilize,t)} u''$, which is possible since $u'.status = seized$. So $u'' = u$ except that $u''.status = stabilized$, $u''.last(seize) = \infty$ and $u''.last(crit) = s.now + a + c$. Since $s.now + a + c$ is greater than any of the time bounds that occur in the condition for $last(crit)$, and $s'.pc_j \neq set$ for all j, we have $f(s', u'')$.
   (b) If $s.pc_j = set$ for some $j \neq i$, then we see that $f(s', u')$ since $s'.pc_j = set$ and $u' = u$ except that $u'.status = seized$, $u'.last(seize) = \infty$, and $u'.last(stabilize) = s.now + a \geq s'.last(set_{j'})$ for all j' such that $s'.pc_{j'} = set$.
2. If $s.x \neq 0$ and $s.pc_j \neq set$ for all $j \neq i$, then let u' be such that $u \xrightarrow{(stabilize,t)} u'$. The state exists because $u.status = seized$, and $u' = u$ except that $u'.status = stabilized$, $u'.last(stabilize) = \infty$, and $u'.last(crit) = s.now + a + c$. Since $s.now + a + c$ is greater than any of the time bounds that occur in the condition for $last(crit)$, and $s'.pc_j \neq set$ for all j, we have $f(s', u')$.
3. If $s.x \neq 0$ and $s.pc_j = set$ for some $j \neq i$, then $f(s', u)$ since $u.status = seized$, and $s'.pc_j = set$.

Our method of proof uses old, familiar techniques (invariant assertions and simulation mappings) in a novel way (on timed automata) to provide rigorous proofs of timing properties. The time bounds established by this simulation are new; there was no clear, rigorous proof of them before. Furthermore, the bounds aren't completely obvious: the extra c is necessary; we can demonstrate executions that need this extra time. We used the same library of LSL traits that we used for the counting process to formalize these automata and simulations, and we used LP to check the entire proof.

4. Conclusions

We have defined, within the Larch Shared Language, a set of abstractions to support proofs of timing properties of timed systems. We have used these abstractions to carry out computer-aided proofs of time bounds for two sample algorithms—a simple counter and Fischer's mutual exclusion protocol—using invariant assertion and simulation techniques.

We see several advantages of this general approach. Because they can be used for proofs of timing properties in addition to ordinary correctness properties, invariants and simulations are very powerful in the real-time setting. The invariants and simulation mappings also serve as "documentation", expressing key insights about a system's behavior (including its timing). Our experience in going from the simple counter to Fischer's algorithm suggests that these methods are scalable to systems of realistic size. They also appear to provide assistance when modifying systems. When we modify a system or its specification only slightly, we expect that LP will be able to recheck most of the original proof automatically, thereby allowing us to concentrate our attention on what has truly changed without having to worry that we have overlooked some important detail.

The first proof we attempted, that of the counter, took many weeks. Making it work successfully required understanding the manual proof better (e.g., that it relied on an invariant of the automaton C), finding LSL formalizations that were easy to reason about using LP, and finding appropriate LP proof strategies (e.g., for dealing with transitivity before LP was enhanced with decision procedures for linear inequalities). As a result of our increased understanding, and of enhancements made to LP in response to our experience, the proof for Fischer's algorithm took much less time—about four days to fill in all the details of the last simulation, from F to I, which was the most difficult. This amount of time does not seem unreasonable, given that we get the added assurance of a machine-checked proof. But we would like to reduce further the amount of time and user guidance required for proofs of this sort. We expect this to happen as we refine our formalizations and our tools, and we believe that practical machine-checked proofs for real-time processes are not such a distant goal.

Finally, we expect to use our methods to prove timing properties for many more examples. We also expect to extend the timed automaton model used in this paper to encompass other timing-based systems that arise in practice. For example, work in [6] on the Generalized Railroad Crossing example uses a slightly more general timed automaton model [10, 12]; nevertheless, the proof uses simulation methods very similar to those in this paper.

REFERENCES

1. Martin Abadi and Leslie Lamport. The existence of refinement mappings. *Theoretical Computer Science*, 2(82):253–284, 1992.
2. Robert S. Boyer and J Strother Moore. *A Computational Logic.* Academic Press, 1979.
3. Robert S. Boyer and J Strother Moore. *A Computational Logic Handbook.* Academic Press, 1988.
4. Stephen J. Garland and John V. Guttag. A guide to LP, the Larch Prover. Technical Report 82, DEC Systems Research Center, December 1991.
5. John V. Guttag and James J. Horning. *Larch: Languages and Tools for Formal Specification.* Springer-Verlag, 1993.
6. Constance Heitmeyer and Nancy Lynch. The generalized railroad crossing: A case study in formal verification of real-time systems. In *Proceedings of the 15th IEEE Real-Time Systems Symposium*, San Juan, Puerto Rico, December 1994. To appear.

7. Leslie Lamport. A fast mutual exclusion algorithm. *ACM Transactions on Computer Systems*, 5(1):1–11, February 1987.
8. P. Loewenstein and David L. Dill. Verification of a multiprocessor cache protocol using simulation relations and higher-order logic. In E. M. Clarke and R. P. Kurshan, editors, *Computer-Aided Verification '90*, number 531 in LNCS, pages 302–311. Springer-Verlag, 1990.
9. Victor Luchangco. Using simulation techiniques to prove timing properties. Master's thesis, MIT Electrical Engineering and Computer Science, 1994. In progress.
10. Nancy Lynch. Simulation techniques for proving properties of real-time systems. Technical Memo MIT/LCS/TM-494, Laboratory for Computer Science, Massachusetts Institute Technology, Cambridge, MA, November 1993.
11. Nancy Lynch and Hagit Attiya. Using mappings to prove timing properties. Technical Memo MIT/LCS/TM-412.e, Lab for Computer Science, Massachusetts Institute Technology, Cambridge, MA, November 1991.
12. Nancy Lynch and Frits Vaandrager. Forward and backward simulations for timing-based systems. In J. W. de Bakker, C. Huizing, and G. Rozenberg, editors, *Proceedings of REX Workshop "Real-Time: Theory in Practice"*, number 600 in LNCS, pages 397–446. Springer-Verlag, 1992.
13. Michael Merritt, F. Modugno, and Mark Tuttle. Time constrained automata. In *CONCUR'91 Proceedings of a Workshop on Theories of Concurrency: Unification and Extension*, Amsterdam, August 1991.
14. Tobias Nipkow. Formal verification of data type refinement. In J. W. de Bakker, W.-P. de Roever, and G. Rozenberg, editors, *Stepwise Refinement of Distributed Systems*, number 430 in LNCS, pages 561–589. Springer-Verlag, 1990.
15. Anna Pogosyants. Incorporating specialized theories into a general purpose theorem prover. Master's thesis, MIT Electrical Engineering and Computer Science, 1994. In progress.
16. F. B. Schneider, B. Bloom, and K. Marzullo. Putting time into proof outlines. In J. W. de Bakker, C. Huizing, W. P. de Roever, and G. Rozenberg, editors, *Real Time: Theory and Practice*, Mook, The Netherlands, June 1991. Springer Verlag.
17. N. Shankar. Verification of real-time systems using PVS. In *Fourth Conference on Computer-Aided Verification*, pages 280–921, Elounda, Greece, June 1993. Springer-Verlag.
18. Jørgen Søgaard-Andersen. *Correctness of Protocols in Distributed Systems.* PhD thesis, Technical University of Denmark, Lyngby, Denmark, December 1993.
19. Jørgen Søgaard-Andersen, Stephen Garland, John Guttag, Nancy Lynch, and Anya Pogosyants. Computer-assisted simulation proofs. In *Fourth Conference on Computer-Aided Verification*, pages 305–319, Elounda, Greece, June 1993. Springer-Verlag.

Position Statements

19

Formal Specification and Analysis of an ISO Communications Protocol

J.R.Rowson
Department of Computer Science, Queen Mary & Westfield College, University of London, Mile End Road, London E1 4NS, United Kingdom

1. INTRODUCTION

This paper demonstrates the practical use of formal methods in a very striking manner by locating an unexpected anomaly in the internationally specified protocol, ISO 8802-2 [1]. This standard describes the service and implementation of the Logical Link Control (LLC) sub-layer protocol constituting the 'upper' portion of the OSI Reference Model's data link layer. In this framework, the peer LLC sub-layer components in different computers communicate with each other by exchanging Protocol Data Units (PDUs) via the Medium Access Control (MAC) sub-layer, which constitutes the remaining 'lower' portion of the data link layer. The LLC sub-layer communicates with its users in the network layer through Service Access Points (SAPs).

This extended abstract includes some results from the full paper's systematic development of progressively more realistic CCS specifications of the service and implementation of the connection phase of the protocol. The purpose of this phase is the establishment of a connection between two user processes in the network layer. The culmination of the development is a full specification of the standard's error recovery methods designed to cope with the unreliability of the MAC. Analysis of this via the University of Edinburgh's Concurrency Workbench provides a verification of the presence of the anomaly.

2. LLC IMPLEMENTATION WITH ERROR FREE COMMUNICATION

For simplicity the CCS specifications are designed for the limited situation of just two network users and two initial assumptions are made in order to postpone consideration of the error recovery procedures. The assumptions are that the MAC provides error free communication (on rec_i and $send_i$) and that all response PDUs are received within the allowed period of time.

Each LLC component starts in the ADM state in which it is ready both to accept a Connect Request message from the local user at the SAP and to receive a SABME PDU from the MAC.

$$ADM \stackrel{\text{def}}{=} Connect\_Request.ADM' + rec(x).\textbf{if } x{=}sabme \textbf{ then } \overline{Connect\_Indication}.CONN$$

If it receives a request for connection from the SAP then it decides locally (the criteria are not given in the standard) whether or not to proceed with the connection. This locally non-deterministic choice is modelled by state *ADM'* using the silent action, τ.

$$ADM' \stackrel{\text{def}}{=} \tau.\overline{Disconnect\_Indication}.ADM + \tau.\overline{send}(sabme).SETUP$$

If the LLC component decides to proceed with the connection then it sends a SABME and enters the SETUP state where it waits for a response via the MAC. Receipt of a UA or DM PDU indicates, respectively, that the remote user has accepted or refused the connection. The third admissible type of PDU expected is a SABME, which indicates the situation of simultaneous connection. In this case UA PDUs are exchanged in both directions.

$$SETUP \stackrel{\text{def}}{=} rec(x).\ \textbf{if}\ x{=}sabme\ \textbf{then}\ \overline{send}(ua).SETUP$$
$$\textbf{else if}\ x{=}ua\ \textbf{then}\ \overline{Connect\_Confirm}\ .NORMAL$$
$$\textbf{else if}\ x{=}dm\ \textbf{then}\ \overline{Disconnect\_Indication}.ADM$$

In state CONN, the LLC component is waiting for a response from the local user that will indicate either acceptance or refusal of the remote request for a connection.

$$CONN \stackrel{\text{def}}{=} Connect\_Response.\overline{send}(ua).NORMAL + Disconnect\_Request.\overline{send}(dm).ADM$$

State NORMAL indicates that connection has been established and transmission of data between the network users may begin. So, $NORMAL \stackrel{\text{def}}{=} \mathbf{0}$.

The two LLC components LLC_i (i =1,2) are defined using appropriate relabelling functions f_i and are combined with the *MAC* in the implementation of the protocol as follows.

$$LLC_i \stackrel{\text{def}}{=} ADM[f_i][rec_i/rec, send_i/send]$$

$$IMPL \stackrel{\text{def}}{=} (LLC_1 \mid MAC \mid LLC_2)\backslash\{rec_1, send_1, rec_2, send_2\}$$

3. LLC SERVICE

The full paper includes the development of two alternative service specifications, each of which captures, to a different degree of abstraction, the behaviour of the protocol. The first service description is expressed as the concurrent composition of two processes, each having the same 'shape' as the LLC agents given above. This service description, which is shown to be observation congruent with the implementation, retains all details of the possible inter-leavings that can arise from the local decisions of the LLC agents over whether or not to proceed with a request to establish a connection. These interleavings, however, concern behaviour that is hidden from the network user processes, or other external observers, via communication at the SAPs. The second version of the service, expressed as a single process, makes use of the testing preorder to abstract away from these hidden details while maintaining the essence of the external behaviour of the protocol from the point of view of its users.

4. IMPLEMENTATION OF ERROR RECOVERY METHODS

It is possible to distinguish in the standard three distinct methods for recovering from the following three types of error which may occur:

(a) the loss of SABME PDUs by the MAC
(b) the loss of response (UA and DM) PDUs by the MAC
(c) the failure by a LLC component to receive a response to a SABME PDU within a predetermined, finite period of time.

From the point of view of the sender of a SABME all three types of error are encompassed by (c) and are dealt with identically by retransmission of the SABME when the wait for a reply is timed out. Both the amount of time that a LLC component waits before retransmission of a SABME and the maximum number of retransmissions are undefined parameters of the standard. The only reference to them is that for "*the proper operation*" of the protocol the amount of time that elapses before expiry of the acknowledgement timer is required to be greater than the maximum time between sending a SABME and receiving a response during "*normal network operation*".

The paper progressively extends the CCS specifications to imitate the error recovery methods. This includes modelling the unreliable MAC and acknowledgement timers (which communicate via *at*) and the extension of the SETUP state to retransmit SABMEs as follows:

$$SETUP \stackrel{\text{def}}{=} (rec(x).\ \textbf{if}\ x{=}sabme\ \textbf{then}\ \overline{send}(ua).SETUP'$$
$$\textbf{else if}\ x{=}ua\ \textbf{then}\ \overline{at}(stop).\ \overline{Connect\_Confirm}.NORMAL$$
$$\textbf{else if}\ x{=}dm\ \textbf{then}\ \overline{at}(stop).\ \overline{Disconnect\_Indication}.ADM\ )$$
$$+\ expired.ADM'$$

$$ADM' \stackrel{\text{def}}{=} \tau.\ \overline{Disconnect\_Indication}.ADM\ +\ \tau.\ \overline{send}(sabme).\ \overline{at}(start).SETUP$$

SETUP now waits either for a PDU or for the timer to expire. If the timer expires first then the decision whether or not to retransmit is imitated by the modified version of *ADM'*. *SETUP'* (given only in the full paper) is a new state introduced to cope with simultaneous connection.

The different types of error demand different recovery methods in the accepting component, however. It is in considering the loss of a UA that a situation of particular interest arises. Here, the sender of the UA that is lost will eventually receive any retransmitted SABME in state NORMAL. The recovery method invoked here forms a small part of the *resetting* phase of the protocol and is exactly that used in normal connection except that a *Reset_Indication* (instead of *Connect_Indication*) is sent to the local user which may then reply with a *Reset_Response* (instead of *Connect_Response*). For the subsequent analysis the artificial action *Ready* is now included in *NORMAL* to indicate that the LLC agent is ready to start transmitting data.

$$NORMAL \stackrel{\text{def}}{=} (rec(x).\textbf{if}\ x{=}sabme\ \textbf{then}\ \overline{Reset\_Indication}.RESET\_CHECK) + Ready.\mathbf{0}$$

$$RESET\_CHECK \stackrel{\text{def}}{=} CONN[Reset\_Response/Connect\_Response]$$

5. DETECTION OF THE ANOMALY

While the CCS specification of this error recovery method was being developed, the suspicion grew that there was an anomaly in the standard's method of dealing with this aspect of type (b) transmission errors. Our concern was that the method seemed to overlook the fact that the decision by the initiating component to stop retransmission of SABMEs may occur when the accepting component has already entered state NORMAL. This would lead to the situation in which the initiator is in state ADM (where no connection is established) and the acceptor is in state NORMAL (where connection is assumed to be established).

To verify this, a *partial* specification of the required behaviour is developed in the process logic of the Concurrency Workbench. The formulation of (half of) the relevant correct behaviour is: "in all reachable states the system is not ready to accept both $Connect\_Request_1$ and $Ready_2$".

$$\nu X.(\neg\ (\langle Connect\_Request_1\rangle \textbf{true}\ \wedge\ \langle Ready_2\rangle \textbf{true})\ \wedge\ [.]X)$$

The Workbench confirmed that the extended version of *IMPL* did *not* have this property, implying that it is possible for just one component to have 'established' a 'connection'. We do stress that this will only occur when the MAC is so unreliable as to lose in succession one UA followed by *n* SABME or UA PDUs, with *n* the given maximum number of retransmissions of SABMEs. We do not know if this case is one that might be considered by the standard as being outside "*normal network operation*". However, since the other recovery methods seem to cater for abnormal operation, it is surprising to us that the standard does not assume the worst here as well. One reason, perhaps, is because there appears to be no obvious cure for this 'bug'.

REFERENCES

1. International Standard ISO 8802-2: 1989. Part 2: Logical link control.

20

Verification Model Reduction through Abstractions

Jean-Charles Grégoire[1]
INRS-Télécommunications, 16, Pl. du Commerce, Verdun, P.Q., CANADA H3E 1H6

1 Introduction

The major problem still facing researchers in the formal verification of finite state systems is the *state space explosion*. This problem occurs partly because the domain of each variable of the system model contributes to the global state and verification typically involves an exhaustive search through the state space, to recognize states corresponding to wanted or unwanted system configurations. In telecommunication or hardware systems, most practical problems have a number of states that exceeds the memory capacity of our computers. New techniques of state-space compression, "on-the-fly" computing, combination of various techniques such as so-called partial order methods or symbolic computing with Binary Decision Diagrams (BDDs) have allowed us to tackle problems several orders of magnitude greater. Yet, it is clear that the "brute force" approach to verification is unlikely to ever be sufficient, and we need to develop not only new techniques but also methods to achieve further practical reduction of the state space. Recent research on this topic has shown promising results (e.g. [Loi94]).

We describe here the salient features of a new verification tool dedicated mainly to the *expression* and the *verification* of communication protocols, but also meant to be suitable to verify distributed applications.

2 Expression

Our modeling language has an algorithmic flavour, and is vaguely related to PASCAL, combined with single-writer/multiple readers synchronous communication primitives and with the nondeterministic repetitive and alternative constructs proposed by Dijkstra [Dij75]. We have found that such a language is quite convenient to express complex problems such as protocols and distributed algorithms in a way that is fairly "natural" for designers. PROMELA[Hol89] is an example of a quite successful modeling language with similar aims. PROMELA and its tool, SPIN, as well some of their limitations, have actually inspired some of our developments.

[1] Prof. Grégoire is supported by the Chaire Cyrille Duquet en Génie logiciel des télécommunications

3 Verification

We have integrated different forms of abstractions which result in a significant decrease in the size of the verification problem. We consider two categories of abstractions: those who reduce the number of transitions, and those who reduce the size of the state information.

3.1 Transition reduction

To achieve a reduction on the number of transitions, we use well-known techniques, namely functional ("compact") transitions, which allows a transition to reflect a complex transformation of the state, *symmetries*[ID93, CFJ93], which restrict the analysis of "similar" behaviours to a unique case, and *partial orders* [GW93], which restrict the amount of interleaving.

3.2 State information compression

State information compression means a reduction of the amount of state information. Such information can be superfluous when it does not contribute to the properties we want to verify, or it can be redundant, when the value of some state variables can be inferred from others.

3.2.1 Value domain abstractions

We make a distinction between two categories of variables in the state space. On the one hand, we have *control variables*, which capture all forms of interaction between processes (i.e. global variables and communication channels) and nondeterministic choices. On the other hand, we have *computation variables*, used mainly for the purpose of computing the values of control variables, or exchanged data.

Computation variables, and their manipulation, are very often removed from verification models, typically through nondeterministic selections where their values might influence behaviour. This may however contribute to generate otherwise unrealistic behaviours, or not capture enough of the model for the purpose of verification.

We manipulate those two types of variables in different ways, to guarantee that computation variables do not contribute to the growth of the state space while still taking their influence into account.

3.2.2 Redundancy elimination

There are cases in practice where state information is redundant, or superfluous. Let us consider for example a process implementing a buffer with a FIFO discipline. We can abstract out the buffer and use only the number of items held as a control variable. This information, combined with the state of a producer and that of a consumer is enough to generate all the data-independent behaviour. Indeed, the behaviour of FIFO buffers does not depend on the value of its content. However, in modeling the buffer this way, we lose the possibility to transmit the information itself to the receiving side.

If the data influences the behaviour of the receiver, we have several courses of action which preserve the expression of the model. Either we keep the buffer in the model, in spite of the redundancy, or we abstract it out and generate all the possible values at the receiver side, a form of nondeterminism which may create more behaviours than the original model would allow. We introduce a mechanism for a third course of action, which is to keep the information behind the scene, i.e. on the evaluation stack rather than in the state information, and give the exact value to the receiver when it is needed, i.e. currently with a read. These so-called "pnambic" techniques can be used with any kind of intermediate buffer structure.

Another technique that we use is the introduction of extra variables in the state space which are evaluated from computation variables but present a better abstraction of the property. One example is a"volume" variable which to count the number of items held in a buffer, and computed from pointer variables in a circular buffer implementation.

3.2.3 Server

In client-server systems, Remote Procedure Calls (RPCs) are very often used as an access mechanism to the operations of the server. This mechanism is a natural extension of function calls. By combining a blocking test mechanism, function calls and persistant (global) computational variables, we can abstract out of the state space (deterministic) servers while preserving their behaviour, and preserving once again the expression of the model.

4 Conclusions

We have presented several techniques of abstraction. These techniques can reduce the number of transitions and the size of the state information. A prototype of new tool integrating these techniques has been realized.

References

[CFJ93] E.M. Clarke, T. Filkorn, and S. Jha. Exploiting symmetry in temporal logic model checking. In *Proc. Conference on Computer Aided Verification*, Lecture Notes in Computer Science, pages 450–462. Springer Verlag, 1993.

[Dij75] E.W. Dijkstra. Guarded commands, nondeterminacy and formal derivation of programs. *Communications of the ACM*, 18(8):453–457, 1975.

[GW93] P. Godefroid and P. Wolper. Using partial orders for the efficient verification of deadlock freedom and safety properties. *Formal Methods in System Design*, 2(2):149–164, April 1993.

[Hol89] G. Holzmann. Algorithms for automated protocol verification. *AT&T Technical Journal*, 68(1):32–44, 1989.

[ID93] C-W. Norris Ip and David L. Dill. Better verification through symmetry. In *International Conference on Computer Hardware Description Languages*, pages 87–100, 1993.

[Loi94] C. Loiseaux. *Vérification symbolique de programmes réactifs à l'aide d'abstractions.* PhD thesis, Université Joseph Fourier de Grenoble, Laboratoire VERIMAG, 1994.

21

Visual Animation of LOTOS using SOLVE

K. J. Turner[a], A. McClenaghan[b]*

[a]Computing Science, University of Stirling, Stirling FK9 4LA (*kjt@compsci.stirling.ac.uk*)

[b]Philips Research Labs., Redhill, Surrey RH1 5HA (*ukrmccl@prl.philips.co.uk*)

SOLVE is an object-oriented approach and toolset based on LOTOS to allow formal requirements capture and visual animation, particularly for interactive systems and digital logic design.

Classification: D.1.5, D.1.7, D.2.1, D.2.2, I.6.2
Keywords: object-oriented programming, visual programming, requirements/specifications, tools and techniques, formal languages, LOTOS (Language Of Temporal Ordering Specification)

1 Overview

The requirements phase in LOTOS-based development has received little attention previously, and was the focus of the project SPLICE I (*Specification and Prototyping for a LOTOS Interactive Customer Environment – Phase I*). Unlike traditional software engineering methods, the SOLVE approach developed by the project allows manipulation of *formal* requirements through *visual animation*. SPLICE studied requirements capture for OSI services, digital logic, neural networks and interactive systems.

SOLVE (*Specification using an Object-oriented, LOTOS-based, Visual language* [1]) uses familiar object-oriented programming concepts translated automatically into LOTOS, thus ensuring a precise basis. The visual animation requires no knowledge of LOTOS, and is suitable for customers/clients, analysts or designers/programmers. SOLVE is partly oriented towards interactive systems, but applications such as digital logic design have also been investigated. SOLVE would benefit from extensions such as more data types, inheritance, dynamic object creation and deletion, dynamic modification of communication paths, and multi-media characteristics.

2 SOLVE Approach

The key concepts in SOLVE are object orientation, interactive animation, and formal specification. SOLVE is designed for those who are not familiar with formal languages (in particular LOTOS). SOLVE allows an analyst to write a requirements specification in an intuitive language that is automatically translated into LOTOS for exploration through visual animation.

A SOLVE specification consists of a number of concurrent objects that communicate via messages. The SOLVE language deliberately avoids the algebraic feel of LOTOS. However, SOLVE descriptions can be automatically translated into LOTOS specifications. An object declaration gives its name, instance parameter sorts and method parameter sorts; an object definition describes inner details. Behaviour within any single simple object is sequential. Method definitions use the following statements:

*The designer of SOLVE and XDILL, supported by the UK Science and Engineering Research Council on SPLICE.

--	comments
Variables *Name : Sort, ...* **EndVariables**	local variables
Assign *(Variable, Value)*	variable bindings
If *Condition* **Then** *Statement* **Else** *Statement* **EndIf**	conditionals
While *Condition* **Do** *Statement* **EndWhile**	loops
AskWaitCall *ObjectName.MethodName (Parameter) (Results)*	blocking calls
TellCall *ObjectName.MethodName (Parameters)*	non-blocking calls

3 SOLVE In Action

SOLVE has been used, for example to define a system for setting the clock of a VCR (*Video Cassette Recorder*). This has an on-screen 24-hour clock that is set using on-screen cursor and control buttons. The cursor can be moved using left or right buttons to select one digit, which can then be adjusted with increment and decrement buttons. The mapping between informal requirements and SOLVE objects, attributes and methods is straightforward. For example, manual operation of a push-button is represented by clicking its icon. A skeleton of the SOLVE description for the VCR clock is given below; the full description is about 250 lines [1], and is translated automatically into about 1700 lines of LOTOS.

```
System VCRclock Is                              -- system name
  PicDecls                                      -- icon declarations
    leftArrow, rightArrow                       -- icon filenames
  ObjectDeclarations                            -- object interfaces
    Object Cursor(Bool) Is                      -- cursor object, flashing parameter
      QueryXPos()(Int)                          -- check cursor x position method
      Left()()                                  -- move cursor left method
  ObjectDefinitions                             -- object internals
    Object Cursor(Bool:flashingOn) Is           -- cursor object
      Method QueryXPos() Is                     -- return cursor x position
        Return(xPos)
      Method Left() Is                          -- move cursor left
        If (xPos Nei 1)                         -- position not 1 (i.e. leftmost)?
          Then
            Assign(xPos,xPos Minus 1)           -- decrement position
            TellCall Interface.SetIcon(xPos,yPos,iconPic)
                                                -- asynchronous call to redisplay icon
          Else                                  -- move attempt ignored
        EndIf
        Return()
```

The SOLVE approach has also been extended to allow interactive visual animation of digital logic specifications written using DILL (*Digital Logic in LOTOS* [3]). DILL provides a library of pre-defined circuit components, combined using a macro language and LOTOS operators. XDILL (*X-based DILL* [2]) visually animates requirements for digital logic circuits. A stage in the visual animation of a D-Latch (a one-bit memory) is illustrated overleaf. In the left-hand animation window the user has clicked on the *F* (False) button to drop the input clock *iC*. The right-hand window shows that the D-Latch has stored its input data *iD*, reflected in the state of output *oQ* and inverted output *oQbar*.

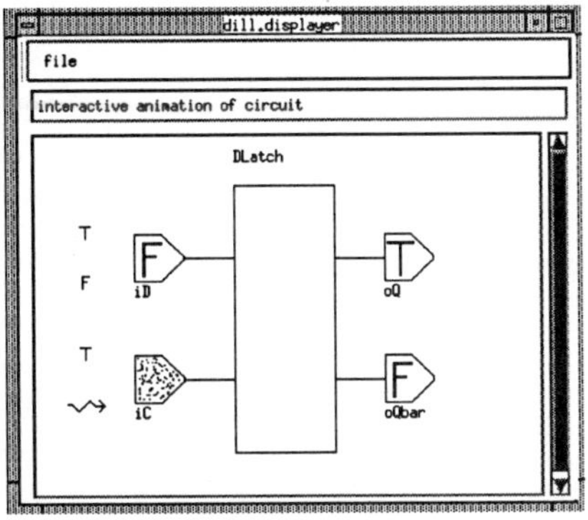

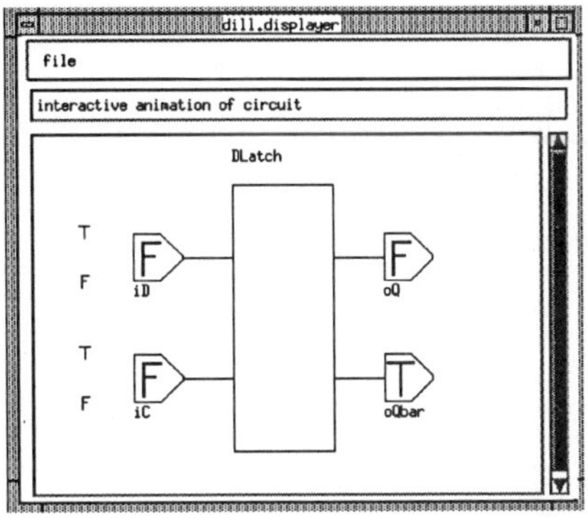

The SOLVE toolset uses the X-Windows environment and is mainly written in C, with some code generated by *yacc* and *X-Designer*. An analyst or customer can use the toolset to explore and assess a requirements specification. Normally automatic selection of events is chosen, allowing direct interaction with the system by mouse clicks and drags.

Front-end tools comprise the editor and parser. SOLVE specifications are textual and can be prepared by a standard editor. However, the syntax-directed editor *syd* was developed with SOLVE in mind; the philosophy of *syd* is to enforce the syntax of the language down to a selected level. The *parser* tool is built using *yacc*, and translates SOLVE to LOTOS fairly straightforwardly [1]. Each object corresponds to a LOTOS process that communicates via a common intermediate process as a communication medium.

Animation tools comprise the simulator, displayer and animator. A modified version of the *hippo* tool is used to carry out LOTOS simulation. Although *hippo* is an old design, its attraction is simple communication with it via standard input/output. Other common tools are used to parse and check the LOTOS specification. To manage an interactive animation, the *animator* tool synchronises *hippo* and *displayer* event offers. The *displayer* displays object icons in response to requests from the *animator*, and passes mouse clicks and drags on object icons to the *animator*. The *animator* manages the interactive animation of a SOLVE description, communicating via Unix pipes to/from the standard input/output of *displayer* and *hippo*.

Overall, SOLVE has gone a long way to achieving the goals of SPLICE. Requirements capture and specification in disparate areas have been undertaken. It has also been possible to demonstrate visual animation of formal requirements specifications without requiring specialist LOTOS knowledge from the tool user.

References

1. A. McClenaghan. SOLVE: Specification using an object-oriented, LOTOS-based, visual language. Technical Report CSM-115, Computing Science, Univ. of Stirling, UK, Jan. 1994.
2. A. McClenaghan. XDILL: An X-based simulator tool for DILL. Technical Report CSM-119, Computing Science, Univ. of Stirling, UK, Apr. 1994.
3. K. J. Turner and R. O. Sinnott. DILL: Specifying digital logic in LOTOS. In R. L. Tenney, P. D. Amer, and M. Ü. Uyar, editors, *Proc. Formal Description Techniques VI*, pages 71–86. North-Holland, Amsterdam, Netherlands, 1994.

Invited Talk

22

How Good is Your Specification Method?

Leslie Lamport[a]

[a]Digital Equipment Corporation, Systems Research Center, 130 Lytton Avenue, Palo, Alto, California 94301, U.S.A.

Abstract

Formal methods abound for specifying concurrent systems. Proponents of a method usually talk about what it does and ignore what it can't do. I will present and explain desiderata for a specification method.

PART SIX

Semantics

23

Relating Maximality-based Semantics to Action Refinement in Process Algebras

J. P. Courtiat and D. E. Saïdouni

LAAS-CNRS 7, av. du Colonel Roche, 31077 Toulouse Cedex (France)
E-mails: {courtiat,saidouni}@laas.fr, Tel: (33) 61336244, Fax: (33) 61336411

This paper extends to process algebras the notion of maximality which has initially been introduced for prime event structures and P/T nets and shows how this notion of maximality may be used for defining an adequate semantics of Basic LOTOS able to support action refinement. Such an approach appears to be more convenient than the classical ST-semantics where non atomic actions are split into start and end sub-actions, as it makes it possible to escape from the potential state space explosion problems induced by action splitting. Main contributions of the paper are related to the definition of a maximality-based operational semantics of LOTOS and to the expression of maximality-based bisimulation equivalences which are shown to be preserved under action refinement.

Keyword Codes: D.1.3; D.3.1; F.3.2

Keywords: Programming Techniques, Concurrent Programming; Programming Languages, Formal Definitions and Theory; Logics and Meanings of Programs, Semantics of Programming Languages

1 Introduction

When applying top-down design approaches, one expects that several good properties, in particular equivalence relations, be preserved under action refinement. It is however well known that the classical interleaving-based bisimulations (either the strong or the weak) are not preserved. For this reason, several authors have proposed new semantics, and by the way new equivalence relations, which have been proven to be preserved under action refinement (look at [16] for an extensive survey).

The interleaving ST-bisimulation (ST-bisimulation in short) without silent moves has been defined on Petri nets [17], and further on prime event structures [15]. The basic idea of the ST-bisimulation is the following: if actions are not atomic, then a system state is determined by (i) the actions that have already taken place and (ii) the actions that have started, but not yet finished their execution. If a bisimulation relates these states, then it has also to relate explicitly these ongoing actions.

Let overlined and underlined action names correspond respectively to the start and the end of these actions, as illustrated below:

$$\underbrace{\overline{a}; \underline{a}; stop}_{S_1} \xrightarrow{\overline{a}} \underbrace{\underline{a}; stop}_{S_2} \xrightarrow{\underline{a}} \underbrace{stop}_{S_3}$$

It can be noted that action a has not yet started its execution in state S_1, that it is being executed in state S_2, and finally that it has terminated its execution in state S_3.

The ST-semantics, using the splitting technique introduced above, has been applied in the literature to process algebras [1, 9]; such an approach is however not completely satisfactory, as the splitting of the actions may significantly contribute to the classical state space explosion problem.

In [18], an alternative definition of the ST-bisimulation has been proposed for prime event structures with silent moves; the main point of this definition is that it does not require anymore to split actions as previously, the partial order relation among events being used instead for determining the set of maximal events in each configuration. The same idea has been used in [8] for defining the maximality preserving bisimulation on labelled P/T nets, and with the hypothesis that all visible actions are non atomic, the maximality preserving bisimulation coincides with the ST-bisimulation.

In this paper, we are therefore interested in extending this notion of maximality to process algebras, using the Basic LOTOS Formal description technique as a target language. The purpose of the paper is therefore twofold: (i) to provide a maximality-based operational semantics of LOTOS in terms of SOS-rules, and (ii) to define maximality-based bisimulations preserved under action refinement, which are the adaptation, within the proposed maximality-based framework, of the classical interleaving-based bisimulations defined by Milner and Park [11, 12].

Current approaches for performing action refinement on a wide spectrum specification language for distributed systems such as CCS, ACP, LOTOS or even Petri nets [1, 10, 7, 13] can be classified into two mains categories: (i) either the action refinement is performed by means of syntactical substitution, i.e. substituting a process for any action to be refined, (ii) or the action refinement is performed by using a new operator of the language.

Performing action refinement by means of a syntactical substitution appears to be very natural and interesting. As already shown in [4], syntactic action refinement implies a careful analysis of synchronization issues, which is far from being obvious when multiway synchronization is to be dealt with, as it is the case for LOTOS. As the purpose of the paper is more oriented towards the demonstration of the interest of the maximality-based semantics, we will extend LOTOS with a refinement operator instead of performing syntactic action refinement as proposed in [14]. A similar approach has been developed in [3, 4] within a causality-based framework.

The paper is organized as follows: Section 2 provides the maximality-based operational semantics of LOTOS in terms of SOS-rules. Section 3 presents the maximality-based operational semantics of LOTOS_ρ which is the language extending LOTOS with a refinement operator. Section 4 investigates behavior equivalence preservation under action refinement, by considering strong and weak maximality-based equivalences, successively. Finally, some conclusions are drawn in section 5.

For the lack of space, proofs are omitted in this paper, and can be found in [5].

2 A Maximality-based Semantics for LOTOS

2.1 Syntax of the language

For illustrating our approach aiming at providing a maximality-based semantics of a concurrent language, we have chosen Basic LOTOS, i.e. the behavioral part of the LOTOS Formal description technique [2]. For convenience reasons, we have slightly modified the standard Basic LOTOS, first by omitting the LOTOS disrupt operator (with the purpose of simplifying the proofs provided in the paper), and second by using a CCS-like syntax for expressing recursive behaviors.

Assuming these slight modifications, the operators to be dealt with are the following:

- ; is the action prefix operator,
- $\gg$ is the process enabling operator,
- [] is the choice operator,
- $|[L]|$ is the general parallel operator, where L represents the finite list of synchronization gates on which the concurrent processes have to synchronize themselves,
- *hide L in E* is the hiding operator, where L represents the finite list of gates which are to be hidden in E,
- $recX.E$ is the fixed-point recursive operator of CCS.

Let $\mathcal{G}$ be a finite set of *gates* ranged over by o, and L stands for an arbitrary subset of $\mathcal{G}$. Let also $Act = \mathcal{G} \cup \{i\}$ be the usual set of actions ranged over by $a, b, \cdots$, where i is the LOTOS notation for the internal action ($i \overset{def}{\notin} \mathcal{G}$). The successful termination action being noted δ, let $\mathcal{G}^+ = \mathcal{G} \cup \{\delta\}$ be the set of the observable actions ranged over by o^+, and let $Act^+ = Act \cup \{\delta\}$ be the set of all actions, ranged over by $a^+, b^+, \cdots$. Finally, let Var be a countable set of variables, ranged over by $\cdots, X, Y, Z$. The syntax of the language may then be defined by the following BNF-like grammar:

$$E ::= stop \mid exit \mid X \mid a; E \mid E \gg E' \mid E[]E' \mid E|[L]|E' \mid hide\ L\ in\ E \mid recX.E$$

Terms generated by this syntax are called *behavior expressions*. Let $\mathcal{L}$, ranged over by $E, F, \cdots$, denote the set of all behavior expressions.

An occurrence of variable X is said to be *bound* in E if it is within a term $recX.E'$ in E, otherwise it is said to be *free*. E is said to be a closed behavior expression if the set of the free variables in E is empty. Let $\mathcal{L}_0$, ranged over by $P, Q, \cdots$, denote the set of all closed behavior expressions, or shortly "*behaviors*". [1]

[1] In order to avoid excessive use of parentheses, the usual priority order for the above operators is assumed. Symbols $|||$ and $||$ denote particular cases of the general parallel composition operator $\_|[L]|\_$, respectively when $L = \emptyset$ and $L = \mathcal{G}$.
Without any loss of generality with respect to the purpose of the paper, only closed behavior expressions are considered.

2.2 The intuition behind the maximality-based semantics

Our motivation for defining a new semantics for LOTOS stems from the need of escaping from the interleaving semantics assumption that actions are atomic. The consequence of such an assumption is that the classical interleaving semantics is not adequate for performing action refinement in a convenient way, as it may be illustrated by the following example:

$$P \equiv a; stop|||b; stop \qquad Q \equiv a; b; stop[]b; a; stop$$

Assuming an interleaving semantics, behaviors P and Q are strongly bisimilar; refining action a by a process offering actions a_1 and a_2 in sequence leads to behaviors P' and Q' which are no more strongly bisimilar, since it is possible for P' to execute action b just after the occurrence of action a_1, which is not the case for process Q':

$$P' \equiv a_1; a_2; stop|||b; stop \qquad Q' \equiv a_1; a_2; b; stop[]b; a_1; a_2; stop$$

Several alternative semantics have then be proposed in the literature, the so-called true concurrency semantics, which distinguish concurrency from a non deterministic sequential composition. Characterizing such semantics for resolving action refinement issues leads to defining behavioral equivalences to be preserved under action refinement, making furthermore as few restrictions as possible on the manner actions are refined into processes.

An interesting semantics in this direction is the ST-semantics, which has first been defined for Petri nets [17], and later translated to other models, like the prime event structures [16, 18]. For prime event structures, the ST-bisimulation has been proven to be the coarsest relation preserved under action refinement [18], and it has also been applied to concrete process algebra [1, 9].

The key point of the ST-semantics consists in splitting each action into its start and its end, as illustrated below. Let $\overline{a}$ and $\underline{a}$ denote the start and the end of action a respectively.

Previous behaviors P and Q may then be written as:

$$P \equiv \overline{a}; \underline{a}; stop|||\overline{b}; \underline{b}; stop \qquad Q \equiv \overline{a}; \underline{a}; \overline{b}; \underline{b}; stop[]\overline{b}; \underline{b}; \overline{a}; \underline{a}; stop$$

It becomes clear that, when using the interleaving transition relation, P and Q can now be distinguished (in other words, they are not anymore strongly bisimilar as previously).

In [18], an alternative definition of the ST-bisimulation has been proposed for prime event structures; the main point of this new definition is that it does not require anymore to split actions as previously, the partial order relation among events being used instead for determining the set of maximal events in each configuration; a configuration characterizes the state of a system, i.e. the set of events which already occurred; maximal events in a configuration are the events which cannot be the cause of any other event present in the configuration. The same idea of maximality has also been applied to P/T nets in [8]. The main interest of the maximality approach, in comparison with the classical ST-semantics approach, is that it is less prone to state space explosion problems, as splitting of actions is no more required.

The purpose of this paper is therefore to show how this notion of maximality may be applied to the definition of a formal semantics of the LOTOS formal description technique, and to study how behavioral equivalences are preserved under action refinement within this framework. The same approach may be translated easily to other languages, like CCS and CSP.

The basic idea of the maximality approach consists in considering that each event corresponds to the start of the execution of an action, instead of being associated with the complete occurrence of an action, as usually assumed in the classical interleaving semantics. Using names for identifying the events, system states are extended into system configurations, in which maximal events are referenced. As stated previously, maximal events in a configuration are events which cannot be the cause of any other event (and consequently action) present in the configuration. In other words, maximal events identify the start of the execution of the actions which may possibly be still executing in the current state of the system. As a consequence, configurations generalize system states as they make it possible (by referencing maximal event names) to identify the set of actions which can concurrently be executing in the system states.

In order to illustrate this notion of configuration, we provide in the sequel some very simple examples. Let us first consider the following derivation which may be applied to previous behavior P:

$$P \xrightarrow{a_x} {}_{\{x\}}stop|||\,{}_{\emptyset}[b; stop]$$

x being the event name identifying the start of action a, nothing can be stated in configuration ${}_{\{x\}}stop|||\,{}_{\emptyset}[b; stop]$ about the termination of action a; as a consequence, x is a maximal event in this configuration (it is in fact the unique maximal event in this configuration). Considering now action b, the following derivation is possible:

$${}_{\{x\}}stop|||\,{}_{\emptyset}[b; stop] \xrightarrow{b_y} {}_{\{x\}}stop|||\,{}_{\{y\}}stop$$

y being the event name identifying the start of action b, nothing can be said about the termination of both actions a and b in configuration ${}_{\{x\}}stop|||\,{}_{\{y\}}stop$; as a consequence, both x and y are maximal events in this configuration.

Let us now consider another derivation which may be applied to previous behavior Q:

$$Q \xrightarrow{a_x} {}_{\{x\}}[b; stop]$$

As previously, x identifies the unique maximal event in configuration ${}_{\{x\}}[b; stop]$; let us now consider the derivation of action b; it should be clear that the derivation of b is only possible once previous action a has indeed terminated its execution (which corresponds to the action prefix semantics). Hence, as event x is no more maximal when the derivation of action b becomes possible, the unique maximal event in the resulting configuration is y, which identifies the start of the execution of action b. The set of the maximal events in the resulting configuration has then been updated (by removing x and adding y), which justifies the following derivation:

$${}_{\{x\}}[b; stop] \xrightarrow{{}_{\{x\}}b_y} {}_{\{y\}}stop$$

${}_{\{x\}}b_y$ will be called an atom; it specifies explicitly the causal relationship which exists between the start of the execution of action b (identified by event name y) and the termination of the action whose start of execution has previously been identified by event name x.

Note finally that configuration ${}_{\{y\}}stop$ is different from configuration ${}_{\{x\}}stop|||\,{}_{\{y\}}stop$, as the first configuration highlights only one maximal event (event name y), whereas two maximal events (event names x and y) are referenced in the second configuration.

Based on the intuition introduced so far, the next paragraph introduces some formal definitions and notations, which will be used for providing the maximality-based operational semantics of LOTOS using the classical SOS style.

2.3 The LOTOS Maximality-based Operational Semantics

Definition 1 *Let $\mathcal{M}$ be a countable set of event names; let $\cdots, x, y, z$ range over $\mathcal{M}$ and let $M, N, \cdots$ range over subsets of $\mathcal{M}$.*

Then, an atom is defined as any triple ${}_M a_x \in 2^{\mathcal{M}} \times A \times \mathcal{M}$ with respect to some set of action A. Let Atm_A denote the set of all atoms with respect to A; the set of all atoms with respect to Act is noted Atm, and we use $\alpha, \beta, \cdots$ to range over it; let Atm^+ be the set of all atoms with respect to Act^+, and we use $\alpha^+, \beta^+, \cdots$ to range over it.

Expressions like ${}_{\{x\}}[c; H|||b; F]$ are called configuration expressions and the set of configuration expressions is noted by C which is ranged over by $\mathcal{E}, \mathcal{F}, \cdots$. $\mathcal{E}$ is a closed configuration expression, shortly a configuration, if for all terms ${}_M[E]$ in $\mathcal{E}$, E is closed. Let finally $\mathcal{C}_0$ denote the set of all closed configurations.

Formally, the syntax of the configuration expressions is characterized by the following BNF-like grammar: $\mathcal{E} \;::=\; {}_M[E] \;\mid\; \mathcal{E} \overset{x}{\gg} \mathcal{E}' \;\mid\; \mathcal{E}\;|[L]|^{\bullet}\;\mathcal{E} \;\mid\; hide^{\bullet}\; L\; in\; \mathcal{E}$.

Symbols $|[L]|^{\bullet}$, $hide^{\bullet}$ and $\overset{x}{\gg}$, as well as $||^{\bullet}$ and $|||^{\bullet}$, are used to distinguish the operators applied to configuration expressions from the corresponding operators applied to LOTOS behavior expressions.

Definition 2 *The set of maximal event names (shortly called names) in configuration expression $\mathcal{E}$ is noted $Max(\mathcal{E})$, and it is defined by induction on the structure of $\mathcal{E}$ as follows:*

$$Max({}_M[E]) = M \qquad Max(\mathcal{E}|[L]|^{\bullet}\mathcal{F}) = Max(\mathcal{E}) \cup Max(\mathcal{F})$$
$$Max(\mathcal{E} \overset{x}{\gg} \mathcal{F}) = Max(\mathcal{E}) \cup (Max(\mathcal{F}) - \{x\}) \qquad Max(hide^{\bullet}\; L\; in\; \mathcal{E}) = Max(\mathcal{E})$$

Definition 3 *The substitution of name y for name x in configuration $\mathcal{E}$, in symbols $y/x\mathcal{E}$, is defined inductively as follows:*

$$y/x\;\, {}_M[E] \equiv {}_{M'}[E] \text{ where } M' = \begin{cases} (M - \{x\}) \cup \{y\} & \text{if } x \in M \\ M & \text{otherwise} \end{cases}$$
$$y/x\;\, \mathcal{E} \overset{z}{\gg} \mathcal{F} \equiv \begin{cases} y/x\mathcal{E} \overset{z}{\gg} y/x\mathcal{F} & \text{if } x \neq z \\ y/x\mathcal{E} \overset{z}{\gg} \mathcal{F} & \text{otherwise} \end{cases}$$
$$y/x\;\, (\mathcal{E}|[L]|^{\bullet}\mathcal{F}) \equiv y/x\;\, \mathcal{E}|[L]|^{\bullet} y/x\;\, \mathcal{F}$$
$$y/x\;\, hide^{\bullet} L\; in\; \mathcal{E} \equiv hide^{\bullet} L\; in\; y/x\;\, \mathcal{E}$$

Definition 4 *Let function $\mathcal{E}\backslash N$ aiming at eliminating a set of names N from configuration $\mathcal{E}$ be defined inductively as follows:*

$${}_M[E]\backslash N \equiv {}_{M-N}[E] \qquad (\mathcal{E}|[L]|^{\bullet}\mathcal{F})\backslash N \equiv \mathcal{E}\backslash N|[L]|^{\bullet}\mathcal{F}\backslash N$$
$$(hide^{\bullet}\; L\; in\; \mathcal{E})\backslash N \equiv hide^{\bullet}\; L\; in\; \mathcal{E}\backslash N \qquad (\mathcal{E} \overset{x}{\gg} \mathcal{F})\backslash N \equiv \mathcal{E}\backslash N \overset{x}{\gg} \mathcal{F}\backslash(N - \{x\})$$

Definition 5 *The maximality-based transition relation* $\longrightarrow \subseteq \mathcal{C}_0 \times Atm^+ \times \mathcal{C}_0$ *is defined as the least relation satisfying the following rules:*

$$(1)\ \frac{}{{}_M[exit] \xrightarrow{M^{\delta x}} {}_{\{x\}}[stop]} \forall x \qquad (2)\ \frac{}{{}_M[a;P] \xrightarrow{M^{a x}} {}_{\{x\}}[P]} \forall x \qquad (3)\ \frac{{}_M[P] \xrightarrow{\alpha^+} \mathcal{P}}{{}_M[P[]Q] \xrightarrow{\alpha^+} \mathcal{P} \qquad {}_M[Q[]P] \xrightarrow{\alpha^+} \mathcal{P}}$$

$$(4.a)\ \frac{{}_M[P] \stackrel{x}{\gg} {}_{\{x\}}[Q] \xrightarrow{\alpha^+} \mathcal{P}}{{}_M[P \gg Q] \xrightarrow{\alpha^+} \mathcal{P}} \qquad (4.b)\ \frac{\mathcal{P} \xrightarrow{M^{a x}} \mathcal{P}'}{\mathcal{P} \stackrel{y}{\gg} \mathcal{Q} \xrightarrow{M^{a z}} z/x\mathcal{P}' \stackrel{y}{\gg} \mathcal{Q}}\ \forall z \notin (Max(\mathcal{P}') - \{x\}) \cup (Max(\mathcal{Q}) - \{y\})$$

$$(4.c)\ \frac{\mathcal{P} \xrightarrow{M^{\delta x}} \mathcal{P}'}{\mathcal{P} \stackrel{y}{\gg} \mathcal{Q} \xrightarrow{M^{i z}} z/y\mathcal{Q}}\ \forall z \notin Max(\mathcal{Q}) - \{y\} \qquad (5.a)\ \frac{{}_M[P] |[L]|^{\bullet} {}_M[Q] \xrightarrow{\alpha^+} \mathcal{P}}{{}_M[P|[L]|Q] \xrightarrow{\alpha^+} \mathcal{P}}$$

$$(5.b)\ \frac{\mathcal{P} \xrightarrow{M^{a x}} \mathcal{P}'}{\mathcal{P}|[L]|^{\bullet}\mathcal{Q} \xrightarrow{M^{a y}} y/x\mathcal{P}'|[L]|^{\bullet}\mathcal{Q}\backslash M \qquad \mathcal{Q}|[L]|^{\bullet}\mathcal{P} \xrightarrow{M^{a y}} \mathcal{Q}\backslash M|[L]|^{\bullet}y/x\mathcal{P}' \text{ if } a \notin L \cup \{\delta\}}$$
$$\forall y \notin (Max(\mathcal{P}') - \{x\}) \cup Max(\mathcal{Q}\backslash M)$$

$$(5.c)\ \frac{\mathcal{P} \xrightarrow{M^{a x}} \mathcal{P}' \qquad \mathcal{Q} \xrightarrow{N^{a y}} \mathcal{Q}'}{\mathcal{P}|[L]|^{\bullet}\mathcal{Q} \xrightarrow{M \cup N^{a z}} z/x\mathcal{P}'\backslash N|[L]|^{\bullet}z/y\mathcal{Q}'\backslash M \text{ if } a \in L \cup \{\delta\}}\ \forall z \notin (Max(\mathcal{P}'\backslash N) - \{x\}) \cup (Max(\mathcal{Q}'\backslash M) - \{y\})$$

$$(6.a)\ \frac{hide^{\bullet}\ L\ in\ {}_M[P] \xrightarrow{\alpha^+} \mathcal{P}}{{}_M[hide\ L\ in\ P] \xrightarrow{\alpha^+} \mathcal{P}}$$

$$(6.b)\ \frac{\mathcal{P} \xrightarrow{M^{a x}} \mathcal{P}'}{hide^{\bullet}\ L\ in\ \mathcal{P} \xrightarrow{M^{a x}} hide^{\bullet}\ L\ in\ \mathcal{P}' \text{ if } a \notin L \qquad hide^{\bullet}\ L\ in\ \mathcal{P} \xrightarrow{M^{i x}} hide^{\bullet}\ L\ in\ \mathcal{P}' \text{ if } a \in L}$$

$$(7)\ \frac{{}_M[[recX.P/X]P] \xrightarrow{\alpha^+} \mathcal{P}}{{}_M[recX.P] \xrightarrow{\alpha^+} \mathcal{P}}$$

Some explanations of the previous rules are provided in the sequel:

Rule (1) characterizes the semantics of the successful termination of a configuration; the set of maximal events in configuration ${}_M[exit]$ is M, and x characterizes the event associated with the start of action δ; notice that, in the resulting configuration, only event x is maximal, as the successful termination ($exit \xrightarrow{\delta}$) depends on the termination of all other actions (i.e. the actions whose start events belong to M.

Rule (2) characterizes the semantics of the action prefix operator; as the start of action a depends on the termination of all the actions associated with the maximal events in M, only x, the event name associated with the start of action a, is a maximal event in configuration ${}_{\{x\}}[P]$.

Rule (3) characterizes the semantics of the choice operator; it is a direct adaptation to configurations of the classical inference rule defined for behavior expressions in the LOTOS operational semantics.

Rules (4.a.b.c) characterize the semantics of the enabling operator and require a more detailed explanation: rule (4.a) relates the enabling operator ($\gg$) applied to behavior expressions to the corresponding operator ($\stackrel{x}{\gg}$) applied to configurations; configuration ${}_M[P] \stackrel{x}{\gg} {}_{\{x\}}[Q]$ means that any action in Q depends on the termination of P, where it is to be noted that x is not a maximal event name (which a posteriori justifies definition 2); rule (4.b) states that any action which may start from $\mathcal{P}$ may also start from $\mathcal{P} \stackrel{y}{\gg} \mathcal{Q}$, where z should be different from the maximal events in $\mathcal{Q}$; rule (4.c) states that, when $\mathcal{P}$ terminates successfully and when the occurrence of δ makes the events in M to be not anymore maximal, then there is a silent move from $\mathcal{P} \stackrel{x}{\gg} \mathcal{Q}$ to $\mathcal{Q}$.

Rules (5.a.b.c) characterize the semantics of the parallel composition operator; rule (5.a) relates the parallel composition operator applied to behavior expressions to the corresponding operator applied to configurations; to explain rules (5.b) and (5.c), let us consider the following example, in which it is assumed that the set of the maximal events in configuration $\mathcal{P}$ is empty:

$$\mathcal{P} \equiv {}_{\emptyset}[a; c; d; stop] \; |[c]|^{\bullet} \; {}_{\emptyset}[b; c; e; stop]$$

In this configuration, actions a and b are enabled and may then start executing; let x (respectively y) be the event name associated with the start of the execution of action a (respectively action b), then the following derivations may be obtained by applying rule (5.b) twice:

$$\mathcal{P} \xrightarrow{\emptyset a_x} {}_{\{x\}}[c; d; stop] \; |[c]|^{\bullet} \; {}_{\emptyset}[b; c; e; stop] \xrightarrow{\emptyset b_y} \underbrace{{}_{\{x\}}[c; d; stop]}_{\mathcal{S}} \; |[c]|^{\bullet} \; \underbrace{{}_{\{y\}}[c; e; stop]}_{\mathcal{S}'}$$

It may now be seen that a synchronization on gate c may take place, as soon as both actions a and b are terminated; performing the synchronization implies starting the execution of action c (identified by event name z) which is consequently the only maximal event in the resulting configuration, as shown when applying rule (5.c):

$${}_{\{x\}}[c; d; stop] \; |[c]|^{\bullet} \; {}_{\{y\}}[c; e; stop] \xrightarrow{\{x,y\}c_z} {}_{\{z\}}[d; stop] \; |[c]|^{\bullet} \; {}_{\{z\}}[e; stop]$$

Note that the causal dependency between action c and both actions a and b is expressed in the inference rule by means of atom ${}_{\{x,y\}}c_z$; note furthermore, that name z could have been chosen equal to either x or y without any ambiguity within the context of the resulting configuration.

Let us now assume that action d might start its execution, implying therefore that action c has terminated its execution, and consequently that z is not anymore a maximal event; in the resulting configuration, name z has then to be eliminated at both sides (see the use of the $\mathcal{E}\backslash M$ function). Applying therefore rule (5.b) leads to the following derivation:

$${}_{\{z\}}[d; stop] \; |[c]|^{\bullet} \; {}_{\{z\}}[e; stop] \xrightarrow{\{z\}d_t} {}_{\{t\}}[stop] \; |[c]|^{\bullet} \; {}_{\emptyset}[e; stop]$$

Rules (6.a.b) and (7) are finally straightforward, as they merely extend to configurations the classical LOTOS inference rules defined on behavior expressions.

Remark 1 *Note that the choice of event names can be made unique; as set $\mathcal{M}$ is countable, then there is a bijection $h : \mathcal{M} \rightarrow \mathbb{N}$, and then the order among elements of $\mathbb{N}$ can be translated to elements of $\mathcal{M}$, by stating that for $x, y \in \mathcal{M}$, $x \leq_{\mathcal{M}} y$ iff $h(x) \leq h(y)$; this order being well founded, then one strategy for choosing an event name consists in selecting the least one satisfying the transition rule.*

3 Extending the language with an action refinement operator

In [14], we have proposed a general solution for performing action refinement as a syntactic replacement of actions by processes. However, as the main emphasis of the paper is on showing the interest of the maximality semantics for performing action refinement in

LOTOS, we will merely consider a refinement approach based on the introduction into the language of a specific refinement operator, leading to language LOTOS_ρ. This approach may be proven to be semantically equivalent to the syntactic refinement proposition formalized in [14].

Definition 6 *The syntax of LOTOS_ρ is characterized by the following BNF-like grammar:* $E ::= stop \mid exit \mid X \mid a; E \mid E \gg E' \mid E[]E' \mid E|[L]|E' \mid hide\ L\ in\ E \mid recX.E \mid \rho(o, E, E')$

Definition 7 *The syntax of the configurations is extended by introducing operator $\rho^\bullet$ which plays for the configurations the counterpart of the ρ operator defined on the behavior expressions:* $\mathcal{E} ::= {}_M[E] \mid \mathcal{E} \overset{x}{\gg} \mathcal{E}' \mid \mathcal{E}|[L]|^\bullet\mathcal{E} \mid hide^\bullet\ L\ in\ \mathcal{E} \mid \rho^\bullet(o, E, \mathcal{E})$

Definition 8 *The set of maximal event names in configuration $\mathcal{E}$, in symbols $Max(\mathcal{E})$, is defined by induction on the structure of $\mathcal{E}$ as in definition 2 to which the following rule is added:* $Max(\rho^\bullet(a, P, \mathcal{E})) = Max(\mathcal{E})$.

Definition 9 *To deal with maximal event name substitution, definition 3 is extended as follows:* $y/x\ \rho^\bullet(a, P, \mathcal{E}) \equiv \rho^\bullet(a, P, y/x\mathcal{E})$.

Definition 10 *Function $\mathcal{E}\backslash N$ is extended to the refinement operator by adding the following rule:* $\rho^\bullet(a, P, \mathcal{E})\backslash N \equiv \rho^\bullet(a, P, \mathcal{E}\backslash N)$.

It is assumed that the refining process P cannot exhibit any derivation sequence like ${}_\emptyset[P] \xrightarrow{\emptyset i_x} \cdots \xrightarrow{M i_y} \xrightarrow{N \delta_x} \mathcal{P}'$. This hypothesis seems to be very acceptable, as there is no practical interest in refining an observable action by a process which can only execute a sequence of internal actions before its successful termination.

The maximality-based operational semantics of LOTOS_ρ may now be formalized.

Definition 11 *The maximality-based transition relation $\longrightarrow \subseteq \mathcal{C}_0 \times Atm^+ \times \mathcal{C}_0$ is defined as the least relation satisfying the rules stated in definition 5 with the addition of the following rules:*

$$(8.a)\ \frac{\rho^\bullet(a,P,{}_M[Q]) \xrightarrow{\alpha^+} \mathcal{P}}{{}_M[\rho(a,P,Q)] \xrightarrow{\alpha^+} \mathcal{P}} \qquad (8.b)\ \frac{\mathcal{Q} \xrightarrow{M b_x^+} \mathcal{Q}'}{\rho^\bullet(a,P,\mathcal{Q}) \xrightarrow{M b_x^+} \rho^\bullet(a,P,\mathcal{Q}') \text{ if } a \neq b^+}$$

$$(8.c)\ \frac{\mathcal{Q} \xrightarrow{M a_x} \mathcal{Q}' \qquad {}_\emptyset[P] \xrightarrow{\emptyset b_y} \mathcal{P}'}{\rho^\bullet(a,P,\mathcal{Q}) \xrightarrow{M b_z} z/y\mathcal{P}' \overset{x}{\gg} \rho^\bullet(a,P,\mathcal{Q}')} \ \forall z \notin Max(\mathcal{Q}') - \{x\}$$

$$(4.d)\ \frac{\mathcal{Q} \xrightarrow{M a_x^+} \mathcal{Q}'}{\mathcal{P} \overset{y}{\gg} \mathcal{Q} \xrightarrow{M a_z^+} \mathcal{P} \overset{y}{\gg} z/x\mathcal{Q}' \text{ if } y \notin M} \ \forall z \notin Max(\mathcal{P}) \cup Max(\mathcal{Q}') - \{x\} \cup \{y\}$$

Rule (8.a) relates refinement operator ρ applied to behavior expressions to the corresponding operator $\rho^\bullet$ applied to configurations.

Rule (8.b) states that whenever action b starts its execution in configuration $\mathcal{Q}$, this action is treated as an ordinary action, as soon as it is not refined.

Rule (8.c) addresses the case of a refined action; if action a starts its execution in configuration $\mathcal{Q}$, then this action is replaced by the behavior of the refining process (process P); note that the concurrency between the refining process and the actions which were concurrent with action a is to be preserved. To illustrate this point, let us analyze the following example: $\mathcal{Q} \equiv {}_{\emptyset}[a; b; stop] \; |||^{\bullet} \; {}_{\emptyset}[c; d; stop]$ with $P \equiv e; f; exit$. Then,

$$\mathcal{Q} \xrightarrow{\emptyset a_x} {}_{\{x\}}[b; stop] \; |||^{\bullet} \; {}_{\emptyset}[c; d; stop] \quad \text{and} \quad {}_{\emptyset}[P] \xrightarrow{\emptyset e_y} {}_{\{y\}}[f; exit]$$

which implies, by applying rule (8.c), that:

$$\rho^{\bullet}(a, P, \mathcal{Q}) \xrightarrow{\emptyset e_y} \underbrace{{}_{\{y\}}[f; exit] \overset{x}{\gg} \rho^{\bullet}(a, P, {}_{\{x\}}[b; stop] \; |||^{\bullet} \; {}_{\emptyset}[c; d; stop])}_{\mathcal{R}}$$

Rule (4.d) can now be applied in order to state that action c is concurrent with action f; the following derivation is then possible:

$$\mathcal{R} \xrightarrow{\emptyset c_z} {}_{\{y\}}[f; exit] \overset{x}{\gg} \rho^{\bullet}(a, P, {}_{\{x\}}[b; stop] \; |||^{\bullet} \; {}_{\{z\}}[d; stop])$$

Definition 12 *A configuration $\mathcal{E}$ is said to be initial, if $Max(\mathcal{E}) = \emptyset$. Given some behavior expression E, the initial configuration associated with E is noted ${}_{\emptyset}[E]$.*

Lemma 1 *For any configuration $\mathcal{E}$, if $\mathcal{E} \xrightarrow{\alpha^+} \mathcal{E}'$ then $Max(\mathcal{E}') \neq \emptyset$.*

Lemma 2 *Given some behavior expression E, and a (possibly empty) sequence of derivations from the initial configuration associated with E, ${}_{\emptyset}[E] \equiv \mathcal{E}_0 \longrightarrow \cdots \longrightarrow \mathcal{E}_n$, then for any $j = 0..n$ and any subterm ${}_{M}[F]$ in $\mathcal{E}_j$, $Card(M) \leq 1$.*

This lemma means that any event which occurs from F has at most one local cause; this point may also be stated as follows:

- if $M = \{x\}$, then x is the event name associated with the start of the action which is the local cause of F;
- if $M = \emptyset$, then the action which was the local cause of F (if any) has already terminated its execution.

4 The Maximality-based semantics and Action Refinement

This section is devoted to the study of behavioral equivalence preservation under action refinement. This issue may be stated as follows: assuming two LOTOS specifications S_1 and S_2 which are equivalent in some sense, and applying the same refinement to both specifications, are the resulting specifications $\rho(a, P, S_1)$ and $\rho(a, P, S_2)$ still equivalent in the same sense ?

Two main equivalences are considered in this section, namely the strong and the weak maximality-based bisimulation equivalences. These bisimulations are the adaptation, within the proposed maximality-based framework, of the classical interleaving-based bisimulation equivalences defined by Milner and Park [11, 12]. They correspond to the ST-bisimulations respectively without (the case of the strong equivalence) and with (the case of the weak equivalence) making abstraction of the silent actions. The notion of maximal event introduced previously is used for defining the equivalences, instead of splitting actions as it has been done in the initial definition of the ST-bisimulations.

4.1 Strong Maximality-based bisimulation

Definition 13 *Let $\Re \subseteq \mathcal{C}_0 \times \mathcal{C}_0 \times \Im$ be a relation between configurations, where $\Im \subseteq 2^{\mathcal{M}\times\mathcal{M}}$ is a set of bijections. Then let $I\!\!F : Rel(\mathcal{C}_0) \longrightarrow Rel(\mathcal{C}_0)$ be a function defined as follows:* [2]
$(\mathcal{P}, \mathcal{Q}, f) \in I\!\!F(\Re)$ *iff:*

1. *whenever* $\mathcal{P} \stackrel{Ma_x^+}{\longrightarrow} \mathcal{P}'$, *there is* $\mathcal{Q} \stackrel{Na_y^+}{\longrightarrow} \mathcal{Q}'$ *such that:*
   (a) *for any* $u \in Max(\mathcal{P})$, *if* $u \notin M$ *and* $f(u) \in Max(\mathcal{Q})$, *then* $f(u) \notin N$;
   (b) *for some* $v, w \in \mathcal{M}$, $(v/x\mathcal{P}', w/y\mathcal{Q}', f\lceil(Max(\mathcal{P}') - \{x\} \cup f^{-1}(Max(\mathcal{Q}') - \{y\})) \cup \{(v, w)\}) \in \Re$;

2. *whenever* $\mathcal{Q} \stackrel{Na_y^+}{\longrightarrow} \mathcal{Q}'$, *there is* $\mathcal{P} \stackrel{Ma_x^+}{\longrightarrow} \mathcal{P}'$ *such that:*
   (a) *for any* $u \in Max(\mathcal{Q})$, *if* $u \notin N$ *and* $f^{-1}(u) \in Max(\mathcal{P})$, *then* $f^{-1}(u) \notin M$;
   (b) *for some* $v, w \in \mathcal{M}$, $(v/x\mathcal{P}', w/y\mathcal{Q}', f\lceil(Max(\mathcal{P}') - \{x\} \cup f^{-1}(Max(\mathcal{Q}') - \{y\})) \cup \{(v, w)\}) \in \Re$.

$\Re$ is called a strong maximality-based bisimulation iff $\Re \in I\!\!F(\Re)$. If $(\mathcal{P}, \mathcal{Q}, f) \in \Re$ for some strong maximality-based bisimulation $\Re$, then configurations $\mathcal{P}$ and $\mathcal{Q}$ are said to be strongly maximality-based bisimilar, in symbols $\mathcal{P} \sim_m \mathcal{Q}$. [3]

In order to later prove some interesting properties of the strong maximality-based bisimulation, let us first define the composition of maximality-based relations.

Definition 14 *Let $\Re \subseteq \mathcal{C}_0 \times \mathcal{C}_0 \times \Im$, then*
$Id_{\mathcal{C}_0} = \{(\mathcal{P}, \mathcal{P}, Id_{\mathcal{M}}\lceil Max(\mathcal{P})) \ : \ \mathcal{P} \in \mathcal{C}_0\}$. $\qquad \Re^{-1} = \{(\mathcal{Q}, \mathcal{P}, f^{-1}) : (\mathcal{P}, \mathcal{Q}, f) \in \Re\}$

Proposition 1 *Assuming that each $\Re_i$ $(i = 1, 2, \cdots)$ be a strong maximality-based bisimulation, then the following relations are all strong maximality-based bisimulations:*

$$(1)\ Id_{\mathcal{C}_0} \qquad (2)\ \Re_i^{-1} \qquad (3)\ \bigcup_{i \in I} \Re_i$$

Definition 15 *Let $\Re_1, \Re_2 \subseteq \mathcal{C}_0 \times \mathcal{C}_0 \times \Im$ such that for any $(\mathcal{P}_1, \mathcal{Q}_1, f_1) \in \Re_1$ and $(\mathcal{P}_2, \mathcal{Q}_2, f_2) \in \Re_2$, $Dom(f_1) \cap Dom(f_2) = \emptyset$ and $Cod(f_1) \cap Cod(f_2) = \emptyset$.*
$\Re = \Re_1\Re_2 = \{(\mathcal{P}, \mathcal{R}, f) : \exists(\mathcal{P}, \mathcal{Q}_1, f_1) \in \Re_1 \wedge (\mathcal{Q}_2, \mathcal{R}, f_2) \in \Re_2\}$ such that for some bijection $h : Max(\mathcal{Q}_1) \longrightarrow Max(\mathcal{Q}_2) : h(\mathcal{Q}_1) \equiv \mathcal{Q}_2$ [4]*, $f = F_1 \cup F_2 \cup F_3 \cup F_4 \cup F_5$ where*

$$
\begin{array}{llllll}
F_1 & = \{(u, v) \in f_1 & : & u \in Max(\mathcal{P}) & \wedge & v \notin Max(\mathcal{Q}_1)\} \\
F_2 & = \{(u, v) \in f_2 & : & u \notin Max(\mathcal{Q}_2) & \wedge & v \in Max(\mathcal{R})\} \\
F_3 & = \{(u, v) & : & u \in Max(\mathcal{P}) & \wedge & v \in Max(\mathcal{R}) \quad \wedge \\
 & & & f_1(u) \in Max(\mathcal{Q}_1) & \wedge & v = f_2(h(f_1(u)))\} \\
F_4 & = \{(u, v) & : & u \in Max(\mathcal{P}) & \wedge & v \in Cod(f_2) - Max(\mathcal{R}) \quad \wedge \\
 & & & f_1(u) \in Max(\mathcal{Q}_1) & \wedge & v = f_2(h(f_1(u)))\} \\
F_5 & = \{(u, v) & : & v \in Max(\mathcal{R}) & \wedge & u \in Dom(f_1) - Max(\mathcal{P}) \quad \wedge \\
 & & & f_1(u) \in Max(\mathcal{Q}_1) & \wedge & v = f_2(h(f_1(u)))\}
\end{array}
$$

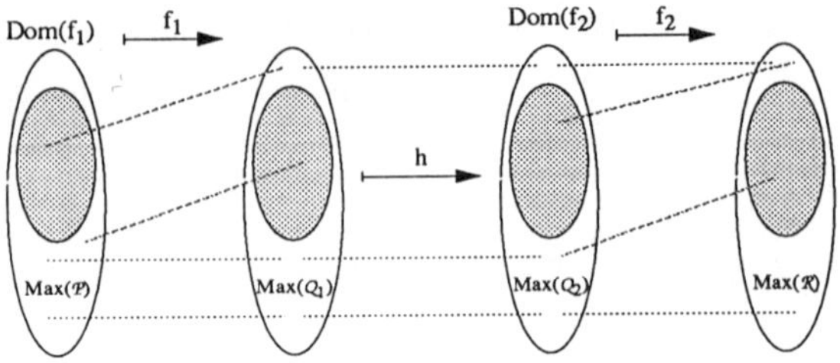

Figure 1: Composition of maximality-based relations

Proposition 2 *Let $\Re_1, \Re_2 \subseteq \mathcal{C}_0 \times \mathcal{C}_0 \times \Im$ be two strong maximality-based bisimulations, such that for any $(\mathcal{P}_1, \mathcal{Q}_1, f_1) \in \Re_1$ and $(\mathcal{P}_2, \mathcal{Q}_2, f_2) \in \Re_2$, $Dom(f_1) \cap Dom(f_2) = \emptyset$ and $Cod(f_1) \cap Cod(f_2) = \emptyset$. Then $\Re_1\Re_2$ is also a strong maximality-based bisimulation.*

The following definition generalizes the strong maximality-based bisimulation between configurations to LOTOS behavior expressions.

Definition 16 *Two behavior expressions P and Q are said to be strongly maximality-based bisimilar, in symbols $P{\sim_m}Q$, if ${}_{\emptyset}[P]{\sim_m}\,{}_{\emptyset}[Q]$, which may also be expressed as follows:*

$$\sim_m = \bigcup\{\Re : \Re \text{ is a strong maximality} - \text{based bisimulation}\}$$

Proposition 3 *$\sim_m$ is an equivalence relation.*

The next proposition states that the strong maximality-based bisimulation is preserved when refining configurations.

Proposition 4 *For any $\mathcal{P}$, $\mathcal{Q} \in \mathcal{C}_0$, $a \in \mathcal{G}$ and $P \in \mathcal{L}_0$, if $\mathcal{P}{\sim_m}\mathcal{Q}$ then $\rho^\bullet(a, P, \mathcal{P}){\sim_m}\rho^\bullet(a, P, \mathcal{Q})$*

The next corollary states that the strong maximality-based bisimulation is also preserved when refining LOTOS behavior expressions.

Corollary 1 *For any P, Q, $R \in \mathcal{L}_0$ and $a \in \mathcal{G}$, if $Q{\sim_m}R$ then $\rho(a, P, Q){\sim_m}\rho(a, P, R)$.*

4.2 Weak Maximality-based bisimulation

As the strong maximality-based bisimulation does not abstract from the internal (silent) actions, its usefulness is limited when one is interested in designing systems based on action refinement. For this reason, a weak version of the maximality-based bisimulation is proposed and also shown to be preserved under action refinement.

[2] $Rel(\mathcal{C}_0) \subseteq 2^{\mathcal{C}_0 \times \mathcal{C}_0 \times \Im}$ is a set of relations

[3] $f\lceil X$ is the usual restriction of the function f to set X.

[4] For $\mathcal{Q} \in \mathcal{C}_0$, $h(\mathcal{Q})$ is the renaming of elements of $Max(\mathcal{Q})$ according to h

Definition 17 *Let $\mathcal{M}_i$ be a countable set of maximal unobservable event names, with the hypothesis that $\mathcal{M}_i \cap \mathcal{M} = \emptyset$. Let furthermore $M, N, \cdots$ range over $\mathcal{M} \cup \mathcal{M}_i$.*

Transition relation $\longrightarrow$ is extended to the new set of maximal event names (i.e. $\longrightarrow \subseteq \mathcal{C}_0 \times 2^{\mathcal{M} \cup \mathcal{M}_i} \times Act^+ \times \mathcal{M} \times \mathcal{C}_0$).

Definition 18 *Let Atm_Λ, ranged over by ν, be the set all atoms with respect to set $\mathcal{G}^+$ of the visible actions. Let us furthermore assume a distinguished symbol $\epsilon \overset{def}{\notin} Atm_\Lambda$ to express an invisible evolution.*

Definition 19 *The weak maximality-based transition relation $\twoheadrightarrow \subseteq \mathcal{C}_0 \times Atm_\Lambda \cup \{\epsilon\} \times \mathcal{C}_0$ is defined as the least relation satisfying the following rules:*

$$\frac{}{\mathcal{E} \overset{\epsilon}{\twoheadrightarrow} \mathcal{E}} \qquad \frac{\mathcal{E} \xrightarrow{M^i x} \mathcal{E}'}{\mathcal{E} \overset{\epsilon}{\twoheadrightarrow} u/x\mathcal{E}'}\ \forall u \in \mathcal{M}_i : u \notin Max(\mathcal{E}') \qquad \frac{\mathcal{E} \overset{\epsilon}{\twoheadrightarrow} \mathcal{E}' \overset{\epsilon}{\twoheadrightarrow} \mathcal{E}''}{\mathcal{E} \overset{\epsilon}{\twoheadrightarrow} \mathcal{E}''} \qquad \frac{\mathcal{E} \overset{\epsilon}{\twoheadrightarrow} \mathcal{E}' \overset{\nu}{\twoheadrightarrow} \mathcal{E}''}{\mathcal{E} \overset{\nu}{\twoheadrightarrow} \mathcal{E}''}$$

In the sequel, the transition $\overset{\epsilon}{\twoheadrightarrow}$ will be noted $\twoheadrightarrow$.

Definition 20 *Function $vis : \mathcal{C}_0 \longrightarrow 2^{\mathcal{M}}$ is defined as: $vis(\mathcal{P}) = Max(\mathcal{P}) \cap \mathcal{M}$.*

Definition 21 *Let $\Re \subseteq \mathcal{C}_0 \times \mathcal{C}_0 \times \Im$ be a relation between configurations. Then, let $\mathbb{F}^i : Rel(\mathcal{C}_0) \longrightarrow Rel(\mathcal{C}_0)$ be a function defined as follows: $(\mathcal{P}, \mathcal{Q}, f) \in \mathbb{F}^i(\Re)$ iff:*

1. *whenever $\mathcal{P} \twoheadrightarrow \mathcal{P}'$, there is $\mathcal{Q} \twoheadrightarrow \mathcal{Q}'$ such that for any $u \in vis(\mathcal{P})$, if $u \in vis(\mathcal{P}')$ and $f(u) \in vis(\mathcal{Q})$, then $f(u) \in vis(\mathcal{Q}')$ and $(\mathcal{P}', \mathcal{Q}', f\lceil vis(\mathcal{P}') \cup f^{-1}(vis(\mathcal{Q}'))) \in \Re$;*
2. *whenever $\mathcal{P} \overset{M o_x^+}{\twoheadrightarrow} \mathcal{P}'$, there is $\mathcal{Q} \overset{N o_y^+}{\twoheadrightarrow} \mathcal{Q}'$ such that:*
   (a) *for any $u \in vis(\mathcal{P})$, if $u \neq x$ and $u \in vis(\mathcal{P}')$ and $f(u) \in vis(\mathcal{Q})$, then $f(u) \neq y$ and $f(u) \in vis(\mathcal{Q}')$;*
   (b) *for some $v, w \in \mathcal{M}$, $(v/x\mathcal{P}', w/y\mathcal{Q}', f\lceil(vis(\mathcal{P}') - \{x\} \cup f^{-1}(vis(\mathcal{Q}') - \{y\})) \cup \{(v, w)\}) \in \Re$;*
3. *whenever $\mathcal{Q} \twoheadrightarrow \mathcal{Q}'$, there is $\mathcal{P} \twoheadrightarrow \mathcal{P}'$ such that for any $v \in vis(\mathcal{Q})$ if $v \in vis(\mathcal{Q}')$ and $f^{-1}(v) \in vis(\mathcal{P})$, then $f^{-1}(v) \in vis(\mathcal{P}')$ and $(\mathcal{P}', \mathcal{Q}', f\lceil vis(\mathcal{P}') \cup f^{-1}(vis(\mathcal{Q}'))) \in \Re$;*
4. *whenever $\mathcal{Q} \overset{N o_y^+}{\twoheadrightarrow} \mathcal{Q}'$, there is $\mathcal{P} \overset{M o_x^+}{\twoheadrightarrow} \mathcal{P}'$ such that:*
   (a) *for any $v \in vis(\mathcal{Q})$, if $v \neq y$ and $v \in vis(\mathcal{Q}')$ and $f^{-1}(v) \in vis(\mathcal{P})$, then $f^{-1}(v) \neq x$ and $f^{-1}(v) \in vis(\mathcal{P}')$;*
   (b) *for some $v, w \in \mathcal{M}$, $(v/x\mathcal{P}', w/y\mathcal{Q}', f\lceil vis(\mathcal{P}') - \{x\} \cup f^{-1}(vis(\mathcal{Q}') - \{y\}) \cup \{(v, w)\}) \in \Re$;*

$\Re$ is called a weak maximality-based bisimulation iff $\Re \in \mathbb{F}^i(\Re)$. If $(\mathcal{P}, \mathcal{Q}, f) \in \Re$ for some weak maximality-based bisimulation $\Re$, then $\mathcal{P}$ and $\mathcal{Q}$ are said to be weakly maximality-based bisimilar, in symbols $\mathcal{P} \approx_m \mathcal{Q}$.

Proposition 5 *Propositions 1 and 2 remain true for the weak maximality-based bisimulations, assuming that Max is replaced by vis in definition 15, and that* $h : Max(\mathcal{Q}_1) \longrightarrow Max(\mathcal{Q}_2)$ *respects* $\mathcal{M}$ *and* $\mathcal{M}_i$.

The following definition generalizes the weak maximality-based bisimulation between configurations to LOTOS behavior expressions.

Definition 22 *Two behavior expressions P and Q are said to be weakly maximality-based bisimilar, in symbols* $P \approx_m Q$, *if* ${}_\emptyset[P] \approx_m {}_\emptyset[Q]$. *This may be equivalent expressed as follows:*

$$\approx_m = \bigcup\{\Re : \Re \text{ is a weak maximality} - \text{based bisimulation}\}$$

Proposition 6 $\approx_m$ *is an equivalence relation.*

The next proposition states that the weak maximality-based bisimulation is preserved when refining configurations.

Proposition 7 *For any* $\mathcal{P}, \mathcal{Q} \in \mathcal{C}_0$, $a \in \mathcal{G}$ *and* $P \in \mathcal{L}_0$, *if* $\mathcal{P} \approx_m \mathcal{Q}$ *then* $\rho^\bullet(a, P, \mathcal{P}) \approx_m \rho^\bullet(a, P, \mathcal{Q}$

The next corollary states that the weak maximality-based bisimulation is also preserved when refining LOTOS behavior expressions.

Corollary 2 *For any* P, Q, $R \in \mathcal{L}_0$ *and* $a \in \mathcal{G}$, *if* $Q \approx_m R$ *then* $\rho(a, P, Q) \approx_m \rho(a, P, R)$.

5 Conclusion

In this paper, we have defined a maximality-based operational semantics for the LOTOS language extended by an action refinement operator. A strong maximality-based bisimulation has been defined and proven to be preserved under action refinement. It corresponds to the interleaving ST-bisimulation but the notion of maximal events has been used instead of the action splitting technique.

This bisimulation relation has been extended by making it possible to abstract from some actions, leading therefore to the weak maximality-based bisimulation. This relation corresponds to the adaptation of the ST-bisimulation, defined on labelled prime event structures, which abstracts from silent actions [18]. This bisimulation relation has also been proven to be preserved under action refinement.

As previously mentioned in the paper, the interest in defining such a maximality-based semantics is to overcome the potential state space explosion induced by the action splitting technique which itself aims at escaping from the action atomicity assumption.

To be convinced, let us consider the behavior graphs presented in figure 2 which correspond both to behavior $P \equiv a; stop \mid\mid\mid b; stop$, the left and right graphs being respectively constructed by means of the splitting and the maximality techniques. Note furthermore that the right graph may be constructed in an unique manner by using an adequate strategy for choosing event names (as discussed in remark 1).

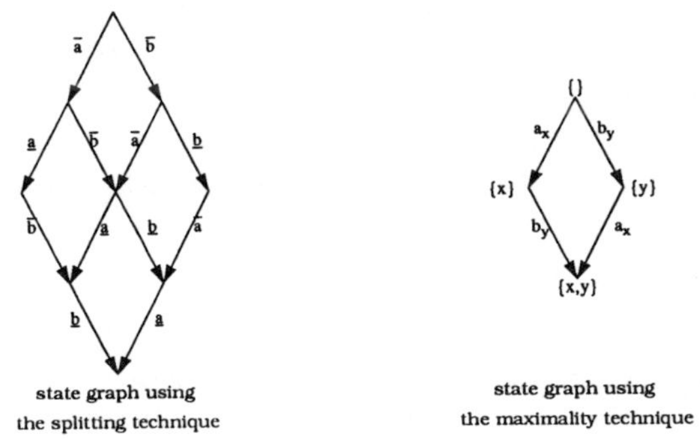

Figure 2: Behavior state graphs

Another interesting point is related to the relationship between the causality-based semantics previously proposed for LOTOS$_p$ in [4] and the maximality-based semantics developed here. In [4], it has been proven that a weak bisimulation relation, namely the *branching causality-based bisimulation*, is preserved under action refinement, as it is the case for the weak maximality-based bisimulation defined in this paper. Comparing these weak bisimulation relations, it may be seen that the weak maximality-based bisimulation is indeed weaker than the branching-causality-based bisimulation, in the sense that it distinguishes less behaviors. As an example, the following behaviors are weakly maximality-based bisimilar, but they are not branching causality-based bisimilar:

$$P = a; stop \; [] \; i; b; stop \qquad \text{and} \qquad Q = P \; [] \; b; stop$$

Current work on this topic deals with the definition of a maximality-based semantics for an extended Basic LOTOS [14] able to support syntactic action refinement, which appears, from a practical point of view, much more interesting than the action refinement operator considered in this paper. Current results suggest that the two approaches are semantically equivalent, as far as action refinement is considered.

References

[1] L. Aceto, M. Hennessy, Adding action refinement to finite process algebra. In proc. of ICALP'91, Volume 510 of LNCS, pages 506-519, Springer-Verlag, 1991.

[2] T. Bolognesi, E. Brinksma, Introduction to the ISO Specification Language LOTOS. Computer Networks and ISDN Systems, 14:25-59, North-Holland, 1987.

[3] R. Coelho Da Costa, J.P. Courtiat, A true concurrency semantics for LOTOS. In Proc. of the IFIP TC6/WG6.1, 5th Int. Conf. on Formal Description Techniques, (FORTE'92), pages 347-362, North-Holland, 1992.

[4] J.P. Courtiat, D.E. Saïdouni, Action Refinement in LOTOS. In Proc. of the 13th IFIP Symposium on Protocol Specification, Testing and Verification, (PSTV'93), pages 341-354, North-Holland, 1994.

[5] J.P. Courtiat, D.E. Saïdouni, Relating Maximality-based Semantics to Action Refinement in Process Algebras (Extended Version). Research Report 94078, LAAS/CNRS, March 1994.

[6] P. Darondeau, P. Degano, About semantic action refinement, Fundamenta Informaticae, 14:221-234, 1991.

[7] P. Degano, R. Gorrieri, Atomic refinement in process description languages. In proc of MFCS'91, Volume 520 of LNCS, pages 121-130, Springer-Verlag, 1991.

[8] R. Devillers, Maximality preservation and the ST-idea for action refinement. In proc. of Advances in Petri Nets 1992, Volume 609 of LNCS, pages 108-151, Springer-Verlag, 1992.

[9] M. Hennessy, Concurrent Testing of Processes. In proc. of CONCUR'92, Volume 630 of LNCS, pages 94-107, Springer-Verlag, 1992.

[10] A. Kiehn, E. Best, R. Devillers and L. Pomello, Concurrent bisimulations in Petri nets. Acta Informatica, 28:231-264, 1991.

[11] R. Milner, Communication and Concurrency. Prentice Hall, 1989.

[12] D. Park, Concurrency and automata in infinite sequences. Volume 104 of LNCS, pages 167-183, Springer-Verlag, 1981.

[13] G. Rozenberg, M. Nielsen, K.S. Thiagarajan, Elementary transition systems and refinement. Acta Informatica, 29: 555-578, 1992.

[14] D.E. Saïdouni, J.P. Courtiat, Syntactic Action Refinement in Presence of Multiway Synchronization. In proc. of the workshop on Semantics of Specification Languages (SoSL'93), pages 289-303, Workshops in Computing, Springer Verlag, 1994.

[15] R.J. van Glabbeek, The Refinement Theorem for ST-bisimulation Semantics. In proc. of IFIP Working Conference on Programming Concepts and Methods, North-Holland, 1990.

[16] R.J. van Glabbeek, Comparative Concurrency Semantics and Refinement of Actions. PhD thesis, Vrije Universiteit Te Amsterdam, 1990.

[17] R.J. van Glabbeek, F.W. Vaandrager, Petri net models for algebraic theories of concurrency. In proc. of PARLE'87, Vol. II (Parallel Languages), Volume 259 of LNCS, pages 224-242, Springer-Verlag, 1987.

[18] W. Vogler, Bisimulation and Action Refinement. In proc. of STACS'91, Volume 480 of LNCS, pages 309-321, Springer-Verlag, 1991.

24

An Abstract Interpreter for the specification language LOTOS *

Franco Fiore and Fosca Giannotti
CNR, Ist. CNUCE, via S.Maria 36 I56126 Pisa Italy
ph. +39-50-593339, Fax: +39-50-904052, email:fosca@cnuce.cnr.it

We present an abstract interpreter for supporting semantic analysis of LOTOS specifications. The tool uses a technique defined in [7], which automatically constructs an abstraction of the LOTOS data domains tailored on the input specification. The abstract interpreter annotates each LOTOS behaviour expression with an abstract environment which keeps trace of the data flow that reaches that point of the specification. The abstract interpretation procedure is proved correct and its application to the detection of starvation and dead code are discussed.

1. Introduction

This work is part of a set of experiments aimed at verifying the practicability of flow analysis techniques such as *Abstract Interpretation* (AI) [1, 5] in new applicative contexts.

Specification languages offer an appealing applicative field for AI even though efficiency is usually not a concern. On one side, program verification starts from the specification phase and therefore early detection or dangerous situations is clear advantage. On the other hand, specification languages are supported by a development methodology based on refinements. So is natural to equip each refinement step with formal statements which establish which properties hold at that step.

This is the setting of the specification language LOTOS [3] for which a collection of semi-automatic tools supporting the different development step have been realized in the Esprit project Lotosphere [10–14]. Moreover LOTOS has been designed for distributed systems which offer a wide set of properties whose early detection is useful, such as deadlock freedom.

This paper continues the experience in [7] where Abstract Interpretation was used for an efficient solution to the problem of gate splitting in LOTOS specifications [8]. The transformation problem is part of a design methodology based on the specification language LOTOS. Abstract Interpretation was used for approximating the sets of possible values which LOTOS expressions can evaluate to. The abstract domain as well as abstract functions were generated *automatically* from the LOTOS specification to be transformed.

In this paper, we exploit the results of [7] as far as the automatic definition of the abstract domain is concerned. On the basis of these results, we define a general tool which uses such abstract domain to abstractly execute the input specification. As a result of such abstract interpretation, we obtain an annotated program. The annotation attached to each point of the program is an *abstract environment* representing information about data flows that will reach that point in all the possible executions. Finally, the annotated

*This work has been carried out within the Progetto Speciale of CNR - ANATRA

program can be used by a transformer which applies those optimizations which are based on dataflow analysis.

We assume the reader to be familiar with LOTOS. The following section describes the basic ideas of Abstract Interpretation.

1.1. Basic ideas of Abstract Interpretation

Abstract interpretation of programs is an analysis technique aimed at gathering static information about the dynamic semantics of a program or specification to be used by compilers, partial evaluators and debuggers, or merely as documentation. It has its roots in imperative languages where it supplied some formal support to correct optimizations. The first formalization is due to P.Cousot and R.Cousot in [4, 5]. In the eighties Abstract Interpretation found fertile ground on declarative formalisms as well, since it fits with the formal framework of such languages and with their need to improve the efficiency of their implementations.

Abstract Interpretation consists in assigning language constructs non- standard (i.e. *abstract*) meaning which approximate standard (i.e. *concrete*) semantics. In other words, an abstract interpretation of a programming language is the homomorphic image of its standard semantics on another (usually simpler) domain. For instance, the set $S_1 = \{[-], [0], [+]\}$ may be used to approximate integers in the obvious way: $[-]$ is an approximation for negative integers, $[0]$ for 0, and $[+]$ for positive integers. Multiplication is re-interpreted as the function $mult : S_1^2 \to S_1$ defined as follows:

a/b	[−]	[0]	[+]
[−]	[+]	[0]	[−]
[0]	[0]	[0]	[0]
[+]	[−]	[0]	[+]

More formally, given a language L, a standard interpretation $\mathcal{M} : L \to D$ and an abstract domain $D^{\mathcal{A}}$ we want define $\mathcal{M}^{\mathcal{A}} : L \to D^{\mathcal{A}}$, such that $\mathcal{M}^{\mathcal{A}}(e)$ approximates $\mathcal{M}(e)$ for every $e \in L$. More precisely, $\mathcal{M}^{\mathcal{A}}(e)$ approximates all possible evaluations of $\mathcal{M}(e)$ (i.e. execution of e with all possible inputs).

Three main requirements have to be fulfilled when defining an abstract interpreter:

- Termination: such property is compulsory for static analysis to make sense. It is generally obtained imposing that the abstract domain be a finite lattice or c.p.o. with monotonic functions defined on it, so that termination of fixpoint computations is guaranteed.
- Safety: such property guarantees that the lack of information due to approximation does not introduce inconsistency with respect to all possible concrete executions.
- Usefulness: the information gathered by the abstract analysis has not to be trivial, i.e. the granularity of the approximation should not be too coarse.

1.2. Plan of the paper

In Section 2 the technique for the construction of the abstract domain is described. In Section.3 are described the abstract interpreter and a discussion about its correctness and termination is presented. In Section 4 two applications are described: namely starvation detection and dead code detection. Section 5 contains just a final remark.

2. Abstract Domains

This section presents the details of the automatic definition of the abstract domain starting from an input specification as defined in [7]. The abstract domain is the sum of several smaller abstract domains corresponding to the sorts of the input specification. Every component will contain finite representations of all possible sets of values of the corresponding sort. These representations give a special status to constants and constructors which appear explicitly in the input specification reflecting the way they are used.

The first step of the construction consists in the definition of a domain of abstract terms equipped with an ordering relation to compare single abstract terms and a comparison relation for sets of abstract terms. Then it is defined when a set of abstract terms is in some sense representative of all other abstract terms. The starting point for the construction of the abstract domains is the notion of a *partial term*.

Definition 2.1 *[Partial term posets:* $< ((\Pi_\Sigma)_s, \preceq_s) | s \in S >$*]*
For all sorts s the set of partial terms *of sort s is defined as the smallest set* $(\Pi_\Sigma)_s$*, such that the following conditions hold:*
1. $\perp_s \in (\Pi_\Sigma)_s$
2. for all constants $c :\rightarrow s, c \in (\Pi_\Sigma)_s$
3. for all constructors $k : s_1, \ldots, s_n \rightarrow s$ *and partial terms* $t_i \in (\Pi_\Sigma)_{s_i}$*, with* $i = 1 \ldots n$*,* $k(t_1, \ldots, t_n) \in (\Pi_\Sigma)_s$*.*
Let relation $\leadsto_s$ *be the smallest set in* $(\Pi_\Sigma)_s \times (\Pi_\Sigma)_s$*, such that the following conditions hold:*
1. for all constants $c :\rightarrow s, \perp_s \leadsto_s c$
2. for all constructors $k : s_1, \ldots, s_n \rightarrow s$*,* $\perp_s \leadsto_s k(\perp_{s_1}, \ldots, \perp_{s_n})$
3. for all constructors $k : s_1, \ldots, s_n \rightarrow s$*,*
$k(t_1, \ldots, t_i, \ldots, t_n) \leadsto_s k(r_1, \ldots, r_i, \ldots, r_n)$ *iff* $t_i \leadsto_{s_i} r_i$ *for exactly one i and* $t_j = r_j$ *for* $j \neq i$
The relation $\preceq_s$ *is defined as the reflexive and transitive closure of* $\leadsto_s$*.*

Notice that $\preceq_s$ is a partial order, $(\Pi_\Sigma)_s, \preceq_s)$ has a bottom element, which is $\perp_s$.

Definition 2.2 *[Partial terms generators:* π_s*]*
For all sorts s $\pi_s : (T_{\Sigma \cup Var})_s \rightarrow (\Pi_\Sigma)_s$ *is defined as follows:*

$\pi_s(c) = c$ *for all constants* $c :\rightarrow s$
$\pi_s(x) = \perp_s$ *for all variables of sort s*
$\pi_s(k(t_1, \ldots, t_n)) = k(\pi_{s_1}(t_1), \ldots, \pi_{s_n}(t_n))$ *for all constructor terms*
$\pi_s(f(t_1, \ldots, t_n)) = \perp_s$ *for all function terms*

Intuitively, given a sort s, a partial term is a finite representation for a subset of $(T_\Sigma)_s$ and is obtained from a term in $(T_{\Sigma \cup Var})_s$ where variables and function terms are replaced with a special symbol, $\perp$, representing *all* terms in $(T_\Sigma)_{s'}$ for some sort s'. Any ground term t simply represents the singleton $\{t\}$. The intuitive meaning of the ordering relation is that $t_1 \preceq t_2$ if t_1 is "less specified", or "more partial" than t_2, or, equivalently, the set of terms represented by t_2 is included in the one represented by t_1.

For instance, if $\{0, s(0), s(s(\perp_{Nat}))\}$ represents the abstraction of sort NAT meaning that the singleton $\{0\}$ represents the number 0, $\{s(0)\}$ represents the number 1 and $\{s(s(\perp_{NAT}))\}$ all the numbers greater than 1, then $\{0, s(s(\perp_{NAT}))\}$ will represent $\{0, s(s(0)), s(s(s(0))), ...\}$.

Definition 2.3 *[Cuts]*
Let $((\Pi_\Sigma)_s, \preceq_s)$ be a partial order and $C, X, Y \subseteq (\Pi_\Sigma)_s$. C is a cut *of $(\Pi_\Sigma)_s$ iff the following conditions hold:*
1. $\forall x, y \in C : x \neq y \Rightarrow x \not\preceq_s y \wedge y \not\preceq_s x$
2. $\forall x \in (\Pi_\Sigma)_s : \exists y \in C$ such that $x \preceq_s y \vee y \preceq_s x$

Definition 2.4 *[Covering]*
X covers Y iff the following holds:
$\forall y \in Y : \exists x \in X : y \preceq_s x$

Intuitively a cut C of $(\Pi_\Sigma)_s$ produces a partition in two sets P_1 and P_2 and then it takes the elements on the border, does not matter if they belong to P_1 or P_2, but they have to be on different chains of the partial order and all chains has to be represented. The *covering* relation simply extend the ordering relation to sets.

In the second step of our construction, with reference to a given LOTOS specification, it is defined a way to extract a set of abstract terms which represent the concrete terms explicitly used into the specification, and the way they are used in such specification.

The following definitions allow to automatically generate a finite representation AB_s for each sort s in the data type part of any given LOTOS specification. All terms textually occurring in the LOTOS specification will be explicitly represented in AB_s.

Definition 2.5 *[Raw Abstract Bases: RAB_s]*
The Raw Abstract Base *of sort s, RAB_s is defined as follows:*

- *if s is a* basic *sort, then RAB_s is the smallest set such that the following conditions hold:*
  1. $\perp_s \in RAB_s$, and
  2. if $t \in (T_{\Sigma \cup Var})_s$ is a term which textually occurs in the LOTOS specification, then $\pi_s(t) \in RAB_s$;
- *if s is non-basic and letting RAB'_s be defined as above, then RAB_s is the smallest set such that the following conditions hold:*
  1. $RAB'_s \subseteq RAB_s$, and
  2. if constructor $k : s_1, \ldots, s_n \rightarrow s$ textually occurs in RAB'_s and $t_i \in RAB_{s_i}$ with s_i basic, then $k(t_1, \ldots, t_n) \in RAB_s$.

Notice that RAB_s is always finite.
Finally, a finite set of abstract terms is defined such that:

- such set is representative of all the concrete domain;
- such set does not contains redudancy with respect to the set of abstract terms extracted by the input specification.

Definition 2.6 *[Abstract Bases: AB_s]*
An Abstract Base *of sort s, AB_s is a finite cut C of $(\Pi_\Sigma)_s$ which covers RAB_s.*

In [7] has been proved that AB_s exists for all sort s. Distinct abstract bases for the same sort s will differ in their "grain", i.e. in what terms are given an explicit representation. The choice between different abstract bases will depend on the implementation of the abstraction procedure as well as on the level of detail required for the particular use of it.

Once chosen an abstract base AB_s, the following definition allows to characterize a finite representation of the powerset of $(T_\Sigma)_s$. It will be made up of finite sets of partial terms which do not suffer from the unpleasant redundance above.

Definition 2.7 *[Abstract Domain:ABS]*
The Abstract Domain *of* sort s, *ABS_s, is defined as 2^{AB_s} with standard set inclusion partial ordering.*
The Abstract Domain, *ABS, for a given data type definition is defined as the* $\perp$−sum *of the abstract domains for all the sorts in the type definition.*

2.1. Abstract Functions

This section shows how to compute the abstract version of any function defined within an abstract data type definition T. We thus need to define what it means for an abstract term to approximate a set of partial terms, which in turn represents a set of possible values. We need some additional definitions. Given a partial order $(D, \preceq), d \in D$,and $W \subseteq D$

Definition 2.8 *[∇]*
∇_d is the partial order with $X = \{x | d \preceq x\}$ as set of elements and $\preceq \cap X \times X$ as ordering relation.

Definition 2.9 *[Approximation]*
W generates *d* iff $\exists\, x \in W : x \preceq d$. *W* represents *d* iff $\exists\, X \subseteq W$ *:X is a* cut *for* ∇_d. *W* approximates *d* iff *W* generates *d or W* represents *d*.

Definition 2.10 *[Approximation Function:$\aleph_s$]*
For all $X \subseteq (\Pi_\Sigma)_s$:

$$
\begin{array}{ll}
\aleph_s(\{\}) & =\{\} \\
\aleph_s(\{\perp_s\}) & = AB_s \\
\aleph_s(\{c\}) & = \bigcup Min(\{a | a \in ABS_s, a \text{ approximates } c\}) \\
 & \textit{for all constants } c \\
\aleph_s(\{k(p_1,\ldots,p_n)\}) & = \aleph_s(\{x | x = k(x_1,\ldots,x_n), \bigwedge_{i=1}^{n} x_i \in \aleph_{s_i}(\{p_i\})\}) \\
 & \textit{for constructor terms} \\
\aleph_s(\{p_1,\ldots,p_n\}) & = \bigcup_{i=1..n} \aleph_s(\{p_i\})
\end{array}
$$

We shall use only *finite, non-empty* sets of partial terms, in which case function $\aleph$ is certainly effectively computable and total.

In the following section we shall describe how to get the abstract version of functions from their definition.

2.2. Function Abstraction Procedure

In order to compute the abstract version of all the functions defined in T, we generate a new system of equations starting from those in T, and let such abstract functions be its minimal solution.

For the sake of simplicity, we shall deal with unary functions only. All definitions and results can be extended to n-ary functions.

For each function $f : s' \rightarrow s$, the set of equations defining its abstract version $f^{\mathcal{A}}$ is composed of two parts. The first one does not depend on the equations which define f. It is as follows where $p_i \in AB_{s'}$:

1. $f^{\mathcal{A}}(\{\}) = \{\}$
2. $f^{\mathcal{A}}(\{p_1, \ldots, p_n\}) = \bigcup_{i=1 \ldots n} f^{\mathcal{A}}(\{p_i\})$

To complete the definition of $f^{\mathcal{A}}$ we have to introduce the equations for $f^{\mathcal{A}}$ when applied to singletons. The right-hand side of $f^{\mathcal{A}}(\{p\})$ is obtained by selecting the equation of f's definition, the left-hand side of which *matches* with the partial term p, if any. Pattern matching is easily extended to partial terms simply by considering $\perp$ as an additional constant. The equation for $f^{\mathcal{A}}(\{p\})$ when no equation for f can be selected is given in case 3) below whereas in case 4) $rhs(p)$ is the right-hand side of the *unique* equation for f which has been selected, and properly instantiated according to the bindings induced by pattern- matching.

3. $f^{\mathcal{A}}(\{p\}) = AB_s$
4. $f^{\mathcal{A}}(\{p\}) = \mathcal{AS}[\![rhs(p)]\!]$

Function $\mathcal{AS}$ is defined as follows:

$$
\begin{aligned}
&\textit{4.1)}\mathcal{AS}[\![c]\!] &&= \aleph(\{c\}), \textit{ for all constants } c :\rightarrow s, \textit{ or } c = \perp \\
&\textit{4.2)}\mathcal{AS}[\![k(e_1, \ldots, e_n)]\!] &&= \aleph(\{x | x = k(x_1, \ldots, x_n), \textstyle\bigwedge_{i=1}^{n} x_i \in \mathcal{AS}[\![e_i]\!]\}), \\
& &&\quad \textit{for all constructor terms} \\
&\textit{4.3)}\mathcal{AS}[\![f(e)]\!] &&= f^{\mathcal{A}}(\mathcal{AS}[\![e]\!]), \textit{ for all function terms.}
\end{aligned}
$$

Note that $f^{\mathcal{A}}(\mathcal{AS}[\![e]\!])$ is a *syntactical* function application, i.e. a *function term*.

Let m_2 a function defined as below:

$$
\begin{aligned}
&m_2(0) = 0 \\
&m_2(s(0)) = s(0) \\
&m_2(s(s(x))) = m_2(x)
\end{aligned}
$$

The relevant equations obtained for m_2 are:

$$
\begin{aligned}
m_2^A(\{0\}) &= \{0\}\\
m_2^A(\{s(0)\}) &= \{s(0)\}\\
m_2^A(\{s(s(\perp_{Nat}))\}) &= m_2^A(\{0, s(0), s(s(\perp_{Nat}))\})
\end{aligned}
$$

3. The Abstract Interpreter

The aim of the abstract interpreter is to annotate each point of a LOTOS specification with an abstract environment representing all possible bindings that the specification variables will have at "execution time". This means that each behaviour expression (bex) will have associated a function that binds variable identifiers to abstract terms containing information about the possible values the variables can assume when executing the associated bex. Therefore, the abstract environment takes trace of all possible data flow reaching a bex.

The following figure sketches the architecture of our system:

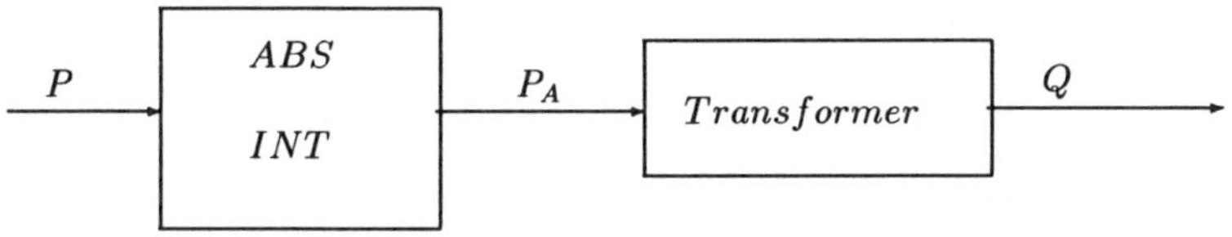

where the module *ABS INT*, given a specification, realizes the construction of a abstract domain tailored on that specification and the annotation phase which uses such a domain. The module *Transformer* analyses the annotated specification, verifies that certain program (specification) properties hold, and realizes the opportune optimizations via program transformation. In our case we explored optimization such as starvation detection and dead code removal. Intuitively this module is a black box for all program transformation which can based on data-flow analysis.

The first *syntactic* step is to associate a unique label to each point of the specification. The labelling function $\mathcal{F}$ performs this task.

Definition 3.1 *[Simple behaviour expression sbex]*
A sbex is a bex Exit, Exit v_s, Stop or $P[g_1, \ldots, g_k](e_1, \ldots, e_n)$.

Definition 3.2 *[Labelled behaviour expression lbex]*
A labelled behaviour expression (lbex) is a pair lbex = (label, B) where B is labex and label *is a label that identifies univocally B in the specification.*

Definition 3.3 *[labex]*
A labex is either a sbex or a structured bex.

Definition 3.4 *[Structured bex]*
A structured bex is a bex defined in terms of lbex.

For example $(label_1, B_1)$ [] $(label_2, B_2)$ is a structured bex derived from Choice.

Definition 3.5 *[Labelling function $\mathcal{F}_l$]*
$\mathcal{F}_l$: bex $\longrightarrow$ Nat $\longrightarrow$ lbex is defined as follow:

$\mathcal{F}_l$ *(Exit) n = (n,Exit)*
$\mathcal{F}_l$ *(Stop) n = (n,Stop)*
$\mathcal{F}_l$ *(g!e; B) n = (n,g!e;* $\mathcal{F}_l$ *B (n+1))*
$\mathcal{F}_l$ *(g?x:S; B) n = (n,g?x:S;* $\mathcal{F}_l$ *B (n+1))*
$\mathcal{F}_l$ $(B_1 [] B_2)$ *n =* $(n,(\mathcal{F}_l\ B_1\ (n+1))\ []\ (\mathcal{F}_l\ B_2\ (max\_lab(\mathcal{F}_l\ B_1\ (n+1))) + 1))$
$\mathcal{F}_l$ $(B_1 >> B_2)$ *n =* $(n,(\mathcal{F}_l\ B_1\ (n+1)) >> (\mathcal{F}_l\ B_2\ (max\_lab(\mathcal{F}_l\ B_1\ (n+1))) + 1))$
$\mathcal{F}_l$ *(sp→ B) n = (n,sp→ (*$\mathcal{F}_l$ *B (n+1))*
$\mathcal{F}_l$ *(g?x:S[sp(x)]; B) n = (n,g?x:S[sp(x)];(*$\mathcal{F}_l$ *B (n+1))*
$\mathcal{F}_l$ $(B_1\ \underline{parop}\ B_2)$ *n =* $(n,(\mathcal{F}_l\ B_1\ (n+1))\ \underline{parop}\ (\mathcal{F}_l\ B_2\ (max\_lab(\mathcal{F}_l\ B_1\ (n+1))) + 1))$
$\mathcal{F}_l$ *(Hide g in B) n = (n,Hide g in (*$\mathcal{F}_l$ *B (n+1)))*
$\mathcal{F}_l$ *(Let x:S=e in B) n = (n,Let x:S=e in (*$\mathcal{F}_l$ *B (n+1)))*
$\mathcal{F}_l$ *(Accept x:S in B) n = (n,Accept x:S in (*$\mathcal{F}_l$ *B (n+1)))*
$\mathcal{F}_l$ *(Choice x:S [] B) n = (n,Choice x:S [] (*$\mathcal{F}_l$ *B (n+1)))*
$\mathcal{F}_l$ *(P[a](e)) n = (n,P[a](e))*

where $max\_label(B)$ *gives the max label in lbex B and* $\underline{parop} \in \{||, |||, |[g_1, \ldots, g_n]|\}$.

Finally, the labelled specification S_l associated to the specification:

S = **Specification** n [gate_list](par_list):(functionality) :=
Behaviour
B
where
...
< *proc_list* >
....
endspec

is a new specification in which bex B is changed in $(\mathcal{F}_l\ (B)\ 0)$ and each bex B_{p_i} of the process definition P_i in < *proc_list* > is changed in $(\mathcal{F}_l\ (B_{p_i})\ k)$ where $k = max\_lab(B_{p_{i-1}})$.

Let *Var* denote the domain of LOTOS variables and let *abs_env* denote the domain abstract environments.

Definition 3.6 *[Abstract environment ρ]*
An abstract environment is a function $\rho : Var \longrightarrow ABS$ that binds variable identifiers to abstract terms.
Let $x :\longrightarrow s$ be a variable of type s. $\rho(x) = [t_1, \ldots, t_k]$ if x is bound to the abstract term $\{t_1, \ldots, t_k\}$ and $t_i \in AB_s$, otherwise $\rho(x) = \emptyset$.

Definition 3.7 *[Annotation Function Γ)]*
The annotation function Γ : label $\longrightarrow$ abs_env, given lbex l associated to labex (l, B) returns the abstract environment ρ associated to B. This environment represents all the possible association between variable identifiers and abstract terms existing before the execution of B.

Definition 3.8 *[Context]*
The tuple $(glist, \beta, exlist, \rho, \Gamma, hlist)$ *is called* context *where:*

- *glist: synchronization gate list;*
- β*: function that binds a gate to an abstract term;*
- *exlist: abstract term list offered in a bex of type* Exit*;*
- ρ*: current environment function: implements the association between variable identifiers and abstract terms in force at a certain moment of the abstract computation;*
- Γ *is the function defined in 3.7;*
- *hlist: hiding gates list;*

Finally we need a function which returns the abstract representation of concrete values.

Definition 3.9 *[Abstract Data Evaluation* $\mathcal{V}$*]*
Let V_{ex_s} *be the set of LOTOS value expression of sort s:*

$$
\begin{aligned}
&\mathcal{V}_s : V_{ex_s} \longrightarrow abs\_env \longrightarrow ABS_s\\
&\mathcal{V}_s(c)\rho = \aleph_s(\{c\}), \ \ \forall \ constant \ c : \longrightarrow s\\
&\mathcal{V}_s(x)\rho = \rho(x), \ \ \forall \ variable \ x \ of \ type \ s\\
&\mathcal{V}_s(k(e_1, \ldots, e_n))\rho = \aleph_s(\{k(x_1, \ldots, x_n) \mid x_i \in \mathcal{V}_{s_i}(e_i)\rho\})\\
&\qquad\qquad\qquad\qquad\quad \forall \ constructor \ k : s_1, \ldots, s_n \longrightarrow s\\
&\mathcal{V}_s(f(e_1, \ldots, e_n))\rho = f^{@}(\mathcal{V}_{s_1}(e_1)\rho, \ldots, \mathcal{V}_{s_n}(e_n)\rho)\\
&\qquad\qquad\qquad\qquad\quad \forall \ functor \ f : s_1, \ldots, s_n \longrightarrow s
\end{aligned}
$$

Let *Context* denote the domain of abstract contexts.
The abstract interpretation of a specification S is performed by the function $\mathcal{PV}$:

$$\mathcal{PV} : B \longrightarrow Context \longrightarrow Context$$

that given a bex B and a context $(glist, \beta, exlist, \rho, \Gamma, hlist)$, returns a new context properly modified from the abstract evaluation of B. $\mathcal{PV}$ is recursively defined on the structure of B.

The followings are few cases of the definition of $\mathcal{PV}$. The whole definition can be found in [9]

Action prefix: g!e; B

$$
\begin{aligned}
&\mathcal{PV}\ (g!e;(label, B))\ (glist, \beta, exlist, \rho, \Gamma, hlist) =\\
&\qquad = \mathcal{PV}\ B\ (glist, \beta, exlist, \rho, \Gamma', hlist) \text{ if } g \in hlist\\
&\qquad = \mathcal{PV}\ B\ (glist, \beta', exlist, \rho, \Gamma', hlist) \text{ if } g \in glist\\
&\qquad = \mathcal{PV} B\ (glist, \beta, exlist, \rho, \Gamma', hlist) \text{ otherwise}\\
&\qquad\quad \text{where}\\
&\qquad\qquad\quad \begin{cases} \Gamma' = \Gamma[\rho \cup_\rho \Gamma(label)\ /\ label] \\ \beta' = \beta[\mathcal{V}(e)\rho \cup \beta(g)\ /\ g] \\ \rho' \cup_\rho \rho'' = \rho[\rho'(x) \cup \rho''(x)/x] \ \ \forall \ variable \ x \end{cases}
\end{aligned}
$$

Action prefix: g?x:S; B

$$\mathcal{PV}\ (g?x : S;(label, B))\ (glist, \beta, exlist, \rho, \Gamma, hlist) =$$
$$= \mathcal{PV}\ B\ (glist, \beta', exlist, \rho', \Gamma', hlist) \text{ if } g \in glist$$
where
$$\begin{cases} \rho' = \rho[\beta(g)/x] \\ \Gamma' = \Gamma[\rho'/label] \end{cases}$$
$$= \mathcal{PV}\ B\ (glist, \beta, exlist, \rho'', \Gamma'', hlist) \text{ if } g \notin glist$$
where
$$\begin{cases} \rho'' = \rho[AB_s/x] \\ \Gamma'' = \Gamma[\rho''/label] \end{cases}$$
$$= \mathcal{PV}\ B\ (glist, \beta, exlist, \rho, \Gamma', hlist) \text{ if } g \in hlist$$
where
$$\Gamma' = \Gamma[\rho \cup \Gamma(label)/label]$$

The interpretation of an action prefix (either in input or in output) consists in modifying Γ in order to augment the environment associated to *label* with the bindings of the current context. Essentially each time that a point is reached by a abstract execution with a certain bunch of data we have to take trace of them.

Choice: $B_1[]B_2$

$$\mathcal{PV}\ ((label_1, B_1)[](label_2, B_2))\ (glist, \beta, exlist, \rho, \Gamma, hlist) =$$
$$=\mathcal{PV}\ (B_1)\ (glist, \beta, exlist, \rho, \Gamma', hlist)$$
$$\uplus$$
$$\mathcal{PV}\ (B_2)\ (glist, \beta, exlist, \rho, \Gamma', hlist)$$
where
$$(g_1, \beta_1, e_1, \rho_1, ge_1, h_1) \uplus (g_2, \beta_2, e_2, \rho_2, ge_2, h_2) =$$
$$(g_1 \cup g_2, \beta_1 \cup_\beta \beta_2, e_1 \cup_{ex} e_2, \rho_1 \cup_\rho \rho_2, ge_1 \cup_{ge} ge_2, h_1 \cup h_2).$$

and:

$$e_1 \cup_{ex} e_2 = \{e_i | e_i = (e_{i_1} \cup e_{i_2}),\ (e_{i_1} \in e_i) \wedge (e_{i_2} \in e_2)\}$$
$$\beta_1 \cup_\beta \beta_2 = \beta \text{ and } \beta(x) = \beta_1(x) \cup \beta_2(x)\ \forall x$$
$$ge_1 \cup_{ge} ge_2 = ge \text{ and } ge(label) = ge_1(label) \cup_\rho ge_2(label)\ \forall label$$

Selection Predicate: $g?x : S\ [sp(x)];\ B$

$$\mathcal{PV}\ (g?x : S[sp(x)];\ (label, B))\ (glist, \beta, exlist, \rho, \Gamma, hlist) =$$
$$= \mathcal{PV}\ B\ (glist, \beta, exlist, \rho', \Gamma', hlist)$$
where
$$\begin{cases} \rho' = \rho[\{y | y \in \rho(x) \wedge sp^{@}(\{y\}) \neq \{ff\}\}/x] \\ \Gamma' = \Gamma[\rho' \cup_\rho \Gamma(label)/label] \end{cases}$$

This last rule allows the reduction of environment because of the filtering of data due to the guard $sp(x)$.

3.1. Correctness and Termination

In this section we describe the procedure for establishing the correctness and the termination of the interpretation process. For further details we refer to [9].

Intuitively, correctness has been proved according this schema: assuming that the context $C = (glist, \beta, exlist, \rho, \Gamma, hlist)$ in which will be executed a bex B is correct with respect to all the possible execution, then the context obtained from the abstract evaluation of B is still valid. The idea is the following: let C be a *concrete context*, that represents a concrete execution in a certain moment, and let $C^{@}$ be an abstract context so that $C \subseteq_C C^{@}$, then applying them respectively the concrete interpretation and the abstract interpretation we obtain two new contexts C' and $C^{@'}$, and the relation $C' \subseteq_C C^{@'}$ is still valid. The relation $A \subseteq_C B$ expresses the fact that the set of partial terms associated to variables in B represents the terms associated to the same variables in A.

In order to prove that the abstract interpretation process terminates, we proved that the operator $\cup_\rho$, which given two abstract environments returns their disjoint union is monotonic and bounded in ABS. In this manner we are granted that the calculus terminates because sooner or later each environment $\Gamma(label)$ reaches a point in which can't never increase and when this happens for all labels we reached the abstract interpretation termination, i.e. the fix_point of the whole computation is reached.

Let Env be the abstract environment domain and $nvar_i$ the number of variable of sort S_i in the specification.

Proposition 3.1 The function $\cup_\rho : Env \times Env \longrightarrow Env$ is monotonic with respect to the operator $\leq$ and is bounded in the following way:

$$0 \leq |\rho' \cup_\rho \rho''| \leq nvar_1 * |AB_{S_1}| + \cdots + nvar_k * |AB_{S_k}|$$

$\forall$ environment ρ', ρ'', k=number of sorts.

□

Proposition 3.2 The abstract interpretation terminates in a finite number of at most p steps:

$$p = l * (nvar_1 * |AB_{S_1}|) * \cdots * (nvar_k * |AB_{S_k}|)$$

with:

$$\begin{cases} l = & \text{highest label value} \\ k = & \text{number of sorts} \\ nvar_k = & \text{Number of variable of sort } S_k \end{cases}$$

□

4. Examples

In this section two simple applications of our interpreter are presented. In the first case the abstract interpretation of a LOTOS specification allows to detect situations of starvation in which the represented system waits for an event that will never happen. The second experiment allows to detect if a part of the specification code is never reached, so that it can be removed without changing the specification semantics.

4.1. Starvation detection

This kind of analysis is able to circumscribe situations in which, when the data flow reaches a specified point of the specification, then the starvation:

1. will surely happen in that point;
2. will possibly happen in that point;
3. will never happen in that point.

The idea is to detect whether a certain sufficient property holds in a point of the program (which guarantees that the property will hold for all executions of that point of the program). Then, if the property holds for all points of the program, we are able to characterize the overall behaviour of the program.

A typical situation that could create starvation is when two processes P_1 and P_2, which evolve in parallel, have to synchronize on a fixed gate g. If P_1 is not able to execute the actions that permit the synchronization, P_2 will wait this event with consequent starvation of P_2 itself.

We can analyze the previous situation in presence of either value matching or value passing synchronization:

1. **Sure starvation**
   Value passing
   Let $(l_1, g!x_1; B_1), \ldots, (l_n, g!x_n; B_n)$ be all the action prefix construct present in the specification. The specification represent a system that gives sure starvation if:

   $$\Gamma(l_i)(x_i) = \emptyset \quad \forall i = 1, \ldots, n$$

   This means that all the synchronization points will never be reached by any control flow. Otherwise if there exists l_j such that $\Gamma(l_j)(x_j) \neq \emptyset$, then it is possible that the synchronization takes place in that point.

   Value matching
   In this situation (output–output synchronization) the system represented reaches starvation if the abstraction of the values offered from the processes which have to synchronize denotes disjoint sets. That is, if process P_1 offers value e_1 by means of $(l_1, g!e_1; B_1)$ and process P_2 offers value e_2 by means of $(l_2, g!e_2; B_2)$, then the starvation is guaranteed in that point if $\mathcal{V}(e_1)(\Gamma(l_1)) \cap \mathcal{V}(e_2)(\Gamma(l_2)) = \emptyset$. If there are more than one synchronization point, in order to decide if the starvation in guaranteed (if the control flow reaches one of these synchronization points) we must verify the previous property for all possible couple of action prefix.

2. **Possible starvation**
   Value passing
   Like the previous situation, starvation is possible in a synchronization point $(l_i, g!x; B)$ only if $\mathcal{V}(x)(\Gamma(l_i)) \neq \emptyset$. This means that some data flow reached the point labelled with l_i during the abstract interpretation.

Value matching
If process P_1 offers value e_1 by means of $(l_1, g!e_1; B_1)$ and process P_2 offers value e_2 by means of $(l_2, g!e_2; B_2)$ then starvation is possible in that point if $\mathcal{V}(e_1)(\Gamma(l_1)) \cap \mathcal{V}(e_2)(\Gamma(l_2)) \neq \emptyset$ because this means that:
$\exists v_i \in \mathcal{V}(e_1)(\Gamma(l_1))$ such that
$v_i \in \mathcal{V}(e_2)(\Gamma(l_2))$, but is also possible to have starvation, i.e. when B_1 offers on g a value $v_j \in \mathcal{V}(e_1)(\Gamma(l_1))$ such that $v_j \notin \mathcal{V}(e_2)(\Gamma(l_2))$. This is possible because $\mathcal{V}(e_1)(\Gamma(l_1)) \cap \mathcal{V}(e_2)(\Gamma(l_2)) \neq \mathcal{V}(e_1)(\Gamma(l_1))$ and $\mathcal{V}(e_1)(\Gamma(l_1)) \cap \mathcal{V}(e_2)(\Gamma(l_2)) \neq \mathcal{V}(e_2)(\Gamma(l_2))$

3. **Synchronization**
   Value passing
   In this situation the synchronization between the usual processes P_1 and P_2 is certain in the value e offered by P_1 on an action prefix $g!e$ don't denotes the empty set.
   Value Matching
   If process P_1 offers value e_1 by means of $(l_1, g!e_1; B_1)$ and process P_2 offers value e_2 by means of $(l_2, g!e_2; B_2)$ then synchronization in certain only if:

   (a) $\mathcal{V}(e_1)(\Gamma(l_1)) = \{p_i\}$ and $\mathcal{V}(e_2)(\Gamma(l_2)) = \{p_i\}$.
   (b) p_i is a "ground" partial term.

   The condition (b) is necessary because the ground partial terms represent a specified element of the starting real domain (for the construction of function π).

4.2. Dead code removal

The abstract execution is able to detect code portion which could be cut off from the specification because of their unreachability from no data flow. For example we could cut off an alternative of a Choice construct. This would be helpful for the run time support that has to decide at run time which alternative of a non deterministic construct should execute.

Example 4.1 *Let consider the labex:* $(label_1, sp(x) \longrightarrow (label_2, B))$
If at the end of the abstract interpretation we have that: $\Gamma(label_2)(x) = \emptyset$ *this means that,* $sp^{@}$ *in* $\Gamma(label_1)(x)$ *is never true so no value filtered by* $sp(x)$ *bounds* x *in* $\Gamma(label_2)$.

5. Final Remarks

The work described in this paper has been implemented under Unix using the functional language Miranda. The Abstract Interpreter has been used in several applications including the Gate Splitting Transformation. The experiment is still in a prototype phase. It confirmed the viability of the approach, but still we have to classify the kind of optimizations suitable for this general paradigm.

Another direction of work is the comparison with other tools for the symbolic execution of LOTOS specifications. In particular, it is interesting to compare our Abstract Interpreter with the Symbolic Executor SMILE [6]. Both tools try to apply technology from other fields, abstract interpretation in the former case, and narrowing the latter. In our case, the main effort is in defining the abstract domain exploiting the data type definition capabilities of LOTOS, then the abstract execution is the standard one extended with few simple operations on sets. In the case of SMILE, the emphasis is on the execution mechanisms which is based on the logic interpretation of the LOTOS specification and executed with narrowing.

The goal of our approach was to verify the usefulness of Abstract Interpretation in the context of specification languages, so that a systematic comparison with SMILE has not been done yet. A very initial impression is that the two approaches are very similar, both with respect to the type of applications and the quality of results they can obtain. The existing differences are probably much more formal than substantial such as the problem of termination which in the case of SMILE is enforced by the implementation while in our case it is guaranteed by the formal framework. In any case, an experimental comparison of accuracy, efficiency and ease of use should be performed.

REFERENCES

1. S. Abramsky, C. Hankin, (eds.): Abstract Interpretation of Declarative Languages. Ellis Horwood, Chichester, U.K. 1987
2. S. Abramsky: Abstract Interpretation, Logical Relations and Kan Extensions, Journal of Logic and Computation, Vol.1, No.1, pp.5-40, 1990
3. T. Bolognesi, E. Brinksma: Introduction to the ISO specification language LOTOS. Computer Network and ISDN Systems - Vol. 14, No 1, pp. 25-59, 1987
4. P. Cousot, R. Cousot: Static determination of dynamic properties of programs. Proc. of the 2nd Int. Symp. on Programming Languages. Dunod, Paris, pp. 106-130, 1976
5. P. Cousot, R. Cousot: Abstract Interpretation: a unified lattice model for static analysis of programs by construction or approximation of fixpoints. Proc. of the 4th ACM Symp. on Principles of Programming Languages, Los Angeles, pp. 238-252, 1977
6. Eertink H., Hankin C.: Simbolic execution of LOTOS Specification. 5th International Conference on Formal Description Technique - FORTE '92, pp. 289-304, Lannion (Francia) 1992
7. F. Giannotti, D. Latella: Gate Splitting in LOTOS Specifications Using Abstract Interpretation. Proceedings of TAPSOFT'93 Ed, Gaudel MC. Jouannaud JB, LNCS 668 Springer Verlag 1993, selected for publication on Science of Computer Programming, special issue on TAPSOFT'93.
8. D. Latella: Correctness of the IPR transformation. ESPRIT Project 2304 - Loto-Sphere, ref. Lo/WP1/T1.2/CNUCE/N0021. University of Twente, Enschede, The Netherlands, 1991
9. F. Fiore: Definizione ed Implementazione di un Interprete Astratto per il linguaggio di specifica LOTOS. Tesi di Laurea, Dipartimento di Scienze dell'Informazione, Universitá di Pisa, 1993 (in italian).
10. L. Pires (Ed.): The LotoSphere Design Methodology - I: Basic Concepts. ESPRIT Project 2304 - LotoSphere, Task 1.1 Deliverable, ref. Lo/WP1/T1.1/N0045/V03. University of Twente, Enschede, The Netherlands, 1992
11. T. Bolognesi (Ed.): Catalogue of LOTOS Correctness Preserving Transformations ESPRIT Project 2304 - LotoSphere, Task 1.2 Deliverable, ref. Lo/WP1/T1.2/N0045/V03. University of Twente, Enschede, The Netherlands, 1992

12. P.H.J. van Eijk, C. A. Vissers, M. Diaz (Ed.): The Formal Description Technique LOTOS. North-Holland, 1989
13. P.H.J. van Eijk: Tool demonstration: The LotoSphere Integrated Tool Environment Lite. Proc. FORTE'91, North-Holland, 1992
14. K. Warkentyne: The Application of Transformations in T3.2 (MiniMail+ISDN). ESPRIT Project 2304 - LotoSphere, ref. WP1/T11/ASCOM/N0015. University of Twente, Enschede, The Netherlands, 1992

An Attempt to Embed a Restricted Version of SDL as a Target Language in Focus*

Eckhardt Holz[a] and Ketil Stølen[b]

[a]Fachbereich Informatik, Humboldt Universität zu Berlin, Lindenstrasse 54a, D-10099 Berlin, Germany

[b]Fakultät für Informatik, TU München, Postfach 20 24 20, D-80290 München, Germany

This paper presents a first attempt to embed a restricted version of SDL as a target language in Focus. Brief introductions to both Focus and SDL are given, and it is shown how both methods can be assigned a denotational semantics based on streams and stream processing functions. A set of Focus specifications, referred to as F-SDL, is characterized whose elements structurally and semantically match SDL specifications to such a degree that an automatic translation is almost straightforward. Finally it is outlined how Focus can be used to develop an SDL specification of a protocol.

1. INTRODUCTION

Focus [1], [2] is a methodology for the formal specification and development of distributed systems. A system is modeled by a network of components working concurrently and communicating asynchronously via unbounded, directed FIFO channels. A number of reasoning styles and techniques is supported. Focus provides mathematical formalisms which support the formulation of highly abstract, not necessarily executable specifications with a clear semantics. Moreover, Focus offers powerful refinement calculi which allow distributed systems to be developed in the same style as for example VDM and Z allow for the development of sequential programs. Finally, Focus is modular in the sense that design decisions can be checked at the point where they are taken, that component specifications can be developed in isolation, and that already completed developments can be reused in new program developments.

SDL [3] has been developed by ITU-TSS and was initially intended for the description of telecommunication systems. However, SDL is also well-suited for more general specification tasks. In SDL the behavior of a system is equal to the combined behavior of its processes. A process is basically a communicating, extended finite-state machine. The processes communicate asynchronously by sending signals via signal routes. SDL provides both a textual and a graphical specification formalism. SDL has received considerable interest from industry and is supported by a large number of tools and environments.

In some sense Focus and SDL are two orthogonal approaches. Because of its very mathematical and abstract notation, Focus has its strength in the area of formal refinement

*This work is supported by the Sonderforschungsbereich 342 "Werkzeuge und Methoden für die Nutzung paralleler Rechnerarchitekturen".

and verification. SDL, on the other hand, due to its graphical notation and many structuring constructs, is well-suited for the formulation of large and complicated "real-life" specifications. It is therefore tempting to try to combine these two approaches into one methodology inheriting the strength of both. This is our motivation!

SDL offers a large number of specification and structuring constructs, and it is important to realize that it is not our intention to transform Focus into a method which allows for the development of any SDL specification. In fact we are only interested in the sublanguage of SDL, which in a natural way corresponds to Focus developments. For example, Focus is not well-suited for the description of dynamic networks — networks where processes can be created and interfaces may change during execution. Thus the full generality of the SDL process creation mechanism is not very important in connection with a Focus development. This does not mean that we consider these additional features of SDL to be of little value. On the contrary, we rather see Focus as a tool or facility which can be used to formally develop and verify certain restricted, critical parts of an SDL specification. Typical examples would be communication protocols, mutual exclusion algorithms or some complicated sorting algorithm.

A development of an SDL specification in Focus is split into three phases: first a requirement specification is formulated in Focus; then in a step-wise fashion this requirement specification is refined into an F-SDL specification; finally the F-SDL specification is translated into SDL.

Section 2 describes the underlying formalism. Then Focus and SDL are introduced in Sections 3 and 4, respectively. It is shown how both Focus and a restricted version of SDL can be assigned the same type of denotational semantics. In Section 5 F-SDL is syntactically characterized, and Section 6 outlines how an SDL specification of a protocol can be formally developed employing the proposed technique. Finally, Section 7 gives a brief summary and discusses possible extensions.

2. UNDERLYING FORMALISM

N denotes the set of positive natural numbers. A stream is a finite or infinite sequence of actions. It models the history of a communication channel by representing the sequence of messages sent along the channel. Given a set of actions D, D^* denotes the set of all finite streams generated from D; D^∞ denotes the set of all infinite streams generated from D, and D^ω denotes $D^* \cup D^\infty$.

Let $d \in D$, $r, s \in D^\omega$, $A \subseteq D$ then ϵ denotes the empty stream; $\#r$ denotes the length of r, which is equal to ∞ if r is infinite, and is equal to the number of elements in r otherwise; $A©r$ denotes the result of filtering away all actions not in A (the projection of r on A); $d \,\&\, s$ denotes the result of appending d to s; $r \frown s$ denotes r if r is infinite and the result of concatenating r with s, otherwise; $r \sqsubseteq s$ holds if r is a prefix of s.

The stream operators defined above are overloaded to tuples of streams in a straightforward way. ϵ will also be used to denote a tuple of empty streams when the size of this tuple is clear from the context. If d is an n-tuple of actions, and r, s are n-tuples of streams, then $\#r$ denotes the length of the shortest stream in r; $d \,\&\, s$ denotes the result of applying $\&$ pointwisely to the components of d and s; $r \frown s$ and $r \sqsubseteq s$ are generalized in the same pointwise way.

A chain c is an infinite sequence of stream tuples $c_1, c_2, \ldots$ such that for all $j \geq 1$, $c_j \sqsubseteq c_{j+1}$. $\sqcup c$ denotes c's least upper bound. Since streams may be infinite such least upper bounds always exist.

A function $\tau \in (D^\omega)^n \rightarrow (D^\omega)^m$ is called a ((n,m)-ary) stream processing function iff it is prefix monotonic and continuous:

$$\text{for stream tuples } i \text{ and } i' \text{ in } (D^\omega)^n : \; i \sqsubseteq i' \Rightarrow \tau(i) \sqsubseteq \tau(i'),$$

$$\text{for all chains } c \text{ generated from } (D^\omega)^n : \; \tau(\sqcup c) = \sqcup\{\tau(c_j) | j \in \mathsf{N}\}.$$

That a function is prefix monotonic implies that if the input is increased then the output may at most be increased. Thus what has already been output can never be removed later on. Prefix continuity, on the other hand, implies that the function's behavior for infinite inputs is completely determined by its behavior for finite inputs.

A stream processing function $\tau \in (D^\omega)^n \rightarrow (D^\omega)^m$ is pulse-driven iff:

$$\text{for all stream tuples } i \text{ in } (D^\omega)^n : \; \#i \neq \infty \Rightarrow \#\tau(i) > \#i.$$

That a function is pulse-driven means that the length of the shortest output stream is infinite or greater than the shortest input stream. This property is interesting in the context of feedback constructs because it guarantees that the least fixpoint is always infinite for infinite input streams.

The arrows $\rightarrow$, $\stackrel{c}{\rightarrow}$ and $\stackrel{cp}{\rightarrow}$ are used to tag domains of ordinary functions, domains of monotonic, continuous functions, and domains of monotonic, continuous, pulse-driven, functions, respectively.

To model timeouts we need a special action $\surd$, called "tick". There are several ways to interpret streams with ticks. In this paper, all actions should be understood to represent the same time interval — the least observable time unit. $\surd$ occurs in a stream whenever no ordinary message is sent within a time unit. A stream or a stream tuple with occurrences of $\surd$'s are said to be timed. Similarly, a stream processing function is said to be timed when it operates on domains of timed streams. Observe that in the case of a timed, pulse-driven, stream processing function the output during the first $n+1$ time intervals is completely determined by the input during the first n time intervals. For any stream or stream tuple i, $\diamond i$ denotes the result of removing all occurrences of $\surd$ in i.

In the more theoretical parts of this paper, to avoid unnecessary complications, we distinguish between only two sets of actions, namely the set D denoting the set of all actions minus $\surd$, and $D_\surd$ denoting $D \cup \{\surd\}$. However, the proposed formalism can easily be generalized to deal with more general sorting, and this is exploited in the examples.

We use one additional function in our examples, namely a function $\propto$ which eliminates repetitions. More explicitly, if d, e are n-tuples of actions, and r is an n-tuple of streams, then $\propto$ is a stream processing function such that the following axioms hold:

$$\propto(d \,\&\, \epsilon) = d \,\&\, \epsilon, \quad \propto(d \,\&\, e \,\&\, r) = \textsf{if } d = e \textsf{ then } \propto(d \,\&\, r) \textsf{ else } d \,\&\, \propto(e \,\&\, r).$$

Note that these axioms together with the monotonicity and continuity constraints determine the semantics also for stream tuples whose stream components are not of the same length.

3. FOCUS AND ITS STREAM SEMANTICS

Depending on the logical concepts they employ, Focus specifications can be divided into a number of subclasses. In this paper, we employ only so-called relational specifications. However, the proposed approach can easily be combined with the other specification techniques in Focus. A relational specification of a component with n input channels and m output channels is written in the form

$$S(i_1 \in D^\omega, \ldots, i_n \in D^\omega \rhd o_1 \in D^\omega, \ldots, o_m \in D^\omega) \equiv R,$$

where S is the specification's name; $i_1, \ldots, i_n$ and $o_1, \ldots, o_m$ are disjoint, repetition free lists of identifiers representing n respectively m streams; R is a formula with the elements of $i_1, \ldots, i_n$ and $o_1, \ldots, o_m$ as its only free variables. Each stream models the communication history of a channel, and R characterizes the allowed relation between the histories of the input and the output channels.

In Focus specifications are modeled by sets of timed, pulse-driven, stream processing functions. In real-time specifications the ticks occur also at the syntactic level. In other specifications they are abstracted away in the sense that they are not allowed to occur explicitly in the specifications. In this paper we consider only specifications of the latter type.

The denotation of the specification S is the set of all (n, m)-ary, timed, pulse-driven, stream processing functions which fulfill R when time-signals are abstracted away and only complete inputs are considered:

$$[\![\, S \,]\!] \stackrel{\mathsf{def}}{=} \{\tau \in (D^\omega_{\surd})^n \stackrel{cp}{\to} (D^\omega_{\surd})^m \,|\, \forall r \in (D^\infty_{\surd})^n : (\diamond r, \diamond\tau(r)) \models R\},$$

where $(\diamond r, \diamond\tau(r)) \models R$ iff R evaluates to true when each input identifier i_j is interpreted as the j'th element of the n-tuple $\diamond r$, and each output identifier o_j is interpreted as the j'th element of the m-tuple $\diamond\tau(r)$.

Networks of specifications are expressed in an equational style. For example, the network S pictured in Figure 1, consisting of the two specifications S_1 and S_2, is characterized as below:

$$S(i \in D^\omega, r \in D^\omega \rhd o \in D^\omega, s \in D^\omega) \equiv (o, y) = S_1(i, x),\ (x, s) = S_2(y, r)$$

The channels represented by x and y are now hidden in the sense that they represent local channels. The comma separating the two equations can be read as an "and". More formally, $[\![\, S \,]\!] \stackrel{\mathsf{def}}{=} \{\tau_1 \otimes \tau_2 \,|\, \tau_1 \in [\![\, S_1 \,]\!] \wedge \tau_2 \in [\![\, S_2 \,]\!]\}$, where for any pair of timed stream tuples i and r, $(\tau_1 \otimes \tau_2)(i, r) \stackrel{\mathsf{def}}{=} (o, s)$ iff (o, s) is the least fixpoint solution with respect to i and r. In fact any dataflow network can be expressed in this equational style. We have already seen an example of a finite network. Infinite networks are expressed using recursion. For a more detailed syntactic and semantic treatment, see [4].

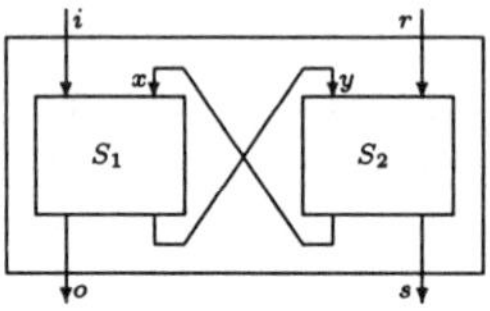

Figure 1. The Network S

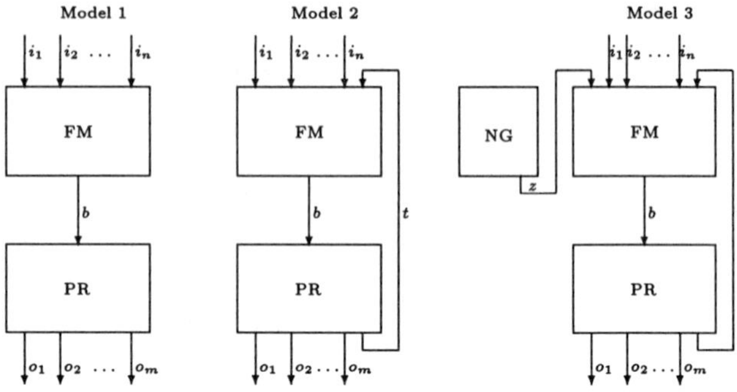

Figure 2. Three Models of an SDL Process

Focus offers a number of refinement concepts with corresponding refinement calculi. The most basic of these is behavioral refinement, which at the semantic level corresponds to set inclusion: a specification S_2 refines a specification S_1, written $S_1 \rightsquigarrow S_2$, if the denotation of S_2 is equal to or contained in the denotation of S_1. More formally: $S_1 \rightsquigarrow S_2$ iff $[\![S_2]\!] \subseteq [\![S_1]\!]$.

4. SDL AND ITS STREAM SEMANTICS

An SDL system specification is a container for a set of blocks. It is separated from its environment by a system boundary. The blocks are connected to one another and to the system environment by channels. Each communication between two blocks or between a block and the environment takes place using signals, which are conveyed by the channels. The transmission of signals can be delaying or non-delaying and uni- or bidirectional. A block can be a container for a set of blocks (block substructure) or it can be a container for sets of processes.

Sets of processes are interconnected by non-delaying signalroutes. Signalroutes are also used to connect processes to the block boundary. A process definition defines a set of processes. Several instances of the same process may exist concurrently and execute asynchronously and in parallel with each other and with instances of other processes in the system. A process instance is a communicating, finite-state machine extended to allow for: a secondary state, represented by local variables, in addition to the ordinary control state; explicit nondeterminism in terms of spontaneous input and non-deterministic decision; deferred consumption of input signals.

Each process has an internal, unbounded buffer in which all incoming signals are inserted in the order of their arrival and thereafter processed. Simultaneously arriving signals are arbitrarily ordered. The set of valid state-transitions is described by a process graph or a service decomposition.

There is a close relationship between a functional core of the SDL language and lazy functional programming languages based on stream communication. It is this functional core that interests us in this paper. Inspired by [5], we sketch how this restricted version of SDL can be given a denotational semantics in terms of streams and stream processing functions. Since Focus is based upon such a semantics this can be achieved by specifying the different SDL constructs in Focus. We consider only the time independent part of SDL. This means that all channels are declared as non-delaying and that only some restricted aspects of the SDL facilities for timers are modeled. Moreover, with the exception of SDL92's statements for explicit nondeterminism, all the constructs considered by us are contained in what [6] calls Basic-SDL.

The semantics of SDL systems and blocks can easily be expressed as finite Focus networks given that the behavior of SDL processes can be described in Focus. This is explained in more detail when we later characterize F-SDL. In this section we concentrate on the modeling of the SDL process construct.

As indicated by Model 1 in Figure 2, if we ignore facilities for process creation, the timer constructs, explicit nondeterminism and that an SDL process can send signals to itself, a Basic SDL process, with n input signal routes and m output signal routes, can be modeled as the sequential composition of two components, namely a fair merge component FM, which merges the streams of signals received on the n input signal routes into a stream b modeling the internal, unbounded buffer of an SDL process, and a processing component PR, which carries out the actual processing.

The component FM is characterized by the following relational Focus specification:

$$\mathrm{FM}(i_1 \in D^\omega, \ldots, i_n \in D^\omega \rhd b \in D^\omega) \equiv \exists o \in \{1, \ldots, n\}^\omega : \wedge_{j=1}^{n} spl_j(b, o) = i_j.$$

Based on an oracle (its second argument) the auxiliary function spl_j extracts (from its first argument) the stream of signals received on the j'th input signal route — mathematically expressed:

$$j = y \Rightarrow spl_j(x \,\&\, b, y \,\&\, p) = x \,\&\, spl_j(b, p), \quad j \neq y \Rightarrow spl_j(x \,\&\, b, y \,\&\, p) = spl_j(b, p).$$

When we later define F-SDL, FM is assumed to be a specification constant with exactly the above characterized semantics. Moreover, FM is overloaded to deal with any number of input channels of any signal sorts. In an SDL process specification the fair merge component FM is hidden in the sense that it is only a part of the semantics of the process. After all, since it is always the case that all incoming input signals are passed on to the internal buffer of the process, this does not have to be stated explicitly at the syntactic level. The visible part of an SDL process specification is basically a (possibly nondeterministic) functional program corresponding to the component PR.

As explained in [5], with respect to the internal, unbounded buffer, the behavior of a deterministic SDL processing component can be modeled by a function

$$g \in D^p \rightarrow D^\omega \xrightarrow{c} (D^\omega)^m,$$

which for any p-tuple of secondary state variables l, returns a stream processing function $g(l)$, which characterizes the behavior of the processing component PR. This means that in the deterministic case the behavior of the component PR can be characterized by a relational Focus specification of the following form:

$$\mathrm{PR}(b \in D^\omega \rhd o_1 \in D^\omega, \ldots, o_m \in D^\omega) \equiv$$

$$\exists l_1, \ldots, l_p \in D : \exists g \in D^p \to D^\omega \xrightarrow{c} (D^\omega)^m : g(l_1, \ldots, l_p)(b) = (o_1, \ldots, o_m) \text{ where } Q$$

The variables $l_1, \ldots, l_p$ represent the secondary state of an SDL process. The existentially quantified function variable g models the behavior of the processing component. The formula Q gives the actual definition of g. Section 5 explains in more detail how Q can be expressed. Based on this outline, we may define the semantics of a simple SDL process as $[\![$ SDL_PROC $]\!]$, where

$$\mathrm{SDL\_PROC}(i_1 \in D^\omega, \ldots, i_n \in D^\omega \rhd o_1 \in D^\omega, \ldots, o_m \in D^\omega) \equiv$$

$$(b) = \mathrm{FM}(i_1, \ldots, i_n), \ (o_1, \ldots, o_m) = \mathrm{PR}(b)$$

As already mentioned, in this paper we are only interested in the time independent part of SDL — time-independent in the sense that all channels are declared as nondelaying. Nevertheless, SDL timers are needed to allow certain weakly time dependent components to be expressed. An example of such a component is the sender specified in Section 6. To allow for the specification of such components, we extend our restricted SDL language with a set-timer command of the following form: set(now, $timer(n)$). The first parameter is fixed as now.

Since the first parameter is fixed as now, according to the SDL semantics the time-out signals are placed in the unbounded, internal buffer in the same order as they are sent. Moreover, they are fairly interleaved with the signals received on the other input channels. Clearly, since we are only interested in the time-independent part of SDL, we do not need reset signals, nor the SDL constructs for checking whether a timer is active or idle.

As indicated by Model 2 in Figure 2, under these restrictions we may model an SDL-process with timers by adding an additional feedback channel t, which allows the processing component PR to send its timer signals back to FM. The latter merges the stream of timer signals with the other streams of input signals in the same way as before.

Unfortunately, as someone familiar with SDL may have observed, a problem has been brushed under the carpet. In SDL a timer signal remains active until it is consumed by the processing component PR. When the processing component sends a timer signal for which there is already an active copy in b, then the already active copy is deleted at the very same moment as the new copy is placed in b. Thus to make sure that we get the intended effect when our Focus specifications are translated into SDL, this problem must somehow be taken into consideration. There are at least two straightforward solutions.

The first alternative is to handle it directly in the specification of PR, namely by for each timer $timer(n)$, to add an additional parameter $m_{timer(n)}$ keeping track of the difference between the number of times $timer(n)$ has been output along t, and the number of times $timer(n)$ has been input from b. Then, whenever a timer signal $timer(n)$ is input from b, if $m_{timer(n)} > 1$, this signal is ignored — otherwise it is processed in the usual SDL way.

Given that T is the set of all timers, then the second alternative is to impose an additional proof-obligation which must be satisfied by the function $g \in D^p \to (D \cup$

$T)^\omega \stackrel{c}{\rightarrow} (D^\omega)^m \times T^\omega$ characterizing the behavior of the processing component PR. More explicitly, to require that

$$g(l_1,\ldots,l_p)(b) = (o_1,\ldots,o_m,t) \Rightarrow \#timer(n)©t \leq \#timer(n)©b + 1 \qquad (*)$$

for all timer signals $timer(n) \in T$.

Clearly, the first alternative allows us to model more SDL-specifications. However, the additional expressiveness we then get is not particularly interesting from a practical point of view. Moreover, it is expensive in the sense that our Focus specifications become more complicated. For this reason we decide in favor of the second alternative. Thus an SDL-process, which behaves in accordance with the additional proof obligation (∗), can be modeled by the set of timed, pulse-driven, stream processing functions characterized by [[SDL_PROC]], where

$$\mathrm{SDL\_PROC}(i_1 \in D^\omega,\ldots,i_n \in D^\omega \rhd o_1 \in D^\omega,\ldots,o_m \in D^\omega) \equiv$$

$$(b) = \mathrm{FM}(i_1,\ldots,i_n,t),\ (o_1,\ldots,o_m,t) = \mathrm{PR}(b)$$

This format can of course easily be generalized to also allow the process to send ordinary signals to itself. It is enough to define t to be of type $(D \cup T)^\omega$.

So far we have considered deterministic processing components only. However, in SDL92 nondeterminism can be expressed explicitly using the constructs for spontaneous input and nondeterministic decision. We now extend our semantic model to deal with these two constructs.

Spontaneous input is in SDL specified using an input symbol with the keyword none. A spontaneous input attached to a state means that the actual transition can be initiated at any time nondeterministically. This construct can for example be used to model unreliable behavior.

To handle spontaneous input, we add an additional component NG to our model, as indicated by Model 3 in Figure 2. The component NG is supposed to output nondeterministically some stream of none's, and is specified by:

$$\mathrm{NG}(\rhd z \in \{\mathsf{none}\}^\omega) \equiv \mathsf{true}$$

The fair merge component must now also take the input from NG into consideration when it generates the stream modeling the internal unbounded buffer b — the signals received along z are of course treated as ordinary input signals.

As a consequence, the denotation of an SDL process with timers (given the stated restrictions) and spontaneous input is characterized by [[SDL_PROC]], where

$$\mathrm{SDL\_PROC}(i_1 \in D^\omega,\ldots,i_n \in D^\omega \rhd o_1 \in D^\omega,\ldots,o_m \in D^\omega) \equiv$$

$$(z) = \mathrm{NG},\ (b) = \mathrm{FM}(z,i_1,\ldots,i_n,t),\ (o_1,\ldots,o_m,t) = \mathrm{PR}(b)$$

The other way of expressing nondeterminism explicitly in SDL — so-called nondeterministic decision — is represented by the keyword **any** inside a decision symbol with no values attached to its outlets. This indicates that the alternative chosen by the process cannot be forecast. To build this into our model we introduce an oracle q in the specification of PR:

$$\mathrm{PR}(b \in (D \cup T)^{\omega} \rhd o_1 \in D^{\omega}, \ldots, o_m \in D^{\omega}, t \in T^{\omega}) \equiv$$

$$\exists q \in \mathsf{N}^{\infty} : \exists l_1, \ldots, l_p \in D : \exists g \in \mathsf{N}^{\infty} \to D^p \to (D \cup T)^{\omega} \xrightarrow{c} (D^{\omega})^m \times T^{\omega} :$$
$$g(q)(l_1, \ldots, l_p)(b) = (o_1, \ldots, o_m, t) \textsf{ where } Q$$

The existentially quantified q represents some infinite stream of positive natural numbers and can be thought of as a seed. If the k'th nondeterministic decision executed by the process has 5 outlets, then outlet $(q[k] \textsf{ mod } 5) + 1$ is chosen where $q[k]$ denotes the k'th natural number in q.

5. THE SYNTAX OF F-SDL

So far we have concentrated on mapping the syntactic expressions of the two formalisms into the very same semantics. In this section we work in the opposite direction. Based on the proposed SDL semantics expressed in Focus, a language called F-SDL is defined. This language characterizes a set of Focus specifications whose elements structurally and semantically match SDL specifications to such a degree that an automatic translation into SDL is almost straightforward. Because of the space constraints only certain restricted aspects can be discussed here. The reader is referred to [7] for more details.

We use the following EBNF convention: **[expr]** — **expr** is optional; **{expr}*** — zero or more repetitions of **expr**; **{expr//sep}*** — zero or more repetitions of **expr** separated by **sep**; **{expr}+** — one or more repetitions of **expr**; **{expr//sep}+** — one or more repetitions of **expr** separated by **sep**; $\textsf{expr}_1|\textsf{expr}_2$ — choice;( ) — grouping.

We start by explaining how SDL system specifications are expressed in F-SDL:

$\langle\textsf{sys\_spec}\rangle$ ::= **system** $\langle\textsf{head}\rangle$ [$\langle\textsf{data\_def}\rangle$] $\langle\textsf{sys\_body}\rangle$ **where** $\langle\textsf{sys\_defs}\rangle$ **end**

$\langle\textsf{head}\rangle$ ::= $\langle\textsf{id}\rangle$ ({ $\langle\textsf{ch\_decl}\rangle$ //,}* $\rhd$ { $\langle\textsf{ch\_decl}\rangle$ //,}*) $\equiv$

$\langle\textsf{sys\_body}\rangle$ ::= { $\langle\textsf{equation}\rangle$ //,}+

$\langle\textsf{sys\_defs}\rangle$ ::= { $\langle\textsf{block\_spec}\rangle$ //,}+

$\langle\textsf{sys\_spec}\rangle$ characterizes a finite Focus network of blocks (i.e. as a set of equations) whose definitions are given in $\langle\textsf{sys\_defs}\rangle$. $\langle\textsf{data\_def}\rangle$ is used to define new datatypes, signals etc. as in SDL. Strictly speaking, the keywords **system** , **where** and **end** have no influence on the stream semantics. They have been included to simplify the implementation of the translation algorithm and to increase the readability of the specifications. There are of course a number of constraints. For example any block identifier occurring in the the right-hand side of an equation must also be defined in $\langle\textsf{sys\_defs}\rangle$.

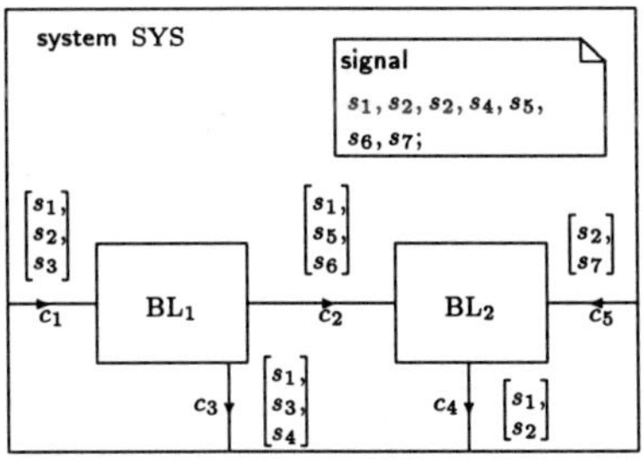

Figure 3. An SDL System Diagram

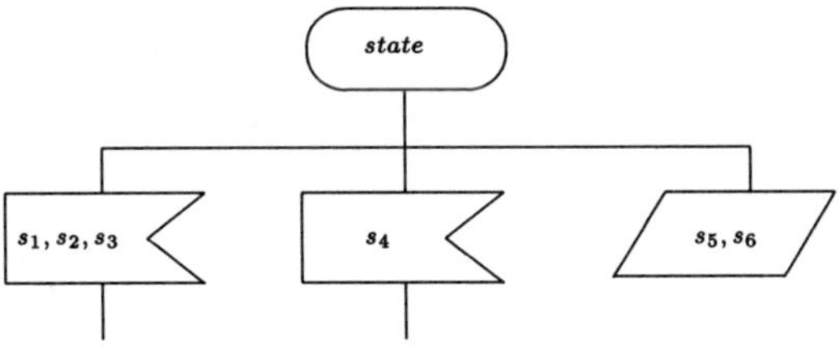

Figure 4. A State Transition Fragment

The SDL system diagram in Figure 3 corresponds to the following F-SDL specification:

$$\textsf{system}\ \mathrm{SYS}(c_1 \in \{s_1, s_2, s_3\}^\omega, c_5 \in \{s_2, s_7\}^\omega \rhd c_3 \in \{s_1, s_3, s_4\}^\omega, c_4 \in \{s_1, s_2\}^\omega) \equiv$$

$$(c_3, c_2) = \mathrm{BL}_1(c_1),\ (c_4) = \mathrm{BL}_2(c_2, c_5)$$

$$\textsf{where}\quad \textsf{block}\ \mathrm{BL}_1 \ldots, \textsf{block}\ \mathrm{BL}_2 \ldots \quad \textsf{end}$$

F-SDL block specifications are expressed in almost the same way as systems:

$$\langle \textsf{block\_spec} \rangle ::= \textsf{block}\ \langle \textsf{head} \rangle\ [\ \langle \textsf{data\_def} \rangle\ ]\ \langle \textsf{block\_body} \rangle\ \textsf{where}\ \langle \textsf{block\_defs} \rangle\ \textsf{end}$$

$$\langle \textsf{block\_defs} \rangle ::= \{\ \langle \textsf{block\_spec} \rangle\ //,\}^+ \mid \{\ \langle \textsf{proc\_spec} \rangle\ //,\}^+$$

$$\langle \textsf{proc\_spec} \rangle ::= \langle \textsf{ord\_proc\_spec} \rangle \mid \langle \textsf{rec\_proc\_spec} \rangle$$

By ordinary processes $\langle \textsf{ord\_proc\_spec} \rangle$ we characterize the SDL processes which do not create other processes. A recursive process $\langle \textsf{rec\_proc\_spec} \rangle$, on the other hand, has process creation and communication as its only responsibility. The reason why we make this distinction is that in F-SDL only some rather restricted aspects of the SDL process creation facilities are modeled, namely those needed to express infinite Focus networks.

We now characterize the syntactic structure of ordinary processes.

$$\langle \textsf{ord\_proc\_spec} \rangle ::= \textsf{ord\_process}\ \langle \textsf{head} \rangle\ [\ \langle \textsf{data\_def} \rangle\ ]\ \langle \textsf{proc\_body} \rangle\ \textsf{where}\ \langle \textsf{pr\_spec} \rangle\ \textsf{end}$$

$$\langle \textsf{proc\_body} \rangle ::= [\ \langle \textsf{ng\_eq} \rangle\ ,\ ]\ [\ \langle \textsf{fm\_eq} \rangle\ ,\ ]\ \langle \textsf{pr\_eq} \rangle$$

$\langle \textsf{ng\_eq} \rangle$, $\langle \textsf{fm\_eq} \rangle$ and $\langle \textsf{pr\_eq} \rangle$ model respectively the NG-equation, the FM-equation and the PR-equation, as explained in Section 4. NG and FM are specification constants and do not have to be explicitly defined in F-SDL. $\langle \textsf{pr\_spec} \rangle$ gives the F-SDL specification of the processing component PR.

$$\langle \mathsf{pr\_spec} \rangle ::= \mathrm{PR}\,(\,[\,\langle \mathsf{ch\_decl} \rangle\,]\,\rhd\,\{\,\langle \mathsf{ch\_decl} \rangle\,//,\}^*\,) \equiv \langle \mathsf{pr\_body} \rangle \text{ where } \langle \mathsf{proc\_graph} \rangle \text{ end}$$

The detailed structure of $\langle \mathsf{pr\_body} \rangle$ corresponds to the similar fragment of the specification PR in Section 4. The process graph $\langle \mathsf{proc\_graph} \rangle$ (the formula Q in Section 4) is basically a functional program based on pattern matching.

$$\langle \mathsf{proc\_graph} \rangle ::= [\,\langle \mathsf{var\_decls} \rangle\,]\,\langle \mathsf{i\_tran} \rangle\,[,\,\{\,\langle \mathsf{tran} \rangle\,//,\}^+\,]$$

$\langle \mathsf{var\_decls} \rangle$ allows us to introduce universally quantified variables needed for the pattern matching. $\langle \mathsf{i\_tran} \rangle$ is used to model the initialization transition — the transition from the start symbol to the first control state. $\langle \mathsf{tran} \rangle$ models any other state transitions of an SDL process graph. $\langle \mathsf{i\_tran} \rangle$ is just a special case of $\langle \mathsf{tran} \rangle$. For an example of a complete F-SDL process graph, see the specification PR in Section 6.

$$\langle \mathsf{tran} \rangle ::= [\,\langle \mathsf{func\_id} \rangle \notin \langle \mathsf{func\_set} \rangle \Rightarrow\,]\,(\,\langle \mathsf{h\_tran} \rangle \,|\, \langle \mathsf{v\_tran} \rangle\,)$$

The optional part of $\langle \mathsf{tran} \rangle$ is used to model an SDL transition from an asterisk state.

A hidden transition $\langle \mathsf{h\_tran} \rangle$ models the way an SDL process consumes input signals for which there is no input or save symbol. A visible transition $\langle \mathsf{v\_tran} \rangle$ models any other transition, namely a save transition or an ordinary state transition.

$$\langle \mathsf{v\_tran} \rangle ::= \langle \mathsf{s\_tran} \rangle \,|\, \langle \mathsf{o\_tran} \rangle$$

A save transition $\langle \mathsf{s\_tran} \rangle$ allows the order of the signals in the internal, unbounded buffer to be permuted. Ordinary state-transitions $\langle \mathsf{o\_tran} \rangle$ model the "real" state-transitions of an SDL process graph.

$$\langle \mathsf{o\_tran} \rangle ::= [\,\langle \mathsf{sig\_id} \rangle \in \langle \mathsf{sig\_set} \rangle \Rightarrow\,]\,\langle \mathsf{left} \rangle = \langle \mathsf{right} \rangle$$

The optional part is used to model that input symbols may contain lists of signals.

Assume that the actual process has $\{s_1, s_2, \ldots, s_7\}$ as its input signal set. Then, the SDL fragment of Figure 4 corresponds to the following F-SDL fragment:

$$a \in \{s_1, s_2, s_3\} \Rightarrow state(a \,\&\, in) = \ \ldots$$

$$state(s_4 \,\&\, in) = \ \ldots$$

$$a \notin \{s_5, s_6\} \wedge \{s_5, s_6\} \copyright in = in \Rightarrow state(in \frown (a \,\&\, in')) = state((a \,\&\, in) \frown in')$$

$$a \notin \{s_1, s_2, s_3, s_4, s_5, s_6\} \Rightarrow state(a \,\&\, in) = state(in)$$

The first two "conditional equations" model the ordinary transitions; the third models the save-transition; the fourth models a hidden transition, namely the implicit consumption of any occurrence of s_7.

6. DEVELOPMENT OF A PROTOCOL SPECIFICATION

In this Section we sketch how Focus can be used to develop an SDL specification of a protocol based on [8]. For more details, see [7]. We refer to the overall protocol network as SP (for Stenning Protocol), and not surprisingly, from an external point of view, it is required to behave as an identity component:

$$\mathsf{system}\ \mathrm{SP}_0(i \in \mathrm{DT}^\omega \rhd o \in \mathrm{DT}^\omega) \equiv o = i.$$

$\mathrm{DT} = \{dt(d) \mid d \in D\}$ is the set of data signals, where D is some nonempty set of data. If we ignore the keyword system, which as explained in Section 5 has no semantics, this is an ordinary Focus specification. However, SP_0 is not a complete F-SDL specification because its internal structure has not yet been fixed in an F-SDL manner. SP is assigned a subscript to allow the different specifications of the same component to be distinguished.

As usual in the case of protocols, the system to be developed consists of a sender, a receiver and a communication medium. Thus, it seems natural to decompose our abstract system specification into three blocks: a block called SND specifying the sender, a block called REC specifying the receiver, and a block called MED specifying the medium. At the block level there are 8 channels altogether; two of these are external; the other six are internal. Thus we want an F-SDL system specification of the following form:

$$\mathsf{system}\ \mathrm{SP}_1(i \in \mathrm{DT}^\omega \rhd o \in \mathrm{DT}^\omega) \equiv$$

$$(sd, sn) = \mathrm{SND}(i, ma),$$

$$(ma, mn, md) = \mathrm{MED}(sd, sn, ra),$$

$$(ra, o) = \mathrm{REC}(md, mn)$$

$$\mathsf{where}\quad \mathsf{block}\ \mathrm{SND}\ \ldots, \mathsf{block}\ \mathrm{MED}\ \ldots, \mathsf{block}\ \mathrm{REC}\ \ldots\quad \mathsf{end}$$

Although the three blocks are unspecified, we already have enough information to generate the corresponding SDL system diagram.

For each data signal $dt(d)$ input on i, the sender SND generates a unique sequence-number signal $sn(n)$ and repeatedly sends these two signals along sd and sn, respectively, until it receives the sequence-number signal $sn(n)$ on ma. Any sequence-number signal input on ma is of course sent by REC to acknowledge that the corresponding data signal has been received. More formally, the sender can be specified as below:

$$\textsf{block}\ \mathrm{SND}(i \in \mathrm{DT}^\omega, ma \in \mathrm{SN}^\omega \rhd sd \in \mathrm{DT}^\omega, sn \in \mathrm{SN}^\omega) \equiv$$

$$\begin{array}{l}
\textsf{let} \\
\quad ma' = \propto(ma) \\
\quad (sd', sn') = \propto(sd, sn) \\
\textsf{in} \\
\quad \#ma' \leq \#i \\
\qquad \Rightarrow \\
\quad \#sd = \#sn \wedge \forall n \in \mathrm{SN} : \#n©sn' \leq 1 \wedge sd' \sqsubseteq i \wedge \\
\quad \textsf{if}\ \#ma' = \#i\ \textsf{then}\ \#sd' = \#i\ \textsf{else}\ \#sd = \infty \wedge \#sd' = \#ma' + 1
\end{array}$$

$\mathrm{SN} = \{sn(n) \mid n \in \mathsf{N}\}$ denotes the set of sequence-number signals. The let-construct defines ma' and (sd', sn') to be equal to mn and (sd, sn) minus consecutive repetitions, respectively.

The antecedent states the environment assumption, namely that the length of ma' is less than or equal to the length of i. This is a sensible assumption since the receiver should only acknowledge the data signals it receives.

The first conjunct of the consequent requires sd and sn to be of the same length. This means that SND never sends a data signal without also sending its corresponding sequence-number — and the other way around.

The second conjunct of the consequent makes sure that each sequence-number has maximum one occurrence in sn'. This means that any data signal in sd that is not equal to its predecessor has a corresponding sequence-number that is different from all its predecessors.

The third conjunct of the consequent requires sd' to be a prefix of i. This means that the sender only sends the data-signals input from i, and moreover that they are sent in the order they are received.

The fourth conjunct of the consequent requires that if ma' is of the same length as i then this also holds for sd'; otherwise sd is infinite and the length of sd' is equal to the length of ma' plus 1. This means that the sender does not send more data signals (when repetitions are ignored) than it receives on i, and that when it never receives an acknowledgement for a data signal, then this signal is sent repeatedly forever.

Note that the second conjunct of the else branch together with conjunct two and three of the consequent make sure that each data signal input on i is assigned a unique sequence-number.

The medium and the receiver can be specified accordingly. The verification of this decomposition boils down to formulating three invariants and showing that six proof-obligations based on these invariants are valid (see [7]).

At the system level our development is now complete. What remains is to transform the two block specifications that we want to implement, namely SND and REC, into F-SDL syntax. The specification of the medium is not refined any further.

The blocks SND and REC have only one process each, and these processes are therefore constrained to behave in exactly the same way as their respective blocks. Thus the verification of this refinement step is trivial. Each process specification is then decomposed into a FM and a PR component, as explained above. In the case of SND an additional feedback channel is introduced to allow for the use of timers. These two decompositions are verified in the same way as the decomposition of SP_0. Then, the specification of the sender's PR component is refined into:

$$\mathrm{PR}(b \in (\mathrm{DT} \cup \mathrm{SN} \cup T)^{\omega} \rhd sd \in \mathrm{DT}^{\omega}, sn \in \mathrm{SN}^{\omega}, t \in T) \equiv$$

$$\exists l \in \mathsf{N} : \exists d \in D : \exists start \in \mathsf{N} \times D \rightarrow (\mathrm{DT} \cup \mathrm{SN} \cup T)^{\omega} \xrightarrow{c} \mathrm{DT}^{\omega} \times \mathrm{SN}^{\omega} \times T^{\omega} : \\ start(l, d)(b) = (sd, sn, t)$$

where $\forall l, n \in \mathsf{N} : \forall d, d' \in D : \forall s \in \mathrm{SN} \cup T : \forall in, in' \in (\mathrm{DT} \cup \mathrm{SN} \cup T)^{\omega}$:

$$start(l, d)(in) = next(1, d)(in)$$

$$next(l, d)(dt(d') \,\&\, in) = \textsf{let } l = l + 1 \textsf{ in } dt(d') \,\&_1\, sn(l) \,\&_2\, ti(l) \,\&_3\, rep(l, d')(in)$$

$$s \in \{sn(n), ti(k) \mid n, k \in \mathsf{N}\} \Rightarrow next(l, d)(s \,\&\, in) = next(l, d)(in)$$

$$rep(l, d)(ti(n) \,\&\, in) = \textsf{if } n = l \textsf{ then } dt(d) \,\&_1\, sn(l) \,\&_2\, ti(l) \,\&_3\, rep(l, d)(in) \\ \textsf{else } rep(l, d)(in) \textsf{ fi}$$

$$rep(l, d)(sn(n) \,\&\, in) = \textsf{if } n = l \textsf{ then } next(l, d)(in) \textsf{ else } rep(l, d)(in) \textsf{ fi}$$

$$s \notin \{dt(d) \mid d \in D\} \wedge \{dt(d) \mid d \in D\} \copyright in = in \Rightarrow \\ rep(l, d)(in \frown (s \,\&\, in')) = rep(l, d)((s \,\&\, in) \frown in')$$

end

The translation algorithm then gives the SDL specification of Figure 5. The PR component of the receiver can be refined accordingly.

7. CONCLUSIONS

It has been outlined how Focus and a restricted version of SDL can be assigned the same kind of stream-based, denotational semantics. Based on the proposed SDL semantics expressed in Focus, a language called F-SDL has been defined. It characterizes a set of Focus specifications whose elements allow an automatic "one-to-one" translation into SDL. The proposed technique was demonstrated on hand of a protocol.

The merge of Focus and SDL into one methodology offers several advantages. Focus with its well-defined formal semantics and very abstract nature allows the use of formal proof techniques for the validation, verification and development of specifications. In particular, employing Focus a certain class of SDL specifications can be developed in a top-down fashion — in other words, in the same style as the SDL protocol specification was developed in Section 6. The SDL specification was formally refined from an abstract Focus specification stating that the overall network should behave as an identity component — a

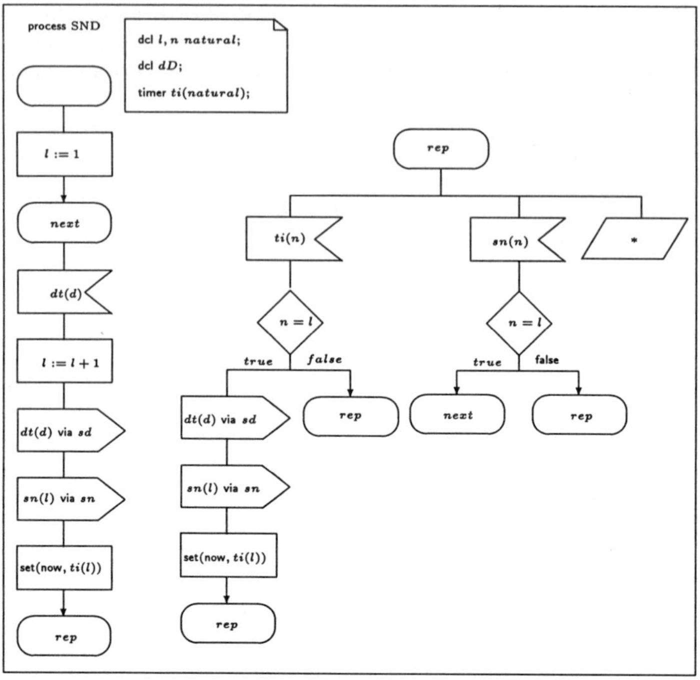

Figure 5. The Sender Process in SDL

requirement specification whose correctness is obvious! Thus the protocol example shows how one can proceed from a simple Focus specification to a non-trivial SDL specification. More complicated protocols can of course be developed accordingly. Another advantage of using Focus is the flexibility with respect to environment assumptions and the treatment of specifications that are not supposed to be implemented. An example of the latter is the medium MED of Section 6. Assumptions about a component's environment can be stated by splitting the specification into an assumption and a commitment part. For example, in the specification SND the antecedent can be seen as an environment assumption, and the consequent is a commitment which must be fulfilled by the component whenever the environment behaves in accordance with the environment assumption.

SDL is a specification language. This means that the readability and structure of SDL specifications is often of crucial importance. For this reason, F-SDL has been designed in such a way that there is a straightforward mapping into SDL. This means that the user has full control of the syntactic structure of the SDL specification he is developing. In SDL there is a lot of syntactic sugar, which can be hard to model directly in Focus. This can be dealt with by defining and implementing user-controlled transformation rules which allow the user to transform the generated SDL specification into its optimal form.

From the Focus user's point of view, the embedding of SDL as a target language means that the many tools and environments already designed for SDL can be used to transform a developed specification into the chosen target architecture. For example, there are SDL tools which allow an automatic translation into C++ [9]. Moreover, since a completely formal development is very resource demanding, the Focus user may concentrate his efforts on the critical parts of a system description and specify the less essential aspects directly in SDL.

Although the considered sublanguage is sufficiently expressive to deal with non-trivial applications, many SDL facilities have been ignored. However, it seems to be relatively easy to extend the proposed approach to handle a much richer part of SDL, including the the SDL timer constructs, procedures (not remote call), and services. On the other hand, the treatment of the more OO-related facilities of SDL, including the full generality of the constructs for process creation, is difficult if at all possible in the context of Focus. This is of course not very surprising since the formal treatment of object-oriented languages is known to be difficult.

8. ACKNOWLEDGEMENTS

Manfred Broy, Christian Facchi, Max Fuchs and Rainer Weber have read preliminary versions of this paper and provided valuable feedback.

REFERENCES

1. M. Broy, F. Dederichs, C. Dendorfer, M. Fuchs, T. F. Gritzner, and R. Weber. The design of distributed systems — an introduction to Focus. Technical Report SFB 342/2/92 A, Technische Universität München, 1992.
2. M. Broy and K. Stølen. Specification and refinement of finite dataflow networks — a relational approach. In H. Langmaack, W.-P. de Roever, and J. Vytopil, editors, *Proc. FTRTFT'94, Lecture Notes in Computer Science 863*, pages 247–267, 1994.
3. ITU-TSS, editor. *Functional Specification and Description Language (SDL), Recommendation Z.100.* International Telecommunication Union, Geneva, 1993.
4. F. Dederichs. *Transformation verteilter Systeme: Von applikativen zu prozeduralen Darstellungen.* PhD thesis, Technische Universität München, 1992. Also available as SFB-report 342/17/92 A, Technische Universität München.
5. M. Broy. Towards a formal foundation of the specification and description language SDL. *Formal Aspects of Computing*, 3:21–57, 1991.
6. F. Belina, D. Hogrefe, and A. Sarma. *SDL with Applications from Protocol Specification.* Prentice Hall, 1991.
7. E. Holz and K. Stølen. An attempt to embed a restricted version of SDL as a target language in Focus. Technical Report SFB 342/11/94 A, Technische Universität München, 1994.
8. V. Stenning. A data transfer protocol. *Computer Networks*, 1:98–110, 1976.
9. J. Fischer, E. Holz, M. van Löwis, and D. Witaszek. A run time library for the simulation of SDL'92-specifications. In O. Færgemand and A. Sarma, editors, *Proc. 6'th SDL Forum*, pages 105–118. North-Holland, 1993.

26

Delayed choice: an operator for joining Message Sequence Charts

J.C.M. Baeten* and S. Mauw*

Dept. of Mathematics and Computing Science,
Eindhoven University of Technology,
P.O. Box 513, 5600 MB Eindhoven, The Netherlands.

We study the extension of a simple process algebra with the delayed choice operator. It differs from the normal non-deterministic choice in that the moment of choice is delayed until all alternatives can be distinguished by their first action. An application is in joining Message Sequence Charts.

1. INTRODUCTION

Message Sequence Charts provide a graphical method for the description of the interactions between system components. The ITU-TS (the Telecommunication Standardization Section of the International Telecommunication Union, the former CCITT) maintains recommendation Z.120 [13] which contains the syntax and an informal explanation of the semantics of Message Sequence Charts. Current developments are the definition of a formal semantics based on branching time process algebra [16, 17] and the extension of the MSC language with operators [12].

The current proposal of the formal semantics only considers single Message Sequence Charts, while the application of Message Sequence Charts often consists of a collection of Message Sequence Charts. Sometimes the intended meaning is the sequential composition of the charts contained, while in other cases they are viewed as alternatives. The semantics of such a collection thus can only be defined if the relation between the contained Message Sequence Charts is explicitly stated. It is proposed to use operators for this purpose.

In [10] several operators are proposed: alternative composition ($+$), sequential composition ($\cdot$), weak sequential composition ($\circ$) and parallel composition ($\|$). We think however that in a branching time setting the non-deterministic alternative composition operator is not always the intention of the user. In branching time the moment of choice is taken into consideration, so, for example, a distinction is made between the processes $a(b+c)$ and $ab+ac$. In the first expression the choice between b and c is made after executing action a, while in the second expression the choice is made before a. Now consider the following typical use of Message Sequence Charts (see Figure 1).

The semantics of the first drawing is that first a message *ICONreq* is received from the environment by the *Initiator*. Thereafter the message *ICON* is being sent to the *Responder*, which sends an *ICONind* to the environment and receives an *ICONresp* from the environment. The *Responder* sends *ICONF* back to the *Initiator*, who sends an *ICONconf* to the environment. The second drawing displays an alternative scenario in

*This research was supported in part by ESPRIT basic research action 7166, CONCUR2.

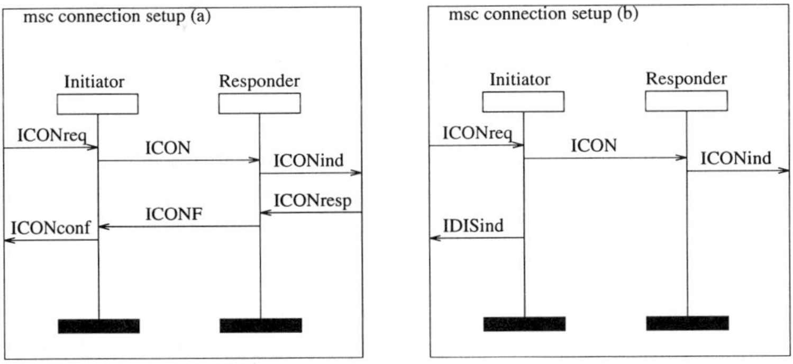

Figure 1. Two Message Sequence Charts

which due to a timeout (which was left out of the picture), a failure *IDISind* is reported to the environment.

If we combine these two Message Sequence Charts by means of the alternative composition operator in a branching time setting, it will mean that the choice between correct operation and faulty operation has been made before reception of the *ICONreq*. This was obviously not the intention.

In order to be able to combine alternative Message Sequence Charts while sharing their initial behavior, we introduce the *delayed choice* operator. Since it behaves as the choice operator in the setting of trace theory, we will denote it by $\mp$ (Trace-+). The most important property is expressed in the following equation (a, b and c are atomic actions, $b \neq c$).

$$ab \mp ac = a(b + c)$$

while, if the two operands do not share an initial action (i.e. $a \neq c$), we have

$$ab \mp cd = ab + cd$$

In this paper we will define this operator within the framework of $BPA_{\delta\varepsilon}$. This is the concurrency theory BPA_{δ} (Basic Process Algebra with deadlock) extended with the empty process ε, which is the neutral element for sequential composition. Introduction of this special process is motivated by the following example. In the expression $a \mp ab$ both operands start with action a. The first operand terminates after execution of a, while the second operand executes b after execution of a. This can only be expressed if we have a special symbol for immediate termination. Using the empty process we obtain $a(\varepsilon + b)$.

Several variants of the alternative composition operator have been defined before, none of which satisfies our needs.

In *TCSP* [8] (Theoretical Communicating Sequential Processes) we have the deterministic choice operator $\Box$ and the non-deterministic choice $\sqcap$. The operator $\Box$ also removes possible non-determinism, but only non-determinism due to internal steps. This operator was defined in a branching time setting in [9], there called τ-angelic choice. The operator we need should remove all non-determinism, also non-determinism due to external, visible actions. Note that TCSP's non-deterministic choice $\sqcap$ was studied in a branching time setting in [2]; there, also a form of choice intermediate between $+$ and $\sqcap$ is studied.

The operational rules we will present for our delayed choice operator also appear in [18] for an operator they call angelic non-determinism. In [18], only linear time semantics is considered, and no axiomatization is presented. We present our delayed choice operator in a more general branching time framework, and moreover provide a complete axiomatization for finite processes.

We prefer not to use the term angelic choice. This is because this term stems from a dualistic world, where specified behavior is distinguished from intended behavior. Transferring this notion to the monistic world of process theory will only cause confusion and differences of opinion.

The remainder of this paper is structured as follows. In Section 2 we will give a short introduction to the theory $BPA_{\delta\varepsilon}$. Section 3 contains an axiomatization and an operational semantics of the delayed choice operator. An example of the use of this operator in the field of Message Sequence Charts is shown in Section 4.

We thank Michel Reniers for proof reading this document and Wim Hesselink for a conversation on angelic choice.

2. BASIC PROCESS ALGEBRA WITH THE EMPTY PROCESS

The process algebra $BPA_{\delta\varepsilon}$ is an algebraic theory for the description of process behavior [5, 6]. This theory is parameterized by a set of unspecified constants A, which are called *atomic actions.* A process is built up from atomic actions, the special constants δ and ε and the operators $\cdot$ and $+$.

The special constant δ denotes the process that has stopped executing actions and cannot proceed. This constant is called *deadlock* or inaction. The special constant ε denotes the process that is only capable of terminating successfully. It is called the *empty process.*

The binary operators $+$ and $\cdot$ are called the *alternative* and *sequential composition.* The alternative composition of the processes x and y is the process that either executes process x or y but not both. The sequential composition of the processes x and y is the process that first executes process x, and upon completion thereof starts with the execution of process y.

The equations of $BPA_{\delta\varepsilon}$ are given in Table 1. Axiom A1, A2 and A3 express that the $+$ is commutative, associative and idempotent. Axiom A4 defines right distributivity of the sequential composition over the alternative composition. Since we consider a branching time theory, we do not have left distributivity. Axiom A5 expresses associativity of the sequential composition. Deadlock is defined by axioms A6 and A7 and the empty process is defined by axioms A8 and A9.

The precedence of the operators is as follows: $\cdot$ binds stronger than all other operators to be introduced, and $+$ binds weaker. The other operators have the same binding power. Brackets are associated to the left. We sometimes use xy to denote $x \cdot y$.

Table 1
Axioms of $BPA_{\delta\varepsilon}$

$x + y = y + x$	A1
$(x + y) + z = x + (y + z)$	A2
$x + x = x$	A3
$(x + y) \cdot z = x \cdot z + y \cdot z$	A4
$(x \cdot y) \cdot z = x \cdot (y \cdot z)$	A5
$x + \delta = x$	A6
$\delta \cdot x = \delta$	A7
$x \cdot \varepsilon = x$	A8
$\varepsilon \cdot x = x$	A9

Table 2 defines a structured operational semantics (S.O.S.) in the style of [19]. Predicate $\downarrow$ expresses that a process has an option to terminate. For every atomic action a, predicate $\xrightarrow{a}$ expresses that the first argument can execute action a and become the second argument. Thus the first rule states that ε has a termination option. The second rule states that whenever x has a termination option then so do $x + y$ and $y + x$, and the third rule states that whenever both x and y have a termination option, then so does $x \cdot y$.

Table 2
Operational semantics for $BPA_{\delta\varepsilon}$

$$\varepsilon\downarrow \qquad \frac{x\downarrow}{(x+y)\downarrow,\ (y+x)\downarrow} \qquad \frac{x\downarrow,\ y\downarrow}{(x\cdot y)\downarrow}$$

$$a\xrightarrow{a}\varepsilon \qquad \frac{x\xrightarrow{a}x'}{x+y\xrightarrow{a}x',\ y+x\xrightarrow{a}x'} \qquad \frac{x\xrightarrow{a}x'}{x\cdot y\xrightarrow{a}x'\cdot y} \qquad \frac{x\downarrow,\ y\xrightarrow{a}y'}{x\cdot y\xrightarrow{a}y'}$$

Using these predicates it is easy to associate an element of the so-called graph model to every process expression. The domain G of this model consists of finitely branching rooted graphs. Edges are labeled with an atomic action and nodes may be attributed with the symbol $\downarrow$ to indicate the option to terminate successfully.

The model $G/\leftrightarrow$ of $BPA_{\delta\varepsilon}$ is the graph domain divided out by bisimulation. Two graphs are bisimilar if there is a bisimulation relating their root nodes. A bisimulation is a binary relation R, satisfying:

- if $R(p, q)$ and $p \xrightarrow{a} p'$, then there is a q' such that $q \xrightarrow{a} q'$ and $R(p', q')$
- if $R(p, q)$ and $q \xrightarrow{a} q'$, then there is a p' such that $p \xrightarrow{a} p'$ and $R(p', q')$
- if $R(p, q)$ then $p\downarrow$ if and only if $q\downarrow$.

3. THE DELAYED CHOICE

3.1. Axioms

The delayed choice between processes x and y, is the process obtained by joining the common initial parts of x and y and continuing with a normal choice between the remaining parts. However, we do not want to remove nondeterministic choices which are internal to x or y. This is expressed in the definition of the delayed choice operator in Table 3. For this definition we need two auxiliary operators. The first is the join operator ($\bowtie$). In $x \bowtie y$ exactly those summands of x and y are selected which have a common initial action. The corresponding summands of x and y are joined with respect to their first common action. The unless operator ($\triangleleft$) has the opposite behavior. In $x \triangleleft y$ only those summands from x are selected which share no initial action with y. Now the definition of the delayed choice can be interpreted as follows. There are three options. The first is that a summand of x and a summand of y have a common initial action. In this case the result is a summand where this action is shared ($x \bowtie y$). The second case is that a summand from x has no initial action with any summand of y in common ($x \triangleleft y$) and the third case is that a summand of y has no initial action with any summand of x in common ($y \triangleleft x$).

Table 3
Axioms for delayed choice

$x \mp y = x \bowtie y + x \triangleleft y + y \triangleleft x$	DC1		
$\varepsilon \bowtie x = \delta$	DC2	$\varepsilon \triangleleft x = \varepsilon$	DC10
$x \bowtie \varepsilon = \delta$	DC3	$x \triangleleft \varepsilon = x$	DC11
$\delta \bowtie x = \delta$	DC4	$\delta \triangleleft x = \delta$	DC12
$x \bowtie \delta = \delta$	DC5	$x \triangleleft \delta = x$	DC13
$ax \bowtie ay = a(x \mp y)$	DC6	$ax \triangleleft ay = \delta$	DC14
$a \neq b \Rightarrow ax \bowtie by = \delta$	DC7	$a \neq b \Rightarrow ax \triangleleft by = ax$	DC15
$(x + y) \bowtie z = x \bowtie z + y \bowtie z$	DC8	$(x + y) \triangleleft z = (x \triangleleft z) + (y \triangleleft z)$	DC16
$x \bowtie (y + z) = x \bowtie y + x \bowtie z$	DC9	$x \triangleleft (y + z) = (x \triangleleft y) \triangleleft z$	DC17

The unless operator $\triangleleft$ that we have here has a filtering behavior very much like the unless operator used in the axiomatization of the priority operator in [3]; this is the reason we use the same name and notation.

Examples (assuming that a, b, c, d, e and f are distinct atomic actions):

$$(ab + ac) \mp ad = a(b + d) + a(c + d)$$

$$(ab + ac) \mp (ad + ae + f) = a(b + d) + a(c + d) + a(b + e) + a(c + e) + f$$

The theory $BPA_{\delta\varepsilon}$ extended with the delayed choice operator and the auxiliary operators is denoted by $BPA_{\delta\varepsilon} + DC$.

Table 4
Operational semantics for Delayed Choice

$$\frac{x \xrightarrow{a} x',\ y \xrightarrow{a} y'}{x \mp y \xrightarrow{a} x' \mp y',\ x \bowtie y \xrightarrow{a} x' \mp y'}$$

$$\frac{x \xrightarrow{a} x',\ y \not\xrightarrow{a}}{x \mp y \xrightarrow{a} x',\ y \mp x \xrightarrow{a} x',\ x \lhd y \xrightarrow{a} x'}$$

$$\frac{x{\downarrow}}{(x \mp y){\downarrow},\ (y \mp x){\downarrow},\ (x \lhd y){\downarrow}}$$

3.2. Structured Operational Semantics

The rules in Table 4 define an operational semantics for the delayed choice operator. In the definition of the $\xrightarrow{a}$ predicate, we make use of so-called negative premises (see [20]). This means that a negation occurs in the condition of an S.O.S. rule. Expression $y \not\xrightarrow{a}$ means that process y cannot execute action a. In Lemma 3.3.4 we prove correctness of this definition. The construction of the graph associated to a process is not as straightforward as in the case without negative premises. Using a layering technique (see [7]) we can derive the so-called *deduction graph.*

We started by formulating the axiomatic laws and afterwards came up with operational rules. This can also be done in reverse order: as [1] shows, there are heuristics on deriving a set of axiomatic laws on the basis of the operational rules. The auxiliary operators $\bowtie$ and $\lhd$, needed for a finite axiomatization, arise by splitting up the operational rules into different cases.

3.3. Soundness and Completeness

In this section we prove soundness and completeness of the axioms with respect to the S.O.S. rules. In order to prove that the extension of $BPA_{\delta\varepsilon}$ with the delayed choice is conservative, we use term rewrite techniques. Consider the term rewrite system consisting of the rules A3–A9 from Table 1, all axioms of Table 3 and the additional axioms from Table 5 ($a \neq b$), oriented from left to right. The additional rules are derivable from the axioms for the delayed choice operator. They are needed because in some derivations axiom A8 ($x \cdot \varepsilon = x$) is used from right to left.

Table 5
Additional rules

$a \bowtie a = a$	$a \bowtie b = \delta$	$a \lhd a = \delta$	$a \lhd b = a$
$a \bowtie ax = a(\varepsilon + x)$	$a \bowtie bx = \delta$	$a \lhd ax = \delta$	$a \lhd bx = a$
$ax \bowtie a = a(\varepsilon + x)$	$ax \bowtie b = \delta$	$ax \lhd a = \delta$	$ax \lhd b = ax$

Lemma 3.3.1 *The term rewrite system defined above is strongly normalizing*

Proof We do this by applying the method of the lexicographical path ordering [14, 15]. There is a complication with this, as in the reduction of $\mp$, the operator $\bowtie$ appears, and in the reduction of $ax \bowtie ay$, it is the other way around. The solution is to weigh these operators with the size of their arguments, as is done in [6]. Thus, we get operators $\mp_n$, $\bowtie_n$, and the ordering is as in Figure 2 (a any atomic action).

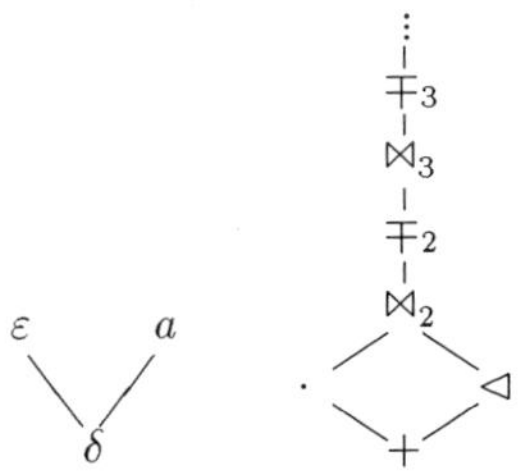

Figure 2. Ordering of operators

Furthermore, we give $\cdot$ the lexicographical status for the first argument, and $\lhd$ the lexicographical status for the second argument. For more information on lexicographical path ordering, we refer to [15]. □

Definition 3.3.2 *Define the class B of basic terms over $BPA_{\delta\varepsilon} + DC$ as the smallest class satisfying*

1. $\varepsilon, \delta \in B$, $A \subset B$
2. $a \in A, t \in B \Rightarrow a \cdot t \in B$
3. $s, t \in B \Rightarrow s + t \in B$

Theorem 3.3.3 *Let t be a closed $BPA_{\delta\varepsilon} + DC$-term. Then there is a basic term s with $BPA_{\delta\varepsilon} + DC \vdash t = s$.*

Proof Using 3.3.1 rewrite t to normal form s. We claim that s must be basic. For, s cannot contain $\mp$ because DC1 can be applied. If s contains $\bowtie$ or $\lhd$, take a smallest sub-term containing one of them, say $s_1 \bowtie s_2$, then we can apply one of DC2–DC9 or the additional rules for $\bowtie$ from Table 5. If this subterm is $s_1 \lhd s_2$ then we can apply one of DC10–DC17 or the additional rules for $\lhd$. It is now easy to finish the proof. □

This is the so-called elimination theorem, saying that the operators newly introduced can be eliminated from closed terms.

Lemma 3.3.4 *Bisimulation is a congruence on the set of deduction graphs generated by the S.O.S. rules.*

Proof The S.O.S. rules for $BPA_{\delta\varepsilon}$ and delayed choice satisfy the *panth* format of [20]. Thus, all we need to do is to provide a stratification for this term deduction system. This is easy: the rank of a step $t \xrightarrow{a} t'$ or a termination option $t \downarrow$ is the number of $\mp$ symbols plus the number of $\lhd$ symbols in t. It is now easy to check that the conditions are met. □

Thus we have that all operators are defined on $G/\leftrightarrows$, the set of deduction graphs modulo $\leftrightarrows$.

Theorem 3.3.5 *Soundness:* $G/\leftrightarrows \models BPA_{\delta\varepsilon} + DC$.

Proof By [4] we have $G/\leftrightarrows \models BPA_{\delta\varepsilon}$. For DC1, consider the relation that relates every closed term of the form $x \mp y$ to the term $x \bowtie y + x \lhd y + y \lhd x$ (and vice versa) and moreover relates every term to itself. On the basis of the definition of a deduction graph, it is easy to verify that this is a bisimulation. All other axioms are equally simple. □

Theorem 3.3.6 $BPA_{\delta\varepsilon} + DC$ *is a conservative extension of* $BPA_{\delta\varepsilon}$.

Proof The operational conservativity follows since our operational rules are in panth format, and pure and well-founded (see [21]). Then we get equational conservativity since the axiomatization of $BPA_{\delta\varepsilon}$ is sound and complete (see [4]) and the axiomatization of DC is sound (Theorem 3.3.5). □

Theorem 3.3.7 $BPA_{\delta\varepsilon} + DC$ *is a complete axiomatization for* $G/\leftrightarrows$.

Proof See [21]. In addition to the ingredients of the previous proof, all we need is the elimination theorem (Theorem 3.3.3). □

3.4. Properties

The main property proven in this section is the commutativity and associativity of $\mp$. This is not derivable from the axioms, but it holds for all closed terms. We need several lemmas to prove this.

Lemma 3.4.1 *The following identities are derivable from the equations.*

1. $ax \mp ay = a(x \mp y)$
2. $a \neq b \Rightarrow ax \mp by = ax + by$
3. $x \mp \delta = x$
4. $x \mp \varepsilon = x + \varepsilon$
5. $(x \lhd y) \lhd z = (x \lhd z) \lhd y$

Proof

1. $ax \mp ay = ax \bowtie ay + ax \lhd ay + ay \lhd ax = a(x \mp y) + \delta + \delta = a(x \mp y)$.

2. If $a \neq b$ then $ax \mp by = ax \bowtie by + ax \lhd by + by \lhd ax = \delta + ax + by = ax + by$.

3. $x \mp \delta = x \bowtie \delta + x \lhd \delta + \delta \lhd x = \delta + x + \delta = x$.

4. $x \mp \varepsilon = x \bowtie \varepsilon + x \lhd \varepsilon + \varepsilon \lhd x = \delta + x + \varepsilon = x + \varepsilon$.

5. $(x \lhd y) \lhd z = x \lhd (y + z) = x \lhd (z + y) = (x \lhd z) \lhd y$. □

In the following we use that every closed term can be written in the form $\sum a_i x_i (+\varepsilon)$, where $(+\varepsilon)$ denotes an optional summand ε. This follows from the elimination theorem (Theorem 3.3.3).

Lemma 3.4.2 *Let a_i and b_j ($i \in I$, $j \in J$, I and J finite) be atomic actions and let x_i and y_j be processes.*

1. $$\sum_i a_i x_i(+\varepsilon) \bowtie \sum_j b_j y_j(+\varepsilon) = \sum_{i,j(a_i=b_j)} a_i(x_i \mp y_j)$$

2. $$\sum_i a_i x_i \lhd \sum_j b_j y_j(+\varepsilon) = \sum_{i(\forall_j a_i \neq b_j)} a_i x_i$$

3. $$\left(\sum_i a_i x_i + \varepsilon\right) \lhd \sum_j b_j y_j(+\varepsilon) = \sum_{i(\forall_j a_i \neq b_j)} a_i x_i + \varepsilon$$

Proof

1. $$\sum_i a_i x_i(+\varepsilon) \bowtie \sum_j b_j y_j(+\varepsilon) = \sum_{i,j} (a_i x_i \bowtie b_j y_j) = \sum_{i,j(a_i=b_j)} a_i(x_i \mp y_j)$$

2. $$\sum_i a_i x_i \lhd \sum_j b_j y_j(+\varepsilon) = \sum_i (a_i x_i \lhd \sum_j b_j y_j(+\varepsilon)) = \sum_{i(\forall_j a_i \neq b_j)} a_i x_i$$

   This last equality can be proven by induction on the number of elements in J.

3. Analogously. □

Define the set of initial actions of a process as in Table 6.

Table 6
Initial actions of a process

$I(\delta) = \emptyset$
$I(\varepsilon) = \emptyset$
$I(ax) = \{a\}$
$I(x + y) = I(x) \cup I(y)$

Lemma 3.4.3 *For closed terms x and y we have*

1. $I(x \bowtie y) = I(x) \cap I(y)$
2. $I(x \lhd y) = I(x) - I(y)$
3. $I(x \mp y) = I(x) \cup I(y)$

Proof This follows directly from the definitions and Lemma 3.4.2. □

The next lemma expresses the fact that in an expression of the form $x \lhd y$, only the initial actions of y matter.

Lemma 3.4.4 *For closed terms x, y and z*

$I(y) = I(z) \Rightarrow x \lhd y = x \lhd z$

Proof Write $x = \sum a_i x_i$, $y = \sum b_j y_j(+\varepsilon)$ and $z = \sum c_k z_k(+\varepsilon)$.
We use the following property: $I(\sum_{j\in J} b_j y_j(+\varepsilon)) = \{b_j : j \in J\}$.

$$\sum_i a_i x_i \lhd \sum_j b_j y_j(+\varepsilon) = \sum_{i(\forall_j a_i \neq b_j)} a_i x_i = \sum_{i(\forall_k a_i \neq c_k)} a_i x_i = \sum_i a_i x_i \lhd \sum_k c_k z_k(+\varepsilon)$$

If x has a summand ε an analogous proof applies. □

Theorem 3.4.5 *For closed terms x, y and z we have*

1. $x \bowtie y = y \bowtie x$
2. $x \mp y = y \mp x$
3. $x \lhd (y \mp z) = x \lhd (y + z) = (x \lhd y) \lhd z$
4. $x \bowtie (y \lhd z) = (x \bowtie y) \lhd z = (x \lhd z) \bowtie y$
5. $x \bowtie (y \bowtie z) = (x \bowtie y) \bowtie z$
6. $x \mp (y \mp z) = (x \mp y) \mp z$

Proof

- (1) and (2) are proven by mutual induction. Again write $x = \sum a_i x_i(+\varepsilon)$, $y = \sum b_j y_j(+\varepsilon)$ and $z = \sum c_k z_k(+\varepsilon)$. First consider (1).

$$\begin{aligned}
& \sum_i a_i x_i(+\varepsilon) \bowtie \sum_j b_j y_j(+\varepsilon) \\
= & \sum_{i,j(a_i=b_j)} a_i(x_i \mp y_j) \\
= & \sum_{i,j(a_i=b_j)} a_i(y_j \mp x_i) \\
= & \sum_{i,j(a_i=b_j)} b_j(y_j \mp x_i) \\
= & \sum_j b_j y_j(+\varepsilon) \bowtie \sum_i a_i x_i(+\varepsilon)
\end{aligned}$$

For (2) we calculate:
$x \mp y = x \bowtie y + x \lhd y + y \lhd x = y \bowtie x + y \lhd x + x \lhd y = y \mp x.$

- (3) follows from Lemma 3.4.3.3, Lemma 3.4.4 and the definitions.
- For (4) write $x = \sum a_i x_i$, $y = \sum b_j y_j (+\varepsilon)$ and $z = \sum c_k z_k (+\varepsilon)$. Then we have:

$$\begin{aligned}
& (x \bowtie y) \lhd z \\
= & \left(\sum_i a_i x_i \bowtie \sum_j b_j y_j (+\varepsilon)\right) \lhd \sum_k c_k z_k (+\varepsilon) \\
= & \left(\sum_{i,j(a_i=b_j)} a_i(x_i \mp y_j)\right) \lhd \sum_k c_k z_k (+\varepsilon) \\
= & \sum_{i,j(a_i=b_j, \forall_k a_i \neq c_k)} a_i(x_i \mp y_j) \\
= & \sum_{i(\forall_k a_i \neq c_k)} a_i x_i \bowtie \sum_j b_j y_j (+\varepsilon) \\
= & (x \lhd z) \bowtie y
\end{aligned}$$

If x has a summand ε, the proof is analogous. The other equation goes similarly.

- For (5) and (6) we use mutual induction. Write $x = \sum a_i x_i (+\varepsilon)$, $y = \sum b_j y_j (+\varepsilon)$, $z = \sum c_k z_k (+\varepsilon)$. Then for (5) we calculate:

$$\begin{aligned}
& x \bowtie (y \bowtie z) \\
= & \sum_i a_i x_i (+\varepsilon) \bowtie \left(\sum_j b_j y_j (+\varepsilon) \bowtie \sum_k c_k z_k (+\varepsilon)\right) \\
= & \sum_i a_i x_i (+\varepsilon) \bowtie \sum_{j,k(b_j=c_k)} b_j(y_j \mp z_k) \\
= & \sum_{i,j,k(a_i=b_j=c_k)} a_i(x_i \mp (y_j \mp z_k)) \\
= & \sum_{i,j,k(a_i=b_j=c_k)} a_i((x_i \mp y_j) \mp z_k) \\
= & \sum_{i,j(a_i=b_j)} a_i(x_i \mp y_j) \bowtie \sum_k c_k z_k (+\varepsilon) \\
= & (x \bowtie y) \bowtie z
\end{aligned}$$

For (6) we have

$$\begin{aligned}
& x \mp (y \mp z) \\
= & x \bowtie (y \mp z) + x \lhd (y \mp z) + (y \mp z) \lhd x \\
= & x \bowtie (y \bowtie z) + x \bowtie (y \lhd z) + x \bowtie (z \lhd y) + \\
& x \lhd (y \mp z) + \\
& (y \bowtie z) \lhd x + (y \lhd z) \lhd x + (z \lhd y) \lhd x \\
= & (x \bowtie y) \bowtie z + (x \bowtie y) \lhd z + (x \lhd y) \bowtie z +
\end{aligned}$$

$$\begin{aligned} & (x \lhd y) \lhd z + \\ & (y \lhd x) \bowtie z + (y \lhd x) \lhd z + z \lhd (x \mp y) \\ = \; & (x \mp y) \bowtie z + (x \mp y) \lhd z + z \lhd (x \mp y) \\ = \; & (x \mp y) \mp z \end{aligned}$$

□

We have established associativity and commutativity for $\mp$. However, the delayed choice operator is not idempotent and it does not satisfy several laws of distributivity.

Proposition 3.4.6 *The following equations are NOT valid in the initial algebra.*

1. $x \mp x = x$
2. $(x + y) \mp z = (x \mp z) + (y \mp z)$
3. $(x \mp y) + z = (x + z) \mp (y + z)$
4. $(x \mp y)z = xz \mp yz$
5. $z(x \mp y) = zx \mp zy$

Proof Let a, b, c, d and e be distinct atomic actions.

1. $(ab + ac) \mp (ab + ac) = ab + a(b + c) + ac$
2. $(ab + cd) \mp ae = a(b + e) + cd$, while
   $(ab \mp ae) + (cd \mp ae) = a(b + e) + cd + ae.$
3. $(ab \mp c) + ad = ab + c + ad$, while
   $(ab + ad) \mp (c + ad) = a(b + d) + ad + c.$
4. $(\varepsilon \mp a)a = a + aa$, while
   $\varepsilon a \mp aa = a(\varepsilon + a).$
5. $(ab + ac)(d \mp e) = ab(d + e) + ac(d + e)$, while
   $(ab + ac)d \mp (ab + ac)e = (abd + acd) \mp (abe + ace) = ab(d + e) + a(bd + ce) + a(cd + be) + ac(d + e).$ □

4. EXAMPLE

Figure 3 shows an example of the use of the delayed choice operator for combining Message Sequence Charts. The complete MSC document consists of four Message Sequence Charts. The first MSC describes the *normal operation* of a system which consists of two instances a and b. Instance a sends a *start* message to instance b. Next a test has to be performed by b, which checks if b operates correctly. This is indicated by a condition named *testing*. Such a condition can be used to denote that there are two possible continuations. The first possibility is expressed in MSC *continuation 1*. It describes the situation where b actually performs a test (indicated by the local action *test*), followed by an acknowledgement to a. The alternative continuation is described in MSC *continuation 2*. In this case, a test is performed and followed by a negative acknowledgment. These two continuations are alternatives, thus expressing that the choice between *ok* and *fail* is non-deterministic. Furthermore, MSC *cancel* expresses yet another scenario. It describes the option that a, immediately after starting b, can *cancel* operation of b. It is clearly not the intention to have a non-deterministic choice between testing and cancellation. So these alternative scenarios are combined with the delayed choice operator.

Using the techniques from [16] we calculate the semantics of these four Message Sequence Charts separately in a simplified notation. Since a Message Sequence Chart describes asynchronous communication, a distinction is made between the output and the input of a message.

$$\begin{array}{ll} normaloperation & = out(start) \cdot in(start) \\ continuation1 & = test \cdot out(ok) \cdot in(ok) \\ continuation2 & = test \cdot out(fail) \cdot in(fail) \\ cancel & = out(start) \cdot \; (in(start) \cdot out(cancel) \cdot in(cancel) \\ & \qquad\qquad\qquad +out(cancel) \cdot in(start) \cdot in(cancel) \\ & \qquad\qquad\qquad) \end{array}$$

Thus the intended semantics of the complete MSC document is

$$normaloperation \cdot (continuation1 + continuation2) \mp cancel$$

This is equal to

$$\begin{array}{l} \quad out(start) \cdot in(start) \cdot (test \cdot out(ok) \cdot in(ok) + test \cdot out(fail) \cdot in(fail)) \\ \mp \; out(start) \cdot \; (in(start) \cdot out(cancel) \cdot in(cancel) \\ \qquad\qquad\quad +out(cancel) \cdot in(start) \cdot in(cancel) \\ \qquad\qquad\quad) \end{array}$$

$$\begin{array}{l} = out(start) \cdot \; (in(start) \cdot \; (test \cdot out(ok) \cdot in(ok) \\ \qquad\qquad\qquad\qquad\qquad +test \cdot out(fail) \cdot in(fail) \\ \qquad\qquad\qquad\qquad\qquad +out(cancel) \cdot in(cancel) \\ \qquad\qquad\qquad\qquad\qquad) \\ \qquad\qquad\quad +out(cancel) \cdot in(start) \cdot in(cancel) \\ \qquad\qquad\quad) \end{array}$$

5. CONCLUSION

We defined the semantics of the delayed choice operator, without proposing a concrete textual or graphical syntax for it in the MSC language. The reason is that the integration of compositional operators and Message Sequence Charts must be uniform for all operators proposed.

We defined operational semantics for the delayed choice operator and gave a sound and complete set of equational laws. We also proved some additional properties of this operator.

If we replace in a process all choices by delayed choices we throw away all branching time information. Thus, on processes that contain only delayed choice, bisimulation semantics

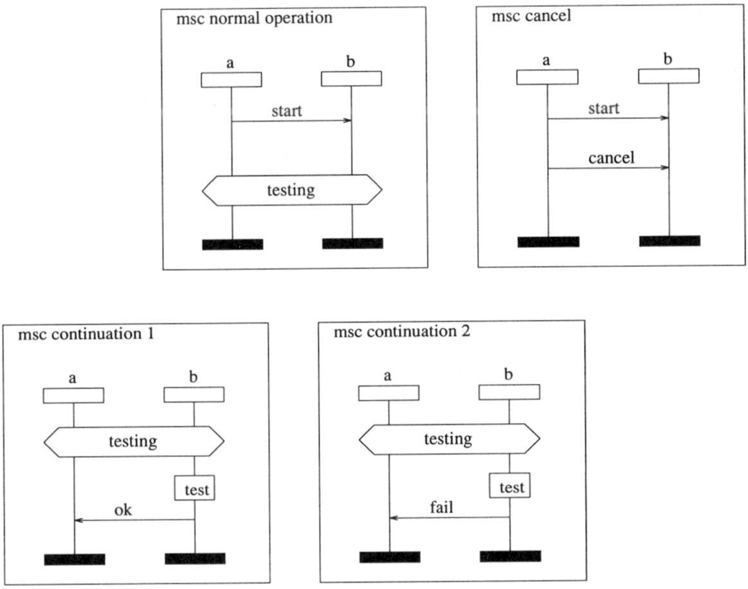

Figure 3. Example Message Sequence Charts

and trace semantics coincide [11]. We leave the formalization of this statement as future work.

The modular approach to process algebra in the style of [5, 6] makes it easy to extend the theory presented here with additional operators. Thus, it is straightforward to achieve extensions with parallel composition, synchronous communication, asynchronous communication, abstraction and other features.

Of course the use of the delayed choice operator is not restricted to the domain of Message Sequence Charts. It is applicable in all cases where we need to express trace-like alternatives in a branching time setting. In general, we prefer the introduction of an extra operator to a total reconsideration of a theory in a different semantics.

REFERENCES

1. L. Aceto, B. Bloom, and F.W. Vaandrager. Turning SOS rules into equations. *Information and Computation*, 111(1):1–52, 1994.
2. J.C.M. Baeten and J.A. Bergstra. Process algebra with partial choice. In B. Jonsson and J. Parrow, editors, *Proc. CONCUR '94*, pages 465–480. Uppsala, Springer Verlag, 1994. LNCS 836.
3. J.C.M. Baeten, J.A. Bergstra, and J.W. Klop. Syntax and defining equations for an interrupt mechanism in process algebra. *Fund. Inf.*, IX(2):127–168, 1986.

4. J.C.M. Baeten and R.J. van Glabbeek. Merge and termination in process algebra. In K.V. Nori, editor, *Proc. FST&TCS 7*, pages 153–172. Pune, Springer Verlag, 1987. LNCS 287.
5. J.C.M. Baeten and W.P. Weijland. *Process Algebra.* Cambridge Tracts in Theoretical Computer Science 18. Cambridge University Press, 1990.
6. J.A. Bergstra and J.W. Klop. Process algebra for synchronous communication. *I&C*, 60(1/3):109–137, 1984.
7. B. Bloom, S. Istrail, and A. R. Meyer. Bisimulation can't be traced: preliminary report. In *Proc. 15th ACM symposium on Principles of Programming Languages*, pages 229–239. San Diego, California, 1988.
8. S.D. Brookes, C.A.R. Hoare, and W. Roscoe. A theory of communicating sequential processes. *Journal of the ACM*, 31:560–599, 1984.
9. P. D'Argenio. τ-angelic choice for process algebra. Technical report, LIFIA, Dpto. de Informàtica, Fac. Cs. Exactas, UNLP, 1994.
10. J. de Man. Towards a formal semantics of Message Sequence Charts. In O. Færgemand and A. Sarma, editors, *SDL'93 Using Objects*, Proceedings of the Sixth SDL Forum, Darmstadt, 1993. Elsevier Science Publishers B.V.
11. J. Engelfriet. Determinacy $\rightarrow$ (observation equivalence $=$ trace equivalence. *TCS*, 36(1):21–25, 1985.
12. O. Haugen. MSC Structural concepts. Experts Meeting SG10, Turin, TD9006, ITU-TS, 1994.
13. ITU-TS. *ITU-TS Recommendation Z.120: Message Sequence Chart (MSC).* ITU-TS, Geneva, 1994.
14. S. Kamin and J.-J. Lévy. Two generalizations of the recursive path ordering. Unpublished manuscript, 1980.
15. J.W. Klop. Term rewriting systems. In S. Abramsky, D. Gabbay, and T. Maibaum, editors, *Handbook of Logic in Computer Science*, volume II, pages 1–116. Oxford University Press, 1992.
16. S. Mauw and M.A. Reniers. An algebraic semantics of Basic Message Sequence Charts. *The computer journal*, 37(4):269–277, 1994.
17. S. Mauw and M.A. Reniers. An algebraic semantics of Message Sequence Charts. Experts Meeting SG10, Turin, TD9009, ITU-TS, 1994. Report CSN94/23, Eindhoven University of Technology, 1994.
18. M.W. Mislove and F.J. Oles. A simple language supporting angelic nondeterminism and parallel composition. In S. Brookes, M. Main, A. Melton, M. Mislove, and D. Schmidt, editors, *Mathematical Foundations of Programming Semantics, 7th international conference*, pages 77–101. Springer Verlag, 1991. LNCS 598.
19. G.D. Plotkin. An operational semantics for CSP. In *Proceedings of the Conference on the Formal Description of Programming Concepts*, volume 2, Garmisch, 1983.
20. C. Verhoef. A congruence theorem for structured operational semantics with predicates and negative premises. In B. Jonsson and J. Parrow, editors, *Proc. CONCUR '94*, pages 433–448. Uppsala, Springer Verlag, 1994. LNCS 836.
21. C. Verhoef. A general conservative extension theorem in process algebra. In *Proc. PROCOMET'94, IFIP 2 Working Conference.* San Miniato, North-Holland, 1994. (to appear), report CSN 93/38, Eindhoven University of Technology 1993.

27

Four Issues Concerning the Semantics of Message Flow Graphs

Peter B. Ladkin[a] and Stefan Leue[b*]

[a]INRIA Lorraine, 615 rue du Jardin Botanique, F-54602 Villers-Lès-Nancy, France
Peter.Ladkin@loria.fr

[b]Institute for Informatics and Applied Mathematics,
University of Berne, Länggassstr. 51, CH-3012 Berne, Switzerland
leue@iam.unibe.ch

Abstract

We discuss four issues concerning the semantics of Message Flow Graphs (MFGs). MFGs are extensively used as pictures of message-passing behavior. One type of MFG, Message Sequence Chart (MSC) is ITU Standard Z.120. We require that a system described by an MFG has global states with respect to its message-passing behavior, with transitions between these states effected by atomic message-passing actions. Under this assumption, we argue (a) that the collection of global message states defined by an MFG is finite (whether for synchronous, asynchronous, or partially-asynchronous message-passing); (b) that the unrestricted use of 'conditions' requires processes to keep control history variables of potentially unbounded size; (c) that allowing 'crossing' messages of the same type implies certain properties of the environment that are neither explicit nor desirable, and (d) that liveness properties of MFGs are more easily expressed by temporal logic formulas over the control states than by Büchi acceptance conditions over the same set of states.

Keyword Codes: F.3.2; D.2.1; 2.
Keywords: Semantics of Programming Languages; Requirements/Specifications; Protocol specification, testing and verification.

1. INTRODUCTION

This paper discusses issues arising from giving a mathematical meaning to Message Flow Graphs (MFGs). Seemingly innocuous syntactic choices may have profound semantic consequences. This highlights the danger of introducing syntactic features without thinking through their semantic consequences, and points to the need for resolution. A precise semantics for MFGs is defined in [19], and was demonstrated for Message Sequence Charts (MSCs) in [18]. MSCs are a well-used class of MFGs, standardised in ITU Z.120 ([16]), which includes some of the syntactic features whose consequences we discuss. Our interpretation is based on the set of partial message-passing states that a system described with MFGs must exhibit.

*This author's work was partly supported by the Swiss National Science Foundation and the Swiss Federal Office for Education and Scientific Research.

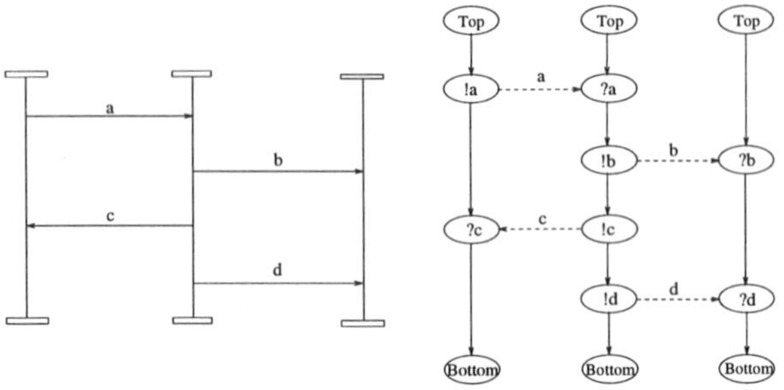

Figure 1. A simple MSC and its corresponding MFG

We present arguments here mainly by means of words rather than in mathematics. It suffices to point to the mathematical formulation of the issues.

What is an MFG? Figure 1 shows an MSC and the same information represented as an MFG. An MFG is a graph with an underlying ontology of message `send` and `receive` events, represented as nodes. Figure 2 presents an MFG which in addition to communication events contains so-called *conditions*, used in Z.120 to allow for branching and iterative behaviour.[2] In a simple MFG (one without conditions, as in Figure 1) the nodes are in one-to-one correspondence with events. In MFGs with conditions, a node may be traversed more than once during a 'run' of an MFG, and therefore does not denote an event *per se*, but rather a 'program statement' indicating that an event of the specified type occurs at this point.

It is generally accepted that the underlying ontology of MFGs (e.g. in their MSC form) is that of events (e.g. [23, 9, 13]). We can summarise features of events as follows: (a) represented events are *message* `sends` and `receives` *only*; (b) the *occurrence* of an event is represented; (c) the *order of occurrence* of events within an individual process is represented; (d) the *types* of the events are represented. These features are represented in the message-passing fragments of other imperative languages such as 'classic'-CSP [14], Esterel [4], Estelle [10] or SDL [6]. The MFG graph is a mathematically more precise formulation of the pictorial information about events and their ordering. Additionally, MFGs (but not CSP, Esterel, Estelle or SDL) connect a `send` to a *unique* `receive` in a different process.[3] A semantics of MFGs must explain how and what this connection is supposed to denote. Here are the issues:

The semantics of MFGs are inherently finite-state. There are only a finite number of control states of all processes described by an MFG, but it might nevertheless be possible, if communication is asynchronous, for there to be an unbounded number of messages 'in the system' (i.e. sent but not received). This is often used as an argument for why MSCs in particular are not finite-state.

[2] We doubt it's a good idea to include conditions without significant restrictions.

[3] MSCs used informally sometimes have `sends` going nowhere, to indicate a lost message. But this isn't in the MSC standard, and it's not significant for the argument made here.

We shall argue in Section 2 that provided that the global state assumption is satisfied (Section 2.4), and traces are interleavings, there are only finitely many global states with respect to the message-passing behavior of the system described by an MFG. Thus we may define a global finite-state automaton, whose accepted language is identical with the set of system traces described by the MFG. Such a semantics was defined in [18, 19].

Besides this argument which is particular to MFGs, justification for a finite-state requirement in general for telecommunications system specifications may be found in [15]. The primary advantage for finite-state interpretations is that reasonable verification and validation techniques may be applied. As of writing, the state-of-the-art in telecommunications system verification and validation is finite-statecraft (see for example [7, 21, 8] and [15]).

A problem with non-local choice. The MSC standard requires 'conditions'. Conditions are global labels such that two MSCs may be 'joined' at this label. We defined conditions, and composition, for MSCs and MFGs in [18, 19]. By having more than one 'joining' possibility, one obtains the effect of non-deterministic choice (conditionals defined on the values of general state predicates are not defined in the standard). Non-terminating-loop-like behavior may be obtained by writing the same condition at the beginning as well as the end of an MFG.

However, unimpeded use of conditions requires unbounded history variables containing the choices in order to resolve non-local choices (Section 3). Such a history of control branching must either be available from the environment or held locally to a process in a history variable of potentially unbounded size. Neither of these options is appealing. We believe that a good specification style should only make explicit assumptions on environment behaviour. Further, they are another potential source of non-finite-stateness.

Crossing message arrows create anomalies. The Z.120 standard allows crossing message arrows as in Figure 7 (a sort of 'message overtaking'). This also leads to implicit requirements on environment behaviour (Section 4).

Büchi acceptance conditions are insufficient to express general liveness properties. A Global State Transition Graph (GSTG) was defined from a given MFG in [18, 19]. However, the GSTG alone does not satisfactorily specify liveness properties for the MFG. We suggested using either Büchi acceptance conditions [26] or Temporal Logic [22] formulae for this purpose. However, in Section 5 we show that one cannot specify every useful liveness property by using Büchi acceptance conditions over the unique GSTG defined from an MFG. We therefore suggest the use of Temporal Logic to specify the desired liveness properties.

Some Other Proposals for MSC Semantics such as [23, 9, 13] do not define the semantics of composed MSCs, therefore cannot observe the semantic problems raised by conditions. These methods provide much less coverage than [19].

2. WHY A FINITE-STATE SEMANTICS?

Finite-state property-checking methods are largely automatic (although controlling state-explosion is a significant problem). This alone is an argument for desiring a finite-state semantics for MFGs. However, there is also an argument from the intuitive meaning of MFGs that they are inherently finite-state. We argue that the *explicit* (as contrasted with hidden) information in an MFG allows only finitely many global system control states.

2.1. What is the Event 'Connection'?

Message-passing connections are shown in the MFG in Figure 1 by means of dotted arrows, in the MSC by means of horizontal (or inclined) arrows, between different processes. What does this symbolic connection correspond to in reality?

`Send` and `receive` events of the same type are connected. Maybe it is intended that *the identical message instance that is sent by the event statement at the arrowtail is received by the event statement at the arrowhead*? However, this is too strong an identity to be useful. Channels, even Ethernet channels, may be lossy. Protocols can try to ensure that if a message of a particular type is sent but not acknowledged, then the contents of the message, along with any message-ID, is regenerated and resent, until successful reception is acknowledged. If message-identity was taken to be message-instance identity, (the actual voltage values raised on the cable), MFGs would be unable to describe higher-level services based on a reliable underlying protocol. So, rather than this strong identity condition, the connection could represent a successful reception of some uncorrupted message instance with a particular message-ID. This is a reasonable interpretation for a higher-layer interaction, such as in a service description (e.g. INRES in [3], JVTOS in [11]), or in Object Models [24, 17]. In some sense, the message-arrow represents the 'same' message sent and received, where 'same message' is a potentially finer individuation than 'message of the same type', and potentially coarser than 'identical message-instance' (while allowing either in the appropriate circumstances). Call such a creature a *message occurrence*. Thus, MSCs represent individual message occurrences, whose properties are assured if necessary by underlying protocols. Unless message occurrences are defined to be message instances in the particular application, one thus cannot discriminate message-instance behavior at the level of description of the MFG.

2.2. In Real Life, There Are a Finite Number of Message Occurrences in a Given MFG Description

The usual means of individuating message occurrences uses an identifier such as a timestamp added at message-generation time. We refer to all such IDs as timestamps. We know of no timestamping mechanism used in real protocols that allows infinite timestamps. Timestamps are often generated by a mechanism formally equivalent to picking numbers in some increasing order from the integers modulo N, for N some large integer. Since there are at most N different timestamps that may be used, only finitely many message occurrences may be individuated.

MSCs are in this regard different from SDL specifications. SDL allows explicit data variables ranging over an unbounded set of values. In principle, one may use a data variable with an infinite range as a timestamp in messages in an SDL specification, allowing distinction of unboundedly many message occurrences. Furthermore, in SDL the communication between processes is by means of explicit unbounded FIFO queues. It follows from these observations that an SDL specification may have unboundedly many states. However, there is nothing in Z.120 that would require MSCs similarly to have unboundedly many states.

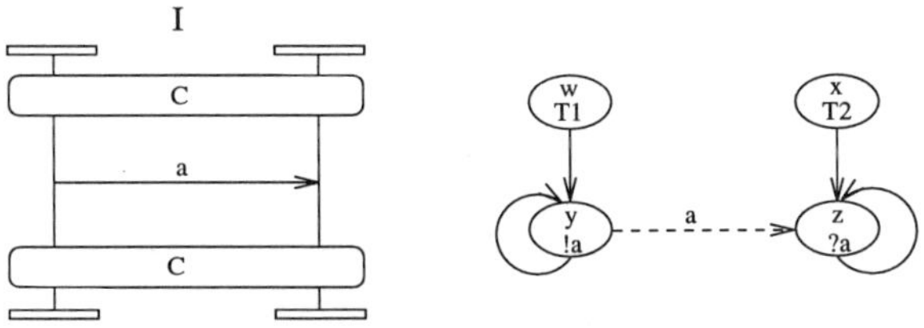

Figure 2. MSC I and corresponding MFG

2.3. Timestamps May Be Eliminated

In Figure 1, the system generates and processes four message occurrences, and each of these message occurrences has a different type. Therefore, the types may be used to individuate messages. However, in Figure 2, an unbounded number of message occurrences is specified, indicated by the condition in the MSC, and the corresponding loop in the MFG obtained by 'joining' the two condition occurrences. Timestamps may be used to individuate the occurrences of messages. Using timestamps modulo N, N iterations through the loop of MSC I in Figure 2 may be individuated. Suppose we duplicate the loop body N times (*loopbody duplication* is defined in [20]). This corresponds to a (maybe much!) larger MFG, in which there are N message arrows, and in which message-occurrence individuation corresponds to different message-occurrence arrows. For MFGs with conditions but without explicit loops, such as MSCs, loopbody duplication corresponds to MFG composition, defined in [19]. We assert without proof that this operation can be carried out for all MFGs with cycles.[4] We call this loopbody operation *timestamp-reduction*. Different message instances generated by the same arrows in the timestamp-reduced graph cannot be individuated. The timestamps individuate precisely the different message arrows in the timestamp-reduced MFG and we may thus remove them. The timestamp-reduced MFG may be much larger than the original MFG, but this increase in size is not semantically relevant. Our concern is only that *in principle* this reduction may be carried out, so that we may identify the arrows in such a reduction with the message occurrences in the MFG. We shall assume from now on that all MFGs have been timestamp-reduced in this way.

2.4. There are Global Control States.

We assume that at any point of time control in each process is located between or at specific message-passing statements or events. Each process $P_i, 0 \leq i \leq n$ contains a finite number of statements defining message-passing events $e_{i_1}, \ldots, e_{i_{n_i}}$ in P. These statements in P correspond with nodes in an MFG. Thus we require that at any point, for any P_i, the Boolean value of the state predicate

$$Last(e)_{P_i} \triangleq \text{the last message event that occurred in } P_i \text{ was one corresponding to node } e$$

is well-defined, where e is one of the e_{i_k}. This entails there is a well-defined vector $\langle e_{p_1}, \ldots, e_{p_n} \rangle$ of the *next message event* for all processes. To handle the startup case, we

[4]Writing out the details from [19] would be tedious but unilluminating.

also include a predicate

$$Last(\texttt{start})_{P_i} \stackrel{\Delta}{\Leftrightarrow} \textit{no message event has yet occurred in } P_i$$

This is the only information available in the MFG concerning the control state of each process P_i; namely, precisely which one of the state predicates $Last(e)_{P_i}$ is true.

2.5. The Different States Engendered by a Message Occurrence

Assuming timestamp-reduction, given the ontology of events, there are three state predicates that a system may satisfy with respect to a given message occurrence m: no `send` or `receive` of a message occurrence m has occurred; a message occurrence m has been sent but not received; a message occurrence m has been sent and received. Since one instance of m may not be discriminated from another instance of m, it follows that there is a finite collection of truth values of these predicates.

2.6. The Total Number of State Predicates is Finite, and There is a Unique Global State Transition Graph

The state predicates of a given state of the MFG are the predicates $Last(e)_{P_i}$ (precisely one of which has the value *true* at any time), and, for each message occurrence m, the truth values of the three state predicates above.[5] The potential global states of the system therefore consist of consistent assignments of truth values to these state predicates. Thus there are only finitely many global states of the MFG.

Since there is a finite collection of global states, it remains to determine the state transition function in order to obtain the *global state transition graph* (GSTG). The nodes of the GSTG are the states. State transitions are represented by edges between pairs of states. State transitions are caused by events (nodes of the MFG), thus an edge of the GSTG may be labelled with the event triggering the transition. Every event causes a change in true value of precisely two predicates of the form $Last(e)_{P_i}$, and corresponding changes in the message-occurrence predicates.

Thus there is a finite number of global states in the timestamp-reduction, and these states and the transitions between them are uniquely determined and may be calculated using the method in [19].[6] We propose this as a *canonical* semantics for MFGs.

2.7. Some General Reasons for Requiring Even Asynchronous Communications to be Finite-State in Telecommunications

There is furthermore a general argument for requiring semantics of message-passing in any real telecommunications protocol or service to be finite-state, even those specified by SDL (which in principle may utilise unboundedly many timestamps).

In a protocol or service definition, each individual process control is usually a finite-state device with respect to `sends` and `receives`. The unboundedly many states are usually attributed to the unboundedly many states a lossless asynchronous channel may have. But consider now system recovery from faults. Regardless of its size, a finite state device can only retain a bounded computation history. Suppose, as the system runs, communication channels are compromised (someone cuts a cable). Assume also that the

[5] We note that not all combinations of values are possible.

[6] We used sets of edges as a code for the collection of truth values of state predicates in [19].

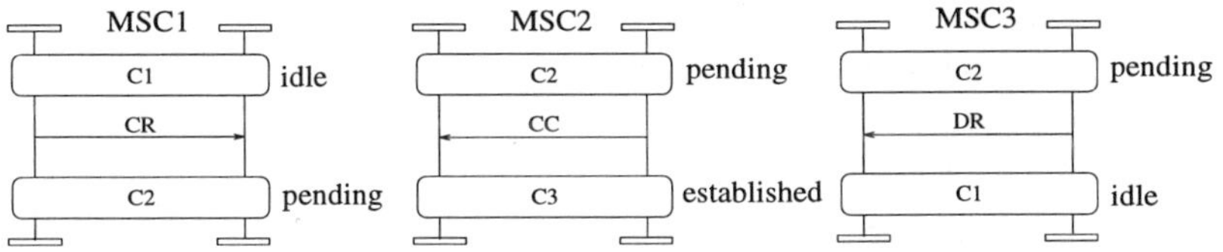

Figure 3. MSC specification with conditions

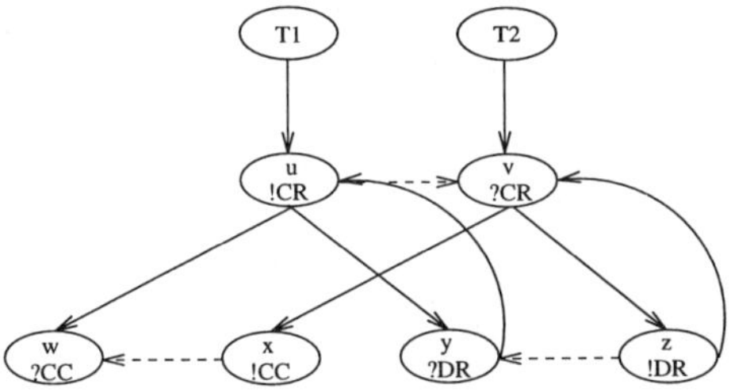

Figure 4. 'Unfolding' an MSC specification into a single MFG

processes themselves are not compromised. A consistent state must be reconstructed. Each process must be asked its state, namely where it thinks its control is, and what it remembers from what it has done. No other information is available. The system itself can have been operating for a very long time, much longer than the bounded memory of any single process. However, what can be reconstructed from the memories of the processes is bounded, no matter how long the system has been running previous to the fault. Two such failures which result in the same local state of each process are therefore equivalent from the point of view of the determinable state of the system. So each such equivalence class can be identified with a *global state* of the system. Since there are finitely many finite-state processes, the global states are some equivalence (probably the identity) relation on a subset of the cartesian product of the state spaces of the individual processes, and thus there are only finitely many global states. A conservative upper bound to the number of these states is the size of this cartesian product.

It is often suggested that asynchronous communication is equivalent to the presence of lossless queues of unbounded capacity on each channel, e.g. in SDL. It is well known that in theory queues may be configured to contain the entire system history information, which is finite at any one point, but unbounded through the history of the system. By the argument above, since the number of practically distinguishable global states is bounded, the contents of these theoretical queues cannot be part of the system, in general, and properties of 'queue' contents may only be inferred from the processes that generated and received those contents – and that information is bounded.

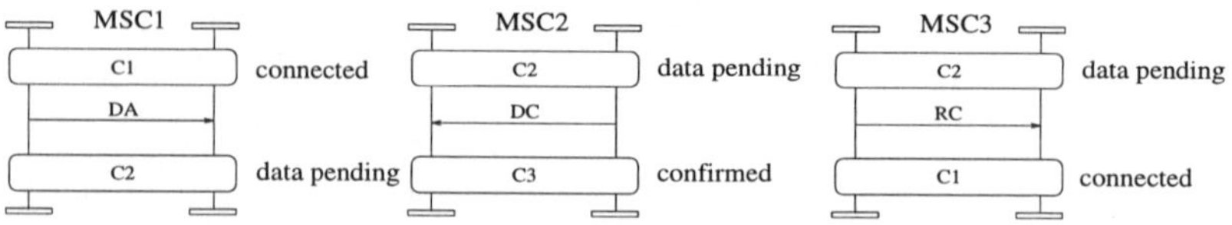

Figure 5. An MSC specification generating non-local control choice

3. CONDITIONS AND NON-LOCAL CHOICE

MSC specifications which include even simple use of conditions, such as in Figure 3, may induce multiple traces. From condition `C2` in `MSC1`, two different possible execution paths are possible, represented by `MSC2` and `MSC3`. This results in branching control, represented by the two next-event out-edges from each of nodes `u` and `v` in the MFG (Figure 4). If one process takes one branch and issues a `send`, the other process is constrained to take its corresponding branch to execute the corresponding `receive`. Thus the control branching must be synchronised. Since an MFG (and the MSCs) represent message exchange, it is consistent with this information that the synchronisation should be accomplished locally by each process, determined from 'observing' what happens during an execution. However, there are cases in which synchronisation cannot be achieved by methods purely local to each individual process (see Figure 6). We show that this non-local decision-making requires that each process have potential access to its complete choice history, a record which is of unbounded size in non-terminating processes.

3.1. Non-Local Choice, and Choice History

We show that history variables or their equivalent are needed to handle control branching that cannot be achieved by local means. The MSC standard Z.120 which considers conditions contains no recognition of the need for history variables in these circumstances. The spirit of MSCs would require that control choice synchronisation between processes which cannot be accomplished by each process acting independently should be accomplished by explicit exchange of messages indicating that a particular control branch is followed. We conclude that either control history variables should be explicitly introduced, or that the use of global initial or final conditions to represent control branching should be limited in some way by the standard. We are content here merely to show the need for a choice, but make no recommendation.

The example is a modification of Figure 3. In the MFG in Figure 4, the control-branch choice may be resolved locally. The first process, equipped with message-type-identification, awaits a signal from the second, and determines whether it is a `CC` or a `DR` message. Indeed, this would be the sensible way to implement it. Thus this example involves no non-local control choice synchronisation. An example in which the control choice must be somehow communicated non-locally is given by the MSC specification in Figure 5, which generates the MFG in Figure 6. In this example, the first and second processes must decide somehow 'together' whether they are going along the `DC` route or along the `RC` route. The MSC specifically disallows that one process could follow its left

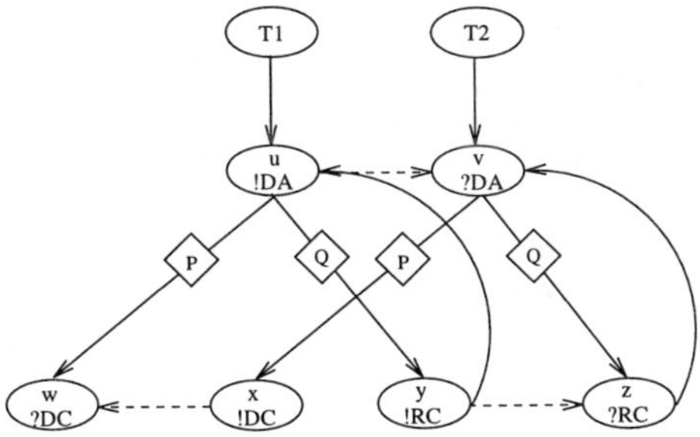

Figure 6. An MFG with non-local-choice nodes

branch and the other its right branch.

We denote this control synchronisation by labeling next-event edges at branches with predicates as in [19]. We assume that there is an unbounded collection of predicate symbols $P_1, P_2, \ldots$, different from all other symbols used. Given a condition symbol for which there exist at least two MSCs starting with that symbol (e.g. the symbol `C2` in both Figures 3 and 5), we label corresponding next-event edges with predicate symbols as we construct the unfolding. For example, in Figure 6 we label edges corresponding to a transition through `C2` as we unfold. The labels are shown in the figure using diamonds on the edges (these diamonds here represent labels, not conditions). The labels used on the next-event edges are predicates from the list $P_1, P_2, \ldots$ that so far have been unused in the unfolding construction, say the first such ones. Labels are the same if each branch of each process with that label arises from the same MSC. In our example, labels `P` and `Q` used in Figure 6 would be respectively P_1 and P_2 according to this scheme, `P` being used to label the branches from MSC2 and `Q` those arising from MSC3. The modifications to the definition of MSC to allow next-event edge-labels, and formal definition of unfolding are straightforward (see [18, 19]).

3.2. Non-Local Choice May Imply Non-Finite-State Control

We show that if unrestricted conditions are allowed in MFGs, then some systems described by MFGs require non-finite-state control within individual processes. Either unbounded history variables are required to keep track of control choices, or very simple MFGs with conditions that require non-local choice, such as those in Figure 5, must be regarded as ill-formed.

The set of MFGs in Figure 5 may represent a higher-level requirement on system behavior. A system implementing this behavior must conform to the requirement. Let us suppose that the implementation may perform some rudimentary amount of error recovery on, say, temporary loss of transmission, such that the system continues to satisfy the MSC requirement.

We refer to the process whose 'line' is on the left [resp. right] in Figures 5 and 6 as the 'left' [resp. 'right'] process. Suppose we label the left branches of both processes in the

MFG in Figure 6 with P, to represent a choice of continuation with MSC2, and the right branches of both processes with Q to represent a continuation with MSC3. Suppose now that the system executes a trace in which the successive choices between continuation with MSC2 and MSC3 are made as follows: 1 P choice, followed by 2 successive Q choices, then 3 successive P, then 4 successive Q, then 5 successive P, ... The semantics of asynchronous communication allows the the left process to fall arbitrarily behind the right process. Somehow, the history of control choices made by the right process must be known by the left process, in order that the left process 'knows' which type of message to receive next. Suppose that the history of the last n control choices in each process is retained, in a 'history variable', which we can assume is an array of length n. After at most $\Sigma_{k \leq n} k = n.(n+1)/2$ branch choices, the history array contains either (a) all P's; (b) all Q's; or (c) at most one change from P's to Q's or *vice versa*. It is easy to see there are $2.(n-1)$ such possibilities for (c), which along with (a) and (b) yields $2.n$ possible configurations of the history variable after $n.(n-1)/2$ branch choices.

Suppose a recoverable fault occurs, and the system is restarted with (i) all message buffers intact as when the fault occurred; (ii) all data (including history variables) intact; (iii) program counters in the same position. Suppose the fault has occurred after somewhat more than $n.(n-1)/2$ branch choices. There are only $2.n$ possibilities for the configuration of the history variables in each process. Hence there are only $(2.n)^2$ total pairs of values of both history variables. Suppose there are k outstanding messages. It is easy to show that there are infinitely many possibilities of the choice history of each process compatible with each given configuration of the two history variables plus the number of outstanding messages. However, because of the construction of the example, unless the two processes start in corresponding places in the choice history, they will not fulfil the requirement expressed by the MSCs in Figure 5.

Not only the control branching history of the current process but also the history of other processes that are 'further ahead' must be known by a process, in order to make the appropriate control branches. To ensure that this condition is fulfilled, under even these mild recovery conditions, either (a) the process must have an implicit means of communicating with the environment in order to access this history, which is held somehow by the environment in a history variable; or if not then (b) the environment must communicate this information implicitly to a process as it happens, and the process must itself retain this entire history in a variable, which as we have seen must be of potentially unbounded size.

These situations are unsatisfactory since they require either greater or lesser roles for the environment as an implicit information-passer, and/or require each process to have potentially unbounded memory to retain control choice history.

A third alternative is to simply consider the requirement as expressed in Figure 5 to be ill-formed. But how to tell which uses of conditions are allowed? Well-formedness should be a matter for syntax, not for semantical analysis. Since non-local choice leads to undesirable requirements, it's necessary to find a syntactic restriction which corresponds to permitting local choice only.

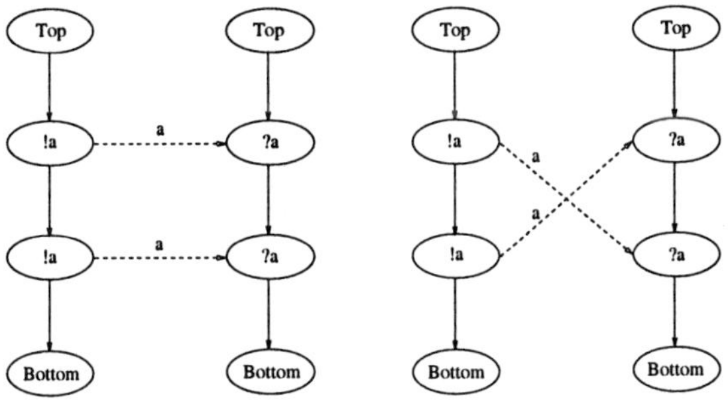

Figure 7. A cross-over MSC example, expressed as MFG

4. A CROSSING ANOMALY

In this section, we point out an anomaly arising from allowing messages to 'cross' in MSCs. We are led to conclude that there are non-trivial but non-obvious properties of the environment implicitly contained in certain types of MSC descriptions. We regard it as infelicitous that such implicit properties should be required.

The MSC standard [16] allows crossing of signals to occur. The two MFGs of Figure 7 representing two simple MSCs describe different system behaviors. In both cases an identical type of signal is transmitted twice. The second case differs from the first in that a 'cross-over' of the messages is specified. The observable behavior of each individual process is identical in the two examples (one sends two 'a' signals, the other receives two 'a' signals), hence code implementing each process will be identical in both examples. However, the two examples have different sets of valid traces. The set of traces (interleaved observable events) of the first MSC is $\{<!a, !a, ?a, ?a>, <!a, ?a, !a, ?a>\}$, and that of the second is is $\{<!a, !a, ?a, ?a>\}$. A system exhibiting behavior $<!a, ?a, !a, ?a>$ satisfies the first specification but not the second. However, a system exhibiting behavior $\{<!a, !a, ?a, ?a>\}$ and no other may satisfy either specification.

Since there is no difference in process code for the two examples, the different trace sets must be accounted for by a difference in the behavior of the environment. Thus, even though the environment is not explicitly represented in this specification, its properties must be invoked implicitly. One may try to resolve this problem by representing the environment explicitly, as a single vertical line like a process axis, provided it engages in message interaction with the processes (cf. [5]). However, this is no solution. Even if the environment is explicitly represented as such a third axis, firstly one can still obtain analogous process behavior by using cross-overs, and secondly this behavior may only be obtained by using a crossover (we leave this as an easy exercise).

One criterion for a good specification method (to distinguish it, say, from the average programming language) is that all asserted properties be represented explicitly, including constraints from the environment. Our example shows that MSCs with cross-over do not pass this test. Further, even if the environment is explicit, cross-over is at best an unintuitive method of enforcing certain orderings on behavior, for which there is no other representation mechanism. Such 'programming tricks' have no place in a good description method.

5. LIVENESS PROPERTIES AND ACCEPTANCE CRITERIA

Liveness Conditions. Given that a system follows a finite state-transition graph, there is nevertheless a question as to whether all traces through this graph are acceptable traces of the system, or whether only a subset of them are. We showed in [19] that general MFGs define a very limited set of liveness properties. In order to facilitate the expression of a wider array of liveness properties, it is necessary to go beyond the Global State Transition Graph (GSTG [19]) to consider which traces through the graph are allowed by the description (along with the liveness properties) and which aren't. A standard way to express these conditions is to consider the GSTG as providing most of the definition of an ω-automaton, lacking only an end-state definition, and to provide that end-state definition.

Büchi- and Other ω-Automata. Since traces may be infinite, a finite-state semantics requires use of a finite-state automaton which accepts infinite strings. The Büchi automaton is probably the most well-known of these, and has been used in the determination of safety and liveness properties of distributed systems [1], [2]. These automata are similar to ordinary finite automata, except for the acceptance condition. Büchi automata include in their definition a set of states called the *end-state set.* A (possibly infinite) string is accepted by a Büchi automaton just in case the automaton passes through an end state unboundedly often on the string (for finite strings, the final state must be an end state).

Given a general MFG specification, involving a family of MFGs with conditions, the GSTG is uniquely determined [19]. From this graph, various different end-state definitions will define various different ω-automata, each of which identifies the set of system traces specified by the MFG with the set of accepted traces of the automaton. The Global-State Transition Graph itself defines a Büchi automaton, namely the one in which the end-states are the set of all states. Even though Büchi automata define a very rich class of trace-sets in order to use them flexibly one must be at liberty freely to design the state set. We are constrained by having to use the global states defined in the GSTG, and we show now that the Büchi acceptance condition does not suffice to define certain natural liveness conditions, given the GSTG states and transitions. Therefore other acceptance conditions may be preferable.[7]

Büchi automata suffice to describe liveness properties of systems, provided that the specifier is free to choose the states of the automaton during the course of designing an automaton to accept precisely the desired traces. However, the global states of an MFG are specified, uniquely, by the MFG. We are not free to choose an alternative state set[8]. If one needs to specify liveness properties of an MFG description, the definition of Büchi acceptance may not suffice. Consider the GSTG on the right hand side of in Figure 8,

[7]We are grateful to Bob Kurshan for this observation.

[8]It may be possible to devise a general transformation of a GSTG into another GSTG G' satisfying identical safety properties such that a different set of of liveness properties may be defined for G' based on Büchi acceptance. However, we don't know of one, or whether such a transformation might have other disadvantages.

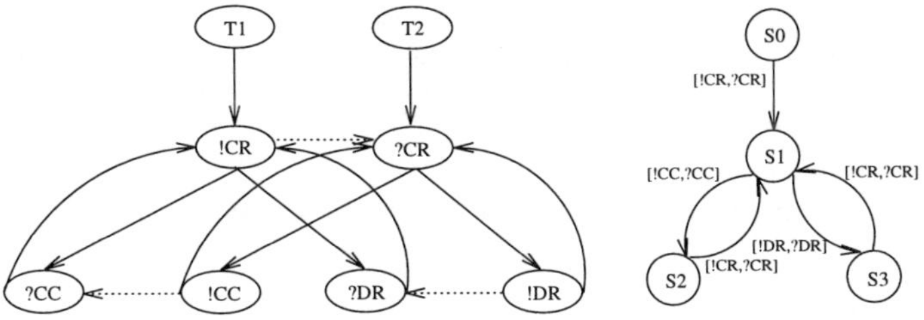

Figure 8. A MFG and the corresponding MFG whose liveness may not be specified by Büchi acceptance

derived from the MFG on the left hand side in the same Figure[9]. It represents a system which when in state $S1$ makes a non-deterministic choice between transiting into two states $S2$ and $S3$, and then returns to state $S1$. A important liveness property may be to require that the system performs a *fair* choice between the subbehaviours $S2$ and $S3$. This is expressed in temporal logic as $(\Box\Diamond at\_S2 \wedge \Box\Diamond at\_S3)$[10]. It is easy to see that there is no set of end-states under which this liveness condition is expressed by Büchi acceptance (just look at the 15 possible non-trivial end-state sets).

The proper definition of liveness properties for the GSTG and therefore for the system described by the MFGs may therefore be accomplished better by temporal logic formulae than by Büchi acceptance. This is because the design of the Büchi automaton is constrained by the necessary selection of a particular transition graph, the GSTG. It remains to be seen whether other acceptance criteria suffice to define automata suitable for all potential liveness criteria. In the meantime, temporal logic appears to be able to define the liveness criteria which users of MFGs may want. We have defined the precise relationship between temporal logic assertions and MFG 'executions' in [19].

6. CONCLUSIONS

We discussed four issues concerning the semantics of Message Flow Graphs (MFGs). We required that a system described by an MFG has global states with respect to its message-passing behavior, with instantaneous transitions between these states effected by atomic message-passing actions. Under this assumption, we argued (a) that the collection of global message-passing states defined by an MFG is finite (whether for synchronous, asynchronous, or partially-asynchronous message-passing); (b) that the unrestricted use of 'conditions' requires processes to have access to control history variables of unbounded size; (c) that allowing 'crossing' messages of the same type implies certain properties of

[9]Note that the signal arrows in this MFG indicate *synchronous* communication, since we can make our point with this GSTG, which is much simpler than that derived from the same MFG with asynchronous communication. MFGs handle synchronous communication without modification. as shown in [19]. This may be another indication that life with synchronous communication is easier.

[10]As described in [22]. at_S2 and at_S3 are state predicates asserting that the system is in state $S2$ or $S3$, respectively.

the environment that are neither explicit nor desirable, and (d) that liveness properties of MFGs are more easily expressed by temporal logic formulas over the control states than by Büchi acceptance conditions over the same set of states.

Acknowledgements
We thank Professor Ken Turner for some helpful discussions on MSCs.

REFERENCES

1. B. Alpern and F. B. Schneider. Recognizing safety and liveness. *Distributed Computing*, 2:117–126, 1987.
2. B. Alpern and F. B. Schneider. Verifying temporal properties without temporal logic. *ACM Transactions on Programming Languages*, 11(1):147–167, 1989.
3. F. Belina, D. Hogrefe, and A. Sarma. *SDL with Applications from Protocol Specification.* Prentice Hall International, 1991.
4. G. Berry and G. Gonthier. The Esterel synchronous programming language: design, semantics, implementation. *Science of Computer Programming*, 19:87–152, 1992.
5. M. Broy. Towards a formal foundation of the specification and description language SDL. *Formal Aspects of Computing*, 3:21–57, 1991.
6. CCITT. Recommendation Z.100: CCITT Specification and Description Language (SDL). CCITT, Geneva, 1992.
7. E. M. Clarke and R. P. Kurshan, editors. *Computer Aided Verification: Proceedings of CAV'90*, volume 531 of *Lecture Notes in Computer Science.* Springer Verlag, 1991.
8. C. Courcoubetis, editor. *Computer Aided Verification: Proceedings of CAV'93*, volume 697 of *Lecture Notes in Computer Science.* Springer Verlag, 1993.
9. J. De Man. Towards a formal semantics of message sequence charts. In *[12]*, pages 157–166. 1993.
10. M. Diaz, J.-P. Ansart, J.-P. Courtiat, P. Azéma, and V. Chari, editors. *The Formal Description Technique Estelle.* North-Holland, 1989.
11. A.J.M. Donaldson. Specification of quality of service measurement points in JVTOS. Master's thesis, University of Stirling, Scotland, U.K., September 1993.
12. O. Færgemand and A. Sarma, editors. *SDL '93: Using Objects.* North-Holland, 1993.
13. P. Graubmann, E. Rudolph, and J. Grabowski. Towards a Petri Net based semantics definition for Message Sequence Charts. In *[12]*, pages 179–190. 1993.
14. C.A.R. Hoare. Communicating sequential processes. *Communications of the ACM*, 21(8):666–677, August 1978.
15. G. J. Holzman. *Design and Validation of Computer Protocols.* Prentice-Hall International, 1991.
16. ITU. Recommendation Z.120: Message Sequence Chart (MSC). Geneva, 1993. Revised Version.
17. I. Jacobson. *Object Oriented Software Engineering.* 1992.
18. P. B. Ladkin and S. Leue. What do Message Sequence Charts mean? In *[25]*. 1994. To appear.
19. P.B. Ladkin and S. Leue. Interpreting message flow graphs. *Formal Aspects of Computing*, 1995. To appear.

20. P.B. Ladkin and B.B. Simons. *Static Analysis of Interprocess Communication.* Lecture Notes in Computer Science. Springer-Verlag, 1994. To appear.
21. K. G. Larsen and A. Skou, editors. *Computer Aided Verification: Proceedings of CAV'91*, volume 575 of *Lecture Notes in Computer Science.* Springer Verlag, 1992.
22. Z. Manna and A. Pnueli. *The Temporal Logic of Reactive and Concurrent Systems: Specification.* Springer-Verlag, 1992.
23. S. Mauw, M. van Wijk, and T. Winter. A formal semantics of synchronous interworkings. In *[12]*, pages 167–178. 1993.
24. J. Rumbaugh, M. Blaha, W. Premerlani, F. Eddy, and W. Lorensen. *Object-Oriented Modeling and Design.* Prentice Hall International, 1991.
25. R. L. Tenney, P. D. Amer, and M. Ü. Uyar, editors. *Formal Description Techniques, VI.* IFIP Transactions C, Proceedings of the Sixth International Conference on Formal Description Techniques. North-Holland, 1994.
26. W. Thomas. Automata on infinite objects. In *Handbook of Theoretical Computer Science*, chapter 4, pages 132–191. Elsevier Science Publishers B. V. (North-Holland), 1990.

Position Statements

28

Towards a Formal Computational Model for Distributed Multimedia Applications*

Andreas Vogel
andreas@dstc.edu.au

CRC for Distributed Systems Technology, Level 7, Gehrmann Laboratories
The University of Queensland Q 4072, Australia

Abstract

A formal computational model for distributed systems and in particular for distributed multimedia applications derived from the Reference Model on Open Distributed Processing is defined and presented. The formal definition is based on the standardized formal description technique LOTOS introducing the LOTOS sub-language approach. This approach is applied to the ODP computational language yielding a sub-language LOTOScomp. In this context, the capability of (standard) LOTOS to model Quality of Service is discussed and compared with approaches based on LOTOS extensions.

1. Introduction

Distributed multimedia applications, e.g. video conferencing systems, on-demand services or multimedia mail, are now available. However, they seem to be mostly ad-hoc solutions implemented in a particular environment. This paper aims a more general approach: distributed multimedia applications are derived from standardized framework, the Reference Model on Open Distributed Processing (ODP). The suitability of the RM ODP for distributed multimedia applications was already seen, e.g. in the multimedia oriented extension of ANSAware [3] or in the *Touring Machine* [1], a software platform for distributed multimedia application based on an ODP oriented architecture. However, these approaches are providing implementation platforms but not a formal framework. Here, such a framework is presented using the standardized formal description technique LOTOS [5].

ODP has been characterized by Kerry Raymond in her tutorial [8] given at the First International Conference on ODP as follows: *"Advances in computer networking have been allowed computer systems across the world to be interconnected. Despite this, heterogeneity in interaction models prevents interworking between systems. Open Distributed Processing describes systems that support heterogeneous distributed processing both within and between organizations through the use of a common interaction model. ISO and ITU-T (formerly CCITT) are developing a Basic Reference Model of Open Distributed Processing (RM ODP) to provide a coordinating framework for the standardization of ODP by creating an architecture which supports distribution, internetworking, interoperability and portability."*

The main concept of the RM ODP is the viewpoint. A viewpoint is an abstraction, focused on parts of an ODP system determined by a particular interest. There are five viewpoints, enterprise, information, computational, engineering, technology. The concepts, rules, and structures appropriate to the specification of an ODP system are given by corresponding description languages, e.g. computational language for the computational viewpoint. The semantics of an ODP infrastructure are defined by ODP functions.

2. Formal computational model

The idea of the LOTOS sub-language approach is to define given properties of a class of specifications, e.g. concepts and structures, within the sub-language. Obviously, this approach is different from the simple reduction of operators as e.g. in the case of Basic LOTOS.

* This work was partly supported by a grant from the Canadian Institute for Telecommunication Research (CITR), under the Networks of Centers of Excellence Program of the Canadian Government.

One reason for the sub-language approach is that the ODP description languages define a class of the specification which can not be defined by one specification (in difference to SDL-92 which provides the necessary genericity) but can be by a sub-language. Another reason is that the sub-language approach provides scalability of prescription, which allows for examples to be so prescriptive to provide interoperability on the specification level.

For ODP's computational viewpoint, the sub-language LOTOScomp has been defined [9]. The sub-language definition includes first the mapping of concepts of the computational language, e.g. computational object or operational interface, in LOTOS templates, e.g. predefined LOTOS processes. Second, the rules of the computational language which define the relationships between the concepts, e.g. a computational object contains one or more computational interfaces are expressed by syntactical constraints of the sub-language. Third, it contains LOTOS specifications of predefined objects which are specified within the Reference Model on ODP, e.g. the trader and the binding-object and specifications of the computational infrastructure and the so called stream object to provide a completely executable computational model*. The same approach has been applied to information and engineering language [10]. The particular suitability for multimedia application is given by the full support of the ODP stream interfaces and according binding mechanisms. Additionally, the modeling of quality of service is investigated in the following section.

3. Modeling of quality of service

The modeling of quality of service is seen in different abstraction levels: (i) specification of QoS parameters, and (ii) specification of behavior depending on QoS parameters.

The considered QoS parameters cover rather than only those of communication protocols also the ones determined by operating systems, file servers, and human end-users. An overview of the different types of QoS parameters are given by Kerherve et al. [6]. In general, two type of parameters can be distinguished, time-independent parameters, eg. loss of cells in a communication network, and time-dependent parameters, e.g. end-to-end delay in a communication network. The specification of QoS parameters can be done straight forward by mapping onto data types incl. a special interpretation, e.g. natural numbers (microseconds).

To model QoS-dependent behavior, LOTOS has no quantitative time concept. However, different approaches has been investigated to express quantitative time. On the one hand, there are approaches using standard LOTOS, e.g. introducing processes modeling clocks [4]. On the other hand, there are approaches extending LOTOS with a time calculus, e.g. as given by Quemada and Fernandez [7]. Within the Tempo project [2], the capability of formal description techniques and in particular LOTOS and its time extensions for multimedia applications has been investigated, leading to a so-called bilingual approach. The idea is to use LOTOS to specify functional behavior and a temporal logic for the nonfunctional behavior, to provide a common semantical basis for both languages.

When considering QoS for distributed multimedia application [6] we identified the delay (caused by file-servers, communication, and operating systems) as the basic time-depended QoS parameter from which other parameters, e.g. jitter, can be concluded and which determine synchronization protocols. Delay is related to data and subject of change when data is processed. This interpretation allows to attach a delay parameter to the data unit which is processed and this parameter can be changed (nondeterministicly) by the processing object, e.g. the communication network. This idea has been applied to different examples*.

In our opinion, the cited and presented approaches are equivalent in respect to their expression power. However, they differ in the level of abstractions. Consequently, the evaluation of the approaches depends on the type of application, the required analysis and validation, available tools, and taste of the specifier. Our approach seems to have its advantage in its simplicity: using the tool and methodological support provided for standard LOTOS. The main disadvantage is seen in the mixing of functional and nonfunctional behavior. This problem was already discussed by Bowmann et al. [2] under the title "*Time*

* The complete language definition, the specification of pre-defined objects, examples and other details can be obtained trough the author's html home-page: http://www.dstc.edu.au/public/staff/andreas-vogel.html.

vs. Abstraction" and consequently the bilingual approach was proposed. However, limitations of this approach are seen in the common semantics which is only defined for Basic LOTOS, and in the lacking of integrated tool support for the temporal logic portion. Finally, LOTOS time extensions seem to be 'oversized' for the specification of QoS, i.e. functionality is provided which is not necessary for QoS specification.

4. Conclusion and further work

First, a formal computational model for distributed applications was defined and its suitability for distributed multimedia applications was shown. The advantages and distinguishing characteristics of this approach are seen in the application of a standardized framework and standardized description technique, introducing of mathematical semantics on the language level of LOTOS, providing executable models, and applicability of already developed analysis, test and verification techniques and accompanying tools. The model has been applied within a CITR (Canadian Institute of Telecommunication Research) project on '*Broadband Services*'.

Second, the modeling of QoS using LOTOS has been reviewed, our own approach has been presented. Consequence of the discussion is that considerations on QoS modeling are not yet mature and future work is required, in particular in the framework of ODP.

References

1. M. Arango, P. Bates, G. Gopal, N. Griffeth, G. Herman, T. Hickey, W. Leland, V. Mak, L. Ruston, M. Segal, M. Vecchi, A. Weinrib, and S.-Y. Wuu, "Touring Machine: a software infrastructure to support multimedia communications," *ACM Computer Communication Review,* 22, pp. 53-54 (Mar. 1992).
2. H. Bowmann, G. S. Blair, L. Blair, and A. G. Chetwynd, "Time Versus Abstraction in Formal Description" in *Formal Desription Techniques VI,* ed. R. L. Tenney, P. D. Amer, and M. Ü. Uyar, pp. 469-484, North Holland (1993).
3. G. Coulson, G. Blair, J. B. Stefani, F. Horn, and L. Hazard, "Supporting the Real-Time Requirements of Contionous Media in Open Distributed Processing," University of Lancaster Technical Report MPG-92-35 (1992).
4. W.H.P van Hulze, P.A. Tilanus, and H. Zuidweg, "LOTOS Extended with Clocks" in *Formal Description Techniques,* ed. S. T. Vuong, pp. 117-132, North Holland (1990).
5. ISO, *LOTOS - A formal description technique based on the temporal ordering of observational behaviour,* ISO 8807, International Standard (1988).
6. B. Kerherve, A. Vogel, G. v. Bochmann, R. Dssouli, J. Gecsei, and A. Hafid, "On Distributed Multimedia Presentational Applications: Functional and Computational Architecture and QoS Negotiation" in *Proc. of Fourth International Workshop on Protocols for High-speed Networks, August 10-12, 1994, Vancouver,* ed. G. Neufeld and M. R. Ito, Chapman & Hall, London (1994).
7. J. Quemada and A. Fernandez, "Introduction of Quantitative Realtime Time into LOTOS" in *Protocol Specification Testing and Verification VII,* North-Holland (1987).
8. K. Raymond, "The Reference Model of Open Distributed Processing: a Tutorial" in *Proceedings of the 1st International Conference on ODP,* ed. J. de Meer, B. Mahr, and O. Spaniol, pp. 3-14, Berlin (1993).
9. A. Vogel, "A Formal Approach to an Architecture for Open Distributed Processing," Technical Report #902, Université de Montréal (1994).
10. A. Vogel, "Entwurf, Realisierung und Test von ODP-Systemen auf der Grundlage formaler Beschreibungstechniken," submitted as PhD thesis, Humboldt-Universität zu Berlin (1993). in German.

29

Specification, Detection and Resolution of IN Feature Interactions with Estelle

J. Bredereke and R. Gotzhein

University of Kaiserslautern, Postfach 3049, D-67653 Kaiserslautern, Germany
{bredereke,gotzhein}@informatik.uni-kl.de

Keyword Codes: C.2.2; D.2.1
Keywords: Network Protocols; Requirements/Specifications

1. Introduction

We present an approach for the treatment of Feature Interactions in Intelligent Networks (IN). It is based on the formal description technique (FDT) Estelle and consists of three steps. 1) A *specification style* supporting the integration of additional features into a basic service is introduced. 2) Feature interactions are *detected* which result from the integration of additional features (before their deployment). 3) Previously detected feature interactions are *resolved*. We emphasize that these three aspects interfere and therefore have to be treated together. In particular, the FDT and the specification style have a strong influence on the detection criteria and on the resolution of feature interactions.

Several authors have proposed classifications of feature interactions in telecommunication systems. In [DaNa93], a distinction is made between *technical interference* and *policy interference*. Technical interference leads to phenomena such as deadlocks, livelocks and congestion. The detection of these phenomena has been studied extensively. Policy interference occurs if features interact such that the policy of one or more of the involved features is violated. Our approach is designed to be general, but we will especially concentrate on policy interference.

The approach sketched in this extended abstract is presented in more detail in [BrGo94], and it is applied to the Basic Call Service and several supplementary interacting features.

2. Specification of IN Features with Estelle

Adding (further) features to a system means extending its specification. The number of spots in the syntax and the semantics which are touched should be as small as possible. The following stylistic rules will allow the use of our detection and resolution procedure below: 1. Extend on a *coarse-grained* level. 2. Modify only by *adding* text.

We will look at a single module instance. For Estelle modules, the coarsest grain of behavioural description is the transition. So if we modify the behaviour, we work on the level of entire transitions only. Furthermore, we only add transitions. Because of this, every possible execution sequence in the hitherto existing module instance remains possible in the extended module instance[1]. The existing behaviour is not touched. (The justification for this proposition will be sketched in Section 3.) Of course there also may

[1]Under some premises which mainly concern the notational abbreviations that are possible in Estelle, and under the premise that there are no priorities defined among the transitions. The latter can be overcome easily by a straightforward extension of our resolution method below.

be cases where the existing behaviour needs to be changed. Then, we don't change the definition of the corresponding transition, but we disable[2] the old transition and add a new one. It is easier to detect problems resulting from a few such transition replacements than those from many small changes on the level of single Pascal-like statements inside the transition bodies. How the disabling of transitions is done will be described in Section 4.

3. Detection of Feature Interactions

We designed our stylistic rules in such a way that the only modification to the automaton which is the semantics of an Estelle module instance is the addition of further transitions (and states). Therefore, every existing execution sequence will remain possible for the extended automaton. It is safe to say that adding to the specification of a system will not lead to any interactions with existing features if the visible behaviour of the system is not changed in any way. Therefore, a change in the visible behaviour of the system is a necessary condition for feature interactions to occur. This leads us directly to the following necessary condition for feature interactions to occur:

Rule: A feature interaction may occur only if there is a (reachable) state in which at least two transitions belonging to different features can be fireable simultaneously, i.e. if there is non-determinism among different features.

For orthogonality, we also denote the basic service as a feature. Note that this formal criterion is a necessary condition for feature interactions to occur, but no sufficient condition. It may be possible that our detection rule points out a spot which proves to be harmless by manual inspection. The advantage of our approach is that it provides us with a relatively short list of critical spots. If we investigate each of them and resolve the non-determinism (as described below), then no unwanted feature interaction will be left.

It has to be stressed that only non-determinism between the transitions of *different* features is an indication of possible feature interactions. Non-determinism among the transitions of a single feature is a common means to express abstraction and has nothing to do with our topic. Therefore, we have to differentiate between these two kinds of non-determinism. This is done by an unambiguous association of each transition to one feature.

4. Resolution of Feature Interactions

One technique to resolve feature interactions is to define, for each pair of features, which one will take precedence over the other. This step can be seen as a high-level design decision that requires background knowledge about policies of the telephone administration, and therefore cannot be automated. The results of this step can be represented in the form of a precedence matrix, with an entry for each pair of features.

The resolution of feature interactions as defined by a precedence matrix can be incorporated into the Estelle specification in three steps:

Step 1. From the precedence matrix, derive a precedence function prec associating a precedence value with each feature such that f takes precedence over f' implies $\text{prec}(f) < \text{prec}(f')$[3].

Step 2. Associate each transition of the Estelle specification with one feature.

---

[2]To be exact, our method will leave the transition *enabled*, but not *fireable*.

[3]In Estelle, smaller values denote higher priorities. We will use the value of the precedence function as the Estelle priority below.

Step 3. For all transitions t: if t is associated with feature f, then add a priority clause "`priority` p_f", where $p_f = \text{prec}(f)$.

The association between transitions and features from Step 2 should also be done syntactically, according to the following provision: for each feature, one constant identifier is defined. In Step 3 we will need a priority constant for each feature anyway, so we will define one priority constant per feature. This way, the priority clause from Step 3 will provide us with the syntactical association between transitions and features.

A simple algorithm for the computation of the precedence function prec from a precedence matrix is presented in [BrGo94].

Sometimes, it is not sufficient to define a precedence between two conflicting features. It may be necessary to introduce completely new behaviour for this special case to handle the complex situation. The addition of this new behaviour is done in our approach formally like the addition of any other feature. This "feature" must be selected exactly if both of the conflicting features are selected, and it must have a (slightly) higher precedence than both of them.

We point out that our formal detection criterion is simple enough to be calculated by tools quickly even for large specifications; the same applies to the algorithm we use in the resolution step which computes priority clause constants automatically from the precedence matrix. Therefore, our approach scales well to the complexity of practical systems.

5. Future Work

The incorporation of further features into the Basic Call Service is currently under investigation. The refinement from the service to a protocol is an obviously promising step to go, introducing more structure into the system architecture. The capacity of our approach should be investigated with respect to the classes of further feature interaction classifications. Furthermore, we will continue the development of supporting tools.

Addition of features and thus extension of the Basic Call Service (BCS) has been achieved by adding specification text, i.e. on a purely syntactical level. We expect that it will be possible to develop a theoretical foundation together with suitable (syntactical) rules for incremental specialization such that the following holds: if the specification of the BCS augmented by additional features is an incremental specialization of the BCS, then it also extends the BCS semantically. Some work in this direction has been reported in [GoBo94], where semantical relations for specializing Estelle module definitions and a notion of incremental specialization based on these relations have been introduced.

References

[BrGo94] Bredereke, J. and Gotzhein, R. *A case study on specification, detection and resolution of IN feature interactions with Estelle.* Tech. Rep. 245/94, Univ. of Kaiserslautern, Dept. of Comp. Sce. (May 1994).

[DaNa93] Dahl, O. C. and Najm, E. *Specification & detection of IN service interference using LOTOS.* In Tenney, R. L. et al., editors, "Formal Description Techniques, VI", Boston, Mass. (26–29 Oct. 1993). North-Holland.

[GoBo94] Gotzhein, R. and von Bochmann, G. *Specialization in Estelle.* In Vuong, S. T. and Chanson, S. T., editors, "Proc. of PSTV, XIV", Vancouver, Canada (7–10 June 1994).

30

Application of Protocol Synthesis Technique to Resolution of the Service Interaction Problem

Yoshiaki KAKUDA†, Hiroyuki ASADA† and Tohru KIKUNO†

†Department of Information and Computer Sciences, Faculty of Engineering Science, Osaka University, 1-3 Machikaneyama-cho, Toyonaka-shi, Osaka 560, Japan

This paper proposes a method for resolving the service interaction problem in the Advanced Intelligent Network [1] [5]. Since this method is based on the protocol synthesis technique, the service interaction problem is resolved in the protocol specification rather than in the service specification. As a result, invalid interactions between communication services can be dynamically avoided.

1. Service Interactions and Specifications

1.1. Service Rules

Since a communication system consists of multiple users, a state of the communication system can be composed by a set of states of such users in the system. In the STR description method [2], a state of each user is represented by a collection of predicates. A service specification is described by service rules, each of which consists of (1) current states of users, (2) an event triggered by a user, and (3) next states of users. Two examples of service rules are shown in Figure 1.

1.2. Service Interactions

For a given state of a communication system (simply, system state) and service rules, if all predicates in current states of users are satisfied by the system state, then the event is said to be executable in the system state. If there exist two service rules including the same executable event in the given system state, then it is said that a service interaction occurs for these two service rules in the system state. If the next states in two service rules for which the service interaction occurs contradict with each other, then the service interaction is invalid and one of them should be chosen.

path(A,B), out-dialtone(C), m-cw(A)	··· current state
dial(C,A):	··· event
out-ringback(C,A), cw-ringing(A,C), path(A,B), m-cw(A)	··· next state

(a) Call Waiting (CW) Service Rule

out-dialtone(C), path(A,B), idle(D), m-cfv(A,D)	··· current state
dial(C,A):	··· event
out-ringing(D,C), out-ringback(C,D), path(A,B), m-cfv(A,D)	··· next state

(b) Call Forwarding (CFV) Service Rule

Figure 1. Examples of Service Rules

Suppose that seven predicates m-cw(A), m-cfv(A,D), path(A,B), path (B,A), out-dialtone(C), idle(D) are satisfied by a certain system state. Then, a service interaction occurs for the CW and CFV service rules in Figure 1 because these two rules include the same event dial(C,A) that is executable in the system state. Since out-ringback(C,A) in the next state of the CW service rule and out-ringback(C,D) in the next state of the CFV service rule contradict each other with single path connection, one of two dial(C,A)s should be chosen.

1.3. Protocol Specification

Service primitives, shortly primitives, are transmitted and received between users in an upper layer and processes in a lower layer through an interface called Service Access Point(SAP). The confirmation, indication, request and response (which are abbreviated as conf, ind, req and resp, respectively) are examples of primitives. A typical example of primitive sequence is req, ind, resp and conf. Messages for synchronization are transmitted and received between processes. A protocol specification is modeled by finite state machines(FSMs), which include sequences of primitives and messages.

2. Resolution of Service Interactions

In this paper, the service interaction problem is defined as follows:

Input ··· A service specification described by STR description method.

Output ··· A protocol specification with such capabilities that resolve the service interaction. For ease of understanding, the protocol specification is represented by sequence charts, which are equivalent to the FSMs.

The proposed method for resolution of the service interaction problem is as follows. First, for each communication service in a given service specification, a sequence of primitives is generated from the service rule. Next, a protocol specification that may include service interactions is derived from all the generated sequences of primitives. This derivation is done by the protocol synthesis technique [4] [6]. Finally, the service interaction is resolved by inserting conditional branch functions to the protocol specification.

As explained in Section 1, the service interaction can be interpreted as a non-deterministic behavior of executable events in the service specification. Since each event in the service specification is translated into a sequence of primitives in the protocol specification, the service interaction is regarded as a non-deterministic execution of primitive sequences. The first half of one primitive sequence includes that of the other, but the latter half is different with each other. Thus the service interaction problem can be basically resolved by inserting, to the protocol specification, conditional branch functions for determining which primitive sequences are selected. The details of this resolution method are shown in [3].

Figure 2 shows a sequence chart representing the protocol specification derived from CW and CFV services which cause service interaction. In this figure, black thick lines represent the sequence in the case that CW service is chosen, and gray thick lines represent the sequence in the case of CFV service.

3. Conclusion

This paper has proposed a method for resolution of the invalid service interaction problem, which is based on the protocol synthesis technique. In this method, non-deterministic

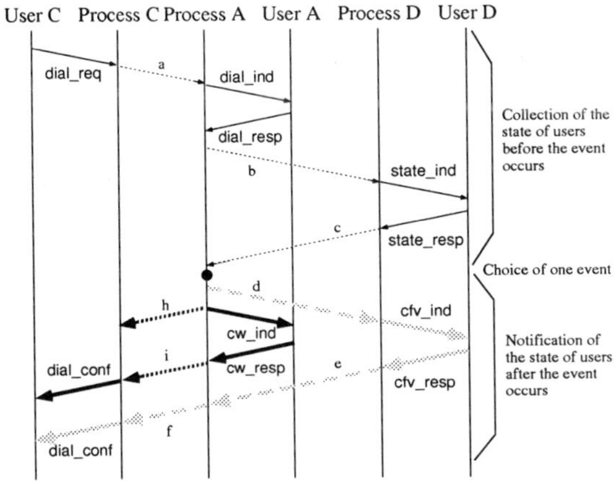

Figure 2. Sequence Chart Representing Protocol Specification Generated by Protocol Synthesis

events in service specification are transformed into deterministic sequences of service primitives with conditional branch functions in the protocol specification. In case of valid service interactions, multiple non-deterministic events only have to be executed simultaneously.

REFERENCES

1. N.Griffeth and Y.-J.Lin(eds.), Special issue on managing feature interactions in telecommunications systems, IEEE COMPUTER, Vol.26, No.8(Aug.1993).
2. Y.Hirakawa and T.Takenaka,"Tecommunication service description using state transition rules," Proc. Sixth Int'l Workshop on Software Specification and Design, pp.140-147(Oct.1991).
3. Y.Kakuda, H.Asada and T.Kikuno,"Protocol synthesis technique and its application to the feature interaction problem," Proc. ATR International Workshop on Communications Software Engineering(Oct.1994) to appear.
4. Y.Kakuda, M.Nakamura and T.Kikuno,"Automated synthesis of protocol specifications from service specifications with parallelly executable multiple primitives," IEICE Trans. on Fundamentals of Electronics, Communications and Computer Sciences(Oct.1994) to appear.
5. Y.-J.Lin and N.Griffeth(eds.), Special issue on managing feature interactions in telecommunications systems, IEEE COMMUNICATIONS MAGAZINE, Vol.31, No.8(Aug.1993).
6. K.Saleh,"Automatic synthesis of protocol specifications from service specifications," Proc. Int'l. Phoenix Conf. on Computers and Communications, pp.615-621(Mar.1991).

Invited Talk

31

The difference between a thing that can break and a thing that can't break is that when the thing that can't break breaks then it can't be fixed.
(Hitchhiker's Guide to the Galaxy, Book 5)

Proving the Value of Formal Methods

Gerard J. Holzmann

AT&T Bell Laboratories, Computer Science Research, Murray Hill, NJ 07974, USA.

The record of successful applications of formal verification techniques is slowly growing. Our ultimate aim, however, is not to perform small pilot projects that show that verification is sometimes feasible in an industrial setting; our aim must be to integrate verification techniques into the software design cycle as a non-negotiable part of quality control.
We take a snapshot of the state of the art in formal verification, and, for inspiration, compare it with other points in history where new technology threatened to replace old technology, only to discover how resilient an established practice can be. To keep a mild sense of perspective, we will also briefly consider how badly mistaken some of the advocates of new technology have sometimes been.

Index Terms: D.2, K.2, K.4

1. *INTRODUCTION*

Despite a number of projects that appear to have demonstrated successfully that the adoption of formal design techniques in a mainstream industrial environment is feasible and beneficial, e.g., [Ch91],[Ru92],[Ho94], in only few cases have such projects led to a permanent change in the design process that they targeted. How, then, can we prove the value of formal methods in such a way that also this second step is taken? To do so, we will have to accomplish two separate objectives. First, we will have to show in a rigorous scientific way (i.e., based on objective measurements and hard data) that the introduction of new formalism can indeed benefit a systems designer. Second, we have to show that adequate formalisms exist today that are ready to be applied in an industrial design environment. Both these objectives have so far proven to be hard to realize. In the following we will examine them in a little more detail.

1.1 *The First Objective: Rationale*

The first objective would appear to be the easier of the two. It is, however, much easier to reason abstractly about the rationale for formal methods, then it is to demonstrate its correctness in a rigorous scientific manner.

Clearly, the basic notion that a sound design has to be based on some type of formal reasoning is not very new. Epictetus (AD 50-c.130), for instance, used the following example in *The Discourses*, to make that point:

> When one of those who were present said, ''Persuade me that logic is necessary,'' he replied: ''Do you wish me to prove this to you?'' The answer was, ''Yes.'' . . . ''How then will you know if I am not cheating you by argument?''

Some form of logic inevitably underlies any design. The main purpose of a formal design method is not to introduce logic for its own sake, but to make the existing logic visible, so that its soundness can be checked. A formal design technique introduces the minimum of formalism that is sufficient to *expose* the logic of the reasoning process that underlies a design, and thus opens up this process to formal scrutiny.

But the above is only an abstract justification that appeals mostly to those of us who are already advocating the use of formal methods. It does not appeal directly to a practitioner, and it fails to provide objective data to support the thesis that formal methods are useful. If presented and justified only informally, the target users will perceive the added formality as, indeed, mere ''formality:'' extra *tasks* to be performed without visible benefit. John Chaves, a regular participant in formal methods projects at AT&T Bell Laboratories, observed that new technology has a much greater chance of being adopted if it is perceived as a *service*, with measurable benefits, rather than a *task*.

At the very minimum we must be able to make the formal method appear to be a service to the user, with a clear and immediate benefit. Even this is often much harder than it seems. It is the sentiment expressed by Dijkstra, when he wrote, in [Dij76]:

> I found this effort at clarification to my amazement (and annoyance) repeatedly rewarded by the accusation that ''I had made programming difficult.'' But the difficulty has always been there, and only by making it visible can we hope to become able to design programs with a high confidence level . . .

If we stay with the informal argument a little longer, we could say that the formal method is a *service* that secures the integrity of the design process, and perhaps more convincingly, the integrity of the designer. In return for satisfying the requirements of a formal method, the designer gains the capability to discover the mistakes that have slipped into the design, and gains the opportunity to repair them before they can tarnish his reputation. As Epictetus wrote, also in *The Discourses*:

> For if any one shall show [that a mistake was made], a man will himself withdraw from that which he does; but so long as you do not show this, do not be surprised if a man persists in his practice; for having the appearance of doing right, he does what he does.

This still does not, however, solve the problem in practice. Being very human, most designers (and researchers) tend to deny the fact that they can err. This brings us to our original objective: to show in a rigorous and scientific way that the introduction of a formal method is beneficial.

In the *NewCoRe* project, a pilot formal methods project we performed at AT&T between 1990 and 1992, our goal was to provide the much needed hard scientific data, [Ch1], [Ho94]. As part of this experiment, a single development project was pursued in two different ways, by two different groups in parallel. The first team used a traditional design method; the second (smaller) team applied formal design and verification techniques. Our aim was to compare the two parallel designs at the end of two years with conventional lab testing techniques, and then determine objectively which design constituted the better product.

The reality the NewCoRe team had to confront, however, was that as soon as it started discovering the first series of logical inconsistencies in the project requirements (after about six months), these errors could not reasonably be kept secret from the parallel team. Our wish for a double blind scientific experiment therefore had to be sacrificied to the realities of a competitive industry. The early success of the NewCoRe project in this case meant that it could no longer prove unambiguously and quantitatively exactly how much better a formal method can be. (Some more details on the project follow in section 2.3.)

1.2 *Opponents*

Formal methods are often opposed by practitioners on purely pragmatic grounds, cf. also [Ha90] and [BoHi94]. The following is from an opponent of formal methods, who shall remain anonymous:

> A formal design method is often overkill. Only a small portion of the errors found by an expensive formal design process would ultimately show up in practice.

And similarly, quoted verbatim from a cautious manager of a formal methods project:

> The paradigm of "zero faults the first time through" is often not as good as an iterative approach to development [i.e., based on testing].

Issues such as these can effectively block the permanent adoption of a formal method, and will need to be addressed seriously. Unfortunately, there are no one-sentence answers.

The first issue is one of economics: can an industry afford to build a higher quality product by retooling the design cycle for a formal method. Any change in an industrial design process involving more than a few engineers is costly, and will be weighed on its immediate merits. The chances of adoption are far greater if there is a measurable interval in which the investment pays for itself. There are no statistics on the types of faults that could be prevented from reaching a delivered product and causing damage, and thus it can be hard to make such a case.

The second issue may be easier to address, merely by properly defining the difference between traditional testing and a design method based on formal verification. Traditional product testing is based on the premise that the design starts with a trusted set of requirements, and is implemented via an untrusted process. Testing is meant to verify that the final product conforms to the original requirements. An example is a craftsperson building a cabinet. When the work is complete, (s)he will test that all the drawers open smoothly and close tightly, and that the cabinet doesn't wobble. There is never any question here that the original design plan for the cabinet as such is in order, and would lead to a proper product if implemented well.

In formal software verification a quite distinct problem is tackled. Unlike the cabinet-maker, we cannot trust our design requirements: they are often incomplete, ambiguous, are even self-contradictory. In a traditional design cycle one expects to discover the ambiguities in the original requirements with testing methods that are not meant, nor capable of, reliably exposing logical inconsistencies in the requirements themselves. In software verification, traditional testing offers too little and it offers it too late in the design cycle (i.e., after the developer has already committed to the implementation of potentially inconsistent requirements).

1.3 *The Second Objective: Methodology*

It is perhaps unfortunate that as long as computers have existed, there have been people who tried to convince their peers that they were powerful enough for deriving or proving significant facts. The reasoning is always persuasive, because in effect there exists no real alternative to mechanical verification. As noted in [ELWW72, p. 120, 143]:

> Perhaps the primary argument against the informal proof technique is that it is as fallible as the person carrying out the proof. . . .
> It is our feeling that a mechanical verifier is the only foolproof approach.
> . . . informal proofs of large programs cannot be handled reliably by a human being, and since the art of informal program proving is not likely to improve, mechanical (or mechanically assisted) verification is the only approach for large programs.

But placing too great a faith in the potential of mechanized verification makes an easy target for the sceptic, since there will always be facts that cannot be proven or disproven by any means, and many of them easily turn up in practice. Edsger Dijkstra wrote, almost twenty years ago [Dij75]:

> We see automatic theorem provers proving toy theorems, we see automatic program verifiers verifying toy programs and one observed the honest expectations that with faster machines with lots of concurrent processing, the life-size problems will come within reach as well.

It could as easily have been written today, even now that our computing machinery runs at least three orders of magnitude faster than in 1975.

It is probably fair to say that we have not yet been able to identify the subdomain where formal methods can be applied without second thoughts, routinely, and with predictable benefit. As Colin West noted in [We93]:

> . . . the FDT's that were available promised the ability to produce technically superior specifications, but failed to satisfy time, resource and ease-of-use constraints.

Overstating the case for formal methods, in this context, may do as much damage as underestimating it. As [BoHi94] phrased it:

> The excessive optimism of the attitude that everything important is provable helps to explain the excessive pessimism of the attitude that nothing important is provable.

Still, there is of course no need to be depressed about the future of our field. Especially in the last few years much progress has been made in the construction of efficient automated verification systems, and in their application to problems of practical significance (e.g., [Ch91], [Ho94]). Where only recently research focused on efficient methods for proving simple properties such as absence of deadlock, a state-of-the-art verification system today handles correctness requirements formalized in temporal logic, and uses sophisticated complexity management techniques to deliver optimal performance. Yet, it would be folly to believe that these tools can be applied blindly to any design problem and deliver instant rewards. System modeling, the formalization of correctness requirements, model verification, and, where necessary, model reduction, are skills that must be learned. Fortunately, many universities are leading the way with courses on formal verification and design techniques. The new skills will therefore inevitably reach their target.

2. *STATE OF THE ART*

For the purposes of this paper the discussion here is restricted to a small sub-field of the general area that is covered by formal methods: the application of automated verification techniques to the proof of logical correctness requirements for concurrent systems. Traditionally, then, there are two approaches to this problem, both first initiated in the early 1980s. The first is usually termed *model checking*, the second is known as *on-the-fly verification*.

Model checking was pioneered by Ed Clarke and his students at Carnegie-Mellon University, e.g., [ClEm81], and by Joseph Sifakis and his students at the University of Grenoble in France, e.g., [Qu82]. It is essentially a two-phase verification process. In the first phase, the reachable state space of a concurrent system is computed and the essential features of it are encoded into a data-structure that is stored in a disk-file, or held in main memory. In the second pass the validity of a correctness requirement, often formalized in a temporal logic formula, is verified for the given global state space structure.

On-the-fly verification has its roots in the mid-seventies, e.g. [Su75], but only began in earnest with the work of Colin West at IBM in the early eighties, e.g., [RuWe82]. The effort here was to prove every property of interest for a concurrent system in a single pass of the state space exploration algorithm. There were many parallel attempts to build verification systems with this property, e.g. [Ho81]. It was not until much later that the first on-the-fly verification systems were built that could verify also temporal logic requirements. As far as we know, the first efficient implementation of such a system was SPIN, which is detailed in [Ho91] and incorporates ideas on temporal logic verification from Costas Courcoubetis and Piere Wolper.

2.1 *Application Domains*

The model checking techniques have proven their strength primarily in their application to hardware circuit verification problems. On-the-fly verification techniques, on the other hand, have their strength primarily in software verification. There are several fundamental differences between these two areas of application, that have led to a quite distinct line of development in the area of model checking and on-the-fly methods. First, hardware systems are typically synchronous, and exhibit a regular, often repetitive, structure (e.g., counters, registers, arrays), that can be exploited successfully by customized verification algorithms. Software systems, such as communication protocols or distributed operating systems, are always (at least logically) asynchronous, not driven by a global clock, and they do not exhibit the same amount of regularity or structure.

The difference in application domain may explain why the application of symbolic data storage techniques, such as binary decision diagrams (BDDs), or symmetry reduction techniques have been more successful when applied in model checking algorithms than when they are applied in on-the-fly verification algorithms. Similarly, it can explain why the application of partial order theories could be shown to be beneficial in on-the-fly systems modeling asynchronous process interleavings, [HoPel94], but (as yet) not in model checking systems modeling synchronous hardware.

One of the advantages of the on-the-fly verification technique is that there is no strict need to construct an accurate global representation of the system behavior as a precondition for performing verifications. The focus of the research in these systems has therefore been to find efficient algorithms that construct the smallest possible portion of the global system that suffices to prove or disprove a temporal logic requirement. It is often sufficient in these cases to compute only an approximate representation of the global state graph, without sacrificing proof accuracy in any way (i.e., by applying coverage preserving reduction techniques).

2.2 *Changes in Limits*

Although the limits have been moved substantially in the last ten to fifteen years, computational complexity still lies at the very heart of the verification problem, and it is likely to remain there for the foreseeable future. Curiously, the improvements in cpu speed of the last decade (roughly three orders of magnitude) have not been matched by an equivalent increase of the available main memory size (roughly one order of magnitude less). The immediate implication of this phenomenon is that what used to be the main bottleneck preventing exhaustive verification of large systems (cpu-time) has disappeared, but has been replaced with another. Today, full-scale verification, both in model checkers and in on-the-fly systems, is primarily limited by the memory requirements of the algorithms. A 100 Mbyte state space can be exhaustively explored in minutes, but this much memory does not always suffice to perform complete verification of larger systems. The challenge in automated verification today, therefore, is to increase the memory efficiency of the search algorithms, and simultaneously to find effective ways to reduce the complexity of validation models *before* they are subjected to automated verification.

2.3 *Some Recent Application Successes*

We first started applying on-the-fly verification techniques to routine development projects at AT&T in 1988 [HoPa89]. Based on the positive results of this first application, it was decided to prepare a custom-made version of a verification system for the AT&T extension of the ITU-T (formerly CCITT) language SDL (Specification and Description Language), which was then the standard for describing the protocol aspects of the AT&T switching systems. This program, called SDLVALID, was completed in 1990, and first applied in the NewCoRe project, mentioned earlier at the end of section 1.1.

The five people who made up the NewCoRe team built a series of validation models for a new ISDN protocol that was in development, the largest model containing 7,500 lines of non-comment SDL source. A total of 145 correctness properties were formalized in temporal logic and verified with SDLVALID. Over a two-year period close to 10,000 verification runs were performed; an average of more than 100 verification runs per week.

The NewCoRe team uncovered a total of 112 serious design errors in the design requirements, by automated verification runs. These errors could be prevented from reaching the product, without requiring field-tests. The NewCoRe effort also successfully demonstrated that close to 55% of the original requirements for one of the designs covered were logically inconsistent, and thus could not have been realized adequately by any implementation. More details on the NewCoRe project can be found in [Ch91] and in [Ho94].

2.4 *Design for Verifiability*

The particular approach to formal design methods that we are pursuing is best summarized in the following principles.

1. A good design method should allow one to
   a. Discriminate between requirements and specifications
   b. Built engineering prototypes of a design
   c. Use these prototypes to predict the essential characteristics of the design
2. A good design method, therefore, minimally requires
   a. An unambiguous notation to specify system behavior
   b. An independent formalism to specify the logical requirements on that behavior
   c. A software tool to verify that the requirements are satisfied for the behavior as specified.

Very few of the existing formal methods standards can fully satisfy these requirements. The specification language SDL, for instance, has no facility for items 1a or 2b, and therefore in its basic form precludes items 1c or 2c. In our applications we therefore first extended the basic language to provide the minimal features necessary for a formal method in the above sense, cf. [HoPa89], [Ch91], [Ho94]. Similarly, few of the existing FDT's were developed with an eye towards the feasibility of 2c. The basic language definition of SDL, to remain with this example, stipulates that all message channels between processes are unbounded. This precise feature makes all correctness requirements formally undecidable, and therefore in our applications too we had to modify this definition and require the use of only explicitly bounded channels.

But modifying an existing FDT to allow for formal verification remains patchwork. Largely as an experiment, the validation modeling language PROMELA [H91] was designed explicitly with efficient verifiability in mind. This PROtocol MEta LAnguage is specifically not meant to represent implementation detail. Being experimental, it undeniably lacks many of the features that would be required for its use as a general FDT, and therefore it can not realistically compete with these. The main limitation of PROMELA, however, is also its strength: it tries to do only one thing (validation modeling) and do it well.

In this context then, the main purpose of a formal method is to enable (empower) the designer to represent design decisions conveniently, accurately, and unambiguously, and, most importantly to enable the designer to verify the logical consistency of those design decisions long *before* the implementation stage is reached. *Design for verifiability*, then, is the new paradigm of design that we hope the formal methods community can advance as the single most important alternative to current software engineering practice.

3. *HISTORICAL PERSPECTIVE*

It is tempting to think that in formal verification we are witnessing the birth of a new methodology of software design. So much of our environment relies on the correct working of software that it is almost inconceivable that we have managed to do without any type of formal method for so long. The very idea that algorithms or protocols can be endorsed by major standardization institutes without any form of formal verification may some day make an unbelievable anecdote of our time. Many of these standards are readily proven incomplete or faulty, simply by running any one of the readily available verifiers.*

We can safely leave it undecided here whether formal methods will ultimately replace the traditional style of software design, and for our own amusement merely assume that this will be so. Fortunately, it is much easier to look back in time than it is to look forward. We can therefore look at points in recent history where people rightfully or mistakenly believed they were in a similar situation. The most appealing examples, then, are at the innovations that we now accept as a quintessential part of our environment. None of them were unopposed, and in some cases the opposition against their adoption caused a delay that lasted for a generation or longer. As examples, we will look at the introduction of the first mechanical computing devices (generously called 'computers' here) by Charles Babbage in ca. 1825, the introduction of the electrical telegraph by Samuel Morse and others in ca. 1837, and the invention of the telephone by Alexander Graham Bell and others in ca. 1876. More details on most of these examples can be found in [HoPehr95].

3.1 *Computers*

One of the motivations for Charles Babbage's work on his analytical engine was to rule out the possibility of human error in repetitive computational tasks, such as the compilation of astronomical and logarithmical tables. Until (relatively speaking) fairly recently such tables were always produced by human computers, and inevitably they contained mistakes. This is how Babbage described how he got the idea to built his first engine, in his autobiography, see [Mo61, p. xiv]:

> Herschel brought in some calculations done by [human] computers for the Astronomical Society. In the course of their tedious checking Herschel and Babbage found a number of errors, and at one point Babbage said ''I wish to God these calculations had been executed by steam.''

When Babbage had succeeded in building his first set of mechanical devices that could perform these tasks mechanically, he had a very hard time convincing the official institutions, such as the Royal Astronomical Society, to change their custom and to use his machines instead of the traditional human computers. Babbage wrote, [Mo61, p. 69]:

> The great importance of having accurate tables is admitted by all who understand their uses; but the multitude of errors really occurring is comparatively little known. . . .
> . . . I have no intention of imputing the slightest blame to the Astronomer Royal, who, like other men, cannot avoid submitting to inevitable fate. The only circumstance which is really extraordinary is that, when it was demonstrated that all tables are capable of being computed by machinery, and even when a machine existed which computed certain tables, that the Astronomer Royal did not become the most enthusiastic supporter of an instrument which could render such invaluable service to his own science.

---

*) E.g., the SPIN system from [Ho91], which is available by anonymous FTP from the server `netlib.att.com` (directory `/netlib/spin`).

3.2 *Telegraphs*

The first patent on an electrical telegraph was issued on 12 June 1837 to William Cooke and Charles Wheatstone in England. It was titled: *Method of Giving Signals and Sounding Alarums at Distant Places by Means of Electric Currents Transmitted through Metallic Circuits.* But electrical telegraphs were being studied by many at this time, especially, of course, also by Samuel Morse. Morse gave his first public demonstration of a so-called ''pendulum'' telegraph at New York University on 3 September 1837, with Alfred Vail in the audience. Vail was a student at N.Y.U. at the time, and joined Morse as a collaborator after this demonstration. The Morse code, and the final version of the Morse–Vail telegraph both date from 1838, and unofficially the first message sent with this device by Vail, in Morse's workshop in Speedwell, New Jersey, was ''A patient waiter is no loser.''

Cooke and Wheatstone's device was based on the deflection of magnetic needles by an electric current. In retrospect, it is easy to see the advantages of the Morse–Vail system over this device: it was simpler, it (inadvertently at first) made the signals both audible and visible, and it could easily be modified to leave a permanent record of the transmissions on strips of paper. Nonetheless, while Cooke and Wheatstone's device found almost immediate practical application in 1837 on a stretch of railway between Euston and Camden Town in England, Morse was unable to convince anyone to use his device for almost 7 years.

In 1837 the United States Congress was considering a plan for the construction of a telegraphic connection between Baltimore and New Orleans, based on optical semaphores (then the standard method for long-distance communications, see [Ho&Pehr94]). Morse submitted a proposal to Congress to use his electrical telegraph instead. Congress, however, could not be persuaded that it was more attractive to use a technique that required not only the construction of hundreds of telegraph stations (as the optical semaphore also required) but also the stringing of expensive copper wires between them. So, Morse's proposal was rejected.

The next year, in 1838, Morse visited France to try his luck there. He demonstrated his telegraph to the French Academy of Sciences, and bid on a contract to install an electrical telegraph line along the railway from Paris to Saint-Germain. In February 1839 Morse left France again, unsuccessful also in this bid. The French government decided that the installation and exploitation of a telegraph network was too important to be trusted to a private individual (and a foreigner at that). It took until 1844 before one of Morse's proposals was finally accepted, and he could speak his famous official 'first' message ''What hath God wrought?'' at the inauguration of the first American electrical telegraph line connecting Washington to Baltimore (24 May 1844).

In France it took still a little longer to switch to electrical telegraphs. In this case it was a prime example of a country being disadvantaged by its leading role. France was the first country in the world to institute a State Telegraph. This happened on 26 July 1793 (not a typo). The telegraph that was used to construct the first ever nation-wide telegraph network was based on an optical semaphore-telegraph invented by Claude Chappe (see [Ho&Pehr95]). By 1840, this network was well established and counted nearly 500 stations, connecting all the main cities of France. The electrical telegraph, therefore, first and foremost had to prove its superiority over the ''traditional'' optical telegraphs. Dr. Jules Guyot took it upon himself to make a study the new devices, traveling to England to see them in operation first-hand. Upon his return to France he published an article in the *Courier Francais* that appeared 5 July 1841. It argued convincingly:

> Every sensible person will agree that a single man in a single day could, without interference, cut all the electrical wires terminating in Paris; it is obvious that a single man could sever, in ten places, in the course of a day, the electrical wires of a particular line of communication, without being stopped or even recognized.

He concluded:

> . . . it [the electrical telegraph] cannot seriously be considered suitable for the needs of the government, and the day is not far distant when this truth will be clearly demonstrated.

The first electrical telegraph lines were installed in France in 1845, but they were not Morse telegraphs. The device that was developed for the new lines, was designed by Alphons Foy, chief administrator of the optical telegraph network at the time, and Louis Breguet, and engineer. It carefully duplicated the positions of the optical Chappe semaphore on the controls of the electrical telegraph (backward compatibility). It took until 1855 before finally these devices were abandoned in favor of the simpler Morse telegraphs.

3.3 *Telephones*

By the 1870s the electrical telegraph was well established. A problem that had been signaled, and that was studied by many, was that a single telegraph wire could only be used for a single communication at a time. If a method could be devised to multiplex Morse signals for two or more connections onto a single wire, the telegraph business could save substantial amounts of money. There was a promise of great riches to the first person who could find a way to accomplish this.

Two of the people who were pursuing this problem were Elisha Gray and Alexander Graham Bell. Gray (1835–1901) was an established inventor with several patents in telegraphy to his name. In comparison, Bell (1847–1922) was an amateur, without formal training in electricity, and with only a scanty knowledge of telegraphy. Both Bell and Gray were working on prototype systems that could transmit a small number of frequencies over a telegraphic wire. Each frequency could, in principle, be used for a separate channel in a multiplexed link. Both Bell and Gray realized at one point or another that if you could transmit sufficiently many tones, you could also transmit sound. Here is for instance what a reporter wrote in The New York Times, May 1874, after witnessing one of the public demonstrations given by Elisha Gray, as quoted in [Ho81]:

> . . . in time [telegraph] operators will transmit the sound of their own voices over the wire, and talk with one another instead of telegraphing.

Gray consider this possibility an amusing side-application, at best, well divorced from the real money-making application: a multiplexing telegraph. Bell, on the other hand, decided that the transmission of sound or voices was the more seminal step, and he dropped the pursuit of a multiplexing telegraph entirely. Gray, of course, couldn't be happier. In a letter to his patent attorney, Elisha Gray wrote in 1875:

> Bell seems to be spending all his energies in [the] talking telegraph. While this is very interesting scientifically it has no commercial value at present, for [the telegraph industry] can do more business over a line by methods already in use than by that system.

A few months later, Gray nonetheless let himself be persuaded by his attorney to file a patent claim on his idea As best as can be determined by coincidence, that same day Bell also filed a patent claim for the transmission of sound over telegraph wires, titled *Improvements in Telegraphy.* After the interference on the claims was discovered, Gray readily granted precedence to Bell, stating in a letter to Bell that though he had had the same idea, he had never attempted to reduce it to practice, and therefore could not rightfully lay claim to the invention. This may seem like a giant mistake on the part of Gray, and we can safely assume that his attorney had recommended against this admission, but things looked very different at this time. The odds seemed to be very much in Gray's favor.

When, for instance, Bell offered his telephone patent to the telegraph company Western Union in

1876 for $100,000 he was turned down. Instead, Western Union bought the rights to Elisha Gray's patents on multiplexing telegraphs. The telegraph industry, like Gray and many others, could not see what possible purpose there could be in having telegraph operators talk to each other over the wires. As we know, Bell went in business for himself and could prove the telegraph industry wrong in due time.

3.4 *Windmills*

The remarkable ability of the human mind to resist technological improvements is, of course, not limited to recent history. For proper measure, we therefore look at one more example, from a few hundred years earlier: improvements in windmill design.

In 1593 the Dutch farmer Cornelis Corneliszn invented a new transmission and gear-mechanism for windmills. The mechanism, the reverse of a modern crankshaft, could translate the rotation of a windmill's main axle efficiently into a lateral movement that could drive, for instance, a powerful saw. Corneliszn received a patent on 15 December 1593, that protected his invention for a period of twelve years. The innovation led to a small revolution in the way that a saw mill was organized, mechanizing much of the duties that could only be performed by hand, and at a much slower pace, until then. The improvement was also timed well, with the great demand for wooden ships that was triggered by the exploits of the East and West Indian Companies in Holland in the beginning of the 17th century.

By 1630 there were 368 conventional mills in Holland (used for milling grain etc.), and 86 wind-driven sawmills. The 368 conventional mills were spread evenly over all districts, including the capital city Amsterdam. But, curiously, not a single wind-driven sawmill could be found inside the gates of Amsterdam. The reason? The Union of Wood Millers successfully fought the mechanization of saw mills, considering it a serious threat to their profession. Although they couldn't stop their construction elsewhere, the union had succeeded in banning the new mills from Amsterdam entirely.

It took almost a century for the ban to be broken. By 1730, finally, there were 78 wind-driven sawmills in Amsterdam, more than in any other city in Holland.

4. *IN CONCLUSION*

With the benefit of hindsight it is of course all too easy to say who was right and who was wrong to reject a new idea. When the events evolve in real-time, the issues are often much less clear. As just one example of someone who was rightfully denied the chance to realize his vision, let use consider the inventor of a telephone predating the ones of Bell and Gray by 80 years: the German professor Gottfried Huth (1763–1818). In 1796 Huth published a translation of a work on acoustics from 1763 by the French physicist J. H. Lambert. Huth gave his book the title *A treatise concerning some acoustic instruments and the use of the speaking tube in telegraphy.* In his own additions to the translation, Huth criticized the Chappe optical telegraphs used in France at this time for being obviously flawed, since they could not be used in bad weather or at night. Huth improvement was to construct telegraph towers at roughly 5km intervals, on which men were positioned equipped with megaphones (''speaking tubes,''). These operators could shout messages to each other, be it sunshine or rain, day or night. Realizing the seminal importance of his idea, Huth knew he needed an original name for his invention, to set it apart from all others. He wrote:

> The fundamental difference therefore deserves, and will ultimately require, a different name for the telegraph based on speaking tubes. But what name would be more appropriate than the word derived from the Greek: ''Telephon,'' or far-speaker [German: *Fernsprecher*].

As we know, Huth did not get his way. Many others shared this fate with him. One of them,

Johann Bergstrasser (1732–1788), started publishing his first works on telegraphy in 1785. He was never very specific in the type of device he had in mind, although he clearly wanted to distinguish his work from his competitors:

> I do not use speaking tubes, and even less electricity or magnets; also no wires or tubes, like Mr. Gauthen [Gauthey]. And still, part of my apparatus can be placed in a closed room, anywhere in a large palace, wherever desired; . . . without it being necessary that the correspondents can see each other, or perform any of the operations themselves . . .

and elsewhere:

> I am completely convinced of an inexpensive construction, and of the feasibility of the scheme. Experiments cannot cost much; they can, however, only be performed in the field; such as the land of a wealthy nobleman who can spend a few thousand thaler on a useful speculation.

When Bergstrasser failed to convince any wealthy noblemen, he became increasingly annoyed at those who were more successful. When, for instance, Claude Chappe had successfully obtained support for the construction of the first optical telegraph lines in France, in 1793, Bergstrasser wrote in anger:

> That my proposal to connect Berlin via Leipzig, Magdeburg, Hanover to Hamburg has never been executed, like the proposal of the gentlemen Chappe to connect Paris to Lille, is not the fault of my method.

Some were even less gracious. When the British lieutenant John Macdonald failed, after 30 years of trying, to obtain support from the British Admiralty for his plans to improve the signaling codes on semaphore telegraphs, he wrote in desperation (in 1819):

> If an angel from heaven were to offer an unexceptionable plan, he would be ordered to walk, or fly, off with it, if he had not an article to offer which is not quite fit for angels to deal in, viz. Parliamentary interest. . . .
> Let it be recollected that a less simple plan, the establishment of mail coaches, was at the time deemed impractical and visionary. Man is a progressive animal; and it is the Press only that points out to him what he ought to do; and assuredly none but drivelling idiots, will oppose the march of knowledge. . .

How Best to Be Wrong

In rare cases even a potentially risky prediction about the potential of a new technology can prove to be a majestic underestimation. Witness this historic statement about the potential of electronic computers, made 45 years ago, in the March 1949 issue of *Popular Mechanics*:

> Where a calculator on the ENIAC is equipped with 18,000 vacuum tubes and weighs 30 tons, computers of the future may have only 1,000 vacuum tubes and perhaps weigh 1.5 tons.'

Wouldn't it be nice if our guess about the potential of formal methods proves to be equally wrong?

REFERENCES

[BoHi94] J. P. Bowen, and M. G. Hinchey, *Seven more myths of formal methods*. Oxford University Computing Laboratory, Rep. PRG-TR-7-94, June 1994, 19 pgs

[Ch91] J. Chaves, *Formal methods at AT&T, an industrial usage report*. Proc. 4th FORTE Conference, Sydney, Australia, 1991, pp. 83–90.

[ClEm81] E. M. Clarke and E. A. Emerson, *Characterizing properties of parallel programs as fixpoints*. 7th Int. Coll. on Automata, Languages and Programming. LNCS 85, 1981.

[Dij75] E. W. Dijkstra, *Correctness concerns and, among other things, why they are resented*. Sigplan Notices, Vol. 10, No. 6, 1975, pp. 546-550.

[Dij76] E. W. Dijkstra, *A Discipline of Programming*. Prentice Hall, Englewood Cliffs, NJ, 1976.

[ELWW72] B. Elspas, K. N. Levitt, R. J. Waldinger, and A. Waksman, *An assessment of techniques for proving program correctness*. Computing Surveys, Vol. 4, No. 2, June 1972, pp. 97–147.

[Ha90] J. A. Hall, *Seven myths of formal methods*. IEEE Software, 7(5):11, 19, Sept. 1990.

[Ho81] G. J. Holzmann, *A theory for protocol validation*. Proc. 1st IFIP/INWG Symp. on Protocol Specification, Testing, and Verification. NPL, Teddington, England, 1981. Reprinted in IEEE Trans. on Computers, Vol C-31, No. 8, pp. 730–738.

[HoPa89] G. J. Holzmann and J. Patti, *Validating SDL specifications: an experiment*. Proc. 9th IFIP WG 6.1 Int. Workshop on Protocol Specification, Testing, and Verification, Twente, Netherlands. North-Holland Publ., Amsterdam, 1989.

[Ho91] G. J. Holzmann, *Design and validation of computer protocols*. Prentice Hall, Englewood Cliffs, NJ, 1991.

[Ho94] G. J. Holzmann, *The theory and practice of a formal method: NewCoRe*. Proc. 13th IFIP World Computer Congress, August 2 - September 2 1994, Hamburg, Germany.

[HoPel94] G. J. Holzmann and D. Peled, *An Improvement in Formal Verification*. This conference: Proc. 7th FORTE Conference, Bern, Sw., 1994.

[HoPehr95] G. J. Holzmann and B. Pehrson, *The Early History of Data Networks*. IEEE Computer Society Press, Los Alamitos, California, 1995, ISBN 0-8186-6782-6.

[Hou81] D. Hounshell, *Two paths to the telephone*. Scientific American, Jan. 1981, pp. 156–163.

[Mo61] P. and E. Morrison, *Charles Babbage and his calculating engines*. New York, Dover, 1961.

[Qu82] J. P. Queille, *Le système César: description, spécification et analyse des applications réparties*. Ph.D. Thesis, June 1982, Computer Science Dept., University of Grenoble, France.

[RuWe82] J. Rubin, J. and C. H. West, *An improved protocol validation technique*. Computer Networks, Vol. 6, Nr. 2, 1982, pp. 65-74.

[Ru92] H. Rudin, *Protocol development success stories: part I*. Proc. IFIP/PSTV Conference, Orlando, Fl., North-Holland, 1992, pp. 149–160.

[Su75] C. A. Sunshine, *Interprocess Communication Protocols for Computer Networks*. Ph.D. thesis 1975, Dept. of Computer Science, Stanford Univ., Stanford, CA.

[We93] C. H. West, *The challenges facing formal description techniques*. Proc. 6th FORTE Conference, Boston, Mass., 1993.

PART SEVEN

Testing and Performance

32

A Structural Analysis Approach to the Evaluation of Fault Coverage for Protocol Conformance Testing*

Mingyu Yao, Alexandre Petrenko and Gregor v. Bochmann

Département d'informatique et de recherche opérationnelle
Université de Montréal, CP. 6128, Succ. Centre-Ville, Montréal (Québec), Canada H3C 3J7

In this paper, we propose a structural analysis approach to the evaluation of fault coverage of protocol conformance testing based on the finite state machine model. The attractiveness of this approach is its low computational complexity. It allows us to calculate the fault coverage of a given test suite by directly analyzing the test suite against the specification machine. Therefore, it avoids the generation and execution of mutants. The approach has been implemented and a number of experiments has been carried out. Some of the experimental results are summarized in this paper to show the accuracy of this approach compared with the mutation analysis technique.

Keywords: I.3, I.6, III,1, IV.3, IV.4.

1. INTRODUCTION

The finite state machine (FSM) model has been widely used in the development of hardware and software systems. Especially in the recent years, it has been extensively used in the conformance testing of communication protocols. Quite a number of methods have been proposed in the literature for generating test suites from protocol specifications given in the form of FSMs. An important issue related to the test suite generation is the *effectiveness* of a test suite which depends primarily on its fault coverage, i.e., its capability of detecting faults in a potential implementation. Approaches based on Monte-Carlo simulation have been proposed in the literature to evaluate the fault coverage of a test suite generated from a given FSM specification [DaSa 88, DDB 91, SiLe 89 and MCS 93]. These approaches require a (large) number of mutants of the given specification machine to be enumerated and executed against the test suite. In our recent work [YPB 94], a different procedure has been developed which, without the need of explicitly generating and then executing a certain (and often large) number of mutant machines, can decide if the given test suite provides full fault coverage (i.e., if it can detect all bad implementation machines). However, this approach does not provide a numeric measure for a test suite which does not have full fault coverage. Consequently, it is impossible to use this approach to compare the fault coverage of two test suites, if none of them provides full fault coverage. In this paper, we will propose a structural analysis approach to evaluate the fault coverage of a given test suite. The basic idea of this approach is, by analyzing the given test suite against the specification machine, to make an estimation on the number of mutants representing potential bad implementations that can be detected by the test suite without the need to generate and execute the mutants.

The rest of the paper is organized as follows. In Section 2, the FSM model is introduced and a framework of software testing based on this model is presented. The structural analysis

---

* This research was supported by the IDACOM-NSERC-CWARC Industrial Research Chair on Communication Protocols at University of Montreal.

approach to the evaluation of fault coverage of a test suite is developed in Section 3. The "Order Coverage", which is based on the structural analysis approach and proposed to deal with large complex specification machines, is then presented in Section 4. Finally, in Section 5, our approach is compared with other related work.

2. A FRAMEWORK OF TESTING BASED ON THE FSM MODEL

We will first introduce the finite state machine model and then present a software testing framework based on this model.

2.1 The FSM Model

A *finite state machine* (FSM), often simply called a *machine* throughout this paper, is essentially an *initialized Mealy machine* which can be formally defined as follows.

Definition 2.1 (finite state machine)

A finite state machine is a 7-tuple $<S, X, Y, S_1, \delta, \lambda, D>$, where
S is a set of n states $\{S_1, S_2, ..., S_n\}$ with S_1 as the initial state;
X is a finite set of input symbols;
Y is a finite set of output symbols;
D is a specification domain which is a subset of S x X;
δ is a transfer function δ: D --> S;
λ is an output function λ: D --> Y. ■

An FSM is said to be *completely specified* (*defined*), iff D = S x X. Otherwise it is said to be *partially* or *incompletely specified* (*defined*). Since δ and λ are required to be functions, this FSM model is *deterministic*. That is, for each $(S_i, x) \in D$, there should be exactly one state $S_j \in S$ and exactly one output symbol $y \in Y$ such that $\delta(S_i, x) = S_j$ and $\lambda(S_i, x) = y$. In this case, we say there is a transition leading from state S_i to S_j with input x and output y. Such a transition is usually written as S_i -x/y-> S_j, or as a triplet < S_i; x/y; S_j >. S_i is said to be the *head* or *starting* state of the transition, while S_j is said to be the *tail* or *ending* state of the transition. An FSM can be given in a graph form, with the states and transitions of the FSM represented by the vertices and arcs of the graph, respectively. As an example, Figure 1 gives a FSM which is partially specified since, at state S_3, no transition is specified for input symbol 1.

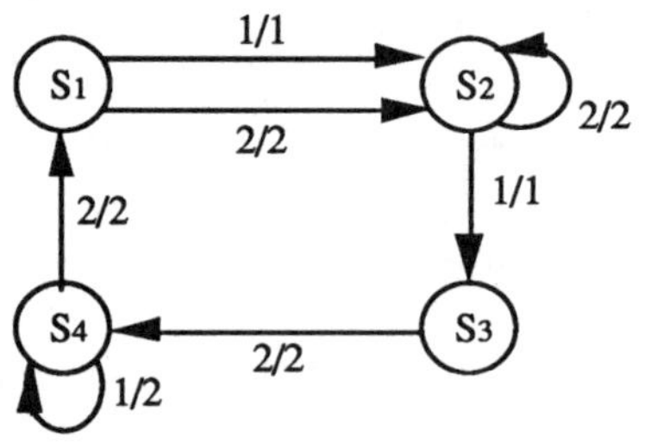

S = { S1, S2, S3, S4 }
X = { 1, 2 }
Y = { 1, 2 }
Initial state is S 1

Figure 1: An example FSM

The following notations will be used throughout the paper. For a given symbol set Z, Z^* is used to represent the set of words constructed on Z and "ε" to represent the *empty* word, i.e., the word consisting of no symbols. Also, the dot "$\cdot$" is used to represent the concatenation operation of two words. However, this dot symbol is often omitted when no ambiguity arises. Furthermore, $|Z|$ is used to represent the cardinality of Z.

Definition 2.2 (defined input sequence)
Let $p = x_1x_2 \cdots x_k \in X^*$. p is called a *defined* input sequence for state $S_i \in S$, if there exist k states $S_{i1}, S_{i2}, ..., S_{ik} \in S$ and an output sequence $q = y_1y_2...y_k \in Y^*$ such that there is a sequence of transitions

$$S_i \text{ -}x_1/y_1\text{-> } S_{i1} \text{ -}x_2/y_2\text{-> } S_{i2} \text{ --> ... --> } S_{ik-1} \text{ -}x_k/y_k\text{->} S_{ik} \quad \textbf{(2-1)}$$

in the finite state machine. ■

We use $\psi(S_i)$ to denote the set of all the defined input sequences for state S_i. A sequence of transitions such as (2-1) can be abbreviated as S_i -p/q-> S_{ik}, which, when we do not care about the output sequence q, can be further simplified as S_i -p-> S_{ik}, with the meaning that the FSM, when in state S_i and given an input sequence p, will enter state S_{ik}. The definitions of the transfer function δ and output function λ can be naturally extended to apply not only to single inputs, but also to sequences of inputs.

Definition 2.3 (extensions of transfer and output functions to input sequences)
Let $p = x_1x_2...x_k \in \psi(S_i)$ and ε be the empty word. Then,

$\delta(S_i, \varepsilon) = S_i, \quad \delta(S_i, p) = \delta(\delta(S_i, p'), x_k)$

$\lambda(S_i, \varepsilon) = \varepsilon, \quad \lambda(S_i, p) = \lambda(S_i, p').\lambda(\delta(S_i, p'), x_k)$

where $p' = x_1x_2...x_{k-1}$. ■

Definition 2.4 (compatible states and distinct states)
We say that S_i and S_j are *compatible* states if for $\forall\ p \in \psi(S_i) \cap \psi(S_j)$, $\lambda_s(S_i, p) = \lambda_s(S_j, p)$. Otherwise, they are called *distinct* states. ■

According to the above definition, if $\psi(S_i) \cap \psi(S_j) = \phi$, then S_i is compatible with S_j. If the FSM happens to be completely specified, then the definition of compatible states given above reduces to the definition of *equivalent states* as found in the literature (see for example, [Gill 62, Koha 78]).

Definition 2.5 (reduced machine)
A FSM is said to be *reduced* if and only if no two states are compatible. ■

It is easy to verify that the FSM given in Figure 1 is reduced.

Definition 2.6 (reachable state and strongly connected FSM)
A state S_i is said to be *reachable* (from the initial state S_1) if there exists an input sequence $p \in \psi(S_i)$ such that S_1 -p->S_i. A machine is said to be *initially connected* if all the states are reachable. ■

All the states of the FSM in Figure 1 can be reached from the initial state and therefore this example FSM is initially connected.

Definition 2.7 (mutant machine)
Let M_1 and M_2 be two given FSMs. M_2 is said to be a *mutant* machine of M_1 if M_2 is obtained by applying to M_1 each of the following four types of operations, in any order, for a certain number of times (including zero times):
Type 1: change the tail state of a transition;
Type 2: change the output of a transition;
Type 3: add a transition; and
Type 4: add an extra state. ■

The following corollary follows directly from the above definition.

Corollary 2.8
A machine is a mutant machine of itself. ■

2.2 A Testing Framework Based on FSM Model

The FSM model was widely used in traditional hardware testing. In recent years, this model has also received much attention in the testing of certain software systems such as communication protocols [PBD 93] and object-oriented programs [HoSt 93, TuRo 92]. Testing based on the FSM model can be formalized as the problem of *testing a FSM implementation*[Ural 91]: given a FSM representation (specification) of a system (denoted henceforth as M_S) and an implementation of the system (denoted henceforth as M_I), we are required to determine if the implementation machine M_I *conforms to* (i.e., is *correct* with respect to) the specification machine M_S by testing M_I as a black-box. This implies that we should generate from M_S a set of input sequences, called a test suite, and the corresponding set of expected output sequences such that M_I conforms to M_S if and only if, when the input sequences in the test suite are applied to M_I, the observed output sequences from M_I are the same as the corresponding expected output sequences. As already pointed out in the literature [Moor 56, Gill 62, YPB 93a, YPB 93b], this problem is not solvable unless it is dealt within a restricted framework. Therefore, some assumptions should be made about the specification machine M_S and the implementation machine M_I. Firstly, the restrictions on the specification machine are summarized in the first assumption.

Assumption 1: (reduced and initially connected specification machine)
The given specification machine M_S is reduced and initially connected. ■
Secondly, testing based on the FSM model is essentially a mutation testing. Therefore, for the given specification machine M_S, an implementation machine M_I is actually a mutant machine (of M_S) obtained from M_S by applying each of the four types of operations listed in Definition 2.7 for a number of times (including zero times). These four types of operations represent the basic types of changes that can be made during the implementation of M_S. However, it should be noted that, in practice, the implementation machine M_I is normally completely defined even though the given specification machine M_S is often only partially specified. Therefore, the following assumption is made throughout this paper.

Assumption 2: (completeness of an implementation machine)
For the given specification machine M_S, an implementation machine M_I is a completely defined mutant machine of M_S. ■
Thirdly, if the number of changes of Type 4 applied to the given specification machine M_S is not limited, the number of mutants of M_S will be infinite and the problem of testing will become intractable. Therefore, in practice, the number of changes of Type 4 is always limited to an upper bound. Throughout this paper, we simply do not allow any change of Type 4 as stated in the next assumption.

Assumption 3: (limited number of states in an implementation machine)
For the given specification machine M_S, any operation of Type 4 is not allowed and therefore the number of states in an implementation machine M_I will not exceed that of M_S. ■

We also note here that additional types of changes, such as changes of inputs, changes of head states and missing states, may be introduced [MiPa 92]. However, these types of changes are not necessary for our discussion as the FSM model is deterministic (Definition 2.1) and implementation machines are assumed to be completely defined (Assumption 2). The following example explains that the same consequences of a missing state can be achieved by changing the tail states of certain transitions (Type 1). Figure 2 (a) shows that, when implementing the FSM given in Figure 1, state S_4 is not implemented (i.e., missing in the implementation) and the transition < S_3; 2/2; S_4 > is changed to < S_3; 2/2; S_1 >. However, we can still think that state S_4 is present in the implementation as shown in Figure 2 (b). The reason is that S_4 is no longer reachable and therefore, during the black-box testing, whether S_4 is present or missing in the implementation makes no difference.

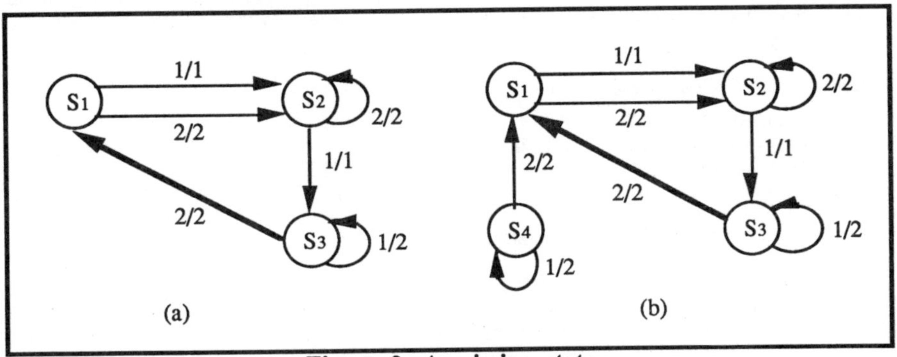

Figure 2: A missing state

Therefore, as a matter of fact, all the n states $S_1, S_2, ..., S_n$ of the specification machine M_S are assumed to be present in an implementation machine M_I. However, some of these states may become unreachable in M_I due to the changes of Type 1 introduced during the implementation. Confusion may arise because the same state names $S_1, S_2, ..., S_n$ are used for both M_S and M_I. It is therefore often helpful, although not necessary, to make things clear by renaming the states $S_1, S_2, ..., S_n$ in M_I to $I_1, I_2, ..., I_n$, respectively. Then without loosing generality, let

$M_S = < \{S_1, S_2, ..., S_n\}, X, Y, S_1, \delta_S, \lambda_S, D_S >$, and

$M_I = < \{I_1, I_2, ..., I_n\}, X, Y, I_1, \delta_I, \lambda_I, D_I >$.

Since M_I is supposed to be completely defined, we know that $D_I = \{I_1, I_2, ..., I_n\} \times X$ and therefore $\psi(I_i) = X^*$ and $\psi(S_j) \subseteq \psi(I_i)$, for any I_i and S_j. Now, we need to introduce some important concepts. The first concept required is the so-called *conformance relation* which essentially defines when M_I is a correct implementation of M_S. This concept is defined through the following two definitions.

Definition 2.9 (equivalence of states in respect to a set of input sequences)

Let I_i be a state of M_I and S_j a state of M_S. V is a set of input sequences such that $V \subseteq \psi(S_j)$. Then

$I_i \cong_V S_j$ if $\lambda_I(I_i, p) = \lambda_S(S_j, p)$, for $\forall\, p \in V$. ■

Definition 2.10 (conformance relation)

M_I conforms to M_S, written M_I **CONF** M_S, if and only if $I_1 \cong_{\psi(S_1)} S_1$, where I_1 and S_1 are the initial states of M_I and M_S, respectively. ■

The above defined conformance relation corresponds to the notion of weak conformance [SaDa 88, SiLe 89 and MiPa 92]. The relationship between the above defined conformance relation and the types of operations listed in Definition 2.7 is established by the following lemma of which the proof is similar to that of Lemma A.1 given in the appendix of [YPB 93a].

Lemma 2.11
For the specification machine M_S and implementation. Then M_I conforms to M_S if and only if there exists a mapping **f**: $\{S_1, S_2, ..., S_n\}$ -> $\{I_1, I_2, ..., I_n\}$, such that
(1) **f** is one-to-one; and
(2) If S_i - x/y -> S_j is in M_S, then I_k - x/y -> I_ℓ is in M_I, where I_k = **f**(S_i) and I_ℓ = **f**(S_j). ■

Since the implementation machine M_I is treated as a black-box, test cases should be generated from the specification machine M_S. The following two definitions formally defines the concepts of test case and test suite.

Definition 2.12 (test case)
A test case is a sequence of inputs which should be of finite length and in $\psi(S_1)$. ■
As is clear from the above definition, a test case always starts from the initial state S_1 of the specification machine M_S. Accordingly, each test case should be applied to the implementation machine M_I when it is in its initial state I_1. Therefore, an important assumption in the testing based on the FSM model is about the availability of the so-called *reliable reset* function and is summarized as our fourth (and final) assumption.

Assumption 4: (availability of reliable reset)
The *reliable reset* is an operation that, when activated, will bring the implementation from any other state back into its initial state. It is assumed to be available in an implementation under test. ■
A special input symbol "r" representing the invocation of the reset operation is added to the beginning of each test case.

Definition 2.13 (test suite)
A test suite is a finite set of test cases. ■

TS = { r.1.1.2, r.2.1.2.1, r.2.2.1 }

(a)

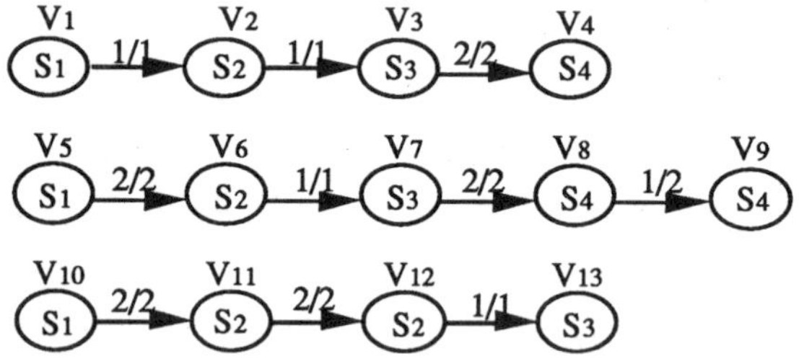

(b)

Figure 3: A test suite generated from the example FSM

Figure 3 (a) lists the test cases of a test suite generated from the machine shown in Figure 1. Each of the test cases is prefixed by the reset symbol "r". Applying the three test cases to the initial state S_1 results in the three sequences of transitions shown in Figure 3 (b) that will be executed.

Definition 2.14 (to pass a test suite)
Let TS be a test suite and $p \in$ TS be a test case. We say that a given implementation M_I passes the test case p, written M_I **pass** p, iff $\lambda_I(I_1, p) = \lambda_S(S_1, p)$. Further, we say that M_I passes the test suite TS, written M_I **pass** TS, iff M_I **pass** p, for $\forall\, p \in$ TS. ■

An implementation machine which cannot pass a given test suite is said to *fail* the test suite or to be *detected* by the test suite.

Let **Impl**(M_S) represent the set of all the implementation machines of M_S, i.e., all the completely defined mutant machines with same number of states as M_S. Then the following lemma follows directly from the above definitions.

Lemma 2.15
Let TS be a given test suite for M_S and $M_I \in$ **Impl**(M_S). M_I does not pass TS implies that M_I does not conform to M_S. ■

3. A STRUCTURAL ANALYSIS APPROACH TO EVALUATING FAULT COVERAGE

In this section, we are going to present a structural analysis approach to the evaluation of fault coverage of a test suite in respect to a given specification machine. This approach avoids the necessity of explicit generation and execution of mutant machines representing possible implementations of the given specification machine M_S. First of all, let us introduce the following notations:

N_1(M_S) - the total number of machines in **Impl**(M_S);
N_2(M_S) - the number of machines in **Impl**(M_S) which conform to M_S;
N_3(M_S) - the number of machines in **Impl**(M_S) which do not conform to M_S;
N_4(M_S, TS) - the number of machines in **Impl**(M_S) which can pass the given test suite TS;
N_5(M_S, TS) - the number of machines in **Impl**(M_S) which do not conform to M_S and cannot pass the given test suite TS;

We have the following lemma whose validity is obvious (see [Gill 62, SiLe 89]).

Lemma 3.1
The total number of implementation machines, that is the cardinality of **Impl**(M_S), is

$$N_1(M_S) = |\mathbf{Impl}(M_S)| = (n|Y|)^{n|X|} \qquad \textbf{(3-1)}$$

Among these implementation machines, there are

$$N_2(M_S) = (n-1)!\,(n|Y|)^{n|X|-|D_S|} \qquad \textbf{(3-2)}$$

implementation machines which conform to M_S, where n, X, Y and D_S are the number of states, the input set, the output set and the specification domain of M_S, respectively. ■

It is also easy to see that **N_3**(M_S) = **N_1**(M_S) - **N_2**(M_S) and **N_5**(M_S, TS) = **N_1**(M_S) - **N_4**(M_S, TS).
Therefore, we can give the following so-called *precise fault coverage* of a given test suite TS in respect to a given specification machine M_S.

Definition 3.2 (precise fault coverage)
The precise fault coverage of a test suite TS in respect to M_S, denoted as $\mathbf{FC_p}(M_S, TS)$, is

$$\mathbf{FC_p}(M_S, TS) = \frac{N_5(M_S, TS)}{N_3(M_S)} = \frac{N_1(M_S) - N_4(M_S, TS)}{N_1(M_S) - N_2(M_S, TS)}$$ ■

As is clear from the above definition, in order to calculate the fault coverage, we still need to find $\mathbf{N_4}(M_S, TS)$, the number of machines in $\mathbf{Impl}(M_S)$ which can pass the given test suite TS. The exact value of $\mathbf{N_4}(M_S, TS)$ is in general too difficult to find without using the exhaustive mutation analysis technique. However, as we have already mentioned, the exhaustive analysis technique is often not feasible in practice due to the high cost. Therefore, in our approach, we will use an estimated value, denoted as $\mathbf{N_4}(M_S, TS)$, of $\mathbf{N_4}(M_S, TS)$. Substituting $\bar{\mathbf{N}}_4(M_S, TS)$ for $\mathbf{N_4}(M_S, TS)$ in the calculation of the fault coverage as defined in Definition 3.2 results in the following estimated fault coverage.

Definition 3.3 (estimated fault coverage)
The estimated fault coverage of a test suite TS in respect to M_S, denoted as $\mathbf{FC_e}(M_S, TS)$, is

$$\mathbf{FC_e}(M_S, TS) = \frac{N_1(M_S) - \bar{N}_4(M_S, TS)}{N_1(M_S) - N_2(M_S, TS)}$$ ■

Definition 3.4 (prefix set of a test suite)
The prefix set $\mathbf{AP}(TS)$ of a test suite TS is the set which consists of all the prefixes of all the test cases in TS, i.e.,
$\mathbf{AP}(TS) = \{ p \mid p \text{ is a prefix of some test case in TS} \}$. ■

Definition 3.5 (transition covered by TS)
A transition $< S_i; x/y; S_j >$ in M_S is said to be covered by TS, if there are α, $\alpha x \in \mathbf{AP}(TS)$ such that $\delta_S(S_1, \alpha) = S_i$ and $\delta_S(S_1, \alpha x) = S_j$. ■

Definition 3.6 (tail state S_j of a transition distinguished from S_k by TS)
The tail state S_j of a transition $< S_i; x/y; S_j >$ in M_S is said to be distinguished from another state S_k by TS if there are α, αx, $\alpha x \gamma$, β, $\beta\gamma \in \mathbf{AP}(TS)$ such that

$\delta_S(S_1, \alpha)=S_i$, $\delta_S(S_1, \alpha x)=S_j$, $\delta_S(S_1, \beta)=S_k$ and $\lambda_S(\delta_S(S_1, \alpha x), \gamma) \neq \lambda_S(\delta_S(S_1, \beta), \gamma)$. ■

We will proceed in two steps to find the value of $\mathbf{N_4}(M_S, TS)$. In the first step, we will make an estimation, denoted as $\mathbf{N_6}(M_S, TS)$, on the number of implementation machines which can pass the given test suite TS, by analyzing the structures of the given specification machine and test suite. In the second step, $\mathbf{N_6}(M_S, TS)$ will be adjusted to $\mathbf{N_4}(M_S, TS)$.

To find the value of $\mathbf{N_6}(M_S, TS)$, we need to classify the transitions of M_S into two classes: the first class includes the transitions covered by TS, while the second class consists of the transitions not covered by TS. Taking the specification machine given in Figure 1 and the test suite TS shown in Figure 3 (a) as an example, we can easily check that, among the seven specified transitions in Figure 1, the six transitions listed in Table 1 (a) are covered by the test suite TS, while the other transition given in Table 1 (b) is not covered.

For a transition $< S_i; x/y; S_j >$ in M_S, we use $\mathbf{Tail\_Dis}(< S_i; x/y; S_j >, TS)$ to denote the set of states from which the tail state S_j of transition $< S_i; x/y; S_j >$ is distinguished. Then,
$\mathbf{Tail\_NDis}(< S_i; x/y; S_j >, TS) = \{ S_1, S_2, ..., S_n \} - \mathbf{Tail\_Dis}(< S_i; x/y; S_j >, TS)$
is the set of states from which the tail state S_j of transition $< S_i; x/y; S_j >$ is not distinguished.

For a transition $< S_i; x/y; S_j >$ covered by TS, we can easily calculate $\mathbf{Tail\_Dis}(< S_i; x/y; S_j >, TS)$ and therefore $\mathbf{Tail\_NDis}(< S_i; x/y; S_j >, TS)$. As an example, let us consider one of the covered transition < S_1; 2/2; S_2 > given in Table 1 (a). As is clear from Figure 3 (b), this transition is covered twice by the test suite. The tail state S_2 at point V_6 is distinguished from state S_4 at point V_8. The same tail state S_2 at point V_{11} is distinguished from state S_3 at point V_7. Therefore, **Tail_Dis**(< S_1; 2/2; S_2 >, TS) = { S_3, S_4 } and **Tail_NDis** (< S_1; 2/2; S_2 >, TS) = { S_1, S_2 }.

< S_1; 1/1; S_2 >	< S_1; 2/2; S_2 >	< S_2; 1/1; S_3 >
< S_2; 2/2; S_2 >	< S_3; 2/2; S_4 >	< S_4; 1/2; S_4 >

(a) transitions covered by TS

< S_4; 2/2; S_1 >

(b) transition not covered by TS

Tail_Dis (< S_1; 1/1; S_2 >, TS) = { S_4 }
Tail_Dis (< S_1; 2/2; S_2 >, TS) = { S_3, S_4 }
Tail_Dis (< S_2; 1/1; S_3 >, TS) = { S_1, S_2 }
Tail_Dis (< S_2; 2/2; S_2 >, TS) = { S_4 }
Tail_Dis (< S_3; 2/2; S_4 >, TS) = { S_1, S_2 }
Tail_Dis (< S_4; 1/2; S_4 >, TS) = ϕ

(c)

Tail_NDis (< S_1; 1/1; S_2 >, TS) = { S_1, S_2, S_3 }
Tail_NDis (< S_1; 2/2; S_2 >, TS) = { S_1, S_2 }
Tail_NDis (< S_2; 1/1; S_3 >, TS) = { S_3, S_4 }
Tail_NDis (< S_2; 2/2; S_2 >, TS) = { S_1, S_2, S_3 }
Tail_NDis (< S_3; 2/2; S_4 >, TS) = { S_3, S_4 }
Tail_NDis (< S_4; 1/2; S_4 >, TS) = { S_1, S_2, S_3, S_4 }

(d)

Table 1: Intermediate calculation results for the example FSM and TS

The related results for the other five covered transitions can be found in Table 1 (c) and (d). Since a state in **Tail_NDis**(<S_i; x/y; S_j>, TS) is not distinguished from the tail state S_j of <S_i; x/y; S_j>, changing the tail state S_j of transition < S_i; x/y; S_j > to any state in **Tail_NDis**(< S_i; x/y; S_j >, TS) will give us an implementation machine which can pass the test suite TS. Therefore, there are |**Tail_NDis**(< S_i; x/y; S_j >, TS)| possible ways to make such a Type 1 change (as defined in Definition 2.7). However, we should note that, as transition < S_i; x/y; S_j > is covered by TS, changing the output symbol "y" to any other output symbol (Type 2 change) will result in an implementation machine which is very likely to be detected by TS. Therefore, to guarantee to generate an implementation machine which can pass TS, we have only one choice of keeping the output "y" of the transition. Consequently, the given transition <

S_i; x/y; S_j > covered by TS gives us |**Tail_NDis**(< S_i; x/y; S_j >, TS)| possible ways of generating an implementation machine which can pass the test suite TS. It can be noted that, with certain test suite generation methods such as the W method [Chow 78], all the transitions in the specification machine will be covered and the tail state of each transition will be distinguished from all other states. Therefore, **Tail_NDis**(< S_i; x/y; S_j >, TS) = { S_j } and |**Tail_NDis**(< S_i; x/y; S_j >, TS)| = 1.

For a transition < S_i; x/y; S_j > not covered by the given test suite TS, its tail state S_j is not distinguished by TS from any state. Therefore, we have **Tail_NDis**(< S_i; x/y; S_j >, TS) = { S_1, S_2, ..., S_n }. Furthermore, since the transition is not covered by TS, we can change the output symbol "y" to any symbol in Y and still get a mutant machine which can pass TS. Combining the possible ways of changing the tail state (Type 1 change) and the possible ways of changing the output (Type 2 change), we can immediately conclude that, for the transition < S_i; x/y; S_j > which is not covered by TS, there are |**Tail_NDis**(< S_i; x/y; S_j >, TS)| x |Y| = n|Y| possible ways to generate an implementation machine which can pass the test suite. As a result, the set of all the transitions not covered by TS gives us $(n|Y|)^m$ choices to generate an implementation machine which can pass TS (where m is the number of transitions not covered by TS).

As we have assumed in Section 2, an implementation machine should be completely defined. Therefore, for the given specification machine M_S which is in general partially specified, we need to apply the Type 3 operation to add an extra transition for each $(S_i, x) \in S \times X - D_S$. Since the tail state of such an extra transition can be any of the n states S_1, S_2, ..., S_n and the output symbol can be any one in Y, we know that there are a total of $(n|Y|)^{n|X| - |D_S|}$ possible ways to generate an implementation machine by adding n|X| - $|D_S|$ extra transitions.

Following from the above discussions, we have

$$\overline{N}_6(M_S, TS) = (n|Y|)^{n|X| - |D_S| + m} \prod_{\substack{< S_i;\, x/y;\, S_j > \\ \text{covered}}} |\textbf{Tail\_NDis}(< S_i; x/y; S_j >, TS)| \quad \textbf{(3-3)}$$

implementation machines in **Impl**(M_S) which are estimated to be able to pass the given test suite, where m is the number of transitions not covered by TS.

However, we should note that, among all the implementation machines in **Impl**(M_S), there are

$$N_2(M_S) = (n - 1)!\,(n|Y|)^{n|X| - |D_S|}$$

implementation machines which conform to the given specification machine M_S and therefore can pass any test suite. As such, we need to adjust the estimated number of implementation machines that can pass the test suite TS and use $\mathbf{N_4}(M_S, TS) = \mathbf{max}(\mathbf{N_2}(M_S), \overline{\mathbf{N}}_\mathbf{6}(M_S, TS))$ (i.e., the maximum of the two)

$$\mathbf{FC_e}(M_S, TS) = \frac{N_1(M_S) - \mathbf{max}(N_2(M_S, TS), \overline{N}_6(M_S, TS))}{N_1(M_S) - N_2(M_S, TS)} \quad \textbf{(3-4)}$$

Let us continue our example with the specification machine given in Figure 1 and the test suite shown in Table 1 (a). For this particular example, $|D_S|$ = 7, m = 1, n = 4, |X| = |Y| = 2. Therefore, we have $\mathbf{N_1}(M_S)$ = 16777216, $\mathbf{N_2}(M_S)$ = 48, $\mathbf{N_6}(M_S, TS)$ = 18432 and finally the estimated fault coverage $\mathbf{FC_e}(M_S, TS)$ = 99.89042%.

Several properties of $\mathbf{FC_e}(M_S, TS)$ given in (3-4) are summarized in the following theorem.

Theorem 3.7
(1) $0 \leq \mathbf{FC_e}(M_S, TS) \leq 1$;
(2) $\mathbf{FC_e}(M_S, TS) = 1$ if $\mathbf{FC_p}(M_S, TS) = 1$;
(3) $\mathbf{FC_e}(M_S, TS) = 0$ if $\mathbf{FC_p}(M_S, TS) = 0$; and
(4) $\mathbf{FC_e}(M_S, TS) \leq \mathbf{FC_e}(M_S, TS')$ if $\mathbf{AP}(TS) \subseteq \mathbf{AP}(TS')$. ■

It is quite straightforward to prove these properties. However, we feel that the meaning of the forth property needs some explanation. We note that **AP**(TS) is the prefix set of TS and that **AP**(TS) ⊆ **AP**(TS') implies that TS' has more or longer test cases than TS. The forth property essentially tells us that the estimated fault coverage for TS and TS' coincide with the intuition that TS' provides better fault coverage than TS.

The calculation of the estimated fault coverage given in (3-4) can be simplified since a common factor $(n\,|Y|)^{n|X|\,-\,|D_S|}$ exists in $\mathbf{N_1}(M_S)$, $\mathbf{N_2}(M_S)$ and $\mathbf{N_6}(M_S, TS)$ as shown in (3-1), (3-2) and (3-3), respectively. After eliminating this common factor from all the items in (3-4), we have the following simplified formula:

$$\mathbf{FC}_e(M_S, TS) = \frac{\mathbf{K}_1(M_S) - \mathbf{max}(\mathbf{K}_2(M_S, TS),\ \mathbf{K}_3(M_S, TS))}{\mathbf{K}_1(M_S) - \mathbf{K}_2(M_S, TS)}, \qquad \textbf{(3-5)}$$

where

$$\mathbf{K}_1(M_S) = (n|Y|)^{|D_S|},$$

$$\mathbf{K}_2(M_S, TS) = (n-1)!,$$

$$\mathbf{K}_3(M_S, TS) = (n|Y|)^m \prod_{\substack{<S_i;\ x/y;\ S_j> \\ \text{covered}}} |\mathbf{Tail\_NDis}(<S_i;\ x/y;\ S_j>, TS)| \cdot$$

From this simplified formula, we can see that only the specified transitions in a given specification machine will contribute to the calculation of the estimated fault coverage and those non specified can be excluded from consideration.

The above structural analysis approach to the evaluation of fault coverage has been implemented under SUN/UNIX and a number of experiments have been done to show how accurately the estimated fault coverage can match the precise fault coverage. Taking the FSM given in Figure 1 as a specification machine, we generated the following nine test suites:

$TS_1 = \phi$
TS_2 = { r.1 }
TS_3 = { r.1.1.2, r.2.1.2.1 }
TS_4 = { r.1.1.2, r.2.1.2.1, r.2.2.1 }
TS_5 = { r.1.1.2, r.2.1.2.1, r.2.2.1.2 }
TS_6 = { r.1.1.2, r.2.1.2.1.1, r.2.2.1.2 }
TS_7 = { r.1.1.2, r.2.1.2.1, r.2.2.1.2.2 }
TS_8 = { r.1.1.2, r.2.1.2.1.1, r.2.2.1.2.2 }
TS_9 = { r.1.1.2.1.1.1, r.1.1.2.1.2.1, r.1.2.1.2.1, r.1.1.2.2.1.1.2.1, r.1.1.2.2.1.2.1, r.1.1.2.2.2.1, r.1.2.2.1, r.2.1.2.1.1, r.2.2.1.2.2 }

Applying our structural analysis approach to these test suites yields the estimated fault coverage listed in the second column of Table 2. To assess the accuracy of these estimated fault coverage values, we need to compare them with the precise fault coverage values for these test suites. Fortunately, for the small specification machine given in Figure 1, we have been able to make an exhaustive mutation analysis. We have written a program which generates and executes one by one all the $(4 \times 2)^{(4 \times 2)} = 16777216$ possible implementation machines against each of the above nine test suites. Therefore, we have been able to calculate the precise fault coverage for these test suites and the results are listed in the third column of Table 2. The differences between the estimated and precise fault coverage are listed in the forth column of Table 2.

TS_i	$FC_e(M_S, TS_i)$	$FC_p(M_S, TS_i)$	**Deviation**
TS_1	0.00000%	0.00000%	0.00000%
TS_2	50.00014%	50.00014%	0.00000%
TS_3	99.34110%	99.25871%	0.08239%
TS_4	99.89042%	99.66497%	0.22545%
TS_5	99.89042%	99.66898%	0.22144%
TS_6	99.94535%	99.77655%	0.16880%
TS_7	99.94535%	99.88084%	0.06451%
TS_8	99.97282%	99.92032%	0.05250%
TS_9	100.00000%	100.00000%	0.00000%

Table 2: Fault Coverage for the FSM in Figure 1

We have also applied this structural analysis approach to some real protocol machines. For instance, the FSM shown in Figure 4 actually represents the control portion of the NBS transport protocol (class 4) [SiLe 89]. It has 15 states, 27 inputs, 46 outputs and 62 specified transitions. For this FSM, the following two test suites are generated in [SiLe 89] with the transition tour method and the UIO method, respectively.

TS_T = { r.x1.x14.x5.x5.x19.x24.x12.x3.x8.x9.x8.x11.x16.x8.x7.x11.x24.x2.x17.x1.x15.
x5.x24.x13.x19.x1.x14.x8.x10.x11.x16.x8.x7.x8.x9.x24.x12.x5.x24.x17. x19.x13.
x18.x21.x23.x24.x1.x15.x6.x24.x12.x22.x4.x19.x24.x17.x2.x27.x1.x14.x23.
x24.x12.x3.x9.x16.x11.x7.x24.x17.x1.x15.x23.x24.x1.x14.x25.x26.x16.x5.x24.
x1.x15.x20.x6.x24.x13.x18.x3.x16.x9.x8.x7.x24.x12.x3.x16.x10.x11.x7.x10.x11.
x9.x24.x1.x14.x16.x8.x9.x11.x7.x24.x1.x14.x16.x23.x24.x1.x14.x16.x.25.x26 }

TS_{UIO} = { r.x1.x14.x5.x5, r.x1.x14.x8.x9, r.x1.x14.x23.x5, r.x1.x14.x25.x8,
r.x1.x14.x26.x8, r.x1.x14.x16.x5.x5, r.x1.x14.x16.x8.x7, r.x1.x14.x16.x23.x5,
r.x1.x14.x16.x25.x8, r.x1.x14.x16.x26.x8, r.x1.x15.x5, r.x12.x3.x8.x9,
r.x12.x3.x9.x8.x16.x7, r.x12.x3.x9.x11.x16.x7, r.x12.x3.x9.x16.x8.x7,
r.x12.x3.x9.x16.x7.x5, r.x12.x3.x9.x16.x11.x7, r.x12.x3.x10.x9,
r.x12.x3.x11.x9, r.x12.x3.x16.x8.x7, r.x12.x3.x16.x7.x8.x9,
r.x12.x3.x16.x7.x9.x5, r.x12.x3.x16.x7.x10.x9, r.x12.x3.x16.x7.x11.x9
r.x12.x3.x16.x9.x8.x7, r.x12.x3.x16.x9.x7.x5, r.x12.x3.x16.x9.x11.x7
r.x12.x3.x16.x10.x7, r.x12.x3.x16.x11.x7, r.x12.x5.x5.x5, r.x12.x5.x19.x5
r.x12.x5.x24.x12, r.x12.x5.x11.x5, r.x12.x21.x3, r.x12.x23.x5, r.x12.x22.x3
r.x12.x4.x5.x5, r.x12.x4.x6.x5, r.x12.x4.x19.x5, r.x12.x4.x23.x5
r.x12.x4.x20.x5, r.x2.x12, r.x17.x1.x14, r.x17.x19.x12, r.x17.x2.x27
r.x17.x27.x12, r.x13.x19.x12, r.x13.x18.x3 }

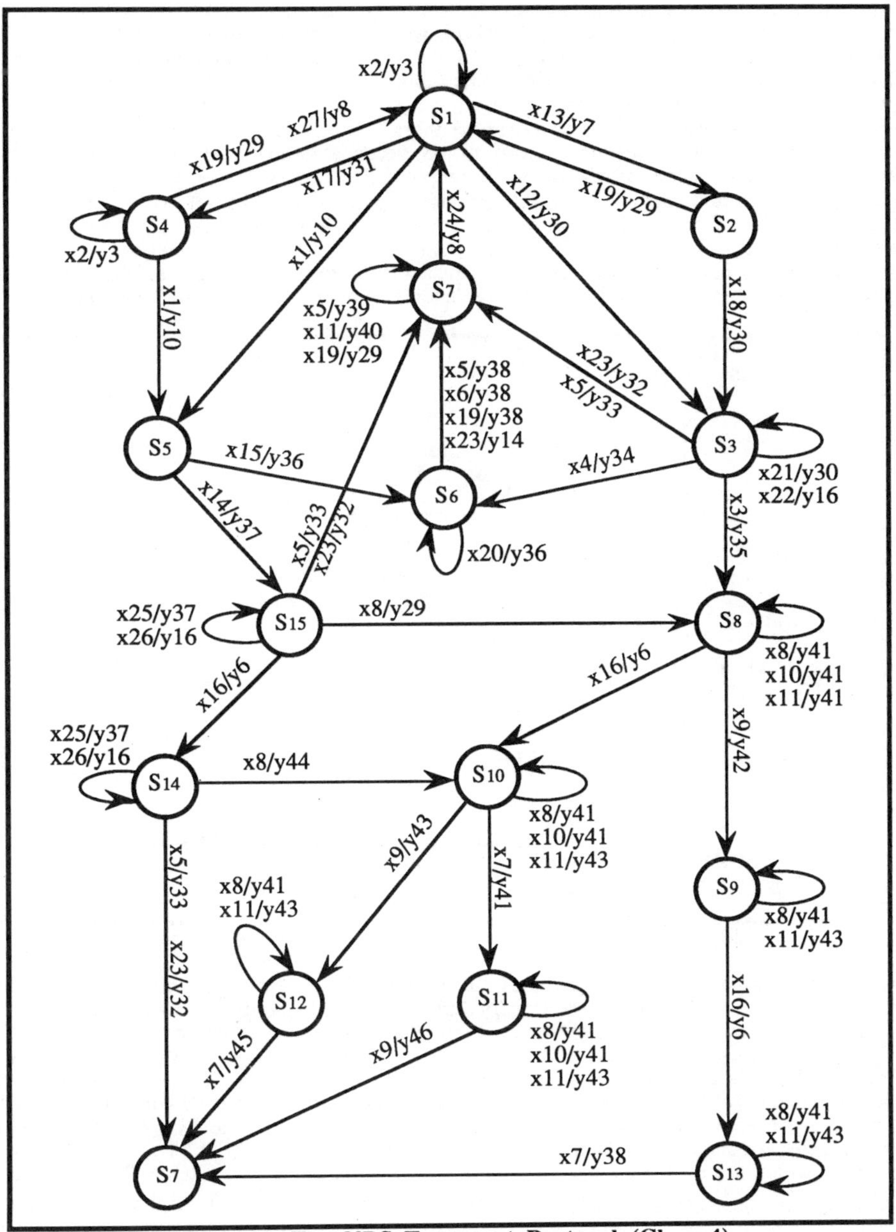

Figure 4: The NBS Transport Protocol (Class 4)

For both these two test suites, our structural analysis approach yields the estimated fault coverage to be 100%. For such a large machine, we are apparently unable to use the exhaustive mutation analysis technique to simulate all the 690^{405} possible implementation machines. Therefore, we cannot compare our estimated fault coverage values with the corresponding precise fault coverage values here. However, we have also used the Monte-Carlo simulation approach to estimate the fault coverage for these two test suites. 10^6 randomly generated mutants, each of which does not conform to the specification machine given in Figure 4 and contains a random number of transfer and/or output faults, have been simulated. It turns out that both test suites can detect all these 10^6 non conforming mutants. Therefore, with the Monte-Carlo simulation approach, we have found that the fault coverage of these two test suites to be also 100%. This coincides with the result obtained with our structural analysis approach. It should be noted that our Monte-Carlo simulation differs from the one used in [SiLe 89] where 10 classes of mutants were defined and only mutants within those classes were simulated.

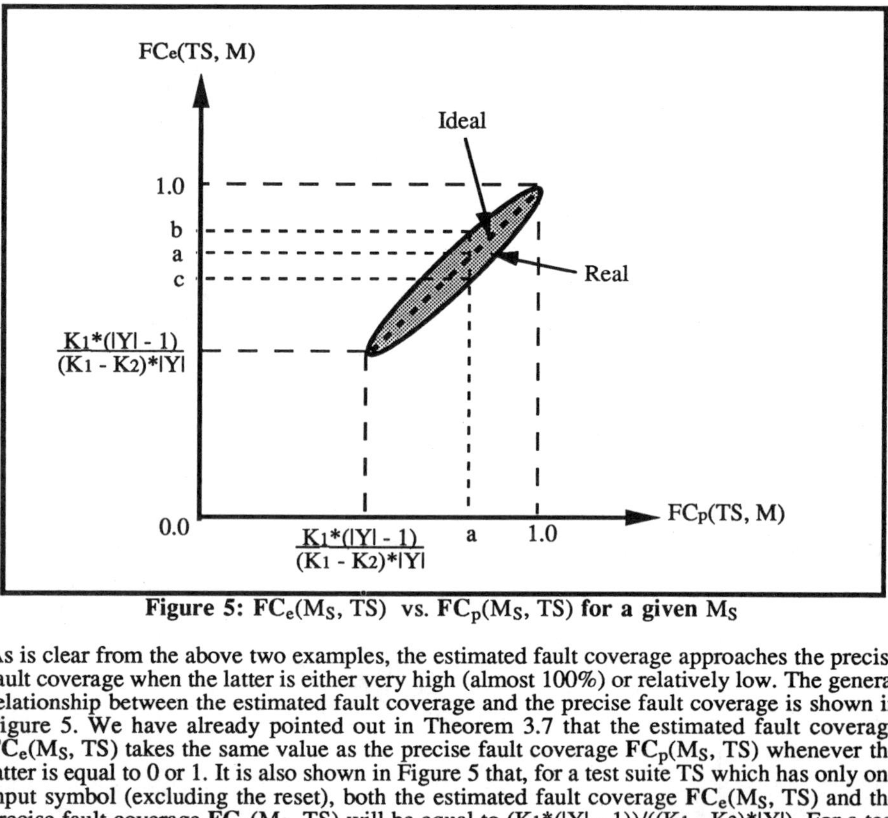

Figure 5: $FC_e(M_S, TS)$ vs. $FC_p(M_S, TS)$ for a given M_S

As is clear from the above two examples, the estimated fault coverage approaches the precise fault coverage when the latter is either very high (almost 100%) or relatively low. The general relationship between the estimated fault coverage and the precise fault coverage is shown in Figure 5. We have already pointed out in Theorem 3.7 that the estimated fault coverage $\mathbf{FC_e}(M_S, TS)$ takes the same value as the precise fault coverage $\mathbf{FC_p}(M_S, TS)$ whenever the latter is equal to 0 or 1. It is also shown in Figure 5 that, for a test suite TS which has only one input symbol (excluding the reset), both the estimated fault coverage $\mathbf{FC_e}(M_S, TS)$ and the precise fault coverage $\mathbf{FC_p}(M_S, TS)$ will be equal to $(K_1*(|Y| - 1))/((K_1 - K_2)*|Y|)$. For a test suite TS which has more than one input, the precise fault coverage $\mathbf{FC_p}(M_S, TS)$ will take a value "a", where $(K_1*(|Y| - 1))/((K_1 - K_2)*|Y|) \leq a \leq 1$. Ideally, we would like the estimated fault coverage $\mathbf{FC_e}(M_S, TS)$ to take the same value "a". In reality, $\mathbf{FC_e}(M_S, TS)$ approximates $\mathbf{FC_p}(M_S, TS)$ by taking a value in the section [c, b], where $c \leq a \leq b$.

4. FAULT COVERAGE AT THE ORDER LEVEL

For a large specification machine M_S (i.e., a machine that has large numbers of states, outputs and specified transitions) and a non trivial test suite TS that covers most or all of the transitions, it can be observed from (3-5) that $\mathbf{K_1}(M_S) >> \mathbf{K_2}(M_S)$ and $\mathbf{K_1}(M_S) >> \mathbf{K_3}(M_S, TS)$ and therefore $\mathbf{FC_e}(M_S, TS) \approx 1$. For example, we have found with formula (3-5) that the estimated fault coverage for the two test suites TS_T and TS_{UIO} of the NBS Transport Protocol shown in Figure 4 to be 100%. However, it is well known in the literature that a test suite generated with the UIO method will in general provide better fault coverage than a test suite generated with the transition tour method. It is therefore necessary to provide some kind of mechanism which can make a distinction between the two. Actually, the following formula, which we call "Order Coverage", can be used for this purpose:

$$\mathbf{FC_o}(M_S, TS) = \frac{\mathbf{Log_{10}}(\mathbf{K_1}(M_S)) - \mathbf{Log_{10}}(\max(\mathbf{K_2}(M_S, TS), \mathbf{K_3}(M_S, TS)))}{\mathbf{Log_{10}}(\mathbf{K_1}(M_S)) - \mathbf{Log_{10}}(\mathbf{K_2}(M_S, TS))} \qquad \mathbf{(3\text{-}6)}$$

The intuition behind this formula is as follows. For a large specification machine and a non trivial test suite such that $\mathbf{K_1}(M_S) >> \mathbf{K_2}(M_S)$ and $\mathbf{K_1}(M_S) >> \mathbf{K_3}(M_S, TS)$, $\mathbf{K_2}(M_S)$ and $\mathbf{K_3}(M_S, TS)$ can be neglected in the calculation with formula (3-5). However, the orders (i.e., the logarithms) of $\mathbf{K_2}(M_S)$ and $\mathbf{K_3}(M_S, TS)$ are still comparable with that of the $\mathbf{K_1}(M_S)$ and therefore will not be neglected in the calculation with formula (3-6). For instance, although $\mathbf{FC_e}(M_S, TS_T) = \mathbf{FC_e}(M_S, TS_{UIO}) = 100\%$, we can still see a difference between the two test suites since $\mathbf{FC_o}(M_S, TS_T) = 63.82301\%$ and $\mathbf{FC_o}(M_S, TS_T) = 66.05067\%$.

The following theorem summarizes the properties of the "order coverage".

Theorem 4.1
(1) $0 \le \mathbf{FC_o}(M_S, TS) \le 1$;
(2) $\mathbf{FC_o}(M_S, TS) = 1$ if $\mathbf{FC_e}(M_S, TS) = 1$;
(3) $\mathbf{FC_o}(M_S, TS) = 0$ if $\mathbf{FC_e}(M_S, TS) = 0$; and
(4) $\mathbf{FC_o}(M_S, TS) \le \mathbf{FC_o}(M_S, TS')$ if $\mathbf{FC_e}(M_S, TS) \le \mathbf{FC_e}(M_S, TS')$. ■

5. CONCLUSIONS

In this paper, we have presented a structural analysis approach to the evaluation of fault coverage of a test suite in respect to a system specification given in the form of a finite state machine. This approach differs from those methods proposed in [DaSa 88, SiLe 89, DDB 91 and MCS 93] as it avoids the necessity of generating and executing a (large) number of mutant machines. Instead, it evaluates the fault coverage of a given test suite by directly analyzing the test suite against the specification machine. It provides a numeric measure for a test suite no matter whether the test suite has full fault coverage (i.e., 100%) or not. This feature makes the metric approach different from one of our previous work [YPB 94] where a test suite is analyzed only to see if it provides full fault coverage or not. As we have seen in Section 3, applications of our approach to a number of examples have shown that the estimated fault coverage of a test suite is quite close to the precise fault coverage, especially when the test suite approaches full fault coverage (100%). Furthermore, this approach has very low computational complexity. Actually, it is not difficult to prove that its complexity is $O(L^2)$, where L is the size of a test suite in terms of the total number of inputs in the test suite. We also proposed the "Order Coverage" based on the structural analysis approach for large complex specification machines. Other related work can be found in [MiPa 92, LoSh 92] where they aimed at generating test suites to achieve full fault coverage (or maximal fault coverage as they called) rather than the evaluation of fault coverage of a given test suite.

We also note that the structural analysis approach has been developed under certain assumptions (Assumptions 1-4 as introduced in Section 2) which are the most relaxed ones compared with other work based on the FSM model [SiLe 89, DaSa 88 etc.]. In particular, we have not assumed the specification machine to be completely specified. Therefore, our approach can be applied to partially specification machines. We believe that this is very important for its practical applications since the real-world systems, such as protocol machines, are normally partially specified.

REFERENCES

[Chow 78] T. S. Chow, "Test Design Modeled by Finite-State Machines", IEEE Trans. SE-4, 3, 1978, pp. 178-187.

[DaSa 88] A. Dahbura and K. Sabnani, "Experience in Estimating Fault Coverage of a Protocol Test", Proc. of the IEEE INFOCOM'88, 1988, pp. 71-79.

[DDB 91] M. Dubuc, R. Dssouli and G. v. Bochmann, "TESTL: A Tool for Incremental Test Suite Design Based on Finite State Model", 4th International Workshop on Protocol Test Systems, Holland, November 1991.

[Gill 62] A. Gill, "Introduction to the Theory of Finite-State Machines", McGraw-Hill Book Company Inc., 1962, pp. 207.

[Koha 78] Z. Kohavi, "Switching and Finite Automata Theory", New York, McGraw-Hill, 1978, pp. 658.

[LoSh 92] F. Lombardi and Y. N. Shen, "Evaluation and Improvement of Fault Coverage of Conformance Testing by UIO Sequences", IEEE Trans. Commun., Vol. COM-40, 8, August, 1992, pp. 1288-1293.

[MCS 93] H. Motteler, A. Chung and D. Sidhu, "Fault Coverage of UIO-based Methods for Protocol Testing", Proc. IWPTS, Pau, France, 1993, pp. 21-33.

[MiPa 92] R. E. Miller and S. Paul, "Structural Analysis of a Protocol Specification and Generation of a Maximal Fault Coverage Conformance Test Sequence", submitted for publication.

[Moor 56] E. F. Moore, "Gedanken-Experiments on Sequential Machines", Automata Studies, Princeton University Press, Princeton, New Jersey, 1956.

[PBD 93] A. Petrenko, G. v. Bochmann and R. Dssouli, "Conformance Relations and Test Derivation", Proc. IWPTS, Pau, France, 1993, pp. 157-178.

[Petr 91] A. Petrenko, "Checking Experiments with Protocol Machines", Proc. of the 4th Int. Workshop on Protocol Test Systems, 1991.

[SiLe 89] D. P. Sidhu and T. K. Leung, "Formal Methods for Protocol Testing: A Detailed Study", IEEE Trans. SE-15, 4, April 1989, pp. 413-425.

[Ural 91] H. Ural, "Formal Methods for Test Sequence Generation", Computer Communications, Vol. 15, No. 5, June 1992, pp. 311-325.

[YPB 93a] M. Yao, A. Petrenko and G. v. Bochmann, "Conformance Testing of Protocol Machines without Reset", Department Publication #861, Département d'informatique et de recherche opérationnelle, Université de Montréal, February 1993, 27 p.

[YPB 93b] M. Yao, A. Petrenko and G. v. Bochmann, "Conformance Testing of Protocol Machines without Reset", Proc. of the 13th IFIP Symposium on Protocol Specification, Testing and Verification, Liege, Belgium, 1993, pp. 241- 253.

[YPB 94] M. Yao, A. Petrenko and G. v. Bochmann, "Fault Coverage Analysis in Respect to an FSM Specification", Proc. of the IEEE INFOCOM'94, Toronto, Canada, 1994.

Test sequence generation using Estelle or SDL structure information

Marc Phalippou[1]
FRANCE TÉLÉCOM - CNET

Abstract : *this paper makes a synthesis of some previous works on automatic test generation from FDT specifications and presents ideas on how a complete framework can be developed - from test theory based on FDTs semantics to a concrete tool built with algorithms that can handle real specifications. The structure information provided by Estelle or SDL descriptions is shown to be a key point both from the theoretical and from the practical point of view.*

Keywords : *I.3, I.4, I.6, II.1, II.3, III.1, IV.3, IV.4.*

1 Introduction

Testing the conformance of final products against their specifications before their introduction in the networks is an important task for telecommunication operators. The cost of this task justifies the attempts to automatize it as much as possible. With the increasing use of formal description techniques in industry [Rudin 92], more and more specifications of real protocols or operational systems are available in Estelle or SDL. It is therefore important to study how to derive automatically test suites from these formal specifications.

This has led to a lot of interesting research in the field of automatic test generation. Trying to balance between their generally admitted advantages (GAA) and their generally admitted drawbacks (GAD), we can mention the following studies:

1. fundamental studies on test theory: what does test mean? What can I prove with a test? Can we design optimal test suites? Does coverage signify anything? [Brinksma 88] [Phalippou 92] [Tretmans 92] [Vuong 91]. GAA: they allow to understand what testers do. GAD: they are generally useless to help designing real test suites.
2. test generation methods based on automata theory: the famous W, UIO, UIOv, DS, Wp (see an overview in [Fujiwara 91]), HSI [Luo 93], UE [Cavalli 92] methods. GAA: they do provide algorithms to develop test suites and we know exactly what such test suites check. GAD: they are generally not applied to any protocol more complex than the alternating bit protocol.
3. heuristic test generation methods: several attempts to obtain test suites of a reasonable size from standard FDTs descriptions [Sarikaya 87] [Phalippou 90] [Grabowski 93] [Chun 90]. Most of these studies underline the importance of FDTs syntactical structures as a basis for efficient test selection. GAA: they can be applied - and they are applied- to real systems. GAD: they often need human assistance and generally nobody knows what such test suites really check.

1. CNET LAA/SLC/EVP, BP-40, F-22301 LANNION Cedex, FRANCE
Tel: 33 96 05 10 51, Fax: 33 96 05 39 45, E-mail: Marc.Phalippou@lannion.cnet.fr

In this paper we show that it is possible to link together several ideas taken from some of the above previous work and to provide a coherent framework which covers most aspects of test generation, from theoretical testing issues on mathematical models to concrete algorithms applied to standard FDTs descriptions in Estelle or SDL. And this is not so surprising, since such work has already been published for other formal models, abstract data types for instance [Gaudel 92].

In order to obtain this unified framework we must introduce a slightly new point of view on the mathematical model (section 2), use previous ideas on test theory in a slightly new context (section 3) and combine previous algorithms in a slightly new way (section 4). New also are the release of our tool TVEDA in which all these ingredients are incorporated (section 5), and maybe our global optimism about the feasibility of automatic test generation (section 6).

2 Semantic model with structure information

2.1 Available semantic models

Currently available test theories (such as the ones which are listed in items 1 or 2 in the introduction) are defined on mathematical models, not on standard FDTs. These models are different kind of automata, such as:

1. Mealy machines (classical W, UIO, etc. test methods),
2. a derived model of Mealy machines called PNFSM (partially specified non-deterministic finite state machines) in [Luo 93] (HSI method),
3. LTS (labelled transition systems) [Brinksma 88],
4. a particular kind of labelled transition systems in which labels are classified into inputs and outputs [Tretmans 92], named IOSM (input-output state machines) with a slightly different communication mechanism in [Phalippou 92].

The differences between these models are the communication semantics defined by the composition of automata (synchronous communication on actions for standard labelled transition systems, inputs and outputs for Mealy machines, PNFSM or IOSM), the finiteness or infiniteness of the state space, and the grouping of events (input and output labelling the same transition for Mealy machines and PNFSM, one event only labelling each transition for labelled transition systems and IOSM).

Their common feature is that they provide a description of an automaton in terms of its interactions with its environment (input-outputs or synchronization events). *Testing being based on a dialogue between the tester and the implementation under test, this kind of description is of prime importance for testing theories.*

The international standards for Estelle and SDL do not provide a semantic model for these languages which has this property. SDL semantics is an operational semantics described in Meta IV which indicates how to execute a complete SDL system. Estelle semantics is an operational semantics which describes the global behaviour of a closed Estelle specification, but the observable behaviour of a stand-alone module cannot be expressed in terms of its input-output interaction with the environment. On the contrary the LOTOS standard does provide a formal semantics which has this property: this explains maybe why work on testing theory historically dealt with LOTOS before Estelle or SDL.

2.2 Extended finite state machines as structured automata

None of the above mentioned models allows to represent information about the syntactic structure of the specification. LTS, PNFSM or IOSM are "flat" automata, in which all states and transitions are individual components of their respective sets: no grouping of states and transitions is a priori provided by the model.

Estelle or SDL are based on an *extended finite state machine (EFSM) model* (this is explicitly stated in the title of Estelle standard). Although this concept has not a single generally admitted precise definition, it means informally a mathematical model of automata which are extended with data, i.e. variables and operations which manipulate these variables. The syntactic structure of an Estelle body or an SDL process consists in factorizing (grouping) the behaviour into syntactic "transitions" which correspond each one to several (often many) transitions of the underlying finite state automaton model.

The link between the EFSM description (i.e. Estelle or SDL syntactic description) and this underlying finite state automaton is done by computing the reachability graph of the specification. But during this transformation the structure information is lost. And we know from practical studies on testing (such as the ones mentioned in the item 3 of the introduction), that this information about the structure is very important for an efficient test selection.

These considerations lead to the following rather unusual definition of EFSM as a pure structure on an automaton, i.e. the grouping of transitions of this automaton into sets that represent the structure. In this paper we illustrate this concept on IOSM, but a similar definition could be provided for Mealy machines, PNFSM or pure LTS.

Definition 1 : an *input-output state machine* (IOSM) is a 4-tuple $M = \langle S, L, T, s_0 \rangle$ where:

1. S is a finite non-empty set of states;
2. L is a finite non-empty set of interactions;
3. $T \subset S \times ((\{?, !\} \times L) \cup \{\tau\}) \times S$ is the transition relation. Each element from T is a transition, from an origin state to a destination state. This transition is associated either to an observable action (input ?a or output !a), or to the internal action τ.
4. s_0 is the initial state of the IOSM.

Definition 2 : let $S = \langle S_S, L, T, s_{0S} \rangle$ be an IOSM. An *extended transition* on S is a non-empty subset $t \subseteq T$. An *extended finite state machine* (EFSM) on S is a finite set of extended transitions on S: $X = \{t_i\}_{1 \leq i \leq n}$.

These concepts are illustrated on figure 1. We show (part of) an Estelle specification, the finite state automaton which represents its semantics in terms of an IOSM, and the extended finite state machine on this automaton. Extended transitions are represented by number labels (❶, ❷, ❸), and as we can see several transitions of the IOSM are grouped into one extended transition. We can interpret the IOSM as being the description of the *semantics* of the Estelle specification (a semantics which has the relevant properties for test study as explained in section 2.1), while the EFSM represents the *syntactic structure* of the Estelle specification: one syntactic Estelle transition is modelled by one extended transition of the EFSM.

Of course the figure shows only an example. The translation between Estelle and the EFSM model shown on this example can be generalized only to a *restricted subset* of Estelle language (a single body with no child, and a limited number of constructs in the transitions: no delay, no priority, etc.). This subset is close to the "Estelle Normal form" used in [Sarikaya 87] or [Chun

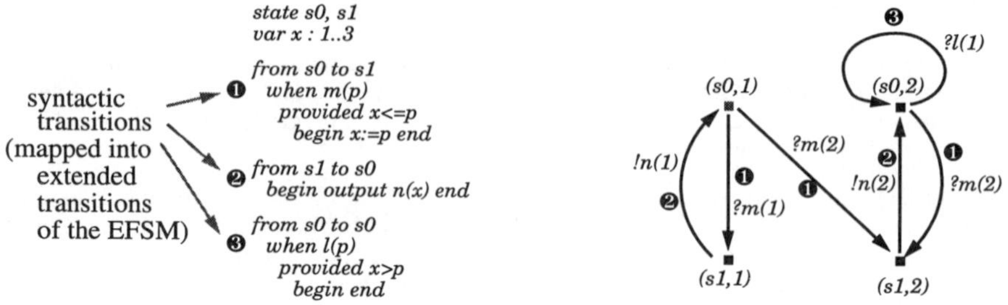

Figure 1 : Estelle description and extended automaton

90]. A very similar translation can be obtained from a comparable SDL subset. We will give in section 5 more details on these subsets, and indicate why we think that we can limit ourselves to them for many applications.

To define precisely the maximal subset of Estelle and SDL which can be interpreted in terms of our EFSM model, and to provide a formal translation (i.e. a formal semantics) would be an important step in the development of a formal test theory for these FDTs, which goes beyond the preliminary ideas which are presented in this paper.

2.3 Generality of the concept

Not only Estelle or SDL, but also programming languages such as C or Pascal, or even informal natural language specifications from ITU or ISO standards can be interpreted with the above EFSM concept. This extends the use of our framework to many other systems.

Figure 2 shows how we can map a sequential program (here a Pascal program) into an EFSM

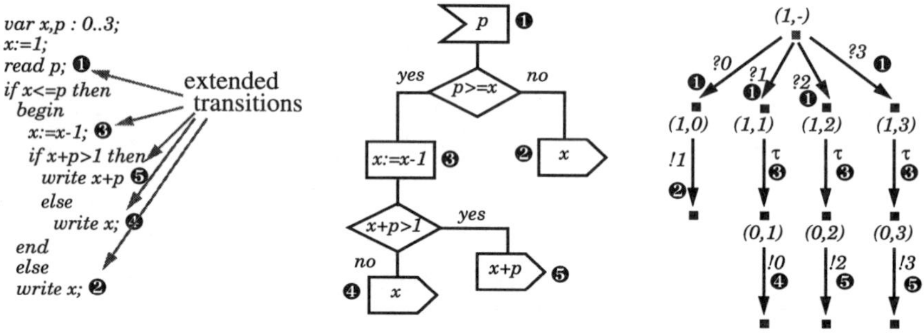

Figure 2 : Pascal description, flowchart, and extended automaton

structure. Read statements are modelled by the inputs of the IOSM, assignments by internal actions (they do not interact with the environment, since they represent the internal computations of the program), and write statements by the outputs of the IOSM. Each statement of the language which manipulates data (read, write, assignments) corresponds to an extended transition of the EFSM. Let us remark that the result is a deterministic IOSM (in testing sense [Phalippou

94], which means that one input leads to one output only), and that the decision instructions do not correspond to any transition, but are used, of course, to compute the structure of the automaton.

Finally, figure 3 shows how we can interpret the structure of a specification written in natural language in terms of EFSM, provided of course that the semantics of this specification can be described by an IOSM. If this is the case, then the sections or subsections of the document, which describe each one part of the behaviour of the automaton, are mapped into extended transitions. Let us remark that a transition of the IOSM can belong to several extended transitions.

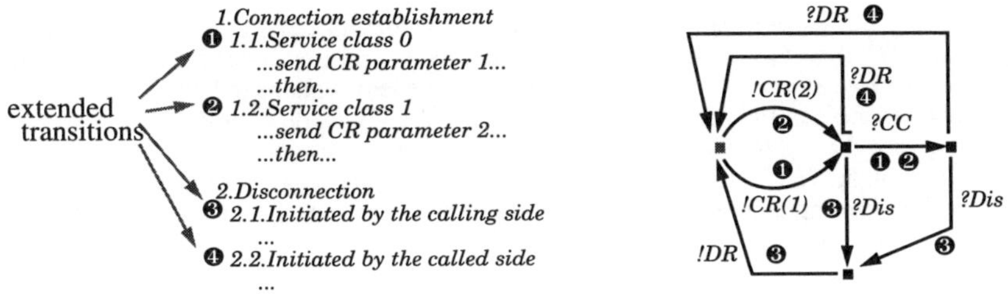

Figure 3 : structured natural language description and extended automaton

3 Uniformity hypothesis is the key for test selection

3.1 Conformance testing

We chose to model conformance by implementation relations, which is a widely used approach [Brinksma 88] [Luo 93] [Phalippou 92]. An implementation relation is a relation between the set modelling the implementations and the set modelling the specifications which indicates which implementations conform to a given specification. In the case in which both specifications and implementations are modelled by IOSM, we have argued [Phalippou 94] that the following implementation relation captures a notion of conformance which corresponds to the needs of practical users:

$$R_5(I, S) \text{ iff } (\forall \sigma \in Tr(S))\ (\sigma \in Tr(I) \Rightarrow (O(\sigma, I) = O(\sigma, S)))$$

(where I is the implementation, S is the specification, $Tr(S)$ is the set of traces of the automaton S and $O(\sigma, S)$ the set of possible outputs of the automaton S after the execution of the trace σ). This implementation relation corresponds to the *quasi-equivalence* relation on PNFSM [Luo 93], and can be related to the *conf* relation on LTS [Brinksma 88]. Formal mappings between these relations have been provided [Phalippou 94].

Testing is modelled by the parallel composition of an automaton called a *tester* with the automaton which represents the implementation. A tester can be either a Moore machine if the implementation is modelled by a Mealy machine [Petrenko 93], a LTS for LTS implementations [Brinksma 88] or an IOSM for IOSM implementations [Phalippou 92]. A definition of the *verdict* of the test, based on the observation of this parallel composition, is provided by each one of the above mentioned studies and completes the theory of testing. We will not need more details in this paper about these theories, and the reader is refered to the referenced papers.

3.2 Test selection

As far as our work is concerned (how to design operational test suites from Estelle or SDL specifications), the important point is that theory has shown that the testers which check directly the above relations (R_5 or *conf* or *quasi-equivalence*) are infinite, i.e. they model infinite test experiments, which is clearly not acceptable in practice. The selection of a reasonable (at least finite) subset of tests is therefore required.

Formulating *test hypotheses* is a convenient way of making such a test selection [Gaudel 92] [Tretmans 92]. A test hypothesis consists in making assumptions on the implementations under test: the more we know about what we are testing, the less remains to be checked in order to assess conformance. In the previous subsection, we already made an implicit test hypothesis, by assuming that the implementations, which are real (physical) objects, can be adequately modelled by automata (LTS, IOSM or Mealy machines). By making additional test hypotheses, we assume that the implementations are modelled by elements of a subset of these automata (see figure 4.)

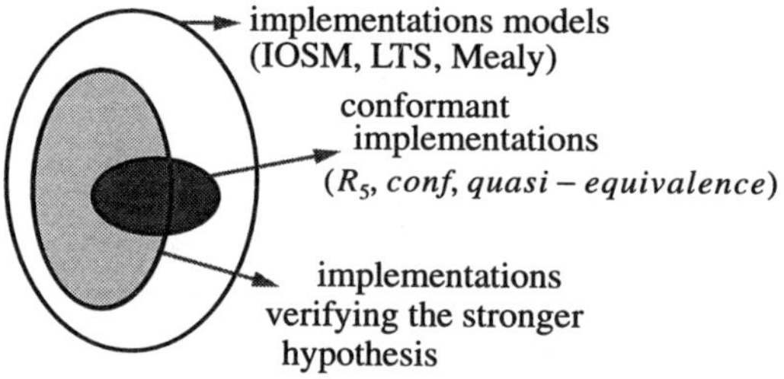

Figure 4 : Stronger test hypothesis

This approach to test selection is often used in test generation methods based on automata theory (W, DS, HSI etc.) where the models of implementations are restricted to Mealy machines or PNFSM with a specified maximum number of states [Fujiwara 91] [Luo 93]. We have called such an hypothesis a *regularity hypothesis* in [Phalippou 94].

3.3 Uniformity hypothesis

Partition testing is a test selection method which has been defined for the test of sequential programs [Rapps 85] [Weyuker 91]. It consists in dividing the input domain of a program into subdomains, and then in making only one test for each one of the subdomains. The assumption on which this method is based is that on a subdomain the program is either correct everywhere either false everywhere. This kind of test hypothesis is formalized in [Gaudel 92] and named *uniformity hypothesis*. In that paper it is applied to the test of sequential specifications, expressed in an abstract data type formalism.

In this paper we study labelled automata, which are a very different kind of formalism. We are interested in the sequences of events which are exchanged between a system and its environment: the implementation relations that we consider are based on traces of observable events. But EFSM structures on such automata, as defined in section 2.2, provide a way of dividing the observable behaviour of an automaton into some behaviour domains which are the extended transitions. This leads naturally to the following test method: test one elementary transition for each of the extended transitions, and assume that if the implementation is correct with respect to these tests, then it is fully conformant. Due to the similarity with the sequential test technique described above, we call this test hypothesis a uniformity hypothesis.

Let us express this formally. Let $S = \langle S_s, L, T, s_{0s}\rangle$ be an IOSM and $t = (s, \mu, s') \in T$ a transition. We denote $dep(t) = s$ the initial state of the transition and $arr(t) = s'$ the final state of the transition.

Definition 3 : let $t_e \subseteq T$ be an extended transition. An *instantiation* of t_e is a sequence of chainable transitions belonging to t_e: $t_1 \ldots t_n$ such that for all $i \in [1, n]$, $t_i \in t_e$ and for $i \in [1, n-1]$, $arr(t_i) = dep(t_{i+1})$.

By definition every extended transition admits at least instantiation. Let $X = \{t_i\}_{1 \le i \le n}$ be an EFSM on S. For every extended transition t_i we chose an instantiation $t_{i1} \ldots t_{in}$. Let $s_i = dep(t_{i1})$. If we assume that every state of the IOSM is reachable from the initial state (this is a common feature of protocol models), we can find a sequence π_i of observable events such that (s_{0S}, π_i, s_i) (this means that there is a sequence of transitions of the IOSM labelled by the elements of π_i and going from the initial state to state s_i): we call such a sequence a *preamble* leading to this state. In a similar way, if we assume that the IOSM is strongly connected (here again this is common for protocol models), then we can find a sequence π'_i of events such that $(arr(t_{in}), \pi'_i, s_0)$, which we call a *postamble* from the state $arr(t_{in})$. Let σ_i be the sequence of observable events labelling the transitions $t_{i1} \ldots t_{in}$ of the instantiation. We say that the sequence (where the product operator means "concatenation")

$$\prod_{t_i \in X} \{\pi_i \sigma_i \pi'_i\}$$

is an *extended transition tour* of the EFSM X. We use this name because this concept is very close to the classical transition tours on automata, but instead of having a sequence which covers every transition, we have a sequence which covers only one instantiation of every extended transition: we use the structure information to reduce the length of the test to be done. In the example of figure 1, if we assume the existence of a reset $?r$ (in addition to the behaviour drawn on the figure) so that the specification is strongly connected, then the sequence

$$?m(2)\,?r?m(2)\,!n(2)\,?r?m(2)\,!n(2)\,!l(1)\,?r$$

is an extended transitions tour obtained with the above method. Notice that we did not try here to optimize the construction of the extended transition tour, by eliminating redundancy (some subtours cover several extended transitions), but this could be done.

Definition 4 : let $S = \langle S_s, L, T, s_{0s}\rangle$ be an IOSM, $X = \{t_i\}_{1 \le i \le n}$ an EFSM on S and σ_{TT} be an extended transitions tour on X. We make a *uniformity hypothesis* (relative to a given σ_{TT}) if we assume that $\sigma_{TT} \in Tr(I) \Rightarrow R_5(I, S)$.

Our definition of uniformity hypothesis *models current practice* in the area of sequential program testing or in the area of functional testing for application layer services of the OSI model. In the case of sequential programs, this hypothesis corresponds to one of the classical techniques for partition testing, called "all nodes" in [Rapps 85], which consists in exercising all statements of the program. In the case of sequential programs, there is implicitly a reset action, which models the fact that we can run the program as many times as we want with different input values. If we optimize the extended transitions tour to eliminate redundancy, then

$$?0\,!1\,?\,r?1!0\,?\,r?2!2\,?\,r$$

is the test sequence corresponding to the example of figure 2 (?0!1 checks the extended transitions ❶ et ❷, ?1!0 the extended transitions ❶, ❸ et ❹, and ?2!2 the extended transitions ❶, ❸ et ❺). In the case of functional testing of application layer services, the current situation is that each section of the natural language specification describes one feature of the service, and often test suites are designed such that each feature is tested once. This test selection strategy corresponds to the optimized test sequence

$$!\,CR\,(1)\,?\,CC?\,DIS!\,DR!\,CR\,(2)\,?\,CC?\,DR$$

obtained from the EFSM of figure 3. Thus our approach provides a unified view on classical sequential program testing and on protocol testing.

3.4 Do we need additional regularity hypothesis?

Of course, we must realize that the uniformity hypothesis is only an assumption: we consider that the implementations under test belong to a particular subset of IOSM in which every element which admits the trace σ_{TT} is a conformant implementation. Specialists of classical test generation methods based on automata (W, DS, etc.) may complain that such a test sequence does not prove much. But we must understand that:

1. in spite of the fact that it is based on a beautiful theorem from automata theory, regularity hypothesis is only an hypothesis too. The resulting test suite proves nothing if the implementations do not satisfy the required property on the number of states.
2. since the test sequences which result from classical methods are in general much longer than the extended transitions tours, the choice of one method rather than another is a matter of trade-off between the efficiency of the test experiment (in terms of detected errors for instance) and the length of the test experiment: we show in section 5 that our method provides test suites of reasonable length for real protocols.

In the case of some protocols (lower layers of the OSI model for instance) for which it is important to test some kind of structure, it is possible to define more complex test techniques which are between these two extreme approaches (extended transition tours and classical methods on finite automata). For this, we use the structure information provided by the EFSM, and build from this structure an upper approximation of the tested IOSM which can be described by a much smaller automaton. Then classical methods (regularity hypothesis) are applied on this small automaton, which results in test suites of still reasonable length. This approach, which is completely formalized in [Phalippou 94], corresponds in particular to what is done when the control flow of a protocol is tested in isolation from its data flow.

4 Symbolic computation on extended automata

We present in this section a method to compute extended transitions tours, and corresponding test cases expressed in a test notation language, such as TTCN. The examples we provide are based on an Estelle description of INRES protocol [Hogrefe 92]. However, our method is restricted to the limited subset of Estelle and SDL that can be interpreted in terms of EFSM (see section 2.2) and is implemented by using a preliminary translation of Estelle and SDL subsets into a common semantic model based on EFSM: therefore what we say in this section is also valid for SDL specifications which belong to the subset. Figure 5 below describes the part of INRES Estelle specification which will be used in the examples (syntax is not strictly respected).

```
state disconnected, wait, connected, sending
var olddata : ...
    counter : 0..4
    number : 0..1

initialize to disconnected begin end;

from disconnected to wait
❶ when user.iconreq
    begin counter := 0 output pdu_access.cr end;

from wait to connected
  when pdu_access.cc
❷   begin number := 1; counter := 0;
    output user.iconconf end;

from wait to same delay(5)
  provided counter<4
❸   begin output pdu_access.cr;
    counter := counter+1 end;

❹ provided otherwise to disconnected
    begin output user.idisind end;

from connected to sending
  when user.idatreq(isdu)
❺   begin output pdu_access.dt(number,isdu)
    olddata := isdu end;

from sending to same delay(5)
  provided counter<4
    begin output pdu_access.dt(number,olddata)
    counter := counter+1 end;

from not_ignore_dr to disconnected
  when pdu_access.dr
    begin output user.idisind end;
```

Figure 5 : INRES specification (partial)

4.1 Why symbolic computation?

Theoretically, there is no problem at all for computing extended transitions tours as soon as we have the EFSM structure and the IOSM automaton: a simple algorithm of graph path calculus is enough. The EFSM structure is directly given by the syntax of Estelle or SDL specifications, as explained in section 2.2. The real problem is to obtain a computer representation of the IOSM, or at least of the part of its transitions graph needed to compute all the subtours.

For small protocols (such as INRES protocol used in this section), this computation is possible by developing the Estelle or SDL specification using reachability analysis with an appropriate environment description. However, for real protocols, this turns out to be infeasible with the currently available tools: the ISDN LAPD protocol, in some configurations, would be modelled by an IOSM of more than 10^6 states and 10^4 interactions [Phalippou 94], which is beyond the capabilities of the tool we tried to use (Veda). And feasibility studies have learnt to us that this would be the case for many protocols with all the tools we heard about.

The technique of symbolic computation has been proposed in [Chun 90] as a way to avoid this problem. It consists in using a compact (symbolic) representation of the states and transitions of the IOSM in order to avoid the storage in the computer of too large transition graphs. The method that we present in the next subsection is close to the one of the above paper. However, it has some additional restrictions (since we do not use a constraint solver we put additional requirements on the allowed inputs), and on the contrary it provides more features which are needed for the computation of test cases (see section 4.3) and can handle some Estelle (or SDL) constructs which are not handled in [Chun 90], such as spontaneous transitions.

4.2 Computing extended transitions tours

Our symbolic computation is based on the representation of the IOSM in a compact way which follows the EFSM structure: what we store in the computer is a representation of the set of extended transitions which compose the structure, together with the procedures which allow to compute the underlying elementary transitions. In a first step we explain how this compact representation is encoded; then we explain how to use it in order to compute the extended transition tours.

Estelle data structures (variables and message parameters) are transformed into *elementary data vectors* which can be of two types: integer interval (including the integer type) or enumerated type (including booleans). For doing this, composed types, such as arrays or records, are first expanded into elementary fields. Each elementary data vector has a *name* (built from the name of the corresponding Estelle data structure field) and can store an *elementary value domain*. An elementary value domain models a set of possible values for the elementary vector. The value domains are represented either by the constant value *any* (modelling all possible values of the given type), either by *empty* (no value) either by a list (of values for enumerated types, of intervals for integer interval types). The product of elementary vectors is called a *vector*, and we call *value domain* the product of elementary value domains. We use a vector to represent a set of states of the IOSM, or a set of events (inputs or outputs). We show on figure 6 the vectors which represent respectively the set of states of the IOSM of INRES after the *initialize* transition, and the set of *when* inputs of transition ❺ of figure 5.

states after initialize

	el_vector1	*el_vector2*	*el_vector3*	*el_vector4*
name	*state*	*olddata*	*counter*	*number*
type	*enum.*	*enum.*	*interv.*	*interv.*
value	*(disconnected)*	*any*	*any*	*any*

when inputs of transition ❺

	el_vector1	*el_vector2*	*el_vector3*
name	*pco*	*pdu*	*isdu*
type	*enum.*	*enum.*	*enum.*
value	*(user)*	*(idatreq)*	*any*

Figure 6 : Vectors and values

For each extended transition of the EFSM structure, we provide a compact representation of the underlying elementary transitions (which define the complete IOSM, if we assume that every elementary transition belongs to at least one extended transition, which is the case for Estelle or SDL derived EFSM structures):

1. the guard of the Estelle transition (from state, when event and provided clause) is transformed into a couple of vectors (*state vector* and *when vector*) which represent the complete domains of this extended transition. Of course this is possible only if some restrictions apply to the provided clauses, such as absence of coupled constraints on several variables or parameters: a vector domain must be the product of the domains of its elementary vectors. But many real protocols satisfy such constraints, as explained in section 5.
2. the body of the Estelle transition (to state, assignment statements and output events) is transformed into a couple of procedures: the first one computes the image domain (final state domain, output domain) of any initial domain (start state domain, input domain) which intersects the *state vector* domain and the *when vector* domain of the Estelle transition. The second one computes the reverse image domain (i.e. a start state domain and an input do-

main, maybe empty) of any final state domain. Of course the definition of these functions is possible only if some restrictions on the allowed constructs apply to the body of the Estelle (or SDL) transition (very limited number of expressions allowed in assignment statements). But many real protocols satisfy such constraints, as explained in section 5.

These concepts are illustrated on figure 7. We claim that these vectors and functions represent the IOSM: in effect all the elementary transitions which belong to a given extended transition can be computed (start state, event, final state) from the vectors and the functions. Moreover, this compact representation allows to compute all the elementary transitions which satisfy some requirements (for instance: whose final state belongs to a given state set), which is useful to compute transition paths in the IOSM graph.

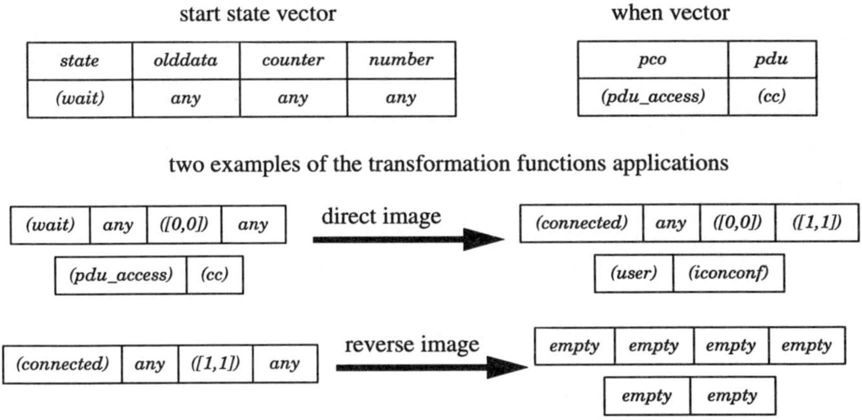

Figure 7 : Compact representation of INRES transition ❷

Finding a transition tour requires finding an instantiation of an extended transition, a preamble and a postamble. These three points can be reduced to a more general problem: how can we find a sequence of transitions of the IOSM which go from a given start domain D_S to a final domain D_F. In order to do this, we compute the shortest sequence of chainable extended transitions such that D_S is included in the start domain of the first extended transition and D_F in the final domain of the last extended transition. For this, we consider successively extended transition sequences of growing length, until the following procedure succeeds:

1. direct propagation of value domains, using the image function mentioned in point 2 above, starting from D_S. If we reach a domain which intersects D_F, we compute the intersection D'_F.
2. reverse propagation of D'_F using the reverse image function mentioned in point 2 above (this indicates from which subset D'_S of D_S we must start and which subsets of the successive when domains of the extended transitions sequence we must use in order to be sure to reach the final domain D'_F).

We apply this method a first time in order to compute a sequence of extended transitions which starts from the initial domain of the Estelle specification and reach the start domain of a given Estelle transition. Then we compute the image domain of this transition, and we apply the method once again to find a sequence going back to the initial state of the specification. These sequences represent sets of extended transition tours as defined in section 3.3: any arbitrary choice of values from the start domain of the sequence and its successive inputs domains will provide a concrete extended transitions tour. Sequences which are obtained by our tool TVEDA (see section 5) for some INRES transitions shown on figure 5 are presented in figure 8.

```
empty preamble
test transition
  state (disconnected) any any any
  when (user) (iconreq)
  output (pdu_access) (cr)
postamble                                   transition ❶
  state (wait) any ([0,0]) any
  when (pdu_access) (dr)
  output (user) (idisind)
  state (disconnected) any ([0,0]) any

preamble
  state (disconnected) any any any
  when (user) (iconreq)
  output (pdu_access) (cr)
test transition
  state (wait) any ([0,0]) any              transition ❷
  when (pdu_access) (cc)
  output (user) (iconconf)
postamble
  state (connected) any ([0,0]) ([1,1])
  when (pdu_access) (dr)
  output (user) (idisind)
  state (disconnected) any ([0,0]) ([1,1])
```

```
preamble
  state (disconnected) any any any
  when (user) (iconreq)
  output (pdu_access) (cr)
  state (wait) any ([0,0]) any
  when ()
  output (pdu_access) (cr)
  state (wait) any ([1,1]) any
  when ()
  output (pdu_access) (cr)                  transition ❹
  state (wait) any ([2,2]) any
  when ()
  output (pdu_access) (cr)
  state (wait) any ([3,3]) any
  when ()
  output (pdu_access) (cr)
test transition
  state (wait) any ([4,4]) any
  when ()
  output (user) (idisind)
postamble empty
  state (disconnected) any ([4.4]) any
```

Figure 8 : Sets of extended transition tours of INRES (partial)

4.3 Test generation

There is still a gap between obtaining transition sequences on the specification automaton and obtaining usable test cases expressed in a language such as TTCN. In particular, two additional ingredients must be taken into consideration:

1. Estelle (and now SDL) allows non-determinism between transitions: in a given state domain and with the same input domain several transitions may be firable. We called this *context non-determinism* in a previous paper [Phalippou 90]. This must be taken into account in the design of test cases. Several solutions are possible here: for instance, in our approach which consists in writing one test case for each extended transition tour, we may consider that the transitions which are non-deterministically firable with the transitions selected for our tour lead to *inconclusive verdicts*.
2. Estelle allows spontaneous transitions, whose firing may disturb the transition tour sequence being tested. Moreover, we have shown in a previous study that when the effect of real test architectures is taken into account, some "when" events in Estelle transitions cannot be controlled from the tester point of view, and must be handled in the same way as spontaneous transitions. We called this *event non-determinism* in [Phalippou 90]. Here again the solution we chose is to consider that these transitions lead to inconclusive verdicts.

The symbolic computation described in the previous subsection provides a convenient means to compute these two kinds of non-determinism. We obtain in such a way complete test cases, which are represented on figure 9 in a TTCN-like tree notation. The empty events which appear on the figure are implemented by timer operations. The *X* which appears in the test of

```
! (user) (iconreq)
  ? (pdu_access) (cr)
    ! (pdu_access) (dr)
      ? (user) (idisind)                    transition ❶
    ? (pdu_access) (cr) <inconc>

! (user) (iconreq)
  ? (pdu_access) (cr)
    ! (pdu_access) (cc)
      ? (user) (iconconf)                   transition ❷
        ! (pdu_access) (dr)
          ? (user) (idisind)
    ? (pdu_access) (cr) <inconc>

! (user) (iconreq)
  ? (pdu_access) (cr)
    ! ()                                    transition ❸
      ? (pdu_access) (cr)
        ! (pdu_access) (dr)
          ? (user) (idisind)
          ? (pdu_access) (cr) <inconc>

! (user) (iconreq)
  ? (pdu_access) (cr)
    ! ()
      ? (pdu_access) (cr)                   transition ❹
        ! ()
          ? (pdu_access) (cr)
            ! ()
              ? (pdu_access) (cr)
                ! ()
                  ? (pdu_access) (cr)
                    ! ()
                      ? (user) (idisind)

! (user) (iconreq)
  ? (pdu_access) (cr)
    ! (pdu_access) (cc)                     transition ❺
      ? (user) (iconconf)
        ! (user) (idatreq) (X)
          ? (pdu_access) (dt) ([1,1]) (X)
            ! (pdu_access) (dr)
              ? (user) (idisind)
              ? (pdu_access) (dt) ([1,1]) (X) <inconc>
      ? (pdu_access) (cr) <inconc>
```

Figure 9 : Test cases obtained for INRES (partial)

transition ❺ is implemented by a test case variable which can be instantiated with any value of the ISDU *undefined* type.

5 Tool, real applications and performance issues

5.1 TVEDA tool

The test generation method which has been described so far is implemented in the latest release of our tool called TVEDA. In addition to what has been described in section 4, TVEDA provides two more features (see figure 10):

1. the tool accepts as inputs either Estelle specifications either SDL specifications, provided that these specifications respect the restrictive conditions mentioned in section 2.2. These specifications are translated into a common semantic model based on EFSM. In this way, most of the tool code is common to both languages Estelle and SDL. The required information to take into account the effect of test architecture (see in the previous section some explanation about non-determinism) is found in an external data file which is given as input to the tool.
2. the tool can produce test cases either in standard TTCN or in a CNET in-house language called Menuet which has some similarities with the future concurrent TTCN (testers are distributed into several distinct test interfaces, which must be explicitly synchronized if necessary). During the translation step from the tests described in section 4.3 to the final test notation, many additional information is computed, with the help of the data file, such as test suite structure, *default* or *otherwise* events in the case of TTCN, splitting of tests into test cases and test steps, etc. (in a word: customization of the test suite).

TVEDA has been developed in Ulysse (an object-oriented language built on top of Le_lisp) using CONCERTO C.A.S.E. environment. It is interfaced with dedicated environments (syntactic editors, parsers and file systems) for the four languages Estelle, SDL, TTCN and Menuet. A previous release of TVEDA, implementing a different (and less complete) test generation strategy has already been presented [Phalippou 90].

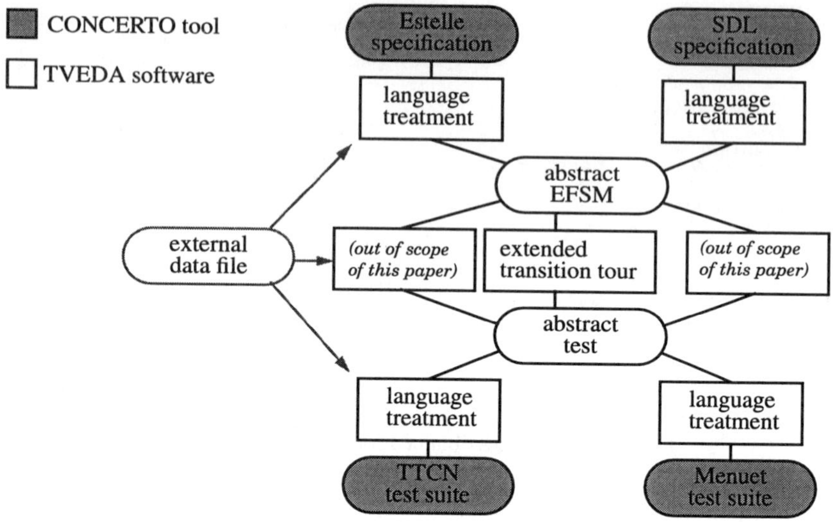

Figure 10 : Architecture of TVEDA

5.2 Main applications

In addition to the treatment of INRES protocol, the test generation method described in this paper has been applied to a given number of real examples in order to assess the quality of the test suites which result from the extended transitions tour strategy and the feasibility of the computation of such test suites with symbolic execution. For the moment, we have reached the following conclusions:

1. the extended transitions tour strategy provides test suites of reasonable length. Since we obtain one test case for each Estelle transition (or SDL *input* construct), it is easy to evaluate this length (there is no need to produce the test cases). The results for some protocols for which we have a formal description are: INRES (43 tests), LAPD (765 tests), D level 3 (user side: 681 tests, network side: 2321 tests), RTSE (320 tests), P1 (a MHS X400 series protocol, 115 tests), MAP (a GSM mobile communication system protocol, 140 tests).
2. comparison with manually-generated test suites shows that the extended transitions tour strategy is close to the test selection method which is actually performed for some protocols, especially the protocols of upper OSI layers, such as P1, or service-oriented protocols, such as MAP (by opposition to automata-oriented protocols, such as LAPD). From points 1 and 2, we conclude that *extended transitions tour test selection strategy is well adapted to generate test suites for real protocols.*
3. several real protocols satisfy the (severe) restrictions on Estelle or SDL constructs which are imposed by our symbolic computation algorithm (this means that it is possible to find a way to describe these protocols - or a significant part of them - which satisfies the restrictions). This is the case for INRES, D level 3, P1 (we did not try up to now with RTSE and MAP). We experimented also successfully on a partial description of FRP (an ATM protocol). LAPD has some constructs that we cannot handle for the moment (sending and reception windows), but we are working on solutions to release the constraints on these constructs.

4. performance is still a key point to be solved. For the moment, only INRES has been successfully processed without any tricky manipulation. We managed to compute LAPD test tours by suppressing the forbidden constructs (see point 3) and by starting from a more abstract model, simpler than the complete specification (TVEDA provides facilities to compute such abstractions): this method allows nevertheless to compute the (real) preambles of more than 2/3 of the transitions of the complete automaton. In the case of P1, the vectors obtained from the complete protocol are very large (more than 1500 elementary vectors in the state vector). Here again nothing can be done without abstractions.

   Performance problems are due to the memory space needed to store large vectors (P1 for instance) or to the length of the graph exploration needed to compute paths (the time needed is an exponential factor of the depth of the graph exploration). We currently investigate two kinds of methods in order to solve these problems: use *abstraction* in a systematic way in order to reduce the size of the symbolic automata, and find *optimization algorithms* or empirical rules to speed up the path search in the graph.

6 Conclusion

We have shown in this paper how to build a complete framework for test generation based on Estelle and SDL languages. This framework covers all steps of test generation: semantic model required for FDTs descriptions, test theory results which give a precise meaning to the defined test suite, algorithms to compute effectively such suites, tool which implements the algorithm, experimentation and assessment of the method on real protocols. The framework is obtained by adapting several existing theories and methods and introducing new concepts where needed.

The final result is the definition of a test selection method which, according to us, fits the needs of test generation for many real applications. Using the structure information of Estelle or SDL descriptions, we obtain test suites of reasonable length which correspond to what human test designers are used to. There are still some problems of algorithm optimization in order to be able to handle successfully with our tool all the real protocols which we took as examples, but research is going on in some promising directions.

Bibliography

[Brinksma 88]
E. Brinksma, *A theory for the Derivation of Tests*, proceedings of Eighth International Conference on Protocol Specification, Testing and Verification, Elseiver Science Publishers, 1988.

[Cavalli 92]
A. Cavalli, Sung Un Kim, *Automated protocol conformance test generation based on formal methods for LOTOS specifications*, proceedings of the 5th International Workshop on Protocol Test Systems, Montréal, September 1992.

[Chun 90]
W. Chun, P. Amer, *Test Case Generation for Protocols Specified in Estelle*, proceedings of FORTE'90, Madrid, november 1990.

[Fujiwara 91]
S. Fujiwara, G. Bochmann, F. Khendek, M. Amalou, A. Ghedamsi, *Test Selection Based on Finite State Models*, IEEE Transactions on Software Engineering, vol. 17, n. 6, juin 1991.

[Gaudel 92]
M.C. Gaudel, *Test Selection Based on ADT Specifications*, proceedings of the 5th International Workshop on Protocol Test Systems, Montreal, September 1992.

[Grabowski 93]
J. Grabowski, D. Hogrefe, R. Nahm, *Test case generation with test purpose specification by MSCs*, proceedings of SDL'93 Using Objects, North-Holland, 1993.

[Hogrefe 92]
D. Hogrefe, *OSI formal specification case study: the Inres protocol and service, revised*, Universit ät Bern, report IAM-91-012, update 1992.

[Luo 93]
G. Luo, A. Petrenko, G.v. Bochmann, *Selecting test sequences for partially-specified nondeterministic finite state machines*, Publication n. 864, Université de Montréal, février 1993.

[Petrenko 93]
A. Petrenko, G. v. Bochmann, R. Dssouli, *Conformance relations and test derivation*, proceedings of IWPTS VI, Pau, September 1993.

[Phalippou 90]
M. Phalippou, R. Groz, *Evaluation of an empirical approach for computer-aided test cases generation*, proceedings of the 3rd IWPTS, Washington, October 1990.

[Phalippou 92]
M. Phalippou, *The limited power of testing*, proceedings of the 5th International Workshop on Protocol Test Systems, Montréal, September 1992.

[Phalippou 94]
M. Phalippou, *Relations d'implantation et hypothèses de test sur des automates à entrées et sorties*, Thèse de l'Université de Bordeaux I, to be published in 1994.

[Rapps 85]
S. Rapps, E. Weyuker, *Selecting Software Test Data Using Data Flow Information*, IEEE Transactions on Software Engineering, vol. 11, n. 4, April 1985.

[Rudin 92]
H. Rudin, *Protocol development success stories, part 1*, proceedings of PSTV XII, Orlando, June 1992.

[Sarikaya 87]
B. Sarikaya, G. v. Bochmann, E. Cerny, *A test design methodology for protocol testing*, IEEE Transactions on Software Engineering, vol. 13, n. 5, May 1987.

[Tretmans 92]
J. Tretmans, *A formal approach to conformance testing*, PhD. thesis, Twente university, 1992.

[Vuong 91]
S. T. Vuong, J. Curgus, *On Test Coverage Metrics for Communication Protocols*, proceedings of the 4th International Workshop on Protocol Test Systems, Leidschendam, October 1991.

[Weyuker 91]
E. Weyuker, B. Jeng, *Analysing Partition Testing Strategies*, IEEE Transactions on Software Engineering, vol. 17, n. 7, July 1991.

34

Derivation of efficient implementations from formal descriptions - issues, methods and conformance

H. Kremer

University of Twente, Fac. Informatica, TIOS Group
P.O. Box 217, NL_7500 AE Enschede, The Netherlands
phone: + 31 53 893703, fax: + 31 53 333815, Email: kremer@cs.utwente.nl

Keyword Codes: D.2.10, D.2.5
Keywords: Software Engineering, Design / Testing and debugging

Abstract: The design and realization of systems is a complex task in which different design issues play a role in different design phases. This paper concentrates on the design issues that play a role during one of the last design phases: the derivation of an efficient implementation for a particular environment. This efficiency, also called performance, is measured in terms of resource usage (CPU and memory) and Quality of Service parameters (e.g. delay and throughput) for the data streams involved.

The design issues that arise during the implementation process are studied using a small system (a duplex 1-slot repeater) and a method for structuring systems that takes these design issues into account is presented. For early assessment of the effect of design decisions on the performance of the realization, Petri-Net based performance analysis techniques are used. A method is given for mapping LOTOS specifications via Extended Transition Systems onto Petri-Nets.

The results of this study include a method for deriving an efficient implementation in a sequential programming language from a LOTOS specification. The usability of the method is demonstrated, and a method is given for LOTOS based performance analysis and new conformance relations.

1. Introduction

The design and realization of systems is a complex task in which different design issues play a role in different design phases. In the first phases, the issues are related to capturing and satisfying the needs of the users, while in the last phases, the issues pertain to making an efficient realization for a particular realization environment. This efficiency (also called performance) is measured in terms of resource usage (CPU and/or memory) and Quality of Service parameter as delay and throughput.

This paper concentrates on the design issues that play a role in the design phase just before mapping a specification in a formal description technique (FDT) onto a realization in software. During this design phase, the issues that play a major role are i) taking characteristics of the procedure calls into account and ii) the removal of non-deterministic choice between interactions. A method for structuring systems that takes these issues into account is presented.

Understanding of the efficiency-effect of design decisions is crucial for making efficient realizations. An inefficient protocol has only inefficient realizations, but an efficient protocol is easily implemented and realized inefficiently. For example, an ill-designed timer package can contribute for more than 50% to the total time needed for processing a packet [3].

This method leaves the designer a lot of freedom in making design decisions. Performance analysis techniques allow the designer to asses the influence of his design decisions on the efficiency of implementations (and realizations). A method is presented for mapping a LOTOS specification onto Petri Nets. These nets are analyzed using the SPNP [2] package whereby estimates for delay and throughput are obtained.

The contribution of this paper is the integration of FDT based implementation with the consideration of efficiency and with performance analysis techniques using Petri Nets.

A running example in the form of a duplex 1-slot repeater is be used throughout the paper for illustration of the design issues and demonstration of the usability of the presented methods.

The following restrictions are put on the specification that is used as a starting point:

- it consists of one or more independent (interleaved) components. This may seems a major restriction, but this isn't the case. Most multiprocessing operating systems do not allow processes to interact directly with each other: all interactions are via the operating system kernel. This kernel is part of the environment of the realization. When independent components of a system are mapped onto different processes, there is no need for direct interaction between these components as that interaction cannot be implemented.
- no dynamic process creation is allowed. No problems are anticipated with an extension of this work that allows dynamic process creation.
- the mapping of observable interactions onto procedure calls is present,
- it may not contain non-deterministic value generation.

This paper uses LOTOS as Formal Description Technique (FDT), but the results can be adapted easily for other process algebraic FDTs. The reader is assumed to be familiar with LOTOS, Petri-Nets [21] and the UNIX operating system.

1.1. Place in design trajectory

The design of a system, being the activity in which user requirements are formulated and translated into a real system, can be structured in a sequence of design phases. The Lotosphere design methodology [18] is taken as the methodology to which the work presented here belongs.

The four phases of this methodology are summarized as follows. The first phase is the *Requirement Capturing* phase in which the user requirement on the system are defined. The second phase is the *Architectural* phase in which a high-level description (called Architecture) is obtained that describes what the system is to do in terms of the future users of the system. In the third phase, the *Implementation* phase, a design is obtained (called Implementation) that can be mapped directly onto a Realization (see below) while preserving the properties of the Architecture. The Implementation describes how the system meets the User Requirements. The fourth and last phase is the *Realization* phase in which the Implementation is mapped onto the Realization. The phases of this methodology are described in more detail in [16].

The Implementation Phase can be divided into two steps. In the first step, the Architecture is transformed into a high-level design that describes how the Architecture is provided and that is realization independent. In the second step, the realization dependent aspects are taken into account The work described in this paper pertains to the second step of the Implementation Phase.

1.2. Outline

The outline of this paper is as follows.

In section 2, the design issues that play are role during the last steps of the second part of the Implementation phase are identified and a method is presented for structuring systems that takes these issues into account.

This method leaves a large design freedom to the designer. The designer can be aided in making decisions by using performance analysis techniques for assessing the influence of his design decisions on the performance of the realization without actually building the realization. In section 3 it is shown how LOTOS descriptions can be mapped onto Petri Nets that are subsequently used for numerical performance analysis. The results of this analysis show the effects the implementation decisions have on the mean delay and throughput.

In section 4 the conformance of the implementations to the specification is addressed. It is shown that, although the implementations are not testing equivalent with the specification, they are to be considered conforming implementations. A new conformance relation is defined that captures precisely the freedom the designer has during the implementation phase.

Section 5 discusses future work and the relationship with other work.

Section 6 contains the conclusions.

1.3. Example system and realization environment

The system used as a running example is a duplex 1-slot repeater. This system consists of two independent simplex 1-slot repeaters, each serving a stream of packets. A simplex 1-slot repeater is an active entity that contains a 1-slot buffer in which packets are received and are subsequently transmitted from. The behaviour of the duplex repeater is best described by the LOTOS specification in Specification 1. Note that the independence of the repeaters is modelled by interleaving of the repeaters.

Apart from the behaviour description there are no other constraints on the realizations.

```
Duplex [r1, w1, r2, w2]
where

  process Duplex[r1, w1, r2, w2] : noexit:=
  Simplex [r1, w1] ||| Simplex [r2, w2]
  endproc (* Duplex *)

  process Simplex[r, w] : noexit:=
  r?p : Packet  ;  w!p  ;  Simplex [r, w]
  endproc (* Simplex *)
```

Specification 1 *Duplex 1-slot Repeater*

Example realizations are made in software in the C programming language [15] on a system running the UNIX operating system. The following mapping of interactions is common practice, so it is presented without motivation. The interactions r_1, w_1, r_2, w_2 are mapped onto the functions calls `read(1,...)`, `write(1,...)`, `read(2,...)` and `write(2,...)` respectively. It is assumed that the write call is 'non-blocking'. This term is explained in detail in section 2.

2. Implementation issues and structure

In this section, some implementation issues that arise during the implementation process are identified, and a method is presented for structuring the specification such that the decisions made regarding the issues are reflected.

2.1. Implementation issues

This subsection contains the identification of implementation issues that arise during the implementation process.

Implementation issue 1 Allocation of independent components onto processes

The specification that serves as starting point consist of a number of independent (interleaved) components. As these components are independent, they can be executed in parallel. The number of processes onto which the independent components are mapped may be smaller than the number of independent components. A decision on the allocation of components onto processes must then be made.

Implementation issue 2 Avoiding choice between blocking call-instances

The default semantics of the read and write calls in UNIX are that they are *blocking calls*: these calls returns only when all activities related to the call have been successfully performed. A read call is successful only when data is read, so this call returns only if some data is actually received.

The notion of 'all activities related to a call' is vague and whether they have completed is not visible to the initiator of the call. A more precise characterization of a blocking call is the following. A call is a *non-blocking call* if the call always returns within a reasonable small period of time; it is called a 'blocking call' otherwise. An instance of a blocking call can be either blocking or non-blocking. It is the latter if it will return within a reasonable small period of time.

In the duplex repeater realizations, it is assumed that the write calls do not block, so only blocking in read calls can occur. If a realization in a single process initiates a read call for one stream and a packet of another stream is ready to be received, then the realization is unable to receive this packet as it is 'blocked' in the read call. It is clear that in this situation, the performance of the realizations is less than optimal as receipt and transmission of a packet is unnecessarily delayed.

This poor performance is caused by the choice the realization is forced to make a between interactions without knowing whether a packet is present that can be received or not.

Implementation issue 3 Executing multiple interactions between scheduling

The execution of poll together with the choice of which interaction is to performed next is called scheduling. Scheduling takes a small amount of time that is generally small compared with the time needed for interactions, but this amount is not negligible. Therefore, it is advantageous to schedule interactions only when necessary: if a non-blocking call can be initiated as an alternative to a blocking call, initiating the non-blocking call is often advantageous.

In the duplex repeater, write calls are never blocking, so after completion of a read call, the related write call can always be performed without need for scheduling, thus improving the performance.

Implementation issue 4 Removal of non-deterministic choice

Specifications generally contain non-deterministic choice between interactions. In realization environments, all processes are fully deterministic. During the implementation process any non-deterministic choice needs to be removed. The way this is done affects the performance of the realizations.

During the execution of the realization, the following sequence of interactions can happen: initially, both slots are empty and a packet is received at one of the interaction points, say $\mathtt{r}_1$.

Immediately after that, a packet is ready to be received at r_2. At this moment, the realization can either i) receive that packet at r_2 and transmit them in arbitrary order or ii) postpone the receipt of the packet at r_2 until after transmission of the packet previously received at r_1.

The differences between both alternatives are that in i) both the maximum amount of memory used and the delay of packets is smaller.

2.2. Structuring systems

The previous subsection has illustrated the effects implementation decisions have on the efficiency of the system. In this subsection, a method is presented for structuring systems that takes the efficiency issues into account.

Rather than taking all issues into account in a single step, the structuring of systems is divided into a sequence of steps, each considering some of the issues. The motivation for the ordering of steps is given in section 2.3. The steps are described in the chronological order in the subsections below.

2.2.1. Step 1 - Allocation of independent components to processes

If multiple processes are available, the designer has the freedom of allocating one or more of the independent components to a process. Allocating each component to a separate process has the disadvantage of the high cost of switching of the CPU between the processes.

In the remainder it is assumed that the duplex repeater is realized using a single process because this is more illustrative of the design issues.

2.2.2. Step 2 - Identification of blocking interactions

In order to be able to reason about blocking of interactions, an attribute is associated with each interaction. The boolean value of this attribute indicates whether the procedure call with which the interaction corresponds is blocking.

In the duplex repeater, the attribute is true for all r interactions and false for w interactions.

2.2.3. Step 3 - Structuring the specification into threads

Implementation issue 3 motivates the execution of behaviour with more than one interaction between instances of scheduling. This can be reflected in the specification by structuring the specification in special behaviours called threads. A thread is a behaviour that starts with a single interaction and that contains at most one blocking interaction. If a blocking interaction is present, only the starting interaction is blocking.

In the duplex repeater, the following threads can be identified: an r_1 followed by w_1 and a r_2 followed by w_2. Another set of threads consists of four threads, each consisting of a single interaction. The choice between the possible sets of threads is a matter that the designer must decide on. Note that in general, a large number of ways for structuring a specification into threads is possible. A structuring into threads of the repeater can be reflected in LOTOS as shown in Specification 2.

```
behaviour
state1 [poll, r1, w1, r2, w2]
where
  process state1[poll, r1, w1, r2, w2] : noexit:=
    r1?p : packet   ;   w1!p   ;   State1 [poll, r1, w1, r2, w2]
       []
    r2?p : packet   ;   w2!p   ;   State1 [poll, r1, w1, r2, w2] )
  endproc (* state1 *)
```

Specification 2 *Threaded specification*

In general, a large number of differently structured specifications can be obtained in this design step.

2.2.4. Step 4 - Addition of poll interactions

Most multiprocessing systems offer a facility that allows a process to determine whether an procedure call instance would block if initiated now. In UNIX, this facility is called poll or select. Using this call, a process can replace the choice between blocking call-instances by a single blocking call, followed by a choice between non-blocking call-instances.

According to Implementation issue 2, no choice should be made between blocking interactions in a single process when it is not certain that they do not block if initiated now. This choice, if present, can be avoided by introducing a special interaction (that is called poll and is mapped onto the poll procedure call) that allows a system to interact with its environment The value established in this interaction allows identification of the interactions whose procedure calls do not block at this moment.

2.2.5. Step 5 - Removal of non-determinism choice

The removal of non-determinism is done in this last step. General techniques for elimination of non-deterministic choice are making random choices and assigning priorities to some interactions over others. The representation of making a random choice in a LOTOS specification is shown in Specification 3.

```
behaviour
state1 [poll, r1, w1, r2, w2]
where
  process state1[poll, r1, w1, r2, w2] : noexit:=
  poll!Set(r1, r2) ?NotBlock : Interaction_denotations  ;
     (let e : interaction_denotation = Random(NotBlock) in
      [e = r1] -> r1?p : packet  ;  w1!p  ;  State1 [poll, r1, w1, r2, w2]
       []
      [e = r2] -> r2?p : packet  ;  w2!p  ;  State1 [poll, r1, w1, r2, w2] )
  endproc (* state1 *)
```

Specification 3 *Random choice between r_1 and r_2*

Notes a) An interaction_denotation is one of a set of constants that identifies one or more interactions
b) The operation Set creates a set of interaction_denotations
c) The operation Random randomly selects an interaction from its parameter.

2.3. Discussion

This section so far has presented a method that was presented "as it is": it was not discussed why this particular method was chosen. It is easy to see that the steps in the method are related to the design issues. The motivation of the ordering of steps is given below.

An important issue is the avoidance of choice between blocking interactions. Thus the identification of which interactions are blocking is one of the first design steps. Allocation of parallel components onto processes was done as a first step, but could be done after step 2) as well, but Step 1 (Allocation of independent components onto processes) must precede step 3) because only within one process the choice between blocking interactions should be avoided.

Step 3 (Structuring the specification into threads) should always precede Step 4 (Addition of poll interactions) because in the reverse order a poll interaction might to be added in a choice between a blocking and a non-blocking interactions. If the non-blocking interaction is always preferred, then the poll interaction never occurs. Hence the addition of the poll interactions was unnecessary. The presence in a specification of an interaction that never happens is a violation of the propriety design criterion [16]. The addition also increases the size of code.

Step 5 (Removal of non-deterministic choice) can be incorporated partly in step 3) (structuring into threads) party in Step 4 (Addition of poll interactions). There are no definitive arguments for preferring either way so for personal preference, the removal of non-determinism is presented as a single design step.

Step 4 (Addition of poll interactions) should be done after Step 3 (Structuring the specification into threads) as, in order to avoid the choice between blocking interactions, knowledge of which interactions are blocking must be available.

3. Performance analysis

The method purposely leaves the system-dependent decisions that effect the performance to the designer: the allocation onto processes, the identification of threads and the removal of non-deterministic choice. This section shows how performance analysis techniques can be used for assessment of the efficiency of the realization without actually building it.

Being able to estimate the performance of a realization without building it is necessary when building realization is expensive (e.g. realizations in hardware), or when realizations are easy to build, but the realizations' environment difficult to influence.

For assessment of the performance of realizations the time needed for the execution of an interaction is an important measure. This time is called the duration of the interaction.

3.1. Approach

The approach taken here involves modelling the system and its environment as stochastic Petri-Nets. The performance of these nets is analyzed using the SPNP package [1]. Petri-Nets were chosen because they are well-known and capable of expressing concurrency and because of the availability of tools for performance analysis of Petri-Nets. SPNP was chosen because it was available locally.

A number of scenarios are defined that represent the outcome of different implementation decisions. These decisions vary in the number of processes and memory locations used and the scheduling of interactions.

The scenarios will be analyzed for different arrival rates of packets. Performance parameters of interest are mean delay and mean throughput of packets. The parameters are obtained for identical arrival rates for both streams that vary from 50 to 700 packets/s.

Performance parameters that could be of interest but that are not studied here are memory usage, processor usage and the variation of delay and throughput.

3.2. Modelling using Petri Nets

The modelling of the system and its environment can be separated into three parts: i) modelling the particular environment in which the system is to operate, ii) modelling the behaviour as described by the specification of the system and iii) modelling the allocation of resources (process and memory) and the scheduling of interactions. The motivation for this division into three parts is that it separates the modelling of the environment from the modelling of the system and that it divides the modelling of the system into a part that depends only on the specification and a part that depends only on the implementation decisions.

For the analysis, it is assumed that the environment acts as a source of packets that are put in a queue from which they are received by the system. The environment also acts as a sink for the packets sent by the system. No queue is necessary here. Both streams of the repeater are independent, so the environment is modelled as two sources of packets. The number of tokens in the environment (10) is such that there are always tokens in the environment. The Petri Net of the environment is shown in Figure 4.

The specification depended part is derived from the specification of the system in two steps: mapping specification is mapped onto an Extended Transition System (ETS) and mapping the ETS onto a Petri-Net.

An ETS is basically a Labelled Transition System extended with variables. A full definition of ETSs and an algorithm for mapping LOTOS onto ETSs is found in [6,7]. The structure of the specification in terms of parallel components is to be preserved, so an ETS is made for each of the parallel components in the specification. The derivation of the ETSs for the duplex repeater is simple, so the ETSs are presented without argumentation in Figure 5.

Due to the nature of Petri Nets, any choice between interactions within the same ETS may not depend on the values established in previous interactions. If such choice is present, it needs to be replaced by a non-deterministic choice whereby the probabilities of the alternatives are known. It would be desirable to have tools for the analysis of Petri Nets that are extended with variables. An example of such nets is found in [9].

The mapping of ETSs onto Petri-Nets consists of two steps: mapping the interactions and mapping the ETS structure.

Transitions in Petri-Nets have no duration so an interaction with a non-zero duration must be mapped onto the Petri-Net |→○→[]. If a token is present in the place then some processor is busy with processing of this interaction. The first transition is an immediate transition and models the start of the interaction. The second transition is a timed transition that models the end of the interaction and the rate of this transition models the duration of the interaction.

The structure of the ETS can be mapped by mapping each state of the ETS onto a set of places, one place for each of the variables present in state, and a set of arc connecting these places with the Petri-Nets onto which the interaction are mapped. The Petri-Net model of the duplex repeater is shown in Figure 6.

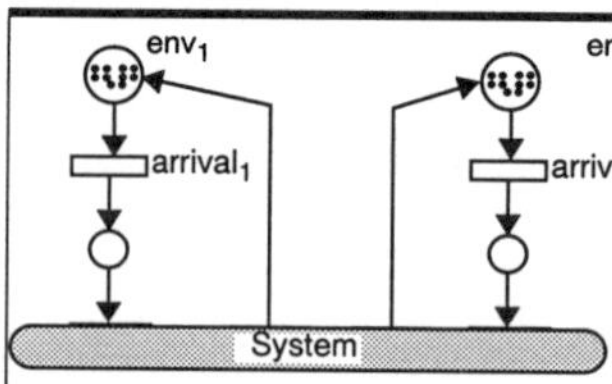

Figure 4 Model of the Environment

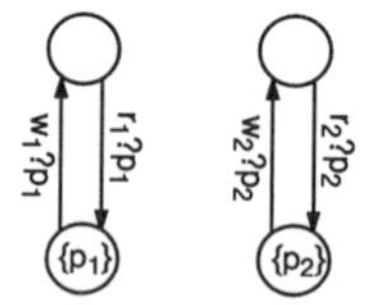

Figure 5 ETS of duplex repeater

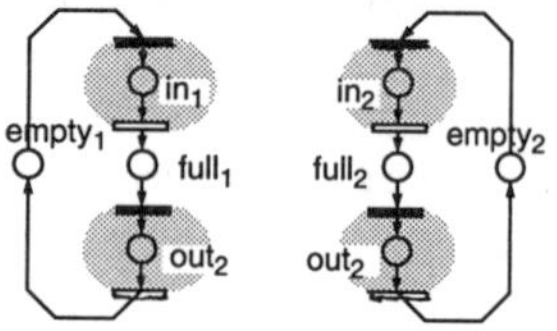

Figure 6 Basic model of the system

Note: The shaded areas in Figure 6 identify the subnets that model the interactions.

For each of the resources (processors and memory), a place is defined that indicates whether the resource is used or free.

Processors

For each processor (or process), a token is defined that, when present in the idle location represent the processor being idle. If a processor is busy executing an interaction, then the corresponding processor token is found at the place in that interaction. The start of a thread is modelled by an input-arc from the processor to the first interaction of the thread. Likewise, the end of a thread is modelled by an output arc from all last interactions of the thread to the processor.

If the processors can be freely allocated to threads, a single place represents all processors. If some processors are only allocated to some threads, then multiple places are needed for representing the processors.

Memory

For each memory location, a token is defined that, when present in the memory location represent the memory location being empty. Arcs present represent the allocation/deallocation of variables to memory. Any memory location to which a variable that is obtained in an interaction is mapped must be allocated before the interaction while any deallocation can only be done at the end of an interaction. The allocation/deallocation is assumed to occur in zero time.

3.3. Scenarios

A number of scenarios is defined in which the effect on the performance of the implementation decisions are analyzed. In Petri nets, only choices can be made between transition that are enabled, so the poll interaction need not to modelled as such.

scenario 1 is the baseline scenario. It involves threads consisting of single interactions. The choice between interactions is a random choice. The Petri Net model of the system and its environment according to this scenario is shown in Figure 7.

In **scenario 2**, the scheduling involves scheduling of threads rather than single interactions. The choice between the interactions is still a random choice.

Scenario 3 analyses the influence of non-negligible time needed for polling. An extra place and a transition are introduced between the place CPU and its outgoing arcs. The rate of the transition is determined by the time needed for scheduling.

Scenario 4 analyzes the effects of making a choice between blocking interactions. In this scenario, interactions are modelled by a different subnet than in the other scenarios. This subnet is shown in Figure 8.

For all of the scenarios non-deterministic choice is removed in two ways: either by random choice or by giving a fixed priority to interactions of stream 2. Note that giving a fixed priority to w interactions coincides in this system with the scheduling of threads consisting of two interactions.

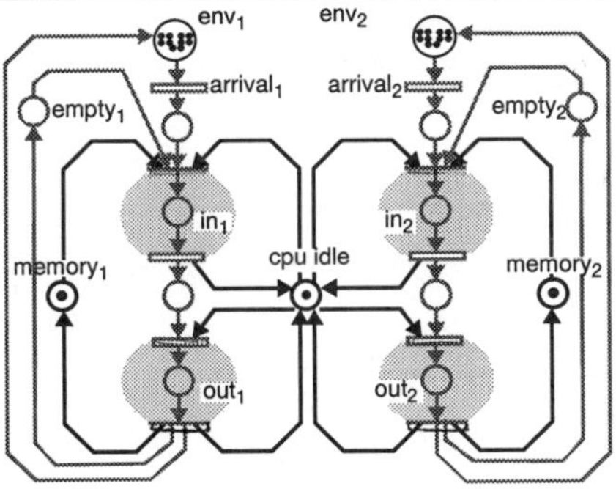

Figure 7 Full Petri Net scenario 1

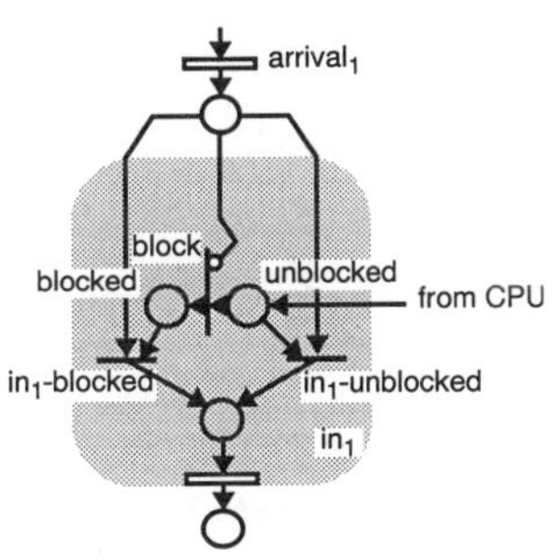

Figure 8 Interaction subnet scenario 4

Note: Figure 7 is obtained by superposition of figures 4, 6 and the allocation of resources

The rates of the timed transitions are based on the actual measurements presented in [3]. The processing time for r and w interactions for a packet with average length 1400 bytes is 926 μs. No estimate is available for the time needed for scheduling, so this time is arbitrarily set to 1/10 of the packet processing time.

For SPNP, only Negative Exponential distributions of parameters are allowed, so it is assumed that the all rates have a Negative Exponential distribution with mean equal to the given time average. In the remainder of this text, when used without and adjective, the terms delay and throughput denote the mean delay and mean throughput.

3.4. Results

All results are obtained using the SPNP version 3.1 on a Sun SPARC station IPX with 32MB memory, running SunOS 4.1.1 at 40 MHz. The total time for determining all results is approximately 13 minutes. This time includes no optimization other than those present within SPNP itself.

The measurements are shown in Figure 9 and Figure 10. The results of assigning priority to stream 2 in scenario 4 are not shown as the stream 1 is permanently starved by stream two.

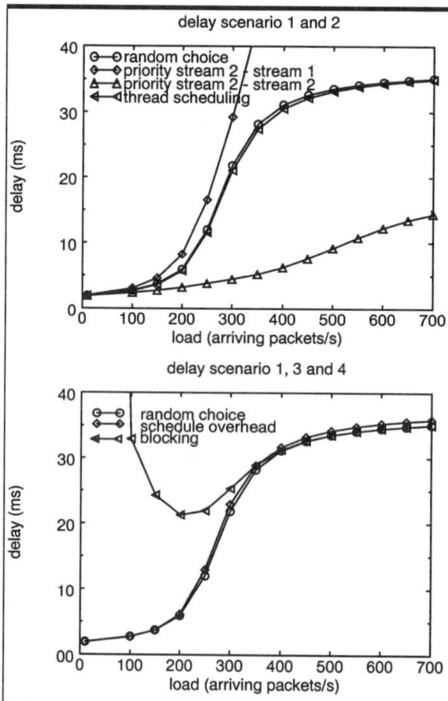

Figure 9 Performance results: mean delay

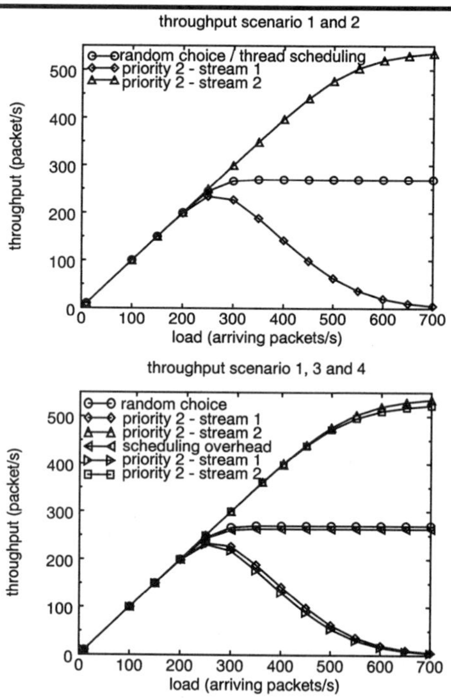

Figure 10 Performance results: mean throughput

Notes: a) Depicted are performance results for varying rates of arriving packets.
b) Unless indicated, measures are the same for both streams.

Based on these measurements the following conclusions can be drawn:

- Under low load, the difference in delay between scenarios 1 through 3 is very large. Under high load, the difference between scenario 4 and 1 is small
- Under low load (≤200 packets/s), the difference in throughput between all scenarios is small. Under high load (≥300 packets/s), assigning priority to stream 2 causes starvation of stream 1.
- The effects of a non-zero scheduling time are small under low load, but not negligible: at 200 packet/s the increase of delay due to scheduling is 3.5%.
- Making a choice between interactions without knowing whether they will block (scenario 4) severely affects the delay, so making avoidance of the choice between interactions a main design issue is justified.

3.5. Discussion

3.5.1. Results

The performance gain by scheduling threads with length greater than 1 rather than single interactions (Implementation issue 3) is a small (3.5%) while a higher gain might be expected, especially for larger systems. This unexpected result may be caused by the simplicity of the repeater system and the particular assumption on the scheduling time.

Current tools are capable of obtaining transition based results (e.g. throughput) and place based results (e.g. average number of tokens or probability of being empty). The mean delay must be calculated using Little Law. This is a drawback as only the mean delay and not its variation can be calculated in this way.

3.5.2. Assuming Negative Exponential distributions

The used analysis package requires that all rates have a Negative Exponential distribution. In practice, this distribution is very seldom observed. This subsection analyses the influence of assuming Negative Exponential distributions.

Negative Exponential distributions are assumed for the load offered by the environment and for the duration of interactions. For the load this is an acceptable assumption as any assumption is equally reasonable in the absence of more information on the nature of the distribution. For the duration, this is less acceptable as more information on the distribution is available: it is a discrete distribution when CPU speed and packet length are constant.

The Negative Exponential distributions were chosen as this choice facilitates numerical analysis of mean delay and throughput without need for simulation. If variation is also important then simulation, e.g. using UltraSAN [4], is necessary. Disadvantages of simulation are the less exact results and much larger analysis time [13].

The assumption relates to the aims of this paper in two ways. One of the aims of this paper was to show how performance analysis can be used during the implementation process. For this aim the precise distribution of load and duration are not relevant at all. Another aim of the analysis was to offer quantitative support for the design issues. Making the assumptions on the distribution of load and duration makes the quantitative results less accurate, but the relative ordering of results can be used to obtain insight in relative effects of the design decisions.

3.5.3. Obtaining the system model

The mapping of the LOTOS specification onto the Petri-Net was simple for the duplex repeater. The mapping consists of two steps: mapping the full LOTOS specification onto an ETS and mapping of the ETS onto a Petri Net. For the latter step, a mapping was given. The first step (LOTOS → ETS) is much more complex. Fortunately, this activity can be performed by a tool (Smile [8])

A mapping of LOTOS onto extended finite state machines (XFSMs) can be found in [14]. In this mapping, a LOTOS behaviour is mapped onto a set of communicating XFSMs. The difference with our mapping onto ETSs is that the mapping onto XFSMs preserve the process-structure of the specification while the mapping onto ETSs flattens the structure. The motivation for performing the flattening is that is facilitates efficient mapping of the ETS onto the finite state model underlying programming languages.

The Petri Net models used are all without data. This disables direct modelling of deterministic choice if this choice depends on the contents of data that is associated with a transition, forcing the designer to make assumptions on the frequency of the choice between interactions.

4. Conformance

The conformance of the implementation to its specification must be ascertained. This is done by i) showing that the implementation of the duplex repeater are not testing equivalent with its specification, ii) motivating why these implementations are, although not testing equivalent with the specification, to be considered conforming implementations, iii) defining a conformance relation that captures precisely the freedom of the designer as identified in section 2 and iv) showing that the implementation conforms to the specification according to the newly defined conformance relation.

Showing that the specification and implementation are not testing equivalent is done by giving a counter-example. If the specification has performed an $\mathtt{r}_1$ interaction, a choice can be made between the interactions $\mathtt{r}_2$ and $\mathtt{w}_1$. A implementation with priority to w interactions will never perform an $\mathtt{r}_2$ interaction. Thus there is an observation of the specification that is never made of the implementation, hence specification and implementation are not testing equivalent. The implementation is a trace reduct of the specification.

There are two different conformance issues related to the conformance of the implementation to its specification: i) the implementation of two independent components in the specification within a single unit of parallelism and ii) the designer's freedom in giving priority to some interactions above others. The motivation for identifying them as separate conformance issues is that they are related to different design issues. The first conformance issue is related to Implementation issue 1 while the second conformance issue is related to Implementation issue 4. These implementation issues are independent from each other, so the corresponding conformance issues are also independent from each other.

The two conformance issues are formalized into two conformance notions that are derived from the well-known testing equivalence notion. Testing equivalence was chosen as starting point because the current design phase is close to realization.

4.1. Implementing independent components within a single process

This conformance issue is explained using scenario 1 with priority to w interactions. The duplex repeater system consists of two independent simplex repeaters that each receive a packet and transmit it with some delay. A realization of a simplex repeater is correct when it exhibits a sequence of alternating r and w interactions, irrespective of the length of the time interval between successive interactions. This implies that a combined realization of two simplex repeaters can delay the execution of a $\mathtt{r}_2$ for any reason, including delaying it because a $\mathtt{w}_1$ interaction is executed first. This execution of $\mathtt{w}_1$ is permissible because $\mathtt{w}_1$ stems from component that is independent of that of $\mathtt{r}_2$.

This can be generalized as follows: for a specification consisting of a number of independent (interleaved) components, a implementation is conformant if it implements the behaviour of each of the components individually correct.

A formal conformance notion called Parallel testing equivalence ($\approx_{|||te}$) is defined below that captures this informal generalization.

Definition: $\approx_{|||te}$ Parallel testing equivalence

Given two (LOTOS) specifications `Imp` and `Spec` with

$$\texttt{Spec} = \texttt{S}_1 \;|||\; \texttt{S}_2 \;|||\; \ldots \;|||\; \texttt{S}_n,$$

`Imp` $\approx_{|||te}$ `Spec` iff $\forall$ `i = 1...n: ` $\texttt{Imp}_{|\texttt{L}(\texttt{S}_i)} \approx_{te} \texttt{S}_i$ whereby

L(S) denotes the set of observable interactions of S and

$\texttt{Imp}_{|X}$ denotes the restriction of the observations of `Imp` to only interactions contained in X. All interactions that are not element of X are simply removed from all observations of Imp. The restriction of the observation $r_1;r_2;w_2;w_1$ to the set $\{r_1, w_1\}$ gives the observation $r_1;w_1$.

This notion is weaker than testing equivalence which is easily inferred from the duplex repeater.

Note that $\approx_{|||te}$ also captures the conformance issue related to the addition of the poll interaction.

4.2. Choice between interactions

The second conformance issue is the way an implementation makes a choice between interactions.

Conformance testing theory (e.g. [23]) involves the notion of refusal of an interaction: an interaction is refused if in a sufficiently long period the interaction has not occurred. Likewise it can be said that a procedure call instance is refused when doesn't return within a sufficiently short period. i.e. the call instance is blocking. This leads to the conclusion that an interaction is refused when the corresponding procedure call instance is blocking.

During the implementation process it was assumed that some procedure calls would never block. This is equivalent with assuming that the corresponding interactions are non-blocking. This implies i) that any test of the realization involving refusal of a non-blocking interaction cannot be performed and ii) that the realization, when given the choice between a set of interactions, it can chose any interaction that is non-blocking.

An implementation that makes such choice in a fixed way is a trace-reduct of its specification (`Imp` $\leq_{tr}$ `Spec`). Rather than allowing any trace reduct of the specification to be a correct implementation, only implementations that a reduct with respect to the choice between non-blocking interactions are considered correct. For the duplex repeater realizations, this means that a realization whose observations consist of sequences of $r_1;w_2$ and $r_2;w_2$ is considered correct while a realization whose observations consist of sequences of only $r_1;w_1$ is not.

This notion of conformance is called Implementation Preorder modulo a set of interactions and is defined as follows.

Definition: $\leq_{imp,X}$ Implementation preorder modulo a set of interactions X

Given two LOTOS specification `Spec` and `Imp` then `Imp` $\leq_{imp,X}$ `Spec` iff
There exists a partitioning of `Traces(Spec)` into `Traces(Imp)` and a set of not observed traces `NOT(Spec,X)` whereby

The set `NOT(Spec, X)` is defined by the following condition:

$\forall$ `v` $\in$ `Traces(Spec)`, $\forall$ `a` $\in$ `L(Spec)`, $\forall$ `w` $\in$ `L(Spec)`*:
(`v;a;w` $\in$ `NOT(Spec,X)` and `v;a` $\notin$ `Traces(Spec)`) implies
$\exists$ `x` $\in$ `X`: `v;x;w` $\in$ `Traces(Spec)`

The `NOT` contains precisely the observations that can be made of the specification but that are not made of the implementation due to a choice between interactions made by the implementation.

Verification that $\leq_{imp,X}$ is indeed a preorder is done by showing that the conformance relation is reflexive (Spec $\leq_{imp,X}$ Spec) and transitive ($Spec_1 \leq_{imp,X} Spec_2$ and $Spec_2 \leq_{imp,X} Spec_3$ implies $Spec_1 \leq_{imp,X} Spec_3$). It is easily verified that this is indeed the case for fixed X.

As both conformance issues are independent, it is conjectured that the definition of $\leq_{imp,X}$ can be extended to the $\leq_{|||impl,X}$ conformance relation. This allows compositional verification of the conformance of an implementation to its specification. A second conjecture is that $\leq_{|||impl,X}$ captures precisely the designer's freedom.

It is easy to see that $\leq_{|||impl,X}$ is weaker than testing equivalence for non-empty X and that $\leq_{|||impl,X}$ and testing equivalence coincide for empty X.

4.3. Verification of conformance

The next thing to do is the verification of the conformance between the implementation of the duplex repeater to their specifications. This verification is done in a rather informal way and done separately for scenarios 1 through 3 and for scenario 4.

In scenarios 1 through 3, the implementations interact with the environment in a desired way: if the environment refuses either r_1 or r_2 or both, all implementations under all scheduling strategies perform some w_i interaction. As the environment is not able to refuse any other interactions, it suffices to conclude that all implementations conform to the specification modulo the set $\{w_1, w_2\}$.

In scenario 4, the situation depends on the scheduling strategy. If random choice is made, the scenario conforms in the same way as the other scenarios. Whether the performance of the scenarios is acceptable is an issue that is not captured by the conformance relation. Under the strategy with priority to interactions of stream 2, any interaction relating to stream 1 is never executed by the implementation, thus the observations of the implementation does not contain the observation r_1, w_2. As this observation is not part of NOT(Specification, $\{w_1,w_2\}$), this implementation is not conformant according to the $\leq_{imp,\{w_1,w_2\}}$ relation.

4.4. Discussion

This section has introduced new conformance notions that are weaker than testing equivalence. Why are these weaker notions used rather than using the well-known testing equivalence? The notion of testing equivalence fails to the design freedom into account that exists when a set of independent and non-deterministic specification components are implemented in a single unit of parallelism.

By making assumption on the environment in which the realization is to operate, the universe of environments is reduced. These assumptions are equally applicable to testers of the implementation because the testers are a particular environment of the implementation. This implies that a test involving refusal of a interaction whose procedure call is assumed to be non-blocking is a test that may not be performed. Performing such test does not discriminate between conforming and non-conforming implementations (which is what it should do), but it discriminates the implementations that do respectively don't make the assumptions on the environment.

5. Relation with other work and future work

5.1. Relation with other work

Mapping of LOTOS onto Petri-Nets is not new. In [20] and [9] nets similar to Petri Nets are used. One of differences between their work and the work in this paper is that in this paper, the mapping is done via an intermediate level (Extended Transition System). The advantage is that by using ETSs, the mapping onto Petri Nets can be adapted more easily for process algebraic languages other than LOTOS. Mapping of other process algebraic languages onto Petri Nets are known from the literature [11,22,10] but the purpose of these mappings is to serve as a vehicle for either the definition of the semantics of the formal language and for simulation, rather than as a vehicle for performance analysis. A drawback of these languages (including LOTOS) is that the duration of interaction is not part of the semantics of the language so the mapping of the languages onto Petri Nets cannot take the duration into account. Recently an extension of process algebra's with durational interactions has been defined [12], but no mapping onto Petri Nets is yet defined for their language.

Other examples of the use of Petri Nets are [17] and [2]. The analysis described in the latter article is similar to the work in this paper, but they don't obtain the Petri Net from a formal specification but from an Ada description.

The work in this paper differs from the approaches taken by many authors when implementing using formal descriptions [24, 19]. Their approaches are more focussed on obtaining a rapid prototype of an arbitrary LOTOS specification than on obtaining an efficient realizations. Their design decisions are made in a pre-defined way, thus ignoring the effect of the implementations issues on the efficiency of the realizations.

5.2. Future work

The current mapping of LOTOS behaviour onto ETSs removes all structure present within the individual components. This sometimes is considered unwanted. Further work is needed that defines a mapping that preserve the process-structure present in the LOTOS behaviour onto some structure present in the C-code.

Insight in the validity of the restriction to Negative Exponential distributions is necessary. This can be gained by simulations or by measurements on realizations.

The need for two new conformance notions was identified and two formal conformance notions were defined. The applicability of these notions for other systems than the duplex repeater is not yet known, so this applicability needs to be studied.

6. Conclusions

This paper has presented a methodology for deriving efficient implementation out of a formal specification that has the mapping of interactions onto procedure calls defined and that contains no non-deterministic value generation.

Aspects of this methodology are i) the identification of implementation issues that affect efficiency, ii) a method for transforming a specification into an efficient implementation in which the designer is assisted by performance analysis, iii) conformance notions that captures the freedom the designer has during this transformation.

The methodology has been shown usable: the method for mapping a specification onto a implementation and the application of performance analysis techniques have been demonstrated using a small system.

Although the validity of the quantitative results in terms of average delay/throughput depend on the validity of the assumptions made on arrival and processing rates, these results can be used for comparing the effect of implementation decisions relative to each other.

Acknowledgments

Mark de Weger and Lex Heerink are thanked for commenting on drafts of this paper.

References

[1] G. Ciardo and J. K. Muppala. *Manual for the SPNP Package Version*. Department of Electrical Engineering, Duke University, Durham, NC, version 3.1 edition, Sept. 1991.

[2] G. Ciardo and J. K. Muppala. Analyzing concurrent and fault-tolerant software using stochastic reward nets. *Journal of Parallel and Distributed Computing*, 15:255–269, 1992.

[3] D. D. Clark, V. Jacobson, J. Romkey, and H. Salwen. An analysis of TCP processing overhead. *IEEE Communications Magazine*, pages 23–29, June 1989.

[4] J. Couvillion, R. Freire, R. Johnson, W. D. Obal II, A. Qureshi, M. Rai, and W. H. Sanders. Performability modelling with UltraSAN. *IEEE Software*, 8(5):69–80, Sept. 1991.

[5] M. Diaz and R. Groz, editors. *Formal Description Techniques — V*. North-Holland Publishing Company, 1993.

[6] H. Eertink. *Simulation Techniques for the Validation of LOTOS Specifications*. PhD thesis, University of Twente, Mar. 1994.

[7] H. Eertink, P. Kars, and H. Wennink. Reducing extended transition systems. Memoranda Informatica 93-09, University of Twente, Feb. 1993.

[8] H. Eertink and D. Wolz. Symbolic execution of LOTOS specifications. In Diaz and Groz [5], pages 295–310.

[9] H. Garavel and J. Sifakis. Compilation and verification of LOTOS specifications. In L. Logrippo, R. L. Probert, and H. Ural, editors, *Protocol Specification, Testing and Verification X*, pages 359–376. IFIP WG 6.1, North-Holland Publishing Company, 1990.

[10] U. Goltz. CCS and Petri nets. In I. Guessarian, editor, *Semantics of Systems of Concurrent Processes*, volume 469 of *Lecture Notes in Computer Science*. Springer-Verlag, 1990.

[11] U. Goltz and A. Mycroft. On the relationship of CCS and Petri nets. In J. Paredaens, editor, *Automata, Languages and Programming*, volume 172 of *Lecture Notes in Computer Science*, pages 196–208. Springer-Verlag, 1984.

[12] R. Gorrieri and M. Roccetti. Toward performance evaluation in process algebras. Technical Report UBLCS-93-16, University of Bologna, July 1993.

[13] B. R. H. M. Haverkort. *Performability Modelling Tools, Evaluation Techniques, and Applications*. PhD thesis, University of Twente, Jan. 1990.

[14] G. Karjoth. Implementing LOTOS specifications by communicating state machines. In W. R. Cleaveland, editor, *Concur'92*, volume 630 of *Lecture Notes in Computer Science*, pages 386–400. Springer-Verlag, 1992.

[15] B. W. Kernighan and D. M. Ritchie. *The C programming language*. Prentice-Hall, 1988.

[16] H. Kremer, J. van de Lagemaat, A. Rennoch, and G. Scollo. Protocol design using LOTOS: A critical synthesis of a standardization experience. In Diaz and Groz [5], pages 231–246.

[17] C. Lindemand and F. Schön. Evaluation sequential consistency in a virtually shared memory system by deterministic and stochastic petri nets. In H. Schwetman, J. Walrand, K. Bagchi, and D. DeGroot, editors, *International Workshop on Modelling, Analysis and Simulation of Computer and Telecommunication systems*, volume 25 of *Simulation Series*, pages 63–68. ACM, 1993.

[18] Lotosphere. The lotosphere design methodology: Basic concepts. Deliverable Lo/WP1/T1.1/N0045/V04, Esprit Project 2034 "Lotosphere", 1992.

[19] J. A. Mañas, T. de Miguel, J. Salvachúa, and A. Azcorra. Tool support to implement LOTOS formal specifications. *Computer Networks and ISDN Systems*, to appear, 1992.

[20] S. Marchena and G. Leon. Transformation for LOTOS specs to galileo nets. In K. Turner, editor, *Formal Description Techniques*, pages 217–230. Elsevier Science Publishers B. V., Sept. 1989. Proceedings of the First International Conference on Formal Description Techniques.

[21] M. A. Marsan, G. Conte, and G. Balbo. A class of generalized stochastic petri nets for the performance evaluation of multiprocessor systems. *ACM Transactions on Computer Systems*, 2(2):93–122, May 1984.

[22] E.-R. Olderog. Operational petri net semantics for CCSP. In G. Rozenberg, editor, *Advances in Petri Nets 1987*, volume 266 of *Lecture Notes in Computer Science*, pages 196–223. Springer-Verlag, 1987.

[23] J. Tretmans. *A Formal Approach to Conformance Testing*. PhD thesis, University of Twente, Dec. 1992.

[24] A. Valenzano, R. Sisto, and L. Ciminiera. Rapid prototyping of protocols from LOTOS specifications. *Software Practice and Experience*, 23(1):31–54, Jan. 1993.

Position Statements

35

Stochastic Process Algebras: Integrating Qualitative and Quantitative Modelling*

J. Hillston[a†], H. Hermanns, U. Herzog, V. Mertsiotakis and M. Rettelbach[b‡]

[a]Dept. of Computer Science, University of Edinburgh, Kings Buildings, Edinburgh, Scotland

[b]IMMD VII, University of Erlangen-Nürnberg, 91058 Erlangen, Germany

1. Introduction

Stochastic process algebras (SPA) have emerged from work over the last five years investigating the use of process algebras for performance modelling. Like other stochastic extensions of existing system description techniques, they offer exciting possibilities for integrating qualitative and quantitative analysis. The use of such languages facilitates the early consideration of the temporal properties of an emerging design. Unlike existing approaches, SPA exploit the compositionality, known from classical process algebras. The algebraic framework of this compositionality has advantages for both model construction and model solution.

The compositional construction provides a systematic approach to the hierarchical development of the models. This means that the structure of the system is clearly reflected in the model. Formal semantics and congruence relations characterising the behaviour of components mean that this structure can be exploited to reason about the model. Abstraction mechanisms allow a subsystem to be modelled in detail but subsequently presented to its environment with only a restricted interface. Moreover the structure imposed on the model may also be used to control both functional and temporal analysis of the system.

2. The Approach

In a SPA, as in a classical process algebra, a system is described as an interaction of *agents*. The behaviour of an agent is defined by the *actions* it can perform or as a composition of smaller agents. The ways such behaviours can be constructed are formally defined by a grammar which specifies how actions and agents may interact. Actions are atomic at the chosen level of abstraction. Each of the published SPA [1, 2, 5, 8, 9] has settled on the same core set of combinators: *prefix, choice, concurrency* and *hiding*. Hiding provides the abstraction mechanism within the language.

In classical process algebras only the relative ordering of events is modelled and branching is non-deterministic. Recent timed and probabilistic extensions to these languages still leave them unsuitable for performance modellers. In a SPA a random variable is associated with every action, representing its *duration*. This is analogous to the association of a duration with the firing

*The cooperation was supported by the German National Research Council (DFG) under *SFB 182*.
†J. Hillston is supported by EPSRC Research Fellowship B/93/RF/1729.
‡The group in Erlangen is supported by SFB 182 and the ESPRIT BRA Project 7269 QMIPS.

of a timed transition in a Stochastic Petri Net.

SPA using generally distributed random variables have been investigated [4], but current work is focussed on languages in which all distributions are assumed to be exponential [1, 2, 8, 9]. This has the advantage that the underlying stochastic process will be a Continuous Time Markov Chain (CTMC), amenable to solution by a variety of techniques. When enabled, an activity will delay for a period determined by its associated exponential distribution. A *race condition* is assumed to exist between concurrently enabled activities — the first completed delay governs which action takes place. Via this race condition branching becomes probabilistic.

The structural operational semantics of the language is given as a labelled transition system containing the stochastic delay of activities as well as their action type. Functional analysis is carried out on a derivation graph representing the transition system; for performance analysis the graph is reduced to the underlying CTMC. The existence of an equilibrium behaviour for this process can be easily checked by conditions on the initial structure of the process term [9]. Numerical techniques are used to extract the steady state distribution. Transient analysis is carried out similarly. From the resulting probability distribution it is straightforward to obtain performance measures such as throughput, utilisation or mean time to failure.

3. Theoretical Results

Theoretical work so far has concentrated on equivalence relations for SPA. The usual notion of equivalence in process algebras is *bisimilarity*, or indistinguishability under observation. In a SPA there are two aspects of behaviour which may be observed, functional and temporal. Bisimulation relations have been developed which capture either or both of these aspects. Most of these relations have been shown to be preserved by the combinators of the languages, i.e. they are congruences reflecting the compositional structure of the model.

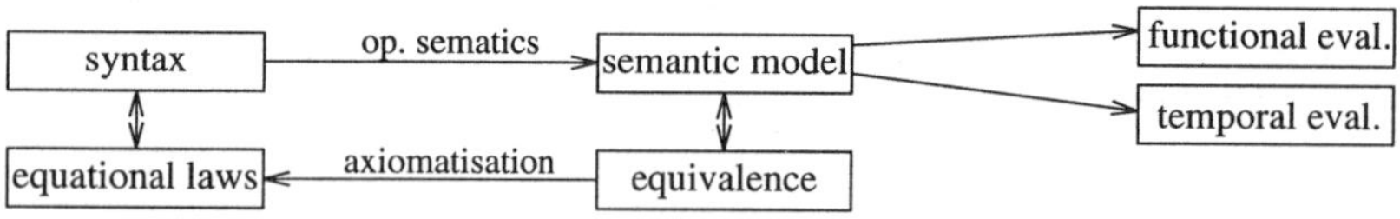

Functional bisimulation captures the notion of agents which are equivalent in terms of the types of actions which they can perform [8]. For example, it may be used to show that a system provides the functionality specified by an abstract specification, or to investigate alternative designs when quantitative aspects of the system are unimportant. *Markovian bisimulation* [8] or *strong equivalence* [9] capture information about the types and rates of actions which an agent can perform. They characterise systems which exhibit exactly the same stochastic behaviour. Additionally they may be used to partition the state space of a single system implying an exact aggregation of the underlying CTMC based on lumpability [9]. Due to the congruence property this aggregation can be systematically applied through the component structure of the model, without the need to construct the state space of the original process. Moreover, term rewriting based on equational laws may be used to carry out the aggregation on the syntactic level. In [8] a set of equational laws is shown to completely characterise Markovian bisimulation on finite behaviour systems. Further notions of equivalence are presented in [9]. For example *weak isomorphism*, in which the observation of internal activities is relaxed, can be used for state space reduction which guarantees the integrity of mean value performance measures.

4. Experiences

The applicability of SPA to the modelling and evaluation of communication and multiprocessor systems has been demonstrated by extensive case studies as well as small examples.

In [5] an example is presented that shows how to model workload mapping to a multiprocessor system. In addition, a simple model of the Send-and-Wait protocol is analyzed. In [7] the SPA TIPP is used to evaluate an alternating bit protocol with a faulty channel.

The description and analysis of several classical and extended polling models using the SPA PEPA are presented in [9]. Moreover, the use of equivalences to aggregate the state space is demonstrated.

An interesting integration of performance and functional analysis using a SPA is shown in [6]. A distributed mail system containing three deadlocks is investigated. Various measures are computed, such as throughput and mean time to deadlock. A refined, deadlock-free version of the mail system is compared with the original version. Calculation of quantitative aspects of qualitative properties, e.g. mean time to deadlock, is an advantage of carrying out both analyses on the same model.

The case studies mentioned above were performed using prototype evaluation tools for PEPA and TIPP respectively; details of the PEPA Workbench can be found in [3].

REFERENCES

1. M. Bernardo, R. Gorrieri, and L. Donatello. MPA: A Stochastic Process Algebra. Technical report UBLCS-94-10, Laboratory for Computer Science, University of Bologna, 1994.
2. P. Buchholz. On a Markovian Process Algebra. Forschungsbericht 500, Informatik IV, Universität Dortmund, 1994.
3. S. Gilmore and J. Hillston. The PEPA Workbench: A Tool to Support a Process Algebra-based Approach to Performance Modelling. In G. Haring and G. Kotsis, eds., *Proc. of the 7th Conf. on Modelling Techniques and Tools for Computer Performance Evaluation*, 1994.
4. N. Götz, U. Herzog, and M. Rettelbach. TIPP – a language for timed processes and performance evaluation. Technical Report 4/92, IMMD VII, University of Erlangen-Nürnberg, November 1992.
5. N. Götz, U. Herzog, and M. Rettelbach. Multiprocessor and Distributed System Design: The Integration of Functional Specification and Performance Analysis using Stochastic Process Algebras. In *Performance'93*, 1993.
6. H. Hermanns, U. Herzog, J. Hillston, V. Mertsiotakis, and M. Rettelbach. Stochastic Process Algebras: Integrating Qualitative and Quantitative Modelling. Technical Report 11/94, IMMD VII, University of Erlangen-Nürnberg, May 1994.
7. H. Hermanns, V. Mertsiotakis, and M. Rettelbach. Performance Analysis of Distributed Systems Using TIPP – a Case Study. In *Proc. of the 10th U.K. Performance Engineering Workshop for Computer and Telecommunication Systems*, 1994.
8. H. Hermanns and M. Rettelbach. Markovian Processes go Algebra. Technical Report 10/94, IMMD VII, University of Erlangen-Nürnberg, 1994.
9. J. Hillston. *A Compositional Approach to Performance Modelling*. PhD thesis, CST-107-94, Department of Computer Science, University of Edinburgh, 1994.

36

A new Approach to Performance Evaluation of Formally Specified Protocols

Stefan Böhmer, Ralf Klafka
Communication Networks
Aachen University of Technology, 52074 Aachen
EMail: sb@dfv.rwth-aachen.de, rkl@dfv.rwth-aachen.de

This report proposes enhanced methods for performance evaluation of formally specified communication protocols. A concept as well as an implementation is shortly described which combines the techniques of formal specification and stochastic simulation. A main focus is on semi–automatic generation of a program for stochastic simulation directly from a given protocol specification in SDL. Further, a new method is introduced to estimate the hardware performance requirements to run a given communication protocol properly. By means of an appropriate hardware emulation system that has been implemented, it becomes as well possible to compare different protocol implementations with regard to the efficiency of hardware utilization.

1. Technical Approach

The importance of using formal specification techniques for the development of communication protocols is generally undisputed. SDL [1], the most popular specification language in the field of telecommunication, is specialized to describe the behaviour of systems. It is widely used to specify telecommunication standards as well as it has been selected in many international projects as the standard description language for the specification of future telecommunication systems. As stated in [2] SDL is now well supported by tools which allow the writing and the correctness analysis of SDL documents. Although SDL has primarily been designed for description purposes, further acceptance and importance of SDL will also be determined through possibilities to integrate formal (SDL-)specifications as a main part in an entire protocol development cycle. This means for example to use SDL in a project as a description and formal programming language from the software requirements specification phase up to final software implementation. Therefore, it isn't sufficient having tools 'just' providing code generation facilities but having objective methods to evaluate the quality of (proposed) protocols in terms of performance, resistance against error failures etc. as well as to compare different SDL development tools with each other. The approach presented here deals with concepts and realizations combining the methods of stochastical simulation and formal protocol specification. Futhermore, concepts and methods will be described which make it possible to determine the computing power of a processor system required to run a protocol. The main improvements with respect to the development cycle of communication protocols can be summarized as follows:

[1]Specification and Description Language

- hardware independent comparison of different communication protocols under defined traffic conditions,
- early insight in dynamic behaviour of newly designed protocols,
- performance evaluation of protocols independent from the existence of a communication network,
- support of hardware selection processes for terminal equipment, switches etc. by estimating the computing power needed by protocols,
- estimation of additional system load caused by protocol options and/or extensions.

Experiences with the methods and appropriate tools described above will also have direct impact on the further development of the SDL language: through the use of SDL as a language for protocol implementation it will be possible to orientate the SDL standardization process closer to user requirements.

2. SPEET: a Toolset for Stochastic Simulation of Formally Specified Protocols

As described above the following approach places formal protocol specifications in SDL in the centre of an entire software development and hardware design process. A main aspect is to consider that the program for a stochastical simulation as well as a final protocol implementation should be generated semi-automatically from a formal specification.

2.1. Simulation model

The development of a tool integrating formal protocol specification and stochastical simulation brought up two major tasks: on the one hand to identify simulation and evaluation specific components as well as to determine essential components of a real protocol environment, and on the other hand to combine them in a suitable manner with the SDL concepts. The resulting model shown in Figure 1 is closely related to the layered structure of a communication system as defined by the ISO/OSI model. It consists of several independent applications supporting a distributed simulation if necessary. The traffic-generator must be seen as the essential part of the simulation environment because modelling of typical workload has central importance for performance analysis. The idea is to describe workload as it would be caused by users, applications and/or protocol layers on the top of the protocol stack to be examined through suitable arrival processes. The simulation control directs all the processes involved in the simulation run by providing a globally unambiguous time-base for asynchronous and synchronous discreet-event simulation. The stochastic evaluation module comprises new algorithms to estimate the stationary distribution function of simulated data as well as to gain a cost effective control of the length of a simulation run. The OSI-layers located below the protocol to be examined and/or the physical transport medium of a communication network are substituted by the channel model. It comprises details about the server processes/ strategies for the substituted protocols, time constraints in view of medium access etc. Moreover, the hardware simulator makes it possible to determine the computing power of a processor system required to execute a protocol properly under given (protocol specific) timing conditions.

2.2. Implementations and Results

A powerful realization of the concept was performed without any limitations to the usage of SDL; we make rather use of the general view that a SDL system is embedded in an

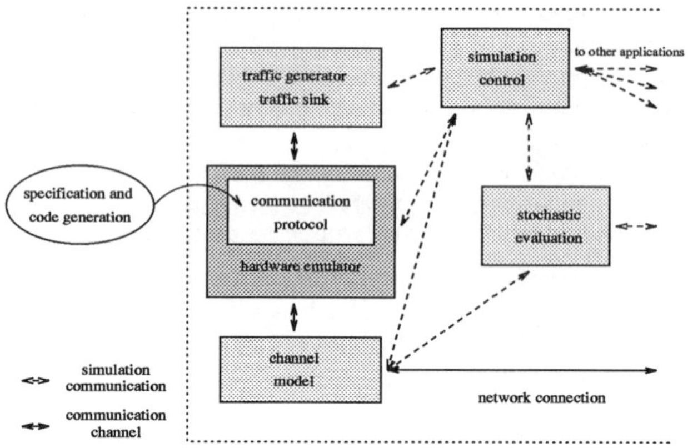

Figure 1. Simulation model

enviroment and communicates with it by means of signals. As shown in Figure 1 the code automatically generated by an SDL tool (e.g. SDT[2]) from a given specification, is used to build the stochastical simulation program. All the modules described in section 2.1 have been implemented as independent processes and have been combined to a toolset called SPEET (SDT Performance Evaluation Toolkit). Practical experiences with SPEET have shown that the demands on a SDL development tool can be reduced to a minor degree in order to use the code generated from a SDL specification directly for the generation of a simulation program. The SPEET tool has also successfully been used for performance analysis of the GSM signalling protocol $LAPD_m$. Appropriate results are published in [1]. Moreover, as a further development of the SPEET tool new measures for hardware dependent protocol performance evaluation have been derived. Hardware emulation results show that a new measure called *CPU–workload* is a suitable number as well for hardware performance evaluation as for hardware dependent protocol analysis. Current work focusses on the design of a SDL-protocol suitable as benchmark and on design criterions for protocol specifications in SDL with regard to an efficient target hardware utilization.

REFERENCES

1. S. Böhmer. *Performance Analysis of GSM Signalling Protocol $LAPD_m$*. In *Creating Tomorrow's Mobile Systems, Proceedings 44th Vehicular Technology Conference, Stockholm, Sweden, June 8-10, 1994*. IEEE, 1994.
2. O. Færgemand, A. Sarma, editors. *SDL '93 Using Objects*. North-Holland, 1993.

[2]SDT is a trademark of TeleLOGIC Malmö AB

Integrated Analysis of Concurrent Distributed Systems using Markovian Process Algebra *

Marco Bernardo[a] and Lorenzo Donatiello[a] and Roberto Gorrieri[a]

[a]Dipartimento di Matematica, Università di Bologna
Piazza di Porta S. Donato 5, 40127 Bologna, Italy

Since the 60's a lot of work has been done in order to devise modeling techniques which are adequate to represent concurrent systems. Process algebras are one of the most relevant results of this work. As a matter of fact, (*i*) process algebras are *abstract languages* conceived for defining formally concurrent systems in a *compositional* way; i.e., they provide the designer with a set of operators by means of which constructing complex systems from simpler ones; hence, the designer is led to use a hierarchical and modular design style. And (*ii*) with process algebras concurrent systems can be given formal descriptions which are more easily understandable and more easily modifiable than the ones obtained by many other models.

Classical process algebras, like LOTOS [2], can be used to describe only the *functional aspect* of the behavior of concurrent systems; neglecting their *temporal aspect* is a remarkable drawback for the expressiveness of classical process algebras because (*i*) time-critical concurrent systems (e.g. real time systems) cannot be modeled in a completely satisfactory manner; moreover, (*ii*) the performance of the concurrent systems cannot be estimated. It often happens that a concurrent system is firstly fully designed and tested for functionality and afterwards tested for efficiency; so, if the performance is detected to be poor, the concurrent system has to be redesigned with negative consequences for both the design costs and the delivery at a fixed deadline.

The need of integrating the performance analysis of a concurrent system into the design process of the system itself has recently stimulated many researchers. In the framework of process algebras, this need is satisfied by introducing the concept of time to express the durations of the activities performed by the concurrent systems being modeled; when such durations are expressed in a deterministic way, the process algebra is said to be a *temporal process algebra* (see, e.g., [6]), whereas when such durations are expressed through random variables, it is said to be a *stochastic process algebra* [7, 8].

In this short abstract we introduce a few ideas underlying a new stochastic process algebra, named *MPA (Markovian Process Algebra)*, which is not actually introduced here for lack of space. Indeed, the purpose of this note is not that of formally defining MPA, rather that of giving an idea about what can be done with MPA. The reader interested in the formal definitions is invited to consult the full version of this paper [1] and the three technical reports (TR-94-10, 94-11 and 94-12) by the same authors.[1]

---

*This reaserch has been partially funded by CNR and MURST.

[1]The technical reports are available via anonymous FTP from the area `ftp.cs.unibo.it:/pub/TR/UBLCS` in compressed PostScript format.

MPA is supplied with an operational interleaving semantics, a markovian semantics and an operational net semantics. Therefore, there are three semantic models associated with each MPA term: an *interleaving model* represented by a labeled transition system, a *stochastic model* represented by the state transition rate diagram of a homogeneous continuous time Markov chain (HCTMC), and a *distributed model* represented by a generalized stochastic Petri net (GSPN) [9].

With a stochastic process algebra like MPA we can exploit a modeling technique for concurrent systems which integrates different views of concurrent systems (*centralized* vs. *distributed*) as well as the different aspects of their behaviour (*qualitative-functional vs. quantitative-temporal*) of concurrent systems. Such a modeling technique can be summarized by the following scheme

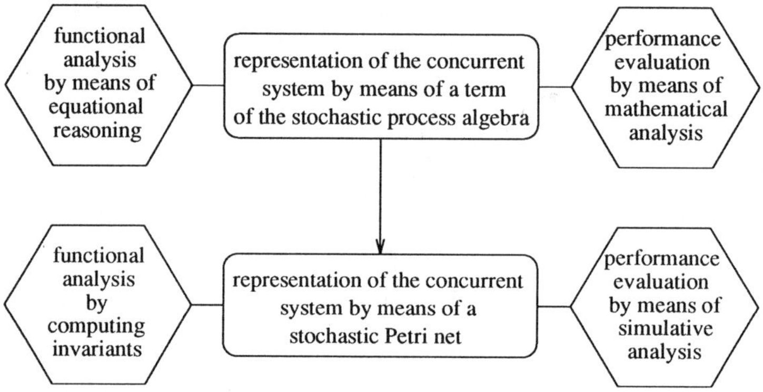

which points out the following integrated specification strategy:

- The first phase consists of specifying the concurrent system as a term of MPA, so as to obtain a first representation, easy to understand and compositional. With this algebraic representation, it is possible to:
  - Perform a functional analysis of the concurrent system by, e.g., equivalence checking or, if the equivalence relation is a congruence, by equational reasoning defined on its axiomatization. Such analyses can detect qualitative properties of the concurrent system (e.g. deadlock-freedom) and can also help in minimizing the state space of the system representation. These analyses can be also computer-aided (see, e.g., [5]).
  - Evaluate the performance of the concurrent system by resorting to the study of its associated HCTMC, which can be assisted by computer (e.g., [11]).
- The second phase consists of translating the term of the stochastic process algebra into a generalized stochastic Petri net (GSPN), giving a distributed representation of the same concurrent system. Such a translation makes explicit the parallelism and the causal dependencies among the activities of the concurrent system itself, but the price to pay is that such a model usually has a remarkable graphic complexity. Moreover, it is not easily compositional, hence making hard, in general, the detection

and the analysis of its subsystems. With this stochastic Petri net representation, it is possible to:

- Perform a qualitative analysis of the concurrent system by, e.g., computing net invariants [3].
- Evaluate the performance of the concurrent systems by resorting to existing tools for simulation on stochastic Petri nets, e.g. [4].

As the various (transformations producing the) semantics we have proposed for MPA can be fully mechanized, future work will be devoted to implement a software tool which associates to each MPA term the collection of its three semantics. On this basis, the next step should consist of integrating this tool with the various ones, already available, tailored for specific purposes. For instance, the interleaving semantics is a labelled transition system which can be input for AUTO [10]; the Markov chain associated to a term can be given as input to SHARPE [11]; its Petri net to PAPETRI [3] and its GSPN to GreatSPN [4]. Therefore, our work opens a perspective to a fully integrated tool which can help in making computer aided – and possibly automatic – the qualitative and quantitative analysis of concurrent distributed systems.

REFERENCES

1. M. Bernardo, L. Donatiello, R. Gorrieri, "Integrating Performance and Functional Analyses of Concurrent Systems with MPA", Technical Report, July 1994.
2. T. Bolognesi, E. Brinksma, "Introduction to the ISO Specification Language LOTOS", *Computer Networks and ISDN Systems*, vol. 14, 1987.
3. G. Berthelot, C. Johnen, L. Petrucci, "PAPETRI: Environment for the Analysis of Petri Nets", in Proc. CAV'90, LNCS 531, Springer, 1991.
4. G. Chiola, "GreatSPN 1.5 Software Architecture", in Proc. of the *Fifth Int. Conf. on Modeling Techniques and Tools for Computer Perf. Eval.*, Turin, 1991.
5. R. Cleaveland, J. Parrow, B. Steffen, "The Concurrency Workbench: A Semantics-Based Tool for the Verification of Concurrent Systems", in ACM *Trans. on Programming Languages and Systems* 15, 1993.
6. R. Gorrieri, M. Roccetti, E. Stancampiano "A Theory of Processes with Durational Actions", to appear in *Theoretical Computer Science*, also available as TR-93-27, Università di Bologna, 1993.
7. N. Götz, U. Herzog, M. Rettelbach, "Multiprocessor and Distributed System Design: the Integration of Functional Specification and Performance Analysis Using Stochastic Process Algebras", in LNCS 729, Springer, 1993.
8. J. Hillston, "A Compositional Approach to Performance Modelling", Ph.D. thesis, CST-107-94, University of Edinburgh (UK), 1994.
9. M. A. Marsan, "Stochastic Petri Nets: an Elementary Introduction", in LNCS, vol. 424, Springer-Verlag, 1990.
10. V. Roy, R. de Simone, "Auto / Autograph", in Proc. CAV'90, LNCS 531, Springer, 1991.
11. R.A. Sahner, K.S. Trivedi, "Reliability Modelling Using SHARPE", *IEEE Trans. Reliabil.* 13 (2), 186-193, 1987.

PART EIGHT

Refinement

38

Modelling Techniques for Evolving Distributed Applications

R. Sekar[a], Y.-J. Lin[a] and C.R. Ramakrishnan[b*]

[a]Bellcore, 445 South Street, Morristown, NJ 07960, U.S.A.

[b]Department of Computer Science, SUNY at Stony Brook, NY 11794, U.S.A.

Several languages and techniques have been proposed for formal specification and validation of concurrent systems. However, these techniques provide no support for modelling incremental changes that take place during software development, such as successive refinements that take place during the design phase or changes that take place later on as a result of software evolution. Consequently any changes to the system model need to be incorporated by manual editing of the system specification, which is cumbersome and error-prone. Moreover, editing being an uncontrolled process, there is no way to automatically carry over (most of the) correctness properties after minor changes to the system. These factors can make formal approaches very expensive for large and evolving systems. To alleviate this problem, we present a language RL (stands for **R**efinement **L**anguage) in this paper that provides syntactic as well as semantic support for modelling incremental changes. Based on the language mechanisms, we then present a method for automatically carrying over properties after refinement. We also present algorithms for compiling RL specifications into finite state automata (FSA) that can be analyzed using traditional algorithms for establishing new properties that hold only after refinement.

Keywords: D.2.1, D.2.4/F.3.1, D.3.3, D.1.3, D.2.2

1. Introduction

Distributed systems are increasingly being used to accomplish complex and critical tasks. However, as compared to centralized systems, distributed systems are inherently harder to program, debug or maintain. Formal modelling and reasoning techniques are therefore being employed to simplify these problems. CSP [8], Estelle [4], Lotos [3], SDL [2] and Petri nets [12] are among many such formal languages. A variety of tools for programming and/or reasoning about specifications in these languages (or variants of these languages) are currently available (e.g., SMV(a tool based on [1] and [5]) , SPIN[9] and COSPAN[10]) and have been used with a good degree of success.

We have been using formal techniques for modelling and reasoning about telecommunication software systems, which are among the largest distributed systems that exist today. Such large systems go through many incremental changes over their life cycle, such as successive refinements that take place during the design phase or changes that take place later on as a result of software evolution. Unfortunately, the techniques mentioned above do not provide support for modelling such changes. Consequently any changes to the system

---

*Work completed while the author was a summer intern at Bellcore.

```
type HookStatus = {idle, offhook};
type Usr2Phone = {liftHandSet, replaceHandSet};
module phone1(hookState, usr2phone) {
   var hookState: HookStatus;
   chan usr2phone: Usr2Phone;
   initial hookState := idle;
   trans
      usr2phone?liftHandSet, hookState = idle → hookState := offhook;
      usr2phone?replaceHandSet, hookState = offhook → hookState := idle;
}
```

Figure 1. A Simple Module Specification in RL

model must be incorporated by manual editing of the system specification. This editing process is cumbersome and error-prone: in the absence of appropriate language support, even conceptually small changes can require editing of the entire system specification and making changes in a number of places. Moreover, since editing is an uncontrolled process, all correctness properties need to be reestablished from scratch, after even minor changes to the specification. These factors can make model-based formal approaches very expensive for large and evolving distributed systems such as telecommunication systems. In order to alleviate these problems, we have developed a new modelling language RL that provides syntactic as well as semantic support for expressing incremental changes. Below, we provide an overview of the language, followed by a summary of results.

1.1. Language Overview

Our overall approach to modelling is similar to CSP [8], in that a system consists of a set of sequential processes that communicate through synchronous channels. The behavior of a process is described by a parameterized module that can be instantiated. A module, in turn, is characterized by a set of state variables and (non-deterministic) transition rules. An example of a simple module specification appears in Figure 1. It gives a simple (incomplete) specification of a telephone that is characterized by a state variable that describes the on-hook or off-hook status of the phone, and two transition rules that correspond to lifting and replacing the handset.

In order to support evolution, we need to model changes that alter the behavior of a module. These changes can take place through modifications to the state component and/or transition rules. In the former case, the change can be addition of new state variables, or expansion of the domain of values assumed by existing state variables. In the latter case, the change can be addition or deletion of transition rules, or selective modifications to existing rules by strengthening or relaxation of the pre and post conditions. These changes can be conveniently described in RL using the following mechanisms.

Module Derivation: All changes are expressed through module derivation, a concept similar to class derivation in object-oriented (OO) languages. By specifying priority relationships among rules in the derived and base modules, the rules in a derived module can override or selectively modify the rules in the base modules. New state variables can also be specified as part of a derived module. More importantly, we can alter the domain of an existing state variable in a structured manner using the following *type derivation* mechanism.

Type Derivation: New types can be derived from existing types either by *refining* a single value into multiple values, or by *augmenting* the type with entirely new values. Even when an existing state variable is redefined to possess a derived type in a derived module, most of the original rules that operate on the base data type can be reused in the derived module. Moreover, the type derivation mechanism serves as the basis of a mapping (between the derived and the original module specifications) that is used in *incremental validation.*

In comparison with OO-languages, modules are analogous to classes, with message transmission taking the role of member function invocation, and (sets of) transition rules taking the role of member function bodies. Addition of new state variables corresponds to having additional data members in the derived class, and addition of new transition rules is similar to adding new member functions. However, for the purposes of supporting refinement and evolution, RL improves upon traditional OO-languages. First, to minimize changes and carry over "general" rules, the semantics of rules that were written for a type continue to remain applicable *even at the new values* added by the derivation process; the inheritance in OO-languages allows the functions on the base class to operate only on components of the base type. Second, priorities provide a more flexible mechanism for refinement; with the override mechanism in OO-languages, a new member function always overrides a function in the base class with the same name.

With regards to our type derivation mechanism, the notion of refinement was first developed in the context of statecharts [7]. Statecharts is a visual formalism, where "boxes" are used to denote states of a machine. In this context, refinement was viewed as a way to decompose a single (super) state into multiple substates. In the context of text-based, typed language such as RL, we needed to develop a different view of refinement. Specifically, we view refinement as operating upon a data type, rather than a state. This view requires us to ascribe much more structure and semantics to refinement, as compared to statecharts. For instance, we need to extend refinement to structured data types, as well as define the semantics of comparisons and assignments over these types. Our notion of type augmentation was not considered in the context of statecharts.

1.2. Summary of Results

Several issues arise in the design of a language for modelling refinement and evolution. In this section, we summarize these issues and outline how we resolve them.

- Design of mechanisms for expressing incremental changes: the design must minimize the potential conflict between expressive power of the mechanism and simplicity of semantics. The former aspect determines the ease with which we can model evolution. The latter aspect defines the ease of understanding or reasoning about incremental modifications. The module and type derivation mechanisms, described in Section 2, have been designed to balance these considerations. We illustrate the use of these mechanisms through examples in Section 3.

- Development of a methodology for incremental validation: In Section 4, we present our approach for *automatically* carrying over properties after incremental modifications.

- Enable reuse of existing techniques and tools: When we cannot use incremental validation, e.g., for properties that do not hold before the refinement, we need to "validate from scratch." Several efficient techniques have been developed for such validation based on FSA representations. In order to reuse these techniques, we develop algorithms in Section 5 for compiling RL specifications into FSAs. The compilation "translates away" type and module derivation mechanisms *without* undue increase in either the state-space or the complexity of transition rules in the resultant automaton.

- Collapsing sequences of incremental modifications: While the language mechanisms to express incremental changes can make it easier to model and validate evolving software systems, they have the disadvantage that after several modification steps, it becomes difficult to understand the specification. Our semantics of module derivation, as well as the compilation algorithms, enable us to collapse a module description that involves a sequence of module derivation steps into a single module description that is much more understandable.

2. Overview of RL

A specification in RL consists of global declarations and module declarations, as shown in the example in Figure 1. A module is characterized by a set of module parameters, declarations, priority specifications and transition rules. Module parameters are analogous to (formal) parameters to a procedure. They are substituted by actual parameters at the time of *module instantiation*, as described later on in the section. Of the three kinds of declarations, type declarations associate names with *type expressions*; variable and channel declarations are used to declare variables and channels respectively.

Transition rules are of the form *Guard* $\rightarrow$ *Effect*, where *Guard* specifies the conditions under which the rule can be executed, and can include tests for presence of messages on channels or comparisons involving state variables. *Effect* includes output operations on channels or assignments to variables. A rule is executable whenever all the conditions in its guard are satisfied. Rule execution is *atomic*, i.e., either all the assignments in its body are executed or no assignment is executed.

Channels in RL are not buffered, and thus provide a rendezvous mechanism for communication. This enables us, in the manner described below, to view channel inputs and outputs as merely syntactic sugar for a particular way of using the state variables. We view each channel as a state variable that can take a value belonging to the channel's type, or be ϵ that denotes an empty channel. Output on a channel is treated as a combination of a test to ensure that the channel is empty, followed by an assignment of the value to be output on the channel. Similarly, an input operation $chan?expr$ is equivalent to the comparison $chan = expr$, followed by the assignment $chan := \epsilon$. We remark that an input operation is viewed in our language as a comparison, as opposed to the more traditional view of an assignment. This is because the effect of an assignment can be achieved using the *pattern-matching* mechanism described below.

A pattern-matching mechanism is supported for equality comparisons through the use of *temporary variables.* The scope of such variables is limited to a single transition rule. When used in an equality comparison, each temporary variable gets *bound* to a value that satisfies the comparison. Subsequent uses of the variable refer to this value. Pattern-matching provides a convenient way to deal with structured data types. For instance,

$$\mathsf{in?}f(\mathsf{X}, \mathsf{Y}) \;\rightarrow\; \mathsf{z1} := \mathsf{X}, \mathsf{z2} := \mathsf{Y}$$

tests for the presence of a message with root symbol f on the channel in, and if present, assigns the children of f to z1 and z2 respectively. (As in the above example, we use the convention that temporary variable names start with a capital letter, whereas state variable names begin in lower case.) In conjunction with module instantiation, pattern-matching provides a mechanism for supporting polymorphism.

2.1. Type Derivation

Type derivation permits us to derive new data types from existing ones. It provides a structured way to extend existing specifications so that existing rules (operating on base types) can be reused whenever they are meaningful in the context of the derived type. The type derivation mechanism is also designed so that properties established for existing specifications can be carried over to refined version of the specification (see Section 4).

New types can be derived from existing types, which can in turn be derived from other types, or be *base types.* The base types in the language include integer subranges (e.g., type t1 below), enumerated lists of flat (type PhoneNum) or structured (type t2) values, as well as arrays (type t3) over these types:

type t1 = 1..20
type PhoneNum = $\{p1, p2, p3\}$
type t2 = $\{\ c, d(\mathsf{t1}), f(\mathsf{HookStatus}, \mathsf{PhoneNum})\ \}$
type t3 = **array** [1..10] **of** PhoneNum

Declaration of t2 refers to the HookStatus type in Figure 1. In type t2, d is a constructor of arity 1, and f has arity 2. Examples of terms in t2 include $c, d(3)$ and $f(idle, p2)$. We note that construction of recursive types is not permitted in the language.

Type derivation in RL is provided by *refinement* and *augmentation.* Refinement allows decomposition (or specialization) of a value into multiple values, whereas augmentation allows us to add a new value to a type. For instance, we can define a new type HookStatus1 by specializing the value *offhook* in HookStatus of Figure 1 as follows:

type HookStatus1 = **refine** HookStatus **at** *offhook* **into** $\{connected, dialtone\}$;

Whenever any of these refined values appears in a context where the HookStatus type is expected, it is treated as if the value *offhook* appeared there. This provides the mechanism for reuse of rules that were designed to operate on HookStatus. For structured values, the refinement declaration specifies a refinement for the root as well as the children:

type t4 = **refine** t2 **at** $f(X, Y)$ **into** $\{g(X, Y, \mathsf{t1}), h(Y, X)\}$;
type t5 = **refine** t2 **at** $f(X : \mathsf{HookStatus}, Y)$ **into** $\{f(X : \mathsf{HookStatus1}, Y)\}$;

The first declaration specifies that any value of the form $g(X, Y, n)$ or $h(Y, X)$, (where X, Y and n are any values in types HookStatus, PhoneNum and t1 respectively) is a refinement of $f(X, Y)$. In the second declaration, the refinement mapping on the type of a child introduces a natural mapping on the type of the parent. A refined type includes all the terms in the base type excluding those terms that were refined, plus the new terms added by the refinement. Thus HookStatus1 contains *idle*, *connected* and *dialtone*.

While type refinement permits us to specialize values, type augmentation provides a mechanism for defining supertypes of existing types by adding entirely new values. The nature of augmentation permits full reuse of rules defined on the base types.

type HookStatus2 = **augment** HookStatus1 **by** {*ring*};

Type augmentation can be used in conjunction with refinement to add new values in a structured manner. In particular, if a type T′ is derived from another type T, then new values can be added to T′ as a refinement of an existing value in T:

type HookStatus3 = **augment** HookStatus2 **at** *offhook* **by** {*wait*(PhoneNum)};

2.1.1. Semantics of Type Derivation

The elements in the types HookStatus through HookStatus3 are shown in Figure 2, where additional elements (of the form $\perp$) have been added to simplify the semantics of type augmentation. The values added in this manner are called $\perp$-values, whereas the other values are said to be *proper*. Type augmentation can now be viewed as refinement at one of the $\perp$-values. Augmentations at the outermost level (as in HookStatus2) can be understood as refinement of the undefined element $\perp$ that is added to every enumerated type. Similarly, augmentation at some value v with new values $v_1, ..., v_n$ has the meaning that $v_1, ..., v_n$ are to be treated as refinements of v. To get this effect, whenever v is refined, we add a new element $\perp_v$ as one of its specializations (see $\perp_{offhook}$ in Figure 2), and later on, refine $\perp_v$ into $v_1, ..., v_n$ to get the effect of augmentation. Thus the $\perp$-elements act as "place-holders" where additional elements can be added in derived types.

With the above view of augmentation as an instance of refinement, we can understand type derivation in terms of the structure of resultant types. Figure 2 shows the associations among the types derived from HookStatus. In particular, a type derivation operation on a type T to get another type T′ defines a (many-to-one) mapping from values in type T′ to values in type T. We use the notation $\mathcal{M}_{\mathsf{T'}\to\mathsf{T}}$ to denote this mapping. For instance,

$$\mathcal{M}_{\mathsf{HookStatus3}\to\mathsf{HookStatus}}(dialtone) = offhook$$

When a value v in T is refined into $v_1, ..., v_n$ in type T1, $\mathcal{M}_{\mathsf{T1}\to\mathsf{T}}$ maps $v_1, ..., v_n$ to v and other values to themselves. We use the notation $\mathcal{M}_{\mathsf{T}}(val)$ to denote the value mapped onto by val in the type T. In order for this mapping to be well-defined, we require that the value val belongs to exactly one type derived from T. If val does not belong to T or any of its derived types, we set $\mathcal{M}_{\mathsf{T}}(val) = \perp$.

Based on the semantics of type derivation, we now specify the semantics of assignments and comparisons. An assignment of the form v := Expr is valid iff the type T′ of Expr is the same as the type T of the variable v, or is derived from it. Moreover, $\mathcal{M}_{\mathsf{T'}\to\mathsf{T}}(\mathsf{Expr})$ must be a proper element of T. The meaning of the assignment is to first evaluate Expr to yield a value V and then assign $\mathcal{M}_{\mathsf{T'}\to\mathsf{T}}(V)$ to the variable v. Since a value in a derived type has a unique image in the base type, we can permit the expression to have a type that is refined from that of the variable, but not vice-versa.

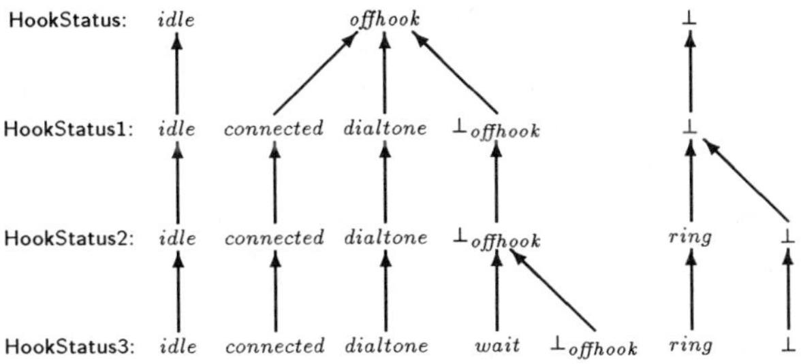

Figure 2. Associations among refinements/augmentations of type HookStatus

All comparisons are of the form *variable op expr*, where *op* can be one of {=, !=, <, <=, >, >=} for integer variables, and one of {=, !=} for others. The semantics of equality is defined as follows. A comparison $\mathsf{v} = \mathsf{Expr}$ is permissible whenever the types T' and T'' of the operands have a least common ancestor type T such that

$$\mathcal{M}_{\mathsf{T}'\to\mathsf{T}}(\mathsf{v}) = \mathcal{M}_{\mathsf{T}''\to\mathsf{T}}(\mathsf{Expr}) = \text{some proper value in type T}$$

For inequality, we require that both operands map to *different proper values* in T.

2.2. Module Derivation

Module instantiation, expressed as a declaration of a module type variable, provides the mechanism for module composition and derivation. Module composition is used to synthesize new modules by *composing* existing modules, whereas module derivation is used to incrementally modify the behavior of a module. The distinction between composition and derivation is in the specification writer's view, but not in the language.

The behavior of an existing module can be modified in a derived module by modifying either the states or the transition rules of the existing module. There are two ways to modify states of a base module: adding new state variables or enlarging the domain of an existing variable through type derivation. In either case, new transition rules are typically specified as part of the derived module that deal with the change in the domain. Priority relationships can be specified between the rules in the base module and those in the derived module. Priorities have the obvious semantics that a lower priority rule (or initialization) is *never* executed when a higher priority rule is executable. Use of module derivation is illustrated through examples in the Section 3. Incremental validation is discussed in Section 4.

The declaration of a variable

var m1: **module** m(a1,...,aN);

within a module m2 creates an instance of the module m. Any variable v inside the module instance m1 can be accessed within m2 using the notation m1.v. In simple terms, this instantiation is equivalent to importing the declarations and the transition rules of m into m2, after substituting the *formal parameters* that appear in the definition of m with the *actual parameters* a1,...,aN. For the substitution to succeed, the types of actual and formal parameters must be compatible. In particular, if the actual parameter is a variable then the type of the formal parameter must be derived from that of the actual parameter or vice-versa. The type of the parameter in the instantiated module is taken to be the more-refined of the formal and actual parameter types. The idea here is that the values in a derived type form a superset of the values in the base type. Hence, a variable of derived type can be thought of as *simultaneously* holding the values in the derived type as well as the base type. This view permits us to reuse most of the transition rules in m that may have been specified in the context of a less-refined type.

2.3. Semantics of RL Specifications

The state of a system specified in RL is given by a function S that maps variables in the specification to values. The behavior of the specification is defined as follows, based on the relation $\mathbf{T}$ that specifies when the system can transition from one state to another.

Definition 1 (History and Behavior) *A* history *is a (potentially infinite) sequence of states $S_0, S_1, \ldots$ such that S_0 is the initial state given by the specification and for any two successive states S_i and S_{i+1}, the condition $\mathbf{T}(S_i, S_{i+1})$ holds. The* behavior *of an RL specification is the set of all histories of the specification.*

Validation of RL specifications is based on finite representations of behavior. On the other hand, simulation of RL programs results in the enumeration of one of the histories.

Due to space limitations, we only provide a brief description of the transition relation $\mathbf{T}$. It is determined by (a) the semantics of rule execution, and (b) the method for identifying a rule that can be executed in a given state. To define rule execution semantics, we need to specify the semantics of expressions, comparisons and assignments, which are all based on the $\mathcal{M}$ function described in Section 2.1.1. For a rule $C \rightarrow A$, we say that the rule is *enabled* in state S if all the comparisons in C hold, with some bindings B for temporary variables. If the rule is enabled, the semantics of its execution is given by the new state S' obtained from S by performing all the assignments in A in sequence, with the temporary variable bindings as in B.

To define when a rule is executable, we can define a *select* function that searches through the rules in a specification and returns all those that are executable. The search starts in the main module and proceeds recursively through the modules that are instantiated. The search proceeds from highest priority rules to the lowest such that (i) once a higher priority rule is enabled, lower priority rules are eliminated from consideration, and (ii) when the search proceeds into an instantiated module, appropriate parameter passing (by reference) is taken into account.

3. Illustration of RL

In this section, we present an example of a telephone service to demonstrate the convenience of the language mechanisms in RL for modelling refinement and evolution. Although we expect these mechanisms to be used mainly on large systems to model incremental changes, for the sake of simplicity of illustration, we use only a small example. The environment for our example is shown in Figure 3. The USER module models actions of a telephone user, whereas NETWORK models the rest of the telephone network. We will not be specifying these two modules, but only the PHONE module. We start with a simple specification that deals only with the interaction between the telephone service and the user of the service. In this simplified view, the phone is either idle (on hook) or off

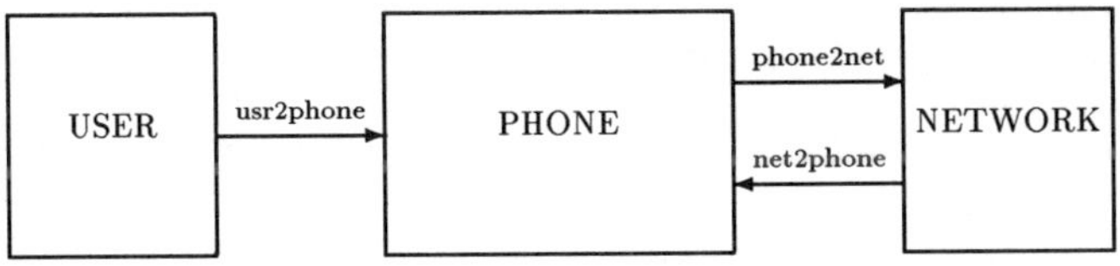

Figure 3. Environment for the telephone service model.

the hook, with the lift-hand-set and replace-hand-set messages from the user controlling the current state of the device. The simplified model is shown in Figure 1.

As the next step of the modelling process, we model the distinction between lifting a hand set to accept an incoming call and lifting it to make an outgoing call. This is done by refining the HookStatus type as shown below. We also need to introduce a new state corresponding to the state when the phone is ringing.

type HookStatus1 = **refine** HookStatus **at** *offhook* **into** $\{connected, dialtone\}$;
type HookStatus2 = **augment** HookStatus1 **by** $\{ring\}$;

Along with these type declarations, we introduce rules for transitions between the new states. At this point, we do not wish to exactly specify the conditions under which hookState can change from *dialtone* to *connected*, or toggle between *idle* and *ring*. However, since we are developing an abstract model of the system, we would like to capture all possible behaviors of the system and so we specify the possibility of making transitions between these states on receipt of some (unspecified) messages from the network. There is also the possibility that some messages may be received in inappropriate states, so that it is possible to simply discard them. (Otherwise, such messages can fill up the message buffers and prevent further messages from being received.) The derived module description is shown below:

```
type FromNetwork = {signal};
module phone2(hookState, usr2phone, net2phone) {
    var hookState: hookStatus2;
    var phone: module phone1(hookState, usr2phone);
    chan net2phone: FromNetwork;
    priority phone < phone2;
    trans
        usr2phone?liftHandSet, hookState = idle  → hookState := dialtone;
        usr2phone?liftHandSet, hookState = ring  → hookState := connected;
        net2phone?signal, hookState = dialtone  → hookState := connected;
        net2phone?signal, hookState = idle  → hookState := ring;
        net2phone?signal, hookState = ring  → hookState := idle;
        net2phone?signal  → ;
}
```

The declaration of the variable phone causes instantiation of the module phone1. Because the module phone2 specifies that rules in phone2 have a higher priority, the first rule in phone1 will never be applicable. However, the second rule (as well as the initialization) continues to be applicable in the two states that correspond to *offhook*.

As a next step, we incorporate interactions corresponding to outgoing calls. To do this, we add the *dial* message to usr2phone, and introduce two new states under *offhook*:

```
type Usr2Phone1 = augment Usr2Phone by {dial(PhoneNum)};
type HookStatus3 = augment HookStatus2 at offhook by {wait, busy};
```

We also refine the message from network to phone, and introduce a channel for communication from the phone to the network. The revised specification is:

```
type ToNetwork = {call(PhoneNumber)};
type FromNetwork1 = refine FromNetwork at signal into {accept, busy};
module phone3(hookState, usr2phone, net2phone, phone2net) {
   chan usr2phone: Usr2Phone1;
   chan phone2net: ToNetwork; net2phone: FromNetwork1;
   var hookState: HookStatus3;
   var phone: module phone2(hookState, usr2phone, net2phone);
   priority phone < phone3
   trans
      usr2phone?dial(X), hookState = dialtone  → hookState := wait, phone2net!call(X);
      usr2phone?dial(X), hookState != dialtone  → ;
      net2phone?accept, hookState = wait  → hookState = connected;
      net2phone?accept, hookState != wait  → ;
      net2phone?busy, hookState = wait  → hookState = busy;
      net2phone?busy, hookState != wait  → ;
}
```

As the next step, we incorporate incoming calls into this specification.

```
type FromNetwork2 = augment FromNetwork1 by {call, hangup};
type ToNetwork1 = augment ToNetwork by {accept, busy, hangup};
type HookStatus4 = augment HookStatus3 at offhook by {disconn};
module phone4(hookState, usr2phone, net2phone, phone2net) {
   chan net2phone: FromNetwork2; phone2net: ToNetwork1;
   var hookState: HookStatus4;
   var phone: module phone3(hookState, usr2phone, net2phone, phone2net);
   priority phone.phone < phone4; /* phone2 < phone4 */
   trans
      net2phone?call, hookState = idle  → phone2net!accept, hookState := ring;
      net2phone?call, hookState != idle  → phone2net!busy;
      net2phone?hangup, hookState = connected  → hookState := disconn;
      net2phone?hangup, hookState = ring  → hookState := idle;
      net2phone?hangup, hookState != ring, hookState != connected  → ;
      user2phone?replaceHandSet, hookState = wait  → phone2net!hangup, hookState := idle;
      user2phone?replaceHandSet, hookState = connected  → phone2net!hangup, hookState := idl
}
```

Note that in this derivation step, the transition rules in this module are not given priority over those in phone3. Thus, all the rules in phone3 are applicable in phone4. In successive refinements, it is easy to add more functionality to this basic model, such as call screening, caller-id, using the type derivation and module derivation operations.

4. Incremental Validation

As mentioned in the introduction, one of the costliest aspects of validation of evolving systems is the need to reestablish all the properties after each modification. We can significantly reduce this cost through *incremental validation*, wherein we spend our effort mainly in establishing the new properties of the modified system, while being able to carry over (most of) the properties established before the modification. In this section, we outline how our language mechanisms can be used to automate incremental validation.

We view validation as a kind of language containment problem, as in the automata-theoretic approach, first proposed in [14]. Specifically, we view the system as an automata A_M, with its *behaviors* (as defined on page 8) corresponding to the set of strings $\mathcal{L}(M)$ (over the alphabet consisting of the set of states of M) accepted by the automaton. In order to verify a property P we construct another automaton A_P that accepts the language $\mathcal{L}(P)$ of all behaviors that *violate* the property. Validation thus amounts to establishing the emptiness of $\mathcal{L}(M) \cap \mathcal{L}(P)$. We remark that currently, RL permits us to express only $*$-automata, whereas most automata-theoretic approaches to verification make use of ω-automata or Büchi automata. However, the discussions in this section on incremental validation are orthogonal to this issue, and hence would be applicable for extensions of RL to support ω-automata as well.

Kurshan [10] proposed a verification method based on *homomorphic reductions* that can be used to deal with step-wise refinement. After modifying an existing specification M to get M', one can go through the following steps to carry over a property P from M to M'. First, specify a mapping h from $\mathcal{L}(M')$ into $\mathcal{L}(M)$. Next, identify a property P' for M' such that $h(\mathcal{L}(P')) \subset \mathcal{L}(P)$. ($P'$ can often be mechanically derived from P, based on inverting h.) It follows that

$$\mathcal{L}(M) \cap \mathcal{L}(P) = \phi \Rightarrow h(\mathcal{L}(M')) \cap h(\mathcal{L}(P')) = \phi \Rightarrow \mathcal{L}(M') \cap \mathcal{L}(P') = \phi$$

The first implication is because $h(\mathcal{L}(M')) \subset \mathcal{L}(M)$ and $h(\mathcal{L}(P')) \subset \mathcal{L}(P)$. The second implication follows from the fact that if the images of $\mathcal{L}(M')$ and $\mathcal{L}(P')$ under h do not have any common elements, then the two sets themselves cannot have common elements.

The above notion of reduction mapping is similar to *protocol projection* [11] and abstraction mappings [6]. However, the difficult aspect of all these approaches is that the mapping h needs to be constructed by the programmer, and is often the most time-consuming and difficult component of the verification approach. In the rest of this section, we discuss how the derivation constructs in RL can be used to automate this process.

4.1. Our Approach

The type and module derivation constructs in RL define a natural mapping h_s from the states of the refined specification M' to the original specification M. Based on this mapping, we can define a mapping h_r from transition rules in M' to those in M such that whenever a transition R' from a state S_1' to S_2' can be taken in M', the transition $h_r(R')$ can be taken to go from $h_s(S_1')$ to $h_s(S_2')$ in M. It is easy to see the existence of the mappings h_s and h_r imply the existence of the mapping h with the desired properties, provided $h_s(S_0') = S_0$, where S_0' and S_0 denote the initial states of M' and M respectively.

The mapping h_s is specified as follows. For each variable v in M' that has a type T'

derived from its type T in M, we replace the value of v by $\mathcal{M}_{\mathsf{T}'\to\mathsf{T}}(v)$. More formally,

$$\begin{aligned} h_s(S_{M'})(v) &= \mathcal{M}_{\mathsf{T}'\to\mathsf{T}}(S_{M'}(v)), \text{ if } \mathcal{M}_{\mathsf{T}'\to\mathsf{T}}(S_{M'}(v)) \text{ is a proper element} \\ &= \text{any proper value derived from same value as } \mathcal{M}_{\mathsf{T}'\to\mathsf{T}}(S_{M'}(v)), \text{ otherwise} \end{aligned}$$

Every rule R that is present in M as well as M' maps onto itself under h_r. For others, we can proceed to check the existence of a h_r satisfying the conditions specified earlier for each state S' of M' — a task that is equivalent to a state-space search of M' and hence defeats the whole purpose of incremental validation. In order to avoid this, when we consider a rule R', we group together all the states that agree on the variables in R'. Typically, the number of such groups is much smaller than the entire state space. Considering each rule R' in M' and each group $\mathbf{S}'_{R'}$ of states that agree on values of variables in R', we identify a rule R in M such that

1. if R' takes $\mathbf{S}'_{R'}$ to some $\mathbf{S}'$ then R takes $h_s(\mathbf{S}'_{R'})$ to $h_s(\mathbf{S}')$.
2. For any R'_1 and R'_2 in M', if they are mapped onto R_1 and R_2 respectively (in M) by the previous step, and R'_1 does *not* have higher priority over R'_2 then R_1 does not have higher priority over R_2.

The second step ensures that whenever a rule R'_2 is enabled, its image in M is enabled.

The above approach yields a procedure for automatically checking for the existence of a mapping h from M' to M. The procedure may have to search through the h_s mappings that correspond to different choices for mapping augmented values. The procedure may also have to consider different mappings between the rules in M' and M. However, we can make the procedures far more efficient by using the mapping between rules to guide the mapping for augmented values and by using heuristics to cut down the search.

4.2. Improvements to the Basic Approach

Two classes of improvements can be made to the basic approach so that the property of interest can be carried over even when the basic approach fails to identify the h_r mapping. In the first class, we exploit the property to be carried over so as to eliminate certain rules in M' that do not map onto rules in M. For instance, we can eliminate rules that can never be taken in any history generated by the property automaton. We can also eliminate rules whose presence or absence does not affect the property. As a concrete example, consider a rule R' in M' that introduces transitions between two refined states. Thus $h_r(R')$ will be a rule that corresponds to a transition from a state onto itself, i.e., a self-loop. Typically, such loops are absent in M, which leads to a failure of the basic approach. However, such self-loops do not affect satisfaction in the context of most properties (liveness properties in the context of certain fairness criteria and safety properties). This is especially true in the context of interleaving semantics, where there is little distinction between making a transition to the same state and not making any transition at all. Therefore, we can often eliminate such rules and establish a mapping only for the others.

In the second case, we can relax the condition on the mappings so that we can look at sets of transitions at a time rather than individual transitions. For instance, if every set of transitions in M' is covered by a corresponding set in M, our correctness properties can still be carried through. When such techniques also fail, it appears that it may still be possible to avoid complete state-space search for certain classes of properties, by exploiting the fact we need only consider those paths corresponding to the rules that were not covered by the mapping. Further research is required to formalize this idea.

4.3. Illustration of Incremental Validation

A class of modifications that often takes place in RL is as follows. We refine some state variables, and strengthen existing rules so that the tests or assignments in them refer to the refined values rather than the unrefined values. Moreover, new transition rules may be added to make transitions between the refined states. For instance, the successive specifications phone1 through phone4 all make use of this kind of transformation. The above technique allows us to carry over all the properties after each such transformation. We illustrate this by defining a mapping h_r from phone3 to phone2. In order to compactly describe the mapping, we first present the image of the new rules in phone3 under a mapping h_1 and then show the existence of a mapping h_2 satisfying the above characteristics from the image to the rules in phone2. (Thus, $h_r(R') = h_2(h_1(R'))$ for every rule R' in phone3.) The mapping h_1 is once again derived from the mapping on states. The mapping of values added through refinement is straight-forward, but there are some choices for the mapping of augmented values. We map a state assigning $dial(X)$ to usr2phone in phone3 to a state that assigns ϵ to the same variable in phone2. Similarly, for the variable hookState, we map *wait* to *dialtone* and *busy* to *connected*. Based on this mapping, we can now map the expressions in the rules. We can also delete the variable phone2net newly introduced in phone3 to get

usr2phone $= \epsilon$, hookState $=$ *dialtone* $\rightarrow$ usr2phone $:= \epsilon$, hookState $:=$ *dialtone*;
usr2phone $= \epsilon$, hookState $!=$ *dialtone* $\rightarrow$ usr2phone $:= \epsilon$;
net2phone?*signal*, hookState $=$ *dialtone* $\rightarrow$ hookState $:=$ *connected*;
net2phone?*signal* $\rightarrow$;

The first four rules in phone3 map to the above four rules, whereas the fifth and sixth rules map onto the third and fourth rules above. Note that when we map the fourth rule, the condition hookState! $=$ *wait* disappears. This is because such a condition would be enabled in states that assign one of the values $\{idle, ring, dialtone, connected, busy\}$ to hookState. Mapping this back to phone2, the condition is enabled in states that assign one of $\{idle, ring, dialtone, connected\}$ to hookState, i.e., for every value of hookState.

From the above image under h_1, we now specify the mapping to rules in phone2. The first and second rules take a state to itself, and hence can be ignored, as discussed before. The third and fourth rules above map onto the third and sixth rules respectively in phone2.

5. Compilation of RL

In this section we describe algorithms for compiling RL specifications into FSMs and present an overview of implementation of our prototype compiler and simulator for RL. The compilation enables us to reuse traditional validation algorithms (based on FSMs) for establishing properties. Our compilation algorithms can also be used to do a source transformation that collapses a module described through a sequence of module derivations into a single module description. As mentioned earlier, this transformation is useful to understand specifications that have gone through several refinements.

Input programs are first parsed and other routine aspects such as elimination of channel variables are taken care of. Following this, module instantiation takes place. To instantiate a module M, we first instantiate its component modules and bring their rules together with those of M. Priority relationships between rules are held in a separate table. Without the module instantiation step, we would have to search through all the module instances to identify the transition(s) that can be made. By bringing together all the rules, we can build efficient automata [13] for quickly testing which rules are executable.

Following module instantiation, we translate the resulting rules into our target language. Our target language, designed to describe FSMs, supports only integer data types and has no notion of priority. It supports several kinds of branching constructs: if-then-else, the **switch** construct similar to that of the C language, the **indeterminate** construct that denotes an n-way non-deterministic branch, and the **ordered** construct, that denotes the fact that the children branches are to be tried in order. Given this target language, the important aspects of code generation are (a) mapping of source language data types into target language (integer) data types, and (b) construction of the transition rule automaton. We describe each of these aspects below.

5.1. Mapping Source Language Data Types into Target Language Data Types

In this phase, we first map structured types onto a set of integer variables, and then encode enumerated values by integers. For the mapping, we associate an array of locations to store a structured variable. For each node of the structured data, the mapping specifies the index where the value of the node is stored. Observe that it is acceptable to have the location corresponding to a node depend upon the values of its ancestor nodes, since we will always examine the ancestor nodes before arriving at this node. However, the location cannot be dependent on values of non-ancestor nodes, since the pattern-matching construct implies that we may not always have inspected such nodes. Consider:

type T1 $= \{a, b, c\}$
type T2 $= \{d(\mathsf{T1}, \mathsf{T1}), e(\mathsf{T1})\}$
type T3 $= \{f(\mathsf{T2}, \mathsf{T1}), g(\mathsf{T1}, \mathsf{T2})\}$

Since T1 is not structured, no mapping is generated. For T2, we map the root and the children values at offsets 0, 1 and 2 respectively. For type T3, the root is stored at offset 0, and the locations of the child nodes will depend on the value of the root node. For instance, in the term $f(e(a), b)$, f is stored at 0, e at 1, a at 2, and b at 4. Note that b is not stored at the location immediately after a, since this location would be occupied by the second child of of d, had d been in the place of e. Similarly, in the term $g(a, d(b, c))$, the values g, a, d, b and c are stored at locations $0, 1, 2, 3$ and 4 respectively.

The encoding of enumerated values becomes complex when we need to support type derivation efficiently. Consider the comparison $\mathsf{v} =$ *offhook* where v has the type HookStatus3 shown on page 6. The test must succeed iff the value of v is one of *connected*, *dialtone* or *wait*. The refinement step thus requires us to make several comparisons in place of one. However, by assigning contiguous codes to all refinements of a value,we can reduce the number of comparisons to two. The table below shows the two comparisons in the above example, where C_{val} denotes the integer code assigned to *val*. In order to assign contiguous codes, after encoding a value V in a type T, we recursively check through the refinements of T, and assign contiguous codes to them. The next value in T gets encoded by an integer that is greater than the code assigned to any refinement of V.

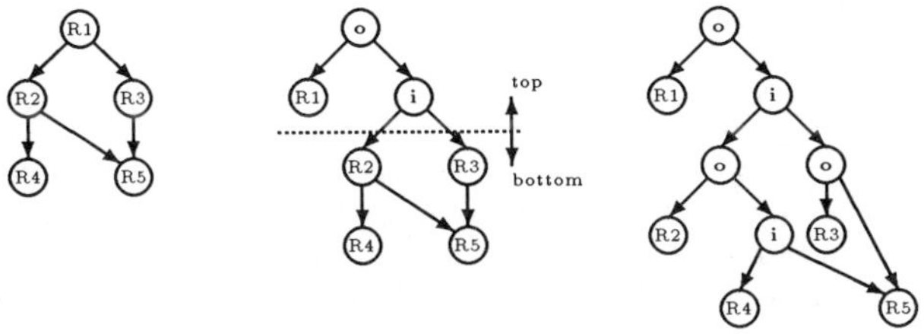

Figure 4. Translation of priorities. **o** and **i** denote **ordered** and **indeterminate** nodes

The table below shows the translation of assignments and comparisons based on the above mappings. In the table, u has the type HookStatus, and the notation M_{var} denotes the memory location for storing a variable var, and $*$ denotes dereferencing.

Source Construct	Target Language Translation
v = *offhook*	$*M_{\mathsf{v}} \geq C_{connected} \wedge *M_{\mathsf{v}} \leq C_{wait}$
v = *wait*(X)	**if** $(*M_{\mathsf{v}}) = C_{wait}$ **then** $M_X := *(M_{\mathsf{v}} + 1)$
u := v	**switch** $(*M_v)$ $C_{connected}, C_{dialtone}, C_{wait}$: $M_u := C_{offhook}$ **default:** $M_u := *M_v$

5.2. Construction of Transition Rule Automaton

In this phase, the priority information is integrated into the rules by using the **ordered** and **indeterminate** constructs. To do this, we start from the roots of the (minimal) directed acyclic graph representing the partial order relationship given by the priorities. At any time during this procedure, we would have converted some "top portion" of the dag into **ordered** and **indeterminate** constructs (see Figure 4). We then select a rule R with maximal priority from among the rules below this top portion. The rules immediately below R are combined using an **indeterminate** node, which is then combined with R using an **ordered** node. Figure 4 shows two such steps.

Note that the automata construction does not lose the structure of the rules. Therefore, if we avoid translating comparisons and assignments, it is possible to generate (from the automaton) a set of rules that closely correspond to the rules in the original specification. However, since the module boundaries have been lost, we have a transformed program wherein the net effect of several module derivation steps is captured in a single specification.

5.3. Discussion of Prototype Implementation

The compiler is written in Standard ML of New Jersey. The whole compiler consists of about 6000 lines of SML code, with about 4000 in the preprocessing phase and 2000 in the code generation phase. All of the components described above have been incorporated into the compiler. The current performance of the prototype compiler is quite good. It is fast and generates compact code. A simulator (about 300 lines of C and SML code) for the resultant FSMs has also been implemented. The simulator first converts the finite state machine into C-code, which is then compiled and run. The compilation of C-code is somewhat slow, but execution speed is already quite fast: about 10^4 transitions per second for some sample specifications on a Sun Sparc LX machine. We are investigating further speed improvements using test-reordering.

6. Concluding Remarks

In this paper, we presented a specification language RL for modelling evolving systems. RL provides two principal mechanisms, namely, type and module derivation, that operate in conjunction to enable modelling of incremental changes to a specification. The design of these mechanisms must minimize the potential conflict between the expressive power of the mechanism and the simplicity of semantics. We illustrated the power and convenience of RL through a set of examples and then developed a methodology for incremental validation. The important aspect of our methodology is that it uses RL mechanisms for refinement and evolution to *automatically* carry over properties, without requiring human intervention. Finally, we presented algorithms and results of a preliminary implementation for compiling RL programs into FSAs. Since such FSAs can be efficiently analyzed for purposes of "validating from scratch," our language support for refinement and evolution is achieved *without* compromising efficiency or requiring rediscovery of previously known algorithms for validation. We are currently investigating means to further enhance the power and convenience of our mechanisms for supporting evolution.

REFERENCES

1. J.R. Burch, E.M. Clarke, K.L. McMillan, D.L. Dill and L.J. Hwang, Symbolic Model Checking: 10^{20} States and Beyond, *Information and Computation, 98(2), 1992.*
2. F. Belina and D. Hogrefe, The CCITT Specification and Description Language SDL, *Computer Networks and ISDN Systems 16(4), 1988.*
3. T. Bolognesi and E. Brinksma, Introduction to the ISO Specification Language Lotos, *Computer Networks and ISDN Systems 14, 1987.*
4. S. Budkowski and P. Dembinski, An Introduction to Estelle: A Specification Language for Distributed Systems, *Computer Networks and ISDN Systems 14, 1987.*
5. E.M. Clarke, E.A. Emerson and A.P. Sistla, Automatic Verification of Finite State Concurrent Systems Using Temporal Logic Specifications, *ACM TOPLAS, 8(2), 1986.*
6. E.M. Clarke, O. Grumberg and D.E. Long, Model Checking and Abstraction, *POPL '92.*
7. D. Harel, Statecharts: A Visual Formalism for Complex Systems, *Sci. of Comp. Prog., 1987.*
8. C.A.R. Hoare, Communicating Sequential Processes, *CACM 21(8), 1978.*
9. G.J. Holzman, Design and Validation of Computer Protocols, *Prentice Hall, 1991.*
10. R.P. Kurshan, Automata Theoretic Verification of Coordinating Processing, *Berkeley Lecture Notes (unpublished), 1991.* Also, Verification of Concurrent Processes: The Automata-Theoretic Approach, *Princeton University Press (to appear), 1994.*
11. S.S. Lam and A.U. Shankar, Protocol Verification via Projections, *IEEE Transactions on Software Engineering, 10(4), July 1984.*
12. J.L. Peterson, Petri Net Theory and the Modelling of Systems, *Prentice Hall, 1981.*
13. R. Sekar, R. Ramesh and I.V. Ramakrishnan, Adaptive Pattern Matching, *ICALP '92.*
14. M.Y. Vardi and P. Wolper, An Automata-theoretic Approach to Automatic Program Verification, *Proc. on Logics in Computer Science, 1986.*

39

A Synthesis Algorithm of a Protocol Model from a Single Entity

Bhed Bahadur Bista,[a] Zixue Cheng,[b] Atsushi Togashi[a] and Norio Shiratori[a]

[a]Research Institute of Electrical Communication, Tohoku University, Japan

[b]Department of Computer Software, University of Aizu, Japan

Protocol entities, in communication protocols, behave under sets of communication rules (protocols). Thus, it is desirable to concentrate on the design of one protocol entity and generate the corresponding protocol entity automatically by exploiting the communication rules. A protocol is desired to be formal, precise and unambiguous, in other words, it is described using FDTs (Formal Description Techniques). In this paper, we propose a protocol synthesis algorithm in which a peer protocol entity is generated from a single given entity. Unlike previous works, where FSMs (Finite State Machines) were used to synthesize partial protocols, we use LOTOS, which is one of FDTs developed by ISO, in our proposed synthesis algorithm. We prove that the resultant protocol consisting of the given entity and the generated peer entity is logical errors free, collectively represented as *deadlock* free, if the given entity is in certain forms which are natural in the context of communication protocols. We also present that the protocol is bisimulated by both the given entity and the peer entity to show that the protocol maintains the behaviours of the entities.

key words:

I. General Keywords: 4. FDT-based software engineering
II. Language: 2. Lotos
III. Semantic models: 3. Process Algebra
IV. Miscellaneous: 2. Protocol Synthesis

1. Introduction

Communication protocols play a central role in the distributed computing and computer networks for their proper operations. It has been always desired to have formal, precise, unambiguous and correct protocols, for which Formal Description Techniques (FDTs) such as Estelle [1], LOTOS [2] and SDL [3] have been developed and standardized.

In order to ensure the correctness of communication protocols, protocol design techniques are provided to protocol designers. Typical techniques to support the design of reliable error free protocols are (1) *analytic method* [4], [5], [6], [7]; (2) *synthetic method* [8], [9], [10], [11], [12],[13]. In (1), a pre-designed protocol is analyzed to detect and correct errors iteratively until the protocol satisfies the properties for correctness. In (2), a partially specified protocol design is completed such that interactions between its entities proceeds without errors and provide a set of services. The advantage of (2) is that no further verification or validation is required as in (1).

Specifications of entities in communication protocols are related to each other as they communicate with each other under sets of communication rules. By exploiting this relationship, it is possible to construct a corresponding peer entity from a given entity [9], thus a protocol designer can concentrate on a single entity.

In this paper, we propose a synthesis algorithm in which from a given entity, specified as communicating processes in LOTOS, we automatically generate a corresponding peer entity, which is also specified as communicating processes, in LOTOS, such that the resulting protocol is *deadlock* free, which will be explained in detail in section 3.2. A communicating process is defined inductively in section 3.1 and is natural for communicating entities providing services to their users. Most of the previous works on synthesis of partial protocol specifications are carried out based on Finite State Machines (FSMs). The main advantages of LOTOS over FSMs are; concurrency [14], which is one of the behaviours in communication system that can be expressed in LOTOS but not in FSMs. States of a system specified in LOTOS are implicit and in FSMs explicit, thus state explosion can occur in the synthesis of protocol specifications of a large system in FSMs.

The idea presented in this paper is the extention of [15]. In [15], a given entity is given a higher priority during the exchange of messages. When both entities send messages, the message sent by the higher priority entity is received by the lower priority entity (i.e. peer entity) and the message sent by the lower priority entity is lost. This is due to the fact that communicating gates are interpreted as half duplex buffer. In this paper, we consider the communicating gates as full duplex buffer thereby relaxing the priorities (no message is lost). Apart from proving the resulting protocol deadlock free, we also prove that the protocol model is bisimulated by both the given entity and the generated peer entity. Thus, it shows that the protocol maintains the behaviour of the entities.

We will briefly review and compare previous related works. In [9], an entity satisfying certain given conditions is specified in Petri nets which is then translated into FSMs. Using transformation rules, a peer FSM is generated which is then transformed into Petri nets. Transformation rules do not consider receiving loops. Considering receiving loops in protocols is very important. In this paper, we consider recursive behaviours which includes receiving loops also. Furthermore, if the given entity is large and complex, the resultant states of the FSMs can be large and difficult to handle. We avoid this by using LOTOS where sub-entities can be designed and combined by operators. [8] extends the transformation rules of [9] including receiving loops as components to synthesize partial protocols consisting of n entities by combining components. Basically, given in [8] is partial specifications of n entities and the result is the completion of the specifications by combining components. Though the approach is different from ours using FSMs, the formal proof of the correctness of the combination of components is not provided and thus, it is not known formally whether the combinations of components always satisfy the proof of correctness. In [16], a protocol service specification and n-1 entities are given and the entity n is constructed. Their limitation is that the notion of correctness does not capture some aspects such as potential deadlocks. Thus, an entity satisfying the automatically generated specification may cause deadlocks.

This paper is organized as follows. Following the Introduction, section 2 gives an outline of LOTOS, section 3 presents protocol model in LOTOS, errors in protocol specifications and some definitions. Section 4 discusses the details of the proposed synthesis algorithm. Section 5 presents a simple application example using the proposed algorithm. Section 6 concludes the paper.

2. LOTOS

LOTOS (Language Of Temporal Ordering Specification) is an FDT (Formal Description Technique) developed by ISO (International Organization for Standardization) [2] for the formal description of distributed systems. A distributed system, specified in LOTOS, consists of a number of *processes* interacting with each other. A process in LOTOS is considered as an abstract entity which is capable of performing *internal actions* and communicates with other processes via *communication actions* by synchronizing at points called *gates*. An action or an *event* is considered atomic (i.e. not divisible in time). Throughout this paper, let $\mathcal{A}$ denote a finite set of actions which are externally observable and controllable. An internal action which can't be externally observed is represented as $\mathbf{i}$ and let $Act = \mathcal{A} \cup \{\mathrm{i}\}$. Act^* denotes the set of all sequences of actions and we write ε as the empty sequence. $\widehat{t}$ is the resulting sequence of actions t where $\mathbf{i}$ is removed.

In LOTOS, the *behaviour expression* of a process is defined by stating the temporal relation of actions (events) of the process. There are various LOTOS operators (eg. *prefix, choice*) to form behaviour expressions and combine behaviour expressions to yield complex behaviour expressions. See [2] for these operators. The description of data structures and data values are based on the algebraic specification of abstract data types (ADTs) for actions involving in exchange of data.

LOTOS has the following two semantic models:

- Label transition systems for behaviour expressions
- Many sorted algebras for data types

In this paper, we are mainly concerned with behaviour expressions, thus label transition systems. A labelled transition system is a state transition graph in which nodes represent states and edges represent state changes. This is formally defined as follows:

Definition 2.1 (Labelled Transition System) A *labelled transition system* L is a quadruple $\langle S, A, T, s_0\rangle$, where
(1) S is a nonempty set of *states*;
(2) A is a subset of *Act*;
(3) $T \subseteq S \times A \times S$ is a *transition relation*;
(4) $s_0 \in S$ is the *initial state* of L. □

The transition relation defines the dynamic changes of states as actions are performed. For $(s, a, s') \in T$, we normally write $s \xrightarrow{a} s'$ and this may be interpreted as "in the state s an action a can be performed and after the action the state moves to s'". If $t = a_1 a_2 \ldots a_n \in Act^*$, then

- $s \xrightarrow{t} s'$ stands for $s \xrightarrow{a_1}\xrightarrow{a_2} \ldots \xrightarrow{a_n} s'$.

- $s \overset{t}{\Rightarrow} s'$ stands for $s \overset{\mathbf{i}^*}{\rightarrow}\overset{a_1}{\rightarrow}\overset{\mathbf{i}^*}{\rightarrow}\overset{a_2}{\rightarrow}\overset{\mathbf{i}^*}{\rightarrow} \ldots \overset{\mathbf{i}^*}{\rightarrow}\overset{a_n}{\rightarrow}\overset{\mathbf{i}^*}{\rightarrow} s'$.
- $s \overset{t}{\rightarrow}$ stands for $s \overset{t}{\rightarrow} s'$ for some s' and $\overset{t}{\nrightarrow}$ denotes the negation of $s \overset{t}{\rightarrow}$.

Based on the operational semantics given by the transition systems, a *bisimulation relation* is defined as follows:

Definition 2.2 (Bisimulation) Let $L_1 = \langle S_1, A_1, T_1, s_{10}\rangle$ and $L_2 = \langle S_2, A_2, T_2, s_{20}\rangle$ be labelled transition systems. A binary relation $R \subseteq S_1 \times S_2$ is a *(weak)bisimulation* if $(s_1, s_2) \in R$ implies that, for all $a \in Act$,

(1) if $s_1 \overset{a}{\rightarrow} s_1'$ for some s_1', then $s_2 \overset{\widehat{a}}{\Rightarrow} s_2'$ and $(s_1', s_2') \in R$ for some s_2';

(2) if $s_2 \overset{a}{\rightarrow} s_2'$ for some s_2', then $s_1 \overset{\widehat{a}}{\Rightarrow} s_1'$ and $(s_1', s_2') \in R$ for some s_1'.

L_1 is *bisimilar* to L_2 (or L_1 is *bisimulated* by L_2) if $(s_{10}, s_{20}) \in R$ for some bisimulation R. □

Define a relation $\approx$ by

$$\approx = \cup\{R \subseteq S_1 \times S_2 \mid R : \text{a bisimulation}\}$$

Then, $\approx$ is an equivalence relation and turns out to be the largest bisimulation. It is clear from the definition that L_1 is bisimilar to L_2 iff $s_{10} \approx s_{20}$.

3. Communication Model and Definitions

3.1. Discussion of Protocol Model in LOTOS

A system in LOTOS is expressed as a set of processes and their gates. We will briefly describe the protocol model in LOTOS with conventional protocol terms and present some notations which will be used throughout the paper.

A *protocol model*, PM, in LOTOS is 5-tuple $\langle P, Q, G, G_p, G_q\rangle$ (see Figure 1). P and Q are entities, processes in LOTOS. G is an abstraction of the lower layers (buffer in usual protocol terms). G, G_p and G_q are the gates at which actual exchange of messages occur between processes and their environments. Another view of looking at the gates is to treat them as buffers in which one process puts in messages in the buffer and another process takes out the messages from the buffer instantly. In fact, such an action is atomic as the putting the messages in the buffer and taking out the messages from the buffer is performed by synchronizing in an action.

In [15], each gate is interpreted (realized) as a half duplex buffer, where synchronization actions for sending and receiving are handled (in a sense) in a single buffer. In such a case, when U_p and U_q send messages at the same time, either the messages will collide causing a deadlock or the priority is given to one message causing another message to be lost as in [15]. We solve that problem by interpreting each gate as a full duplex buffer, the intuitive meaning of which is as shown in Figure 1. For example, G is interpreted in such a way that g_{in} is the gate at which P sends messages ($\overrightarrow{\alpha}$) to Q (and the gate at which Q receives messages ($\overrightarrow{\alpha}$) from P) and g_{out} is the gate at which Q sends messages

($\overleftarrow{\alpha}$) to P (and the gate at which P receives messages ($\overleftarrow{\alpha}$) from Q). Similarly G_p and G_q are interpreted. The communication between processes at the gate (interpreted as full duplex buffer) is by interleaving the sending processes and receiving processes which will be discussed in detail in section 4.

In conventional protocol terms, G_p and G_q are the service access points for the users U_p and U_Q, respectively. In LOTOS, G_p and G_q are the sets of gates in which P and Q interact with the environments represented as the users U_p and U_q, respectively. The environment of P is U_p and Q and the environment of Q is U_q and P. In LOTOS, the PM is formally expressed as,

specification PM$[G_p, G_q]$:=
behaviour hide G **in** $P|[G]|Q$

...

endspec

In this paper, we use a process and an entity in the same meaning. Gates and actions (events) are used as synonyms in LOTOS if data types are not used and we treat them interchangeably.

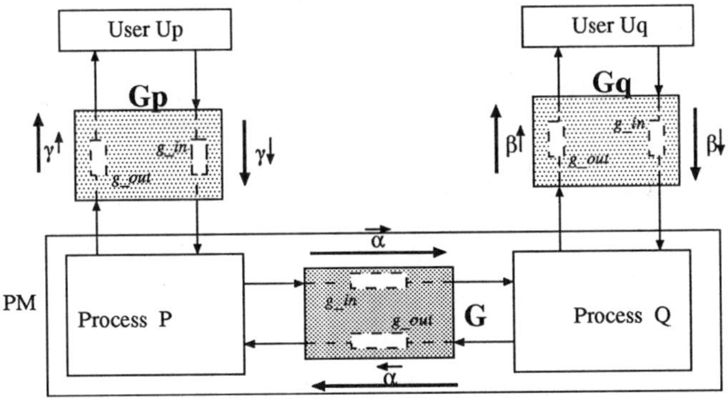

Figure 1. LOTOS protocol model

In the behaviour expression of a system without data type in LOTOS, when processes synchronize, there is no notion of messages flow between processes. In this paper, we consider synchronizing actions as distinct messages and attach arrows to them to represent the directions of messages flow. Note that it is possible to replace each action by a single gate with message attached to it. For instance, $\gamma\downarrow$ by $g?x$ where x is any data variable. See [2] for details. We use the following notations to the rest of the discussion (see Figure 1).

P :	a given entity
Q :	the peer entity
$\gamma_1\downarrow, \gamma_2\downarrow, \ldots \in G_p$:	actions of type '$\downarrow$' representing messages from U_p to P
$\gamma_1\uparrow, \gamma_2\uparrow, \ldots \in G_p$:	actions of type '$\uparrow$' representing messages from P to U_p
$\beta_1\downarrow, \beta_2\downarrow, \ldots \in G_q$:	actions of type '$\downarrow$'representing messages from U_q to Q
$\beta_1\uparrow, \beta_2\uparrow, \ldots \in G_q$:	actions of type '$\uparrow$' representing messages from Q to U_q
$\overrightarrow{\alpha}_1, \overrightarrow{\alpha}_2, \ldots \in G$:	actions of type '$\rightarrow$' representing messages from P to Q
$\overleftarrow{\alpha}_1, \overleftarrow{\alpha}_2, \ldots \in G$:	actions of type '$\leftarrow$' representing messages from Q to P

Note that in the OSI context, actions occurring at the gate G_p and G_q can be interpreted as Service Primitives (SPs) and the actions occurring at the gate G as Protocol Data Units (PDUs).

A protocol entity specification (e.g. P) in Figure 1 describes the behaviour of that entity with respect to its upper interface (G_p) and with respect to its lower interface (G). As far as the model in Figure 1 is concerned, we assume that a given entity P is expressed in terms of communicating processes. We assume that a communicating process is deterministic and the same synchronization action does not occur more than once. To define a communicating process, it is assumed that we have a denumerable set X of *process variables*.

Definition 3.1 (communicating process) A *communicating process* for a given entity is defined inductively as follows:

(1) **stop**, **exit**, and a process variable $x \in X$ are communicating processes.

(2) If P_i $(i \in I)$ and P_j $(j \in J)$ are communicating processes, then a process of the form $(\sum_{i \in I} \gamma_i\downarrow; \overrightarrow{\alpha_i}; P_i) \mid\mid\mid (\sum_{j \in J} \overleftarrow{\alpha_j}; \gamma_j\uparrow; P_j)$ is a communicating process [1], where $\gamma_i\downarrow; \overrightarrow{\alpha_i}$ and $\overleftarrow{\alpha_j}; \gamma_j\uparrow$ are called *primitive components* .

(3) If P is a communicating process, then $\mathbf{rec}\, x.P$ is a communicating process, where $x \in X$.

In the same way, a *communicating process* for the peer entity is defined inductively as follows:

(1) **stop**, **exit**, and a process variable $x \in X$ are communicating processes.

(2) If Q_i and Q_j are communicating processes, then a process of the form $(\sum_{i \in I} \overrightarrow{\alpha_i}; \beta_i\uparrow; Q_i)$ $\mid\mid\mid(\sum_{j \in J} \beta_j\downarrow; \overleftarrow{\alpha_j}; Q_j)$ is a communicating process where $\overrightarrow{\alpha_i}; \beta_i\uparrow$ and $\beta_j\downarrow; \overleftarrow{\alpha_j}$ are called *primitive components*.

(3) If Q is a communicating process, then $\mathbf{rec}\, x.Q$ is a communicating process, where $x \in X$. □

A communicating process $\mathbf{rec}\, x.P$, sometimes called a *recursive* communicating process, is interpreted as a process x whose meaning is defined by the recursive equation $x \stackrel{\text{def}}{=} P$ where any occurrence of x in P may be replaced by P itself. In a recursive process $\mathbf{rec}\, x.P$, every occurrence of x in P is called *bound*. We say P is a scope of $\mathbf{rec}\, x$ in $\mathbf{rec}\, x.P$. An occurrence of process variable which is not within any scope of $\mathbf{rec}\, x.P$ is called *free*. If every occurrence of any process variable in P is bound, P is called *closed*.

[1]For example, if $I = \emptyset$, $\sum_{i \in I} \gamma_i\downarrow; \overrightarrow{\alpha_i}; P_i$ can be interpreted as **exit**

Otherwise it is called *open*. For simplicity, we treat only communicating processes without mutual recursions as above. However, it is easy to extend it for the processes with mutual recursions.

In [15], a given process is classified into *sending process*, i.e. $\sum_{i\in I}\gamma_i\downarrow;\overrightarrow{\alpha_i};P_i$, *receiving process*, i.e. $\sum_{j\in J}\overleftarrow{\alpha_j};\gamma_j\uparrow;P_j$ and *sending-receiving process*, i.e. $(\sum_{i\in I}\gamma_i\downarrow;\overrightarrow{\alpha_i};P_i)$ [] $(\sum_{j\in J}\overleftarrow{\alpha_j};\gamma_j\uparrow;P_j)$. They are the special cases of the communicating process of the form $(\sum_{i\in I}\gamma_i\downarrow;\overrightarrow{\alpha_i};P_i)\ |||\ (\sum_{j\in J}\overleftarrow{\alpha_j};\gamma_j\uparrow;P_j)$ and can be obtained from it except the sending-receiving process which is implied here by interleaving sending and receiving processes. For instance, a sending process can be obtained when $(\sum_{j\in J}\overleftarrow{\alpha_j};\gamma_j\uparrow;P_j)$ is empty in the above communicating process. The main advantages of the new formulation of a communicating process in this paper are that it is more general than in [15], peer entity is also a communicating process whereas in [15] the peer entity is not a communicating process and the priority of entities is relaxed and thus no loss of messages during messages exchange between the entities.

3.2. Logical Errors in Protocol Specification

Communication between processes in LOTOS is by means of synchronization in actions. If processes cannot synchronize in an action, then the resultant behaviour of processes is *deadlock*, a situation in which no further synchronization is possible in remaining actions and communication between processes stops forever. We will explain the intended meaning of deadlock using simple examples before defining it formally in Definition 3.2.

Generally speaking, there are four types of logical errors which are well discussed specially in FSM based model; 1) deadlock, 2) unspecified reception, 3) non-executable transitions and 4) buffer overflow. We do not consider the buffer overflow in our LOTOS model, because the buffer does not exist explicitly. We use the same example as in [15] to discuss the other 3 errors. Let P, Q, R, T, and S be processes.

$P = a; b; c; \mathbf{stop}$, $Q = a; b; c; \mathbf{stop}$, $R = a; c; b; \mathbf{stop}$
$T = a; b; d; \mathbf{stop}$, $S = a; b; c; \mathbf{stop}\ []\ d; \mathbf{stop}$

Now, consider the followings:

$PM_1 = P|[a,b,c]|R \approx a; \mathbf{stop}$, $PM_2 = P|[a,b,c,d]|T \approx a; b; \mathbf{stop}$
$PM_3 = P|[a,b,c,d]|S \approx a; b; c; \mathbf{stop}$

PM_1 is deadlock after executing the action a, because there are actions b, c and c, b in P and R, respectively, still to be executed, i.e. P and R cannot synchronize in the actions b and c. PM_2 can be considered unspecified reception, because after $a; b$ are executed, P provides the c, but the peer process T refuses the c. Similarly T provides d but P refuses it. PM_3 can be considered to be non-executable transition because the action d is never executed.

Under the LOTOS model, every action has to be executed in a synchronous way. That is both processes have to provide the same action at the same time. All above 3 errors are due to the fact that one process provides an action but another process refuses to do that. So, from the "action execution" point of view, we collectively call the 3 errors deadlock.

Definition 3.2 (Deadlock) Let $PM = P|[G]|Q$ be a protocol model. PM is *deadlock* if there exist processes $P'(Q')$ and $a \in Act$ such that

(1) $P \xrightarrow{t} P' \xrightarrow{a}$ ($Q \xrightarrow{t} Q' \xrightarrow{a}$) for some $t \in Act^\star$
(2) For any $Q''(P'')$ such that $PM \xrightarrow{t'} P'|[G]|Q''$ ($PM \xrightarrow{t'} P''|[G]|Q'$), for some $t' \in Act^\star$ we have $P'|[G]|Q'' \not\xrightarrow{a}$ ($P''|[G]|Q' \not\xrightarrow{a}$).

In this case, we say that deadlock *occurs* at P' (Q') and a is the *non-executable* action[2].□

Now, we can formalize our protocol synthesis problem as follows:

Given: The specification of an entity P, which is a given entity and is expressed as communicating processes.

Problem: Generate the specification of the peer entity Q, which is expressed as communicating processes also, such that

(1) $PM = P|[G]|Q$ is *deadlock* free.
(2) P with an action of the form $\overleftarrow{\alpha}$ replaced by **i**; $\overleftarrow{\alpha}$ is bisimilar to **hide** $\tilde{\beta}$ **in** $P|[G]|Q$, where $\tilde{\beta}$ denotes the sequence of actions in Q of the form $\beta\uparrow$ and $\beta\downarrow$.
(3) Q with an action of the form $\overrightarrow{\alpha}$ replaced by **i**; $\overrightarrow{\alpha}$ is bisimilar to **hide** $\tilde{\gamma}$ **in** $P|[G]|Q$, where $\tilde{\gamma}$ denotes the sequence of actions in P of the form $\gamma\uparrow$ and $\gamma\downarrow$.

In order to realize (2) and (3) we have to consider the semantical meaning of entities' behaviours. For instance, the behaviour of P in (2) may be expressed in different form (i.e. $\overleftarrow{\alpha}$ is replaced by **i**; $\overleftarrow{\alpha}$) because for P, any messages sent by user U_q to Q seems invisible thus non-deterministic, i.e. it is not known to P which message is sent to Q by U_q and hence to P. In order to realize this non-determinism, we replace receiving actions of P, $\overleftarrow{\alpha}$ by **i**;$\overleftarrow{\alpha}$. (See Theorem 4.2 for detail).

4. Synthesis Algorithm

In this section, we present the basic idea of the synthesis algorithm in order to generate the peer entity from a given entity. Before stating the algorithm in section 4.2, we give synthesis rules, which are used in the algorithm to generate the peer entity.

4.1. Synthesis Rules

We give here simple synthesis rules to generate the peer entity Q from a given entity of the simple form, $p_1; P_1 \,[\,]\, p_2; P_2$, to improve the understanding of the synthesis algorithm. In the rules, we assume that Q_1 and Q_2 are the peer entities recursively generated from P_1 and P_2, respectively.

Synthesis Rules:

Rule 1: If $p_1 = \gamma_1\downarrow; \overrightarrow{\alpha_1}$, $p_2 = \gamma_2\downarrow; \overrightarrow{\alpha_2}$ then $Q = \overrightarrow{\alpha_1}; \beta_1\uparrow; Q_1 \,[\,]\, \overrightarrow{\alpha_2}; \beta_2\uparrow; Q_2$

Rule 2: If $p_1 = \overleftarrow{\alpha_1}; \gamma_1\uparrow$, $p_2 = \overleftarrow{\alpha_2}; \gamma_2\uparrow$ then $Q = \beta_1\downarrow; \overleftarrow{\alpha_1}; Q_1 \,[\,]\, \beta_1\downarrow; \overleftarrow{\alpha_2}; Q_2$

In Rule 1, when P receives a message from its user and wants to send the message to the peer entity Q, Q receives the message and delivers the message to its user. In Rule 2, when P receives a message from Q and delivers the message to its user, then make Q such that it receives the message from its user and sends it to P. In rule 1, p_1 and p_2 are the

[2]Note that a must be a synchronization action.

sending primitive components of P and generated $\overrightarrow{\alpha_1}; \beta_1\uparrow$ and $\overrightarrow{\alpha_2}; \beta_2\uparrow$ are the receiving primitive components of Q. Similarly p_1 and p_2 in rule 2 are the receiving primitive components of P and $\beta_1\downarrow; \overleftarrow{\alpha_1}$ and $\beta_1\downarrow; \overleftarrow{\alpha_2}$ are the sending primitive components of Q. In the synthesis rules, for each $\gamma\downarrow$ and $\gamma\uparrow$ there is $\beta\uparrow$ and $\beta\downarrow$ respectively. This is to show that the communication is end-to-end, i.e. U_p to U_q and vice versa. The communication between P and Q and vice versa only is also possible by extending the rules.

4.2. Synthesis Algorithm

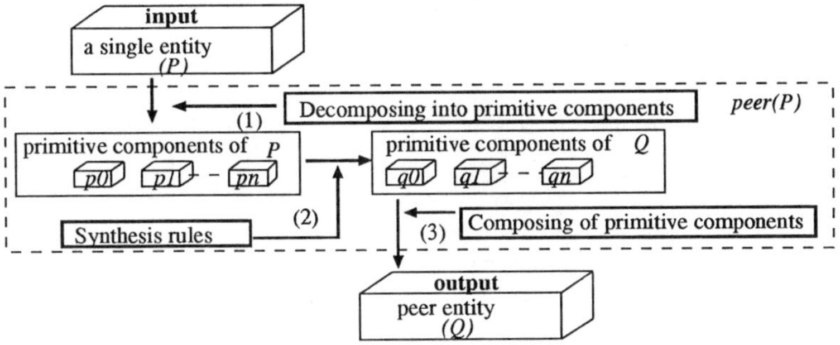

Figure 2. Outline of Proposed Protocol Synthesis Method

The synthesis algorithm we will propose is based on the communication patterns discussed in the previous section. The algorithm generates a peer entity from a given entity. Our algorithm takes care of any recursive behaviour of a given entity.

The outline of the algorithm can be briefly sketched in the Figure 2. Basically, it is a three steps procedure. In step (1), the communication patterns of the primitive components of the given entity are checked and decomposed into primitive components. In step (2), the corresponding communication patterns of the primitive components of the peer entity are obtained by using synthesis rules. In step (3), the primitive components of the peer entity are composed together to obtain the complete peer entity. In the algorithm *peer* is the function (procedure) whose input is a given entity P. *peer* checks the communication pattern of primitive components of P recursively and generates the corresponding primitive components of Q according to synthesis rules. For example, $peer(P)$ checks (see Figure 2) the communication pattern of the primitive component p_0, and generates the primitive component q_0 of Q. $peer(P_1)$ checks p_1 and generates q_1 and so on until the whole structure (behaviour) of P is checked. The relationship between or composition of $q_0, q_1, \ldots$ are handled by *peer* according to the relationship between $p_0, p_1, \ldots$.

Algorithm 4.1 (Synthesis Algorithm)

procedure $peer(P)$;
begin
  case P **of**
    stop: **return stop**; (case 1)
    exit: **return exit**; (case 2)
    x: **return** x; (case 3)
    $(\sum_{i\in I}\gamma_i\downarrow; \overrightarrow{\alpha_i}; P_i) \mid\mid\mid (\sum_{j\in J}\overleftarrow{\alpha_j}; \gamma_j\uparrow; P_j)$: (case 4)
      return $(\sum_{i\in I}\overrightarrow{\alpha_i}; \beta_i\uparrow; peer(P_i)) \mid\mid\mid (\sum_{j\in J}\beta_j\downarrow; \overleftarrow{\alpha_j}; peer(P_j))$;
    rec $x.P$: **return rec** $x.peer(P)$; (case 5)
end

Note that any synchronized action appearing in the entity P also appears in the peer entity $Q = peer(P)$.

The synthesis algorithm has many advantages, e.g. (1) the peer entity is generated automatically, thus the protocol designer can concentrate on the design of the given entity; (2) the synthesis algorithm is easy to extend by considering other communication patterns; (3) safety properties of the protocol generated by the synthesis algorithm are satisfied (Theorem 4.1).

4.3. Properties of the Synthesis Algorithm

The peer entity is derived from a given entity by the Algorithm in such a way that the resulting protocol model is guaranteed to be deadlock free (Theorem 4.1) and is bisimulated by both the given entity and the generated peer entity (Theorem 4.2) stated as follows.

Lemma 4.1 *If $P_i|[G]|Q_i$ are deadlock free for all $i \in I$, then the protocol model, $PM = (\sum_{i\in I}\gamma_i\downarrow; \overrightarrow{\alpha_i}; P_i)|[G]|(\sum_{i\in I}\overrightarrow{\alpha_i}; \beta_i\uparrow; Q_i)$ is deadlock free.*

Proof. Suppose that $P_i|[G]|Q_i$ are deadlock free for all $i \in I$. The resulting protocol model is equivalently expanded as

$$PM = \sum_{i\in I}\gamma_i\downarrow; \overrightarrow{\alpha_i}; (P_i|[G]|\beta_i\uparrow; Q_i)$$

Since the process $\beta_i\uparrow; Q_i$ can always execute $\beta_i\uparrow$ in $P_i|[G]|\beta_i\uparrow; Q_i$ (P_i may be able to execute some non-synchronized actions in $P_i|[G]|\beta_i\uparrow; Q_i$), if $P_i|[G]|\beta_i\uparrow; Q_i$ is deadlock for some i then $P_i|[G]|Q_i$ must be deadlock that contradicts the assumption. Thus, $P_i|[G]|\beta_i\uparrow; Q_i$ is deadlock free for all i. Therefore, we can conclude that PM is deadlock free. □

Lemma 4.2 *Let P_1 and P_2 be processes having no synchronized actions in common with processes Q_2 and Q_1, respectively. If protocol models $P_1|[G]|Q_1$ and $P_2|[G]|Q_2$ are deadlock free, then $PM = (P_1 \mid\mid\mid P_2)|[G]|(Q_1 \mid\mid\mid Q_2)$ is deadlock free.*

Proof. If PM is deadlock, then either $P_1|[G]|Q_1$ or $P_2|[G]|Q_2$ must be deadlock. This contradicts the assumption. Thus, we get the lemma. □

Note: it is assumed that no synchronization actions are common between P_1 and Q_2, and P_2 and Q_1.

Theorem 4.1 (syntactic property) *Let $Q = peer(P)$ be the peer entity of a given entity P obtained by the synthesis algorithm. Then $PM = P||[G]||Q$ is deadlock free.*

Proof. We show that $PM = P||[G]||Q$ is deadlock free by structural induction on P, where $Q = peer(P)$. In the base cases, i.e. $P =$ **stop**, **exit**, or x for some variable $x \in X$, the proof is trivial.

Now suppose $P = (\sum_{i \in I} \gamma_i\downarrow; \overrightarrow{\alpha_i}; P_i) \,|||\, (\sum_{j \in J} \overleftarrow{\alpha_j}; \gamma_j\uparrow; P_j)$. In this case, the deadlock freeness of PM is derived by Lemma 4.1 and Lemma 4.2.

Suppose $P = \mathbf{rec}\, x.P'(x)$. We will show that $PM = \mathbf{rec} x.P'(x)||[G]||\mathbf{rec}\, x.Q'(x)$ is deadlock free, where $Q'(x) = peer(P'(x))$. As the induction hypothesis, we assume that $P'(x)||[G]||Q'(x)$ is deadlock free. Let $P'^{(n)}(x)$ be the entity $P'(x)$ where the variable x is simultaneously unfolded n times according to the defining equation $x \stackrel{\text{def}}{=} P'(x)$. Let $Q'^{(n)}(x) = peer(P'^{(n)}(x))$. The claim can be proved by showing that $P'^{(n)}(x)||[G]||Q'^{(n)}(x)$ is deadlock free by induction on n.

For $n = 0$, it is clear from the structural induction hypothesis. As for the inductive case, deadlock-freeness of $P'^{(n+1)}(x)||[G]||Q'^{(n+1)}(x)$ is proved by contradiction. Suppose deadlock occurs at R in $P'^{(n+1)}(x) = P'(P'^{(n)}(x))$ and a is a non-executable action. Recall that a is a synchronized action and any synchronized action appearing in $P'(x)$ also appears in $Q'(x)$. From these basic properties of the algorithm, if R with the transition $R \xrightarrow{a} R'$ for some R' is in $P'(x)$, then $P'(x)||[G]||Q'(x)$ is deadlock. This contradicts the (structural) induction hypothesis. Otherwise, R turns out to be in $P'^{(n)}(x)$, thus it is implied that $P'^{(n)}(x)||[G]||Q'^{(n)}(x)$ is deadlock. This contradicts the induction hypothesis for $n \geq 0$. In the same way, occurrence of deadlock in $Q'^{(n+1)}(x)$ is denied.

If $PM = \mathbf{rec} x.P'(x)||[G]||\mathbf{rec}\, x.Q'(x)$ is deadlock, then it naturally implies that $P'^{(n)}(x)||[G]||Q'^{(n)}(x)$ is deadlock for some sufficiently large $n \geq 0$. Therefore, it is concluded that PM is deadlock free. □

Theorem 4.2 (Semantic Property) *Let $Q = peer(P)$ be the peer entity of a given entity P obtained by the synthesis algorithm. Then the following semantic properties are satisfied, where $\tilde{\beta}$ denotes the sequence $\beta_1\uparrow, \ldots, \beta_n\uparrow, \beta_1\downarrow, \ldots, \beta_m\downarrow$ and $\tilde{\gamma}$ denotes the sequence $\gamma_1\uparrow, \ldots, \gamma_n\uparrow, \gamma_1\downarrow, \ldots, \gamma_m\downarrow$ of non-synchronized actions appearing in P and Q respectively.*

(1) $\hat{P} \approx$ **hide** $\tilde{\beta}$ **in** $P||[G]||Q$, *where $\hat{P}$ is the resulting P with an action of the form $\overleftarrow{\alpha}$ replaced by* $\mathbf{i}; \overleftarrow{\alpha}$, *i.e.* $\hat{P} = P\{\mathbf{i}; \overleftarrow{\alpha} / \overleftarrow{\alpha}\}$

(2) $\hat{Q} \approx$ **hide** $\tilde{\gamma}$ **in** $P||[G]||Q$, *where $\hat{Q}$ is the resulting Q with an action of the form $\overrightarrow{\alpha}$ replaced by* $\mathbf{i}; \overrightarrow{\alpha}$, *i.e.* $\hat{Q} = Q\{\mathbf{i}; \overrightarrow{\alpha} / \overrightarrow{\alpha}\}$.

Proof.

We will just outline the proof of the theorem. We will just prove the property (1) by structural induction on P with the help of Figure 4 for simplicity. The property (2) can be proved similarly. For the base cases, i.e. $P =$ **stop**, **exit** or x for some variable $x \in X$, the proof is trivial.

Now suppose $P = (\sum_{i\in I}\gamma_i\downarrow;\overrightarrow{\alpha_i};P_i)\,|||\,(\sum_{j\in J}\overleftarrow{\alpha_j};\gamma_j\uparrow;P_j)$. As the induction hypothesis, we assume that
$\hat{P}_i \approx$ **hide** $\tilde{\beta}$ **in** $P_i|[G]|Q_i$ for all $i \in I$, and $\hat{P}_j \approx$ **hide** $\tilde{\beta}$ **in** $P_j|[G]|Q_j$ for all $j \in J$, where $Q_i = peer(P_i)$ and $Q_j = peer(P_j)$.

The graphs (a) and (b) in Figure 4 illustrate the initial parts of the labelled transition systems of P and Q respectively, where

$$P^{(s)} = \textstyle\sum_{i\in I}\gamma_i\downarrow;\overrightarrow{\alpha_i};P_i, \qquad P^{(r)} = \sum_{j\in J}\overleftarrow{\alpha_j};\gamma_j\uparrow;P_j$$
$$Q^{(r)} = \textstyle\sum_{i\in I}\overrightarrow{\alpha_i};\beta_i\uparrow;Q_i, \qquad Q^{(s)} = \sum_{j\in J}\beta_j\downarrow;\overleftarrow{\alpha_j};Q_j.$$

For example, the node labelled with (o_{ij}) in (a) in the Figure 4 indicates the process $P_i\,|||\,P_j$ and the path $(a_{00}) \to (b_{io}) \to (e_{ij}) \to (h_{ij}) \to (o_{ij})$ indicates the transitions

$$P \overset{\gamma_i\downarrow}{\to} \overrightarrow{\alpha_i};P_i\,|||\,P^{(r)} \overset{\overleftarrow{\alpha_j}}{\to} \overrightarrow{\alpha_i};P_i\,|||\,\gamma_j\uparrow;P_j \overset{\gamma_j\uparrow}{\to} \overrightarrow{\alpha_i};P_i\,|||\,P_j \overset{\overrightarrow{\alpha_i}}{\to} P_i\,|||\,P_j.$$

The parts of the transition systems of $\hat{P}$ and $PM = P|[G]|Q$ are given by (c) and (d) respectively in Figure 4.

Now, we will prove the equality, i.e. (1), for the inductive case, by showing that any process in (c), e.g. (e_{ij}), is bisimilar to the corresponding process inside the rectangle in (d), e.g. (e_{ij}, f_{oj}), where all $\beta_i\uparrow$ and $\beta_j\downarrow$ are interpreted as **i**.

To this end, we need the following claims.

claim 1:
$(\hat{P}_i\,|||\,\hat{P}_j) \approx$ **hide** $\tilde{\beta}$ **in** $(P_i\,|||\,P_j)|[G]|(Q_i\,|||\,Q_j)$ (i.e. $(o_{ij}$ in $\hat{P} \approx (o_{ij}, o_{ij})$ in $PM)$.

claim 2a:
hide $\tilde{\beta}$ **in** $(P_i\,|||\,P^{(r)})|[G]|(Q_i\,|||\,Q^{(s)}) \approx$ **hide** $\tilde{\beta}$ **in** $(P_i\,|||\,P^{(r)})|[G]|(\beta_i\uparrow;Q_i\,|||\,Q^{(s)})$.
(i.e.$(d_{io}, d_{io}) \approx (d_{io}, b_{io})$).

claim 2b
hide $\tilde{\beta}$ **in** $(P_i\,|||\,P^{(r)})|[G]|(Q_i\,|||\,\overleftarrow{\alpha};Q^{(s)}) \approx$ **hide** $\tilde{\beta}$ **in** $(P_i\,|||\,P^{(r)})|[G]|(\beta_i\uparrow;Q_i\,|||\,\overleftarrow{\alpha};Q^{(s)})$
(i.e. $(d_{io}, g_{ij}) \approx (d_{io}, e_{ij})$).

claim 2c:
hide $\tilde{\beta}$ **in** $(P_i\,|||\,\gamma_j\uparrow;P_j)|[G]|(Q_i\,|||\,Q_j) \approx$ **hide** $\tilde{\beta}$ **in** $(P_i\,|||\,\gamma_j\uparrow;P_j)|[G]|(\beta_i\uparrow;Q_i\,|||\,Q_j)$.
(i.e.$(g_{ij}, o_{ij}) \approx (gij, h_{ij})$).

claim 2d:
hide $\tilde{\beta}$ **in** $(P_i\,|||\,P_j)|[G]|(Q_i\,|||\,Q_j) \approx$ **hide** $\tilde{\beta}$ **in** $(P_i\,|||\,P_j)|[G]|(\beta_i\uparrow;Q_i\,|||\,Q_j)$.
(i.e.$(o_{ij}, o_{ij}) \approx (o_{ij}, h_{ij})$).

Claim 1 can be verified by the induction hypothesis and the fact that $P_i(P_j)$ and $Q_i(Q_j)$ have no actions in common. The proofs of claim 2a to claim 2d are easy, so omitted.

It remains to check the other inductive case, i.e. the case for an entity of the form **rec** $x.P'$. For this case, in the same style as we have done for the recursive case in the proof of the previous theorem, the equality can be proved by showing that $P'^{(n)}(x) \approx$ **hide** $\tilde{\beta}$ **in** $P'^{(n)}(x)|[G]|Q'^{(n)}(x)$ holds by induction on n.

For $n = 0$, it is clear from the structural induction hypothesis. As for the inductive case, the equality $P'^{(n+1)}(x) \approx$ **hide** $\tilde{\beta}$ **in** $P'^{(n+1)}(x)|[G]|Q'^{(n+1)}(x)$ is implied by the (structural) induction hypothesis, the induction hypothesis for $n \geq 0$, and basic properties for the operation **hide**, where we use the same notations as in the previous proof.

Now, suppose $\mathbf{rec}\,x.P' \not\approx \mathbf{hide}\,\tilde{\beta}\,\mathbf{in}\,\mathbf{rec}\,x.P'(x)|[G]|\mathbf{rec}\,x.Q'(x)$, then from the definition of bisimilarity we have that $P'^{(n)}(x) \not\approx \mathbf{hide}\,\tilde{\beta}\,\mathbf{in}\,P'^{(n)}(x)|[G]|Q'^{(n)}(x)$ for some sufficiently large n. But, this contradicts the above statement. Therefore, the equality must be satisfied. □

4.4. An Application Example of the Protocol Synthesis Algorithm

As an example, we apply the synthesis algorithm to the small portion of simplified version of the protocol handler of the Transport Layer [17]. We have omitted the detail specifications and are contained in [17]. The protocol handler has two entities, a *calling* entity and a *called* entity. Entities exchange messages in three phases, *connection phase*, *data phase* and *disconnect phase*. After the connection is established successfully, the entities enter data phase. During the data phase the entities can be disabled by the disconnect phase. We just state the sequential composition ($\gg$) and the disabling ([>) LOTOS operators, which are used to combine the different phases, due to the lack of space. We use notations given in section 3.1 to denote primitives (service primitives (SPs) and protocol data units (PDUs)) used in the Transport Layer. We use the following abbreviations:

For SPs, $ConReq{\downarrow} : \gamma_1{\downarrow}, ConCnf{\uparrow} : \gamma_2{\uparrow}, DisIn{\uparrow} : \gamma_3{\uparrow}, DatReq{\downarrow} : \gamma_4{\downarrow}, DatIn{\uparrow} : \gamma_5{\uparrow}, DisReq{\downarrow} : \gamma_6{\downarrow}$.

For PDUs, $\overrightarrow{CR} : \overrightarrow{\alpha_1}, \overleftarrow{CC} : \overleftarrow{\alpha_2}, \overleftarrow{DR} : \overleftarrow{\alpha_3}, \overrightarrow{DT} : \overrightarrow{\alpha_4}, \overleftarrow{DT} : \overleftarrow{\alpha_5}, \overrightarrow{DR} : \overrightarrow{\alpha_6}$.

The given entity, P, is *calling* entity. Due to the space, we do not write the full descriptions of the entities i.e. reserved words, gate names etc., instead we write it in equation forms. The phases of the given entity are as follows (Note, P = Connect_Phase_P $\gg$ (Data_Phase_P [> Disconnect_Phase_P) where) ;

$$\begin{aligned} \text{Connect\_Phase\_}P &= \gamma_1{\downarrow}; \overrightarrow{\alpha_1}; (\overleftarrow{\alpha_2}; \gamma_2{\uparrow}; \mathbf{exit}[]\overleftarrow{\alpha_3}; \gamma_3{\uparrow}; \text{Connect\_Phase\_}P) \\ \text{Data\_Phase\_}P &= \gamma_4{\downarrow}; \overrightarrow{\alpha_4}; \text{Data\_Phase\_}P \;|||\; \overleftarrow{\alpha_5}; \gamma_5{\uparrow}; \text{Data\_Phase\_}P \\ \text{Disconnect\_Phase\_}P &= \gamma_6{\downarrow}; \overrightarrow{\alpha_6}; \mathbf{exit} \;|||\; \overleftarrow{\alpha_3}; \gamma_3{\uparrow}; \mathbf{exit} \end{aligned}$$

The algorithm is applied to each phase of P to generate the corresponding phase of peer (called) entity, Q. Here, we apply the algorithm to the Connect_Phase_P of the given entity and generate the Connect_Phase_Q of the peer entity. The application of the algorithm is shown as follows;

$$\begin{aligned} \text{Connect\_Phase\_}Q &= peer(\text{Connect\_Phase\_}P) \\ \text{from (case 5)} &= (\mathbf{rec}\,x.peer(\gamma_1{\downarrow}; \overrightarrow{\alpha_1}; (\overleftarrow{\alpha_2}; \gamma_2{\uparrow}; \mathbf{exit}[]\overleftarrow{\alpha_3}; \gamma_3{\uparrow}; x))) \\ \text{from (case 4)} &= \mathbf{rec}\,x.\overrightarrow{\alpha_1}; \beta_1{\uparrow}; peer((\overleftarrow{\alpha_2}; \gamma_2{\uparrow}; \mathbf{exit}[]\overleftarrow{\alpha_3}; \gamma_3{\uparrow}; x)) \\ \text{from (case 4)} &= \mathbf{rec}\,x.\overrightarrow{\alpha_1}; \beta_1{\uparrow}; (\beta_2{\downarrow}; \overleftarrow{\alpha_2}; peer(\mathbf{exit})[]\beta_3{\downarrow}; \overleftarrow{\alpha_3}; peer(x)) \\ \text{from (case 2)} &= \mathbf{rec}\,x.\overrightarrow{\alpha_1}; \beta_1{\uparrow}; (\beta_2{\downarrow}; \overleftarrow{\alpha_2}; \mathbf{exit}[]\beta_3{\downarrow}; \overleftarrow{\alpha_3}; peer(x)) \\ \text{from (case 3)} &= \mathbf{rec}\,x.\overrightarrow{\alpha_1}; \beta_1{\uparrow}; (\beta_2{\downarrow}; \overleftarrow{\alpha_2}; \mathbf{exit}[]\beta_3{\downarrow}; \overleftarrow{\alpha_3}; x) \end{aligned}$$

The result is: Connect_Phase_Q = $\overrightarrow{\alpha_1}; \beta_1{\uparrow}; (\beta_2{\downarrow}; \overleftarrow{\alpha_2}; \mathbf{exit}[]\beta_3{\downarrow}; \overleftarrow{\alpha_3};$ Connect_Phase_Q)
Other phases of the peer entity can be generated similarly.

We show the *Labelled Transition Systems* of Connect_Phase_P and Connect_Phase_Q in Figure 3.

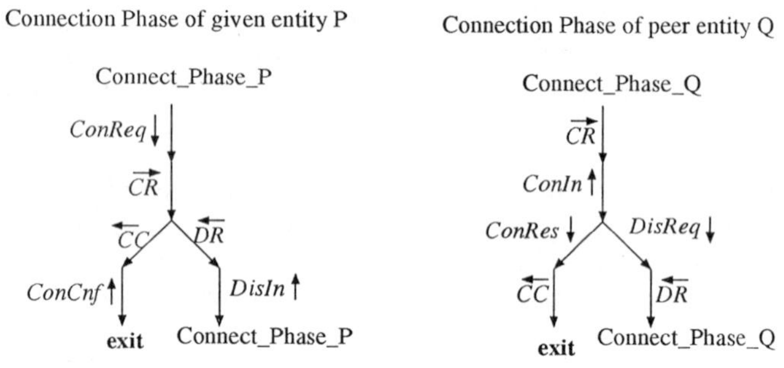

Figure 3. *Labelled Transition Systems* of Connection Phases of P and Q

5. Conclusions

We proposed a protocol synthesis algorithm, which is an extention of [15], to generate the peer entity from a given single entity. The specifications of entities are in LOTOS. We also proposed the definition of *deadlock* which is different from the deadlock defined in the protocol synthesis based on FSMs. The synthesized protocol is proved to be *deadlock* free if the given entity is expressed as communicating processes, which are natural for communicating entities. We also presented a *bisimilarity* between the given entity and the synthesized protocol and also the generated peer entity and the synthesized protocol. It shows that the generated protocol coincides with the behaviour of the given entity and the generated peer entity. The protocol synthesis procedure we have presented is applicable to every structure of protocol layer which provides services to its upper layer. To show the effectiveness of the proposed synthesis algorithm, we have applied it to a simplified version of the protocol handler of the Transport Layer.

The proposed synthesis algorithm checks the structure or communication patterns of the given entity and generates the corresponding structure of the peer entity using synthesis rules. Thus, the proposed synthesis procedure can be extended by considering other more complicated communication patterns and thus making the protocol design a flexible one.

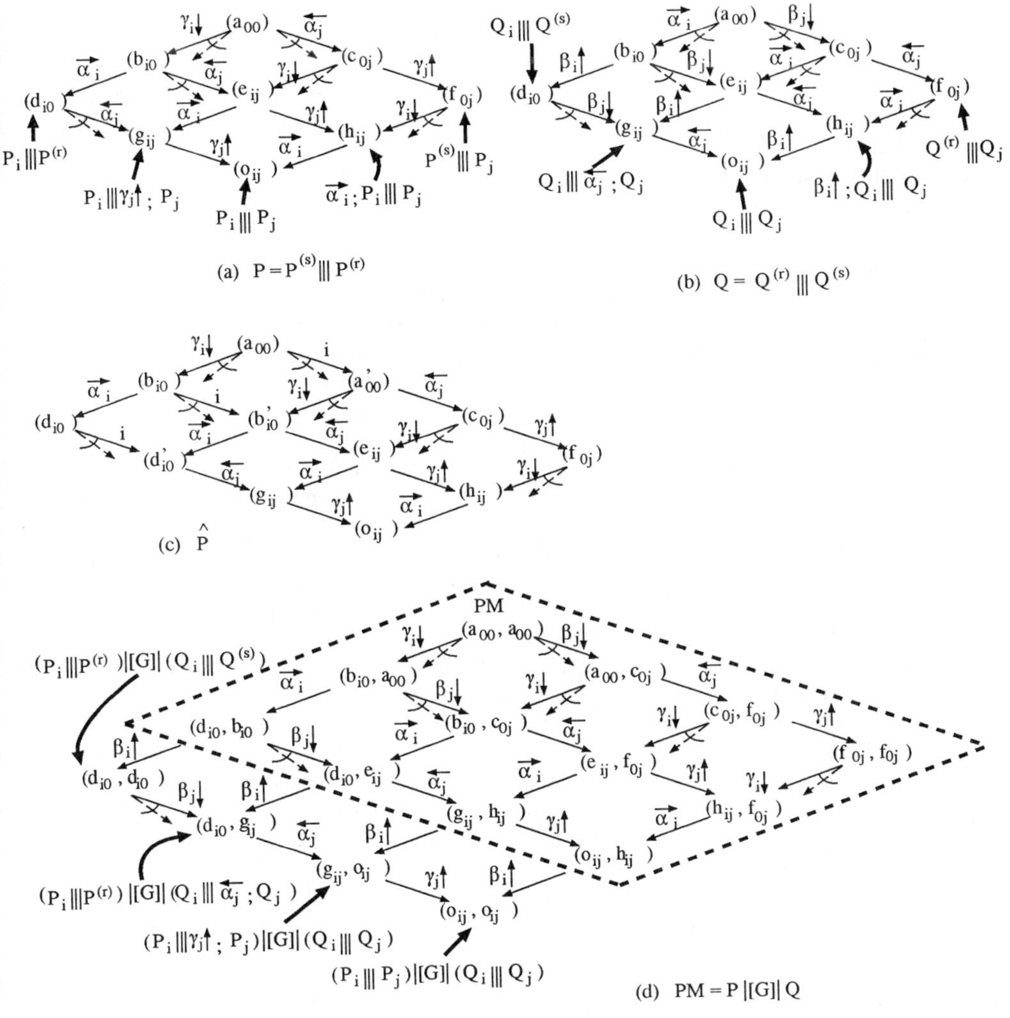

Figure 4. Proofs of Theorem 4.2

REFERENCES

1. ISO. Estelle: A formal Description Technique based on an Extended State Transition Model. ISO 9074, 1989.
2. ISO. Information Processing Systems — Open Systems Interconnection — LOTOS — A formal description technique based on the temporal ordering of observational behaviour. IS 8807, 1989.
3. CCITT. SDL: Specification and Description Language. CCITT Z.100, 1988.
4. Y. Kakuda, Y. Wakahara, and M. Norigoe. An acyclic expansion algorithm for fast protocol validation. *IEEE Transactions on Software Engineering*, 14(8):1059–1070, August 1988.
5. G.J. Holzmann. Automated protocol validation in argos: Assertion proving and scatter searching. *IEEE Transactions on Software Engineering*, 13(6):683–695, June 1987.
6. Y. Kakuda and H. Saito. An integrated approach to design of protocol specifications using protocol validation and synthesis. *IEEE Transactions on Computers*, 40(4):459–467, April 1991.
7. N. Shiratori, H.Kaminaga, K. Takahashi, and S. Noguchi. A verification method for LOTOS specifications and its application. In C.A. Vissers E. Brinksma, G. Scollo, editor, *Protocol Specification, Testing, and Verification, IX*, pages 59–70. North-Holland, 1990.
8. Y. Kakuda and Y. Wakahara. Component-based synthesis of protocols for unlimited number of processes. In *Proc. IEEE COMPSAC'87*, pages 721–730, October 1987.
9. C. V. Ramamoorthy, S. T. Dong, and Y. Usuda. An implementation of an automated protocol synthesizer(APS) and its application to the X.21 protocol. *IEEE Transactions on Software Engineering*, SE-11(9):886–908, 1985.
10. B.V. Bochmann and R. Gotzhein. Deriving protocol specification from service specification. In *Proc. SIGCOMM'86*, volume 14, pages 144–156, 1986.
11. Rom Langerak. Decomposition of functionality : a correctness preserving LOTOS transformation. In Luigi Logrippo, Robert L. Probert, and Hasan Ural, editors, *Proceedings of the Tenth International IFIP WG 6.1 Symposium on Protocol Specification, Testing, and Verification*, pages 203–218, Ottawa, Ontario, Canada, June 1990. North-Holland.
12. Yao-Xue Zhang, Kaoru Takahashi, Norio Shiratori, and Shoichi Noguchi. An interactive protocol synthesis algorithm using a global state transition graph. *IEEE Transactions on Software Engineering*, 14(3), March 1988.
13. B.B. Bista, Zixue Cheng, and Norio Shiratori. A LOTOS based synthesis method for protocol specification. Technical Report 2, Multimedia Communication and Distributed Processing Workshop, IPS of Japan, Nov 1993.
14. R. Milner. *Communication and Concurrency*. Prentice-Hall, 1989.
15. B.B. Bista, Zixue Cheng, Atsushi Togashi, and Norio Shiratori. A new approach for protocol synthesis based on LOTOS. To appear in *IEICE Trans. on Fundamentals of Electronics, Communications and Computer Sciences*, E77-A(10), Oct. 1994.
16. P. Merlin and G.V. Bochmann. On the construction of submodule specifications and communication protocols. *ACM Trans. Program. Language Systems*, 5(1), Jan 1983.
17. A. Tanenbaum. *Computer Networks*. Prentice-Hall International Editions, 1989.

40

A New Distributed Algorithm for Implementation of LOTOS Multi-Rendezvous*

Zixue Cheng[a], Tongjun Huang[b], and Norio Shiratori [c]

[a]Department of Computer Software,
The University of Aizu
Tsuruga, Ikki-machi, Aizu-Wakamatsu City, Fukushima, 965-80 Japan
z-cheng@u-aizu.ac.jp

[b]Information Systems and Technology Center,
The University of Aizu
Tsuruga, Ikki-machi, Aizu-Wakamatsu City, Fukushima, 965-80 Japan
t-huang@u-aizu.ac.jp

[c]Research Institute of Electrical Communication,
Tohoku University
2-1-1, Katahira, Aoba-ku, Sendai, 980 Japan
norio@shiratori.riec.tohoku.ac.jp

LOTOS[1] is a high level specification language which incorporates the Multi-Rendezvous. Multi-Rendezvous is a powerful communication mechanism that allows a set of processes to execute an event in synchronous way. Besides the multi-synchronous property, Multi-Rendezvous has nondeterministic and exclusion properties. It is not a trivial problem to implement Multi-Rendezvous in a network system. The implementation has to maintain the properties mentioned above, be fair and make progress.

So far many researches have been developed for implementation of Multi-Rendezvous. Among them, [2] requires $O(N^2)$ message passes, [3] is not fully distributed because it is based on a binary tree, and [4][5] uses auxiliary resources such as event manager, coordinator, forks or tokens.

In this paper, we propose a new efficient distributed algorithm, based on the concept of coterie, for implementing Multi-Rendezvous with $O(N\sqrt{N})$ message passes. Our algorithm is fully distributed and does not use manager processes and auxiliary resources.

Coterie is an information structure which has been used in solutions for many distributed problems such as mutual exclusion and replica control problems. However, our algorithm is the first one to use coterie for Multi-Rendezvous.

*This work has been supported by Grant-in-Aid for Encouragement of Young Scientists from the Ministry of Education, Science and Culture, Japan

1. Introduction

LOTOS is a high level specification language for OSI communication protocols and services and other distributed systems [1]. LOTOS incorporates Multi-Rendezvous Communication mechanism, as an extension of binary rendezvous that has been suggested for CSP and Ada, to enhance the power of the languages.

Multi-Rendezvous has the following properties

1. Synchronous property:
   A set of processes that interact in an event participate in its execution at the same time.

2. Nondeterministic selection property:

   A process may have more than one possible event, but it cannot participate in these events simultaneously. It has to select one of the possible events nondeterministically.

3. Exclusion property:

   No more than one process-set execute their events simultaneously if these process-sets share a common process.

Multi-Rendezvous has to be implemented in a network environment which only has one-to-one message passing mechanism. The implementation of Multi-Rendezvous in a network environment is not a trivial problem.

The problem is to design a distributed algorithm that decides which processes-set should execute their event in order to guarantee the above properties. The decision made by the distributed algorithm should be fair and make progress.

Fairness means if a process-set tries to participate in an enabled event infinitely often, the event will eventually be executed.

Progress means when an event is enabled, either the event or some conflicting event will eventually be executed.

There are several researches on implementation of Multi-Rendezvous in a network environment. The distributed algorithm proposed by Gao and Bochmann[2] requires $O(N^2)$ message passes. The solution by Sisto et al.[3] requires only $O(N)$. However it uses a binary tree structure and messages are sent to the root of the tree for making decision. Therefore this approach is not fully distributed. Chandy and Misra[4] first pointed out that the problem of implementing Multi-Rendezvous is associated with synchronization and exclusion problems . They described the Multi-Rendezvous as committee coordination problem and proposed an elegant algorithm to implement it. In their algorithm, auxiliary resources are used to implement synchronization and exclusion. That is, for every event in the system, a special process called coordinator is postulated and a fork is used between two neighbor sets of processes.

Following Chandy and Misra's approach, Bagrodia[5] proposed two algorithms (Event Manager Algorithm and Modified Event Manager Algorithm). And in Bagrodia's paper, he also created a simulation approach to evaluate a distributed algorithm for implementation of Multi-Rendezvous. It is the first study that use the simulation method to evaluate an implementation algorithm of Multi-Rendezvous. However the Event Manager Algorithm uses a circulating token to solve the exclusion problem, so it is not a complete distributed one. Same as Chandy and Misra's research Modified Event Manager algorithm uses a manager process for every event and auxiliary resources(forks).

In this paper, we propose a distributed algorithm with $O(N\sqrt{N})$ message passes. Our algorithm is completely distributed and it does not use manager processes and auxiliary resources. The proposed algorithm use the coterie and randomized method to solve the implementation problem of Multi-Rendezvous. Coterie is a kind of information structure which has been used in many distributed problem such as mutual exclusion [6], replica control [7] and distributed consensus [8]. However there is no research which use coterie to solve implementation problem of Multi-Rendezvous.

Section 2 gives a precise description of the problem and the network model used in this paper. Section 3 presents our solution to the problem. In Section 4, properties and complexity of the distributed algorithm are given. Section 5 contains our conclusions.

2. Network Model and Problem Description

2.1. Network Model

In this paper, we consider a connected, undirected, and complete graph, representing a network, with N processes and an link between each pair of processes.

Each process executes the same algorithm, which consists of sending messages over adjoining links, waiting for incoming messages, and processing them. Messages can be transmitted independently in both directions on a link and arrive after an unpredictable but finite delay, without error and in sequence.

2.2. The Problem

For the definition of the Multi-Rendezvous problem, we follow [5]. Let $\mathbf{P} = \{p_1, p_2, ..., p_n\}$ be a set of **n** processes and $\mathbf{E} = \{e_1, e_2, ..., e_m\}$ be a set of **m** events. Each process participates in zero or more events. The set of events in which a process p_i participates is called its event-set and is referred to as E_i; each E_i is a subset of $\mathbf{E}$. Each event, say e_k, is associated with a set of one or more processes; this set is called the process-set of the event and is referred to as P_k; each P_k is a subset of $\mathbf{P}$. Two events e_k and e_j are said to be in conflict if $P_k \cap Pj \neq \{\}$. The process-set of an event and the event-set of a process are both static sets.

A process is either idle or active. An active process autonomously makes the transition to become idle. An idle process p_i waits to commit to any one of the events in E_i. A process commits to an event e_k only when it determines that all other processes that belong to P_k will also do so. An idle process may commit to at most one event at any time. Every process in the system satisfies the following conditions:

1) An idle process remains idle until it commits to some event.

2) An idle process becomes active if it commits to some event.

An event is either enabled or disabled. An event e_k is enabled if and only if all processes that belong to P_k are idle and these processes satisfy synchronous conditions; e_k is disabled if there exists some $p_i \in P_k$ such that p_i is active or synchronous conditions on the event are not satisfied. There are three kinds of synchronous conditions shown in Table 1.

Table 1
Synchronous conditions

p_i	p_j	Synchronous conditions	Interaction kinds
$e!V_1$	$e!V_2$	value(V_1) = value(V_2)	value matching
e !V	e ?x:t	value(V) in domain(t)	value passing
e ?x:t	e ?y:u	t=u	value generation

For any pair of $p_i, p_j \in P_k$ the above synchronous conditions need to be satisfied, in order to commit to e_k. Here e!V is a structured event. "e" is an event name. "!V" is a parameter of the event, representing a value. Similarly, "?x:t" is a parameter representing a variable. In general, the number of values or/and variables added to an event is more than one (e.g. e_k = e!5;!6;?x:integer).

We use the phrase "an event e_k is executed" to mean that each process that belongs to P_k has committed to e_k. We are required to devise an algorithm which allows an idle process to commit to an enabled event such that the following properties are satisfied.

Synchronous property:

A disabled event can not be committed.

Nondeterministic selection property:

A process may have more than one enabled event, but only commits to at most one event of the enabled events at any time. The process has to select one enabled event in nondeterministic way.

Exclusion property:

If a process p_i commits to an event e_k, then $\forall p_j \in P_k$, process p_j cannot commit to another event. In other words, conflicting events cannot be executed simultaneously.

Fairness property:

if a process-set try to participate in an enabled event infinitely often, it will eventually execute the event.

Progress property:

If there are enabled events, at least one of them will be committed by its process-set.

3. The Distributed Algorithm

3.1. Communication Structure

(1)Coterie

A coterie is a family C of subsets of a process-set such that any pair of subsets in C has at least one process in common and no subset in C contains any other subset in C. A subset in C is called a quorum.

We give here a method for creating coterie for a process-set P_k. First, create a complete graph that has n_k node where $n_k \geq \frac{1+\sqrt{8N_k+1}}{2}$ and N_k is the number of processes in P_k.

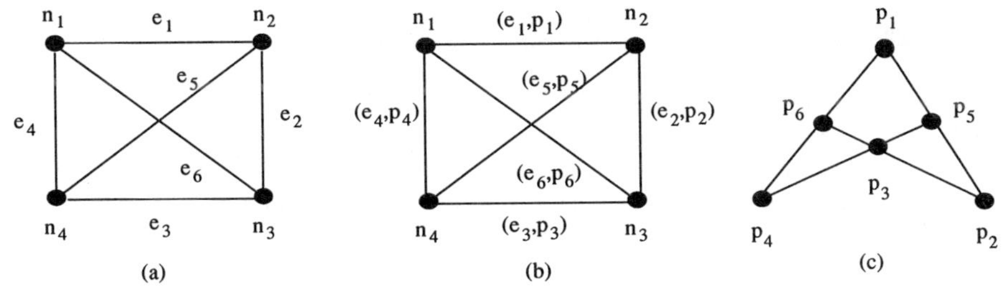

Figure 1. Example of coterie

To differ from the original network system, we call the created graph auxiliary graph

Link one edge in the auxiliary graph to one process in P_k. In this paper we only consider the case where $n_k = \frac{1+\sqrt{8N_k+1}}{2}$, that is, the number of edges is $\frac{n_k(n_k-1)}{2} = N_k$. Hence the one-to-one mapping is possible. We do not consider $n > \frac{1+\sqrt{8N+1}}{2}$ case in this paper. Our method can be easily extended to that case by adding some virtual process to P_k, similar to the method in [8].

Make a dual graph of the auxiliary graph. That is, let a node in the auxiliary graph be a line, let a edge be a node. From above discussion, a node in the dual graph represents a process in P_k.

Obviously, the dual graph has the following properties

1) Each line consists of exactly $n_k - 1$ nodes.

2) Each node is on exactly two lines.

3) Every two lines cross at one node.

4) There are n_k lines in the graph.

5) There are $\frac{n_k(n_k-1)}{2}$ nodes in the graph.

If we take a line as a quorum, then, the set of lines in the graph form a coterie.

Example:

Suppose $P_k = p_1, p_2, \cdots, p_6$, $N_k = 6$, an auxiliary graph is shown in Fig 1(a). The processes are linked to the edges as shown in Fig 1(b). The dual graph is shown in Fig 1(c).

(2) Our Communication Structure

A process(node) only sends messages to processes on the two lines where the process is on. And a process receives messages only from the processes on the two lines where the process is on.

A set of processes which a process sends messages to is called the **sending set** of the process. A set of processes which a process receives messages from is called the **receiving set** of the process.

3.2. Basic ideas

To be easily understood, let us first discuss a simple case where only one process-set participates in the execution of one event, and then discuss the general case where there exist more than one process-set, and a process may be shared by several process-sets.

Case 1: The simple case

Because there is only one process-set, we do not need to consider nondeterministic property, exclusion, and fairness. We only consider how to maintain the synchronous property. Suppose the number of processes in the process-set is N and $N = \frac{n(n-1)}{2}$ (i.e. $n = \frac{1+\sqrt{8N+1}}{2}$), we use the communication structure described above. To make the multi-synchronization, two rounds (steps) of communication are needed.

Every process executes the following steps after it becomes idle.

Step 1

(1) Sending messages
Send a REQUEST message to every process in its sending set. The REQUEST message carries parameters on the event which the process participates in.

(2) Receiving messages
Receive all messages from its receiving set.

(3) Local calculation
Do the value matching described in Section 2, using the parameters carried by the messages received.

The result of the value matching will be a matched value or unmatched.

Step 2

(1) Sending messages
Send a MATCHED message or an UNMATCHED message to every process in its sending set, depending on the result of value matching.

(2) Receiving messages
Receive messages from its receiving set.

(3) Local calculation

IF all the received messages are MATCHED messages.
THEN execute the event; exit.
ELSE (some received message is UNMATCHED message).
exit.

Case 2: General case

Our basic idea is that a communication structure is used for every process-set. That is, every process-set does the same thing (processing) as described in **case 1**. However, there is some process participating in more than one event, that is, more than one process-set share a common process. In order to maintain the nondeterministic and exclusion properties we use a random-number based method.

First in each process, for every event the process participates in, the process assigns a random number (from 0 to 1) to the event. And then the process sends a REQUEST message carrying the random number to every process of its sending set in Step 1. The maximum random number of a process-set will be obtained by every process in the process-set in Step 2(3).

If more than one process-set which shares a common process satisfies synchronous conditions, to maintain the nondeterministic property, the process shared by more than one process-set, selects only one process-set to participate in, according to the following strategy.

Selection strategy:

Suppose mrn(k) is a maximum random number of process-set P_k which is obtained using the method described above, and id(k) is an identifier of process-set P_k given in advance, select the process-set which has the largest (mrn(k), id(k)), and reject other process-sets.

$$(mrn(k), id(k)) > (mrn(j), id(j)) \quad \text{iff}$$
$$mrn(k) > mrn(j) \quad \textit{or}$$
$$mrn(k) = mrn(j) \quad \textit{and} \quad id(k) > id(j)$$

Moreover, a process which participates in only one process-set will select the process-set.

□

To maintain the exclusion property, after selecting a process-set, a process needs to let other processes in process-sets, which share the process, know its selection. To this end, two rounds (Step 3 and Step 4) of communication are needed. We use the same communication structure in Step 3 and Step 4 as in Step 1 and Step 2. Step 3 and Step 4 are similar to Step 1 and Step 2, except instead of sending messages carrying synchronous conditions, SELECTED or UNSELECTED messages are sent and received.

3.3. The formal description of the distributed algorithm

A process p_i holds the following variables.

E_i : The event-set which p_i participates in.

S_k : A sending set for event $e_k \in E_i$.
Here, $S_k \subset P_k$. P_k is a process-set which participate in e_k.

R_k : A receiving set for event $e_k \in E_i$.
Here, $R_k \subset P_k$.

mrn(k): A variable for holding the maximum random number in P_k.

id(k): An integer which represents the identifier of process-set P_k.

Every process executes the following steps, after it becomes idle.

Step 0

For each $e_k \in E_i$, assign a random number.

Step 1

(1) Sending messages
For $\forall e_k \in E_i$, send a REQUEST message to every process $p_j \in S_k$. The REQUEST message carries parameters of e_k and the random number assigned to e_k.

(2) Receiving messages
For $\forall e_k \in E_i$, receive all messages from R_k.

(3) Local calculation
(a) For $\forall e_k \in E_i$, do the value matching described in Section 2, using the parameters carried by the messages received.

The result of the value matching will be a matched value or unmatched.

(b) For $\forall e_k \in E_i$, choose the maximum random number in the all messages received and set mrn(k) to the number.

Step 2

(1) Sending messages
For $\forall e_k \in E_i$, send a MATCHED message or a UNMATCHED message to every process $p_j \in S_k$, depending on the result of value matching. The MATCHED or UNMATCHED message carries the random number chosen.

(2) Receiving messages
For $\forall e_k \in E_i$, receive messages from R_k.

(3) Local calculation
For $\forall e_k \in E_i$, check whether all the received messages from R_k are MATCHED messages

IF there are more than one event (say $e_k \in E_i$ $e_j \in E_i \cdots$), such that all messages received from their receiving sets are MATCHED messages.

THEN For $\forall e_k \in E_i$, choose the maximum random number in the all messages received and set mrn(k) to the number. Select only one $e_k \in E_i$, which has the largest (mrn(k),id(k)). Reject other events.

ELSE IF there is only one $e_k \in E_i$, such that all the received messages from R_k are MATCHED messages.

THEN select the e_k.

ELSE reject all $e_k \in E_i$.

Step 3

(1) Sending messages
For $\forall e_k \in E_i$, send a SELECTED message or a UNSELECTED message to every process $p_j \in S_k$, depending on the result of whether e_k has been selected.

(2) Receiving messages
For $\forall e_k \in E_i$, receive all messages from R_k.

(3) Local calculation
For $\forall e_k \in E_i$, check whether all the received messages from R_k are SELECTED messages
The result will be YES and NO.

Step 4

(1) Sending messages
For $\forall e_k \in E_i$, send a DO message or a UNDO message to every process $p_j \in S_k$, depending on the result in the last step (i.e YES or NO).

(2) Receiving messages
For $\forall e_k \in E_i$, receive all messages from R_k.

(3) Local calculation

IF there is a $e_k \in E_i$, such that all the received messages from R_k are DO messages

THEN execute the event ; exit

ELSE (For $\forall e_k \in E_i$, some received message is UNDO)
exit

4. The Properties and Complexity of the Algorithm.

4.1. Properties

We now outline a proof that the algorithm is correct, that is, we show that the algorithm maintains the properties of Multi-Rendezvous described in Section 2.

Synchronization:

Suppose an event is not enabled, we show the process-set corresponding to the event cannot execute the event.

Obviously if some process has not become the idle state, the process-set cannot execute the corresponding event. Suppose all processes in the process-set become the idle state, they will communicate with each other in Step 1 and Step 2. Since the event is not enabled, that means the synchronous conditions are not met, every process of the process-set will receive an UNMATCHED message in Step 2(2). A process that receives UNMATCHED does not select the process-set to participate in, so all the processes of the process-set do not execute the corresponding event.

Nondeterministic property:

Suppose a process participates in more than one event and all these events are enabled, then the process is shared by more than one process-set. Note that the process communicates with the processes in these process-sets in Step 1 and Step 2 and obtains matched values and maximum random numbers of these process-sets. Since, in Step 2 (3) the process selects only one process-set to participates in and the selection is decided in a randomized way (using the random numbers), the nondeterministic property is maintained.

Exclusion property:

A process shared by more than one process-sets selects one process-set to participate in (see Step 2 (3)).

The process sends SELECTED messages to the sending set of the selected process-set and UNSELECTED messages to the sending sets of unselected process-sets which share the process in Step 3(1).

After receiving an UNSELECTED message, a process will send UNDO messages to its sending set corresponding to the unselected process-set (see Step 3(2)(3) and Step 4(1)).

Thus, all processes in a process-set will receive an UNDO message in Step 4, if some process (even one process) in the process-set sends UNSELECTED message. The processes received UNDO cannot execute the corresponding event. Thus, no more than one process-set execute their events simultaneously if these process-sets share a common process.

Note that the exclusion property cannot be achieved in Step 3(3). Because a process sends an UNSELECTED only to every process in the sending set of an unselected process-set, not to all processes in an unselected process-set, some process in an unselected process-set receives an UNDO message in Step 4(2) instead of directly receiving an UNSELECTED in Step 3(2).

Fairness:

Suppose a set of processes try to participate in an enabled event infinitely often, all processes in the process-set communicate with each other in Step 1 and Step 2. In Step 2 (3), every process of the process-set will commit on a matched value and obtain a maximum random number.

It is possible that the random number is the largest one in all process-sets. In that case, every process in the process-set will select the process-set in the Step 2(3). Then all processes in the process-set communicate with each other to conform they have selected the process-set in Step 3 and Step 4. In Step 4 (3) all the processes in the process-set will eventually execute the event.

Progress:

Suppose there are events which are enabled, we show at least one event will be executed.

We have introduced a total order into the family of process-sets. That is for any two process-sets

$$(mrn(k), id(k)) > (mrn(j), id(j)) \quad \text{iff}$$
$$mrn(k) > mrn(j) \quad or$$
$$mrn(k) = mrn(j) \quad and \quad id(k) > id(j)$$

In Step 2 (4), a process which is shared by more than one process-sets will select one process-set according to the total order. Hence, among the process-sets which is enabled, all processes of one process-set which has the largest (mrn, id) will select the process-set so that the process-set will execute the corresponding event.

4.2. Complexity

This section shows how many messages are used during an execution of the distributed algorithm. First, we give some notations. Let N be the number of processes in a network. Generally, the N processes form m process-sets which participate in m events. Let N_k be the number of processes of process-set P_k. A communication structure based on coterie is used in P_k. A sending set of a process in P_k has $2(n_k - 2)$ processes, where $n_k = \frac{1+\sqrt{8N_k+1}}{2}$.

The number of message used in one process-set (say P_k) is calculated as follows:

In Step 1 (1), Step 2 (1), Step 3 (1) and Step 4 (1), every process in the process-set sends a message only to every process in its sending set. The number of processes in the sending set is $2(n_k - 2)$, thus

$$M_k = 4 \times N_k \times 2(n_k - 2) = 8N_k(n_k - 2)$$

Because of $n_k = \frac{1+\sqrt{8N_k+1}}{2}$,

$$O(n_k) = O(\sqrt{N_k})$$

Thus we obtain

$$O(M_k) = O(N_k\sqrt{N_k})$$

Next let us calculate the total number of messages used during the execution of the algorithm. Since there are m process-sets, and the largest N_k is equal or less than N.

$$O(M_{total}) = O(\sum_{k=1}^{m} M_k) \leq mO(N\sqrt{N}) = O(N\sqrt{N})$$

5. Conclusions

A new efficient distributed algorithm was developed for implementation of LOTOS Multi-Rendezvous. The algorithm maintains synchronous, nondeterministic and exclusion properties. The implementation made by the algorithm is fair and deadlock free.

Our algorithm is based on an information structure: coterie. Although coterie was used in solutions for many distributed problems such as mutual exclusion, replica control problems, there was no paper that described solution using coterie for Multi-Rendezvous.

Our algorithm requires $O(N\sqrt{N})$ messages, as opposed to $O(N^2)$ messages needed in [2]. And our algorithm is fully distributed in the sense it does not use binary-tree [3] or event manager which is used in [4][5]. Another merit of the algorithm is that it do not use auxiliary resources such as tokens or forks [4][5].

Generally, an event does not need participation of all processes in its process-set. It can be committed by only a subset of its process-set. How to extend our method to the general case is left as a future work.

Acknowledgment

The authors thank Prof. K. Naik for his helpful suggestions. We also wish to thank the anonymous referees whose comments on the manuscript help us improve this work substantially.

REFERENCES

1. ISO, "Information Processing Systems-Open System Interconnection – LOTOS – A Formal Description Technique Based on the Temporal Ordering of Observational Behavior," ISO 8807,(Feb. 1989).
2. Q.Gao and G.v.Bochmann, "A Virtual Ring Algorithm for the Implementation of Multi-Rendezvous," Tech. Rep. 675, Universite de Montreal, Dept. IRO.
3. R.Sisto, L.Ciminiera and A.Valenzano, "A Protocol for Multi-rendezvous of LOTOS Processes," IEEE Trans. on Computer, Vol.40, No.1, pp.437-447(Apr.1991).
4. K.M.Chandy, J.Misra, Parallel Program Design: A Foundation, Reading, MA: Addison-Wesley, 1988.
5. R.Bagrodia, "Process Synchronization: Design and Performance Evaluation of Distributed Algorithms," IEEE Trans. on Software Engineering, Vol.15, No. 9, (Sept. 1989).

6. M.Maekawa, "A $\sqrt{N}$ algorithm for mutual exclusion in decentralized systems," ACM Trans. Comput. Syst., vol. 3, no. 2, pp. 145-159, (May. 1985).
7. A.Kumar, "Hierarchical Quorum Consensus: A new method for managing replicated data," IEEE trans. on Computers, Vol. 40, no. 9, pp. 996-1104, (Sept. 1991).
8. T.V.Lakshman and A.K.Agrawala, "Efficient decentralized consensus protocols," IEEE Trans. Softw. Eng., Vol.SE-12, no. 5, pp. 600-607, (May. 1986).

41

Stepwise Transformations for Fault-Tolerant Design of CCS Processes

Tomasz Janowski *
Department of Computer Science, University of Warwick,
Coventry CV4 7AL, United Kingdom,
tomasz@dcs.warwick.ac.uk

This paper provides an approach to the formal design of distributed, fault-tolerant processes, using the language of CCS and the theory of bisimulations. The novel feature of the method is the language by which hypotheses about faults can be specified and also combined. The development of a fault-tolerant process, under a fault hypothesis, makes use of the structure of this hypothesis. This allows to first design a process which does not tolerate any faults and then to stepwise transform this process to tolerate an increasing variety of faults. We illustrate this approach designing a protocol which ensures a reliable transmission for weak assumptions about the faults of the underlying medium.

1. INTRODUCTION

It is widely believed that a retrospective proof of correctness, for programs of realistic size, is infeasible in practice. Much effort is thus spent to ensure that proof systems support the stepwise development of programs, allowing to reason about them compositionally: a program is correct if only its components can be proven correct. The resulting program is then correct by construction. However, the absence of (software) design faults does not guarantee that the program will behave properly in practice. There is a class of faults that is not due to the wrong design decisions but to the malfunction of hardware components upon which the program relies. Such faults are often due to the physical phenomena and only occur under specific environmental conditions e.g. electromagnetic perturbation. Because they do not exist before a system is put into practice, they cannot be removed beforehand but have to be tolerated.

Clearly, it is not possible to tolerate arbitrary faults. Deciding which faults are anticipated (and thus should be tolerated) and which are not (such faults are catastrophical) is the role of the fault hypothesis. The task is then to design a program which behaves properly in the presence of the anticipated hardware faults. A feasible approach to this task is to separate concerns about the correctness of the program in the absence of faults (the functional correctness) and in the presence of the anticipated faults (fault-tolerance). This can be done in two stages:

1. Designing a program which is correct but fault-intolerant.
2. Transforming this program into the one which is correct and fault-tolerant.

*Supported by the University of Warwick, under its Scholarship Scheme for East Europe, and by an Overseas Students Award from the Committee for Vice-Chancellors and Principals.

However, while the standard techniques (e.g. the stepwise development) can be used to support the first development stage, the author is not aware of any technique to handle complexity of the second stage. This complexity stems from the fact that the more faults are anticipated, the more difficult is the task to ensure that all these faults are tolerated. The lack of the appropriate support can be seen as a major obstacle in the formal development of highly-reliable, fault-tolerant systems.

This paper presents a development method for concurrent processes, using the language of CCS [1], which attempts to fill this gap. The idea is to use a language where hypotheses Ψ about faults can be specified and also combined. The development of a fault-tolerant process, under a given fault hypothesis, makes use of the structure of this hypothesis. This allows to first design a process which does not tolerate any faults and then to stepwise transform this process to tolerate an increasing variety of faults.

We use the standard semantic model for concurrent process description languages, the labelled transition system $(\mathcal{P}, \mathcal{A}, \longrightarrow)$ where $\mathcal{P}$ is the set of processes, $\mathcal{A}$ is the set of actions and $\longrightarrow \subseteq \mathcal{P} \times \mathcal{A} \times \mathcal{P}$ is the labelled transition relation. In order to verify correctness in the presence of the specified faults, we need to represent the effect of these faults on the behaviour of processes. We do this first semantically, providing transition relation $\underset{\Psi}{\longmapsto} \subseteq \mathcal{P} \times \mathcal{A} \times \mathcal{P}$ for Ψ-affected processes Q, and then syntactically, applying process transformation $\mathcal{T}(Q, \Psi)$ [2, 3]. We show that both methods coincide, i.e. that the semantics of Q in the Ψ-affected environment (performing transitions $\underset{\Psi}{\longmapsto}$) is 'the same' as the semantics of the transformed process $\mathcal{T}(Q, \Psi)$ in the fault-free environment (performing transitions $\longrightarrow$). This allows to reason about fault-tolerance using the standard verification techniques e.g. applying bisimulation equivalence [1, 4]:

$$Q \text{ is the } \Psi\text{-tolerant implementation of } P\text{, according } \approx\text{, iff } P \approx \mathcal{T}(Q, \Psi). \tag{1}$$

For multiple faults Ψ and Φ we provide transition relation $\underset{\Psi,\Phi}{\longmapsto}$ and show that it coincides with transition relation $\underset{\Psi\oplus\Phi}{\longmapsto}$ for the combined fault $\Psi \oplus \Phi$. The same is the case when we compare transformations $\mathcal{T}(\mathcal{T}(Q, \Phi), \Psi)$ and $\mathcal{T}(Q, \Psi \oplus \Phi)$. As a result, if $P \approx \mathcal{T}(Q, \Psi \oplus \Phi)$ then Q is an implementation of P which tolerates (simultaneous) faults Ψ and Φ. This enables to develop such Q in two steps:

$$P \approx \mathcal{T}(\mathcal{R}_1(P), \Psi) \approx \mathcal{T}(\mathcal{R}_2(\mathcal{R}_1(P)), \Psi \oplus \Phi) \tag{2}$$

where $\mathcal{R}_1$ and $\mathcal{R}_2$ are fault-tolerant transformations which may employ various techniques to detect (e.g. coding of data), confine (e.g. atomic actions) and recover (e.g. backward and forward recovery) from erroneous states of P. Thus P is transformed to tolerate $\Psi \oplus \Phi$ in stages, one for each sub-hypothesis Ψ and Φ.

The rest of this paper is as follows. Both specification languages, of processes and of faults are defined in Sections 2 and 3. The verification theory, based on bisimulations and transformations of processes is introduced in Section 4. In Section 5 we present the development method for fault-tolerance by the stepwise transformations of processes and exemplify this method designing a protocol which ensures a reliable transmission for weak assumptions about the faults of the underlying medium. Finally, in Section 6, we draw some conclusions and comment on the directions for future work.

2. SPECIFICATION: PROCESSES

In this section we present the language for describing concurrent processes. The language is a version of CCS which is slightly modified for our purposes. It is given the structured operational semantics [5] in terms of the labelled transition relation $\longrightarrow$. We proceed describing first finite then recursive and finally value-passing processes.

2.1. Finite Processes

The language $\mathcal{E}$ of finite processes is based on a set $\mathcal{A}$ of actions. We have $\tau \in \mathcal{A}$ where τ is internal and represents the outcome of a joint activity (interaction) between two processes. The set $\mathcal{L} =_{def} \mathcal{A} - \{\tau\}$ is partitioned between actions a and their complements $\overline{a}$ where the function $\overline{\cdot}$ is bijective and such that $\overline{\overline{a}} = a$. We extend $\overline{\cdot}$ into $\mathcal{A}$ by allowing $\overline{\tau} = \tau$. Let $\alpha \in \mathcal{A}$, $L \subseteq \mathcal{L}$ and $f : \mathcal{A} \rightarrow \mathcal{A}$ where $f(\tau) = \tau$, $f(a) \neq \tau$ and $f(\overline{a}) = \overline{f(a)}$. $\mathcal{E}$, ranged over by E, is defined by the grammar:

$$E \;::=\; \mathbf{0} \mid \alpha.E \mid E+E \mid E|E \mid E\setminus L \mid E[f] \qquad (3)$$

Informally, $\mathbf{0}$ denotes a process which is incapable of any actions. $\alpha.E$ is a process which performs action α and then behaves like E. $E+F$ behaves either like E or like F where the choice may be nondeterministic. $E|F$ is the parallel composition of E and F where E and F can proceed independently but also synchronise on complementary actions (performing action τ). $E \setminus L$ performs all actions of E except actions in L and their complements. Finally, $E[f]$ behaves like E with all actions a renamed into $f(a)$.

The formal semantics of $E \in \mathcal{E}$ is defined by induction on the structure of E, in terms of the labelled transition relation $\longrightarrow \subseteq \mathcal{E} \times \mathcal{A} \times \mathcal{E}$. If $(E, \alpha, E') \in \longrightarrow$ then we write $E \xrightarrow{\alpha} E'$ and say that E performs α and evolves into E' (we also use $E \xrightarrow{s} E'$ for $s \in \mathcal{A}^*$). Following [1], $\longrightarrow$ is defined as the least set which satisfies all inference rules in Figure 1.

$$\frac{}{\alpha.E \xrightarrow{\alpha} E} \qquad \frac{E \xrightarrow{\alpha} E'}{E+F \xrightarrow{\alpha} E'} \qquad \frac{F \xrightarrow{\alpha} F'}{E+F \xrightarrow{\alpha} F'}$$

$$\frac{E \xrightarrow{\alpha} E'}{E|F \xrightarrow{\alpha} E'|F} \qquad \frac{F \xrightarrow{\alpha} F'}{E|F \xrightarrow{\alpha} E|F'}$$

$$\frac{E \xrightarrow{a} E' \quad F \xrightarrow{\overline{a}} F'}{E|F \xrightarrow{\tau} E'|F'}$$

$$\frac{E \xrightarrow{\alpha} E'}{E\setminus L \xrightarrow{\alpha} E'\setminus L}, \;\; \alpha, \overline{\alpha} \notin L \qquad \frac{E \xrightarrow{\alpha} E'}{E[f] \xrightarrow{f(\alpha)} E'[f]}$$

Figure 1. Operational semantics of finite processes $\mathcal{E}$

We use one derived operator, $E \frown F$, where E and F proceed in parallel with actions *out* of E and *in* of F 'joined' and restricted (*mid* is not used by E or F):

$$E \frown F =_{def} (E[mid/out]|F[mid/in])\setminus\{mid\} \qquad (4)$$

2.2. Recursion

It is most common to model hardware devices by cyclic, possibly nonterminating processes. In order to specify such processes consider a set $\mathcal{X}$ of process identifiers and, with slight abuse of notation, a new grammar for $\mathcal{E}$ which extends (3) by $X \in \mathcal{X}$:

$$E \ ::= \ \ldots \mid X \tag{5}$$

We call such E a process expression and use $\mathcal{X}(E) \subseteq \mathcal{X}$ for the set of identifiers occurring in E. We also use $E\{Y/X\}$ for process expression E where all identifiers X are replaced by Y. The semantics of E is well-defined by $\longrightarrow$ (Figure 1). Such a relation however treats X like $\mathbf{0}$, as incapable of any actions. In order to interpret X, we will use declarations of the form $X \triangleq E$. If $X \in \mathcal{X}(E)$ then such X is defined by recursion. Given $X \triangleq E$ and $Y \triangleq F$ where $X \in \mathcal{X}(F)$ and $Y \in \mathcal{X}(E)$, such X and Y are defined by the mutual recursion. In the sequel we will often need to manipulate declarations for mutually recursive identifiers. Then, it will be helpful to use a simple language $\mathcal{D}$ for specifying collections of such declarations. $\mathcal{D}$, ranged over by Δ and ∇, is defined by the following grammar:

$$\Delta \ ::= \ [\,] \mid \Delta[X \triangleq E] \mid \Delta \oplus \nabla \tag{6}$$

Informally, $[\,]$ is an empty declaration and X is defined as E in $\Delta[X \triangleq E]$ and as $E + F$ in $(\Delta[X \triangleq E]) \oplus (\nabla[X \triangleq F])$. Formally, Δ is assigned a denotation $[\![\Delta]\!]$ which is a partial function from $\mathcal{X}$ to $\mathcal{E}$. We use $dom(\Delta)$ and $ran(\Delta)$ for the domain and the range of $[\![\Delta]\!]$ respectively, and we define $[\![\Delta]\!]$ in Figure 2, by induction on the structure of Δ.

$$
\begin{array}{lll}
dom([\,]) & =_{def} & \emptyset \\[2ex]
[\![\Delta[Y \triangleq E]]\!](X) & =_{def} & \begin{cases} E & \text{if } X = Y \\ [\![\Delta]\!](X) & \text{if } X \neq Y,\ X \in dom(\Delta) \end{cases} \\[2ex]
[\![\Delta \oplus \nabla]\!](X) & =_{def} & \begin{cases} [\![\Delta]\!](X) & \text{if } X \in dom(\Delta) - dom(\nabla) \\ [\![\Delta]\!](X) + [\![\nabla]\!](X) & \text{if } X \in dom(\Delta) \cap dom(\nabla) \\ [\![\nabla]\!](X) & \text{if } X \in dom(\nabla) - dom(\Delta) \end{cases}
\end{array}
$$

Figure 2. Denotational semantics of declarations $\mathcal{D}$.

We say that the declaration Δ is closed if all identifiers of right-hand side expressions of Δ are interpreted by Δ: $\bigcup_{E \in ran(\Delta)} \mathcal{X}(E) \subseteq dom(\Delta)$. A useful abbreviation is to take $[X \triangleq E, Y \triangleq F]$ instead of $[\,][X \triangleq E][Y \triangleq F]$ and $[X \triangleq E \mid p]$ for all definitions $X \triangleq E$ such that the predicate p holds.

A process $P \in \mathcal{P}$ is finally the pair $\langle E, \Delta \rangle$ of the process expression E and the closed declaration Δ for all identifiers of E: $\mathcal{X}(E) \subseteq dom(\Delta)$. The semantics of $\langle E, \Delta \rangle$ is defined, with slight abuse of notation, by transition relation $\longrightarrow \ \subseteq \mathcal{P} \times \mathcal{A} \times \mathcal{P}$ which is the least set which satisfies all inference rules in Figure 3. It is easy to see that Δ persists through the transitions of $\langle E, \Delta \rangle$ i.e. if $\langle E, \Delta \rangle \stackrel{a}{\longrightarrow} P'$ then there exists $E' \in \mathcal{E}$ such that $P' \equiv \langle E', \Delta \rangle$ ($\equiv$ is the syntactic identity).

$\dfrac{\langle E_i, \Delta\rangle \xrightarrow{\alpha_i} \langle E_i', \Delta\rangle,\ \ i \in I}{\langle E, \Delta\rangle \xrightarrow{\alpha} \langle E', \Delta\rangle}$	for	$\dfrac{E_i \xrightarrow{\alpha_i} E_i',\ \ i \in I}{E \xrightarrow{\alpha} E'}$	in Figure 1
$\dfrac{\langle [\![\Delta]\!](X), \Delta\rangle \xrightarrow{\alpha} \langle E, \Delta\rangle}{\langle X, \Delta\rangle \xrightarrow{\alpha} \langle E, \Delta\rangle}$	for	$X \in dom(\Delta)$	

Figure 3. Operational semantics of processes $\mathcal{P}$.

Example 1 Consider $n \in \mathbb{N} \cup \{\infty\}$ and let R_n be a process which performs actions a and b; the first at any time; the second no more than n times in a row (R_0 can never perform b and R_∞ can perform b at any time). We have:

$$\begin{array}{lll} R_n & =_{def} & \langle X_0, \Delta \oplus \nabla\rangle \\ \text{where } \Delta & =_{def} & [X_i \mathrel{\hat{=}} a.X_0 \quad | \quad 0 \leq i \leq n] \\ \nabla & =_{def} & [X_i \mathrel{\hat{=}} b.X_{i+1} \quad | \quad 0 \leq i < n] \end{array}$$

□

Operators (3) will be also used to combine processes, under syntactic restriction that processes involved have disjoint sets of identifiers. If $dom(\Delta) \cap dom(\nabla) = \emptyset$ then:

$$\begin{array}{lll} \alpha.\langle E, \Delta\rangle & =_{def} & \langle \alpha.E, \Delta\rangle \\ \langle E, \Delta\rangle \backslash L & =_{def} & \langle E \backslash L, \Delta\rangle \\ \langle E, \Delta\rangle[f] & =_{def} & \langle E[f], \Delta\rangle \\ \langle E, \Delta\rangle + \langle F, \nabla\rangle & =_{def} & \langle E + F, \Delta \oplus \nabla\rangle \\ \langle E, \Delta\rangle \mid \langle F, \nabla\rangle & =_{def} & \langle E \mid F, \Delta \oplus \nabla\rangle \end{array} \tag{7}$$

2.3. Value-Passing

In the language defined so far, processes interact by synchronising on complementary actions a and $\overline{a}$. There is no directionality or value which passes between them. In practice however, we may find it convenient to use a value-passing language for the set V of values (we assume, for simplicity, that V is finite).

To this end we introduce value constants, value variables x, value and boolean expressions e and p built using constants, variables and any function symbols we need. These include ε as the empty sequence; $\sharp s$ as the length of the sequence s; s_0, its first element; s', all but the first element; $s : x$, the sequence s with x appended. We also introduce parameters into process identifiers: $X(e1, .., en)$ for X of arity n. Then we extend $\mathcal{E}$ by input prefixes $a(x).E$, output prefixes $\overline{a}(e).E$ and by conditionals **if** p **then** E **else** F. The semantics of the resulting language is defined by translation into the basic one [1].

Example 2 Consider a buffer of capacity $m > 0$, Buf_m, which receives (by action in) and subsequently transmits (by action $\overline{out}$) all values x unchanged, in the same order and with at most m of them received but not sent. We have:

$$\begin{array}{lll} Buf_m & =_{def} & \langle X(\varepsilon), \Delta \oplus \nabla\rangle \\ \text{where } \Delta & =_{def} & [X(s) \mathrel{\hat{=}} in(x).X(s:x) \quad | \quad 0 \leq \sharp s < m] \\ \nabla & =_{def} & [X(s) \mathrel{\hat{=}} \overline{out}(s_0).X(s') \quad | \quad 0 < \sharp s \leq m] \end{array}$$

□

3. SPECIFICATION: FAULTS

We treat transition relation $\longrightarrow$ as the semantics of processes in the idealised, fault-free environment. The primary effect of faults however is that processes no longer behave according to $\longrightarrow$. We use $\longmapsto$ to model the fault-affected semantics and assume that $\longmapsto\ \supseteq\ \longrightarrow$. In the first part of this section we show how to specify such relations. The idea is to use process identifiers as 'states' which can be potentially affected by faults. Each process $\langle E, \Delta\rangle$ provides its own declaration Δ for process identifiers. In contrast to this, 'normal' declaration, we specify faults by an alternative, 'faulty' declaration Ψ.

For certain faults, fault-tolerance cannot be ensured in full. In these circumstances, we may be satisfied with its conditional version, for certain assumptions about the quantity of faults. Even when the 'full' fault-tolerance is (in theory) possible, we may choose its conditional version to first design a process for restricted assumptions about faults and then to stepwise transform this process to ensure fault-tolerance for more relaxed assumptions. For these reasons, it is important that in addition to possible 'faulty' transitions $\dashrightarrow\ =_{def}\ \longmapsto - \longrightarrow$, we can also specify assumptions about the quantity of such transitions. Such assumptions are introduced in the second part of this section.

3.1. Qualitative Assumptions

In order to specify faults we will use declarations $\mathcal{D}$. As defined in Figure 3, transitions of the process $\langle X, \Delta\rangle$ are determined by the transitions of $\langle [\![\Delta]\!](X), \Delta\rangle$ where $[\![\Delta]\!](X)$ is the process expression assigned to X by Δ. Consider $\Psi \in \mathcal{D}$ which assigns yet another expression $[\![\Psi]\!](X)$ to X. Such Ψ specifies the following transitions $\underset{\Psi}{\longmapsto}$ of $\langle X, \Delta\rangle$:

if $\langle [\![\Psi]\!](X), \Delta\rangle \overset{\alpha}{\underset{\Psi}{\longmapsto}} \langle E, \Delta\rangle$ then $\langle X, \Delta\rangle \overset{\alpha}{\underset{\Psi}{\longmapsto}} \langle E, \Delta\rangle$.

Ψ is not assumed to be closed. Instead, we assume that all process identifiers in the right-side expressions of Ψ are declared in Δ, so that $\underset{\Psi}{\longmapsto}$ does not lead from the well-defined process to the ill-defined one: $\bigcup_{F \in ran(\Psi)} \mathcal{X}(F) \subseteq dom(\Delta)$. We use $\mathcal{P}_\Psi$ for the set of all such $\langle E, \Delta\rangle$. Given Ψ which specifies anticipated faults, we can define a new, Ψ-affected semantics of processes, in terms of transition relation $\underset{\Psi}{\longmapsto}\ \subseteq \mathcal{P}_\Psi \times \mathcal{A} \times \mathcal{P}_\Psi$ which is the least set which satisfies inference rules in Figure 4. If $(P, \alpha, P') \in\ \underset{\Psi}{\longmapsto}$ then we write $P \overset{\alpha}{\underset{\Psi}{\longmapsto}} P'$ (we also use $P \overset{s}{\underset{\Psi}{\longmapsto}} P'$ for $s \in \mathcal{A}^*$).

$\dfrac{\langle E_i, \Delta\rangle \overset{\alpha_i}{\underset{\Psi}{\longmapsto}} \langle E_i', \Delta\rangle,\ \ i \in I}{\langle E, \Delta\rangle \overset{\alpha}{\underset{\Psi}{\longmapsto}} \langle E', \Delta\rangle}$	for	$\dfrac{E_i \overset{\alpha_i}{\longrightarrow} E_i',\ \ i \in I}{E \overset{\alpha}{\longrightarrow} E'}$	in Figure 1
$\dfrac{\langle [\![\Delta]\!](X), \Delta\rangle \overset{\alpha}{\underset{\Psi}{\longmapsto}} \langle E, \Delta\rangle}{\langle X, \Delta\rangle \overset{\alpha}{\underset{\Psi}{\longmapsto}} \langle E, \Delta\rangle}$	for	$X \in dom(\Delta)$	
$\dfrac{\langle [\![\Psi]\!](X), \Delta\rangle \overset{\alpha}{\underset{\Psi}{\longmapsto}} \langle E, \Delta\rangle}{\langle X, \Delta\rangle \overset{\alpha}{\underset{\Psi}{\longmapsto}} \langle E, \Delta\rangle}$	for	$X \in dom(\Psi)$	

Figure 4. Operational semantics of processes $\mathcal{P}$ affected by Ψ.

Observe that each of the three inference rules in Figure 4 determines a set of rules: one for each rule in Figure 1 and one for each process identifier $X \in dom(\Delta)$ and $X \in dom(\Psi)$ respectively. It is easy to see that $\longrightarrow \subseteq \underset{\Psi}{\mapsto}$. We will write:

$$P \overset{\alpha}{\underset{\Psi}{\dashrightarrow}} P' \text{ iff } P \overset{\alpha}{\underset{\Psi}{\mapsto}} P' \text{ and } P \overset{\alpha}{\not\rightarrow} P'$$

Example 3 Consider the following declarations which specify various communication faults of the bounded buffer Buf_m, creation (Ψ_e), corruption (Ψ_c), omission (Ψ_o), replication (Ψ_r) and permutation (Ψ_p) of messages:

$$\begin{array}{lll}
\Psi_e =_{def} [X(s) \triangleq \tau.\overline{out}(\surd).X(s) & \mid & 0 \leq \sharp s \leq m] \\
\Psi_c =_{def} [X(s) \triangleq \tau.\overline{out}(\surd).X(s') & \mid & 0 < \sharp s \leq m] \\
\Psi_o =_{def} [X(s) \triangleq \tau.X(s') & \mid & 0 < \sharp s \leq m] \\
\Psi_r =_{def} [X(s) \triangleq \tau.\overline{out}(s_0).X(s) & \mid & 0 < \sharp s \leq m] \\
\Psi_p =_{def} [X(s) \triangleq \tau.\overline{out}((s')_0).X(s_0 : s'') & \mid & 1 < \sharp s \leq m]
\end{array}$$

Because we are not interested in the particular value of the corrupted or created messages, we represent them all by the same, distinguished message $\surd$. An implicit assumption is that corruption and creation can be easily detected (although not easily distinguished). We also assume that when permuted, only one message is delayed. □

Consider assumptions Ψ and Φ about faults which affect the semantics of $\langle X, \Delta\rangle$. Suppose that $X \in dom(\Psi) \cap dom(\Phi)$. In addition to the 'normal' transitions of $\langle X, \Delta\rangle$, following transitions of $\langle [\![\Delta]\!](X), \Delta\rangle$, two kinds of faulty transitions are also possible for $\langle X, \Delta\rangle$, according to the 'faulty' declarations $[\![\Psi]\!](X)$ and $[\![\Phi]\!](X)$ of X. We use transition relation $\underset{\Psi,\Phi}{\mapsto}$ for the semantics of $\mathcal{P}$ affected by both Ψ and Φ (provided all process identifiers in right-side expressions of Ψ and Φ are declared). Such a relation is defined by inference rules in Figure 5.

$$\frac{\langle E_i, \Delta\rangle \overset{\alpha_i}{\underset{\Psi,\Phi}{\mapsto}} \langle E_i', \Delta\rangle,\ \ i \in I}{\langle E, \Delta\rangle \overset{\alpha}{\underset{\Psi,\Phi}{\mapsto}} \langle E', \Delta\rangle} \quad \text{for} \quad \frac{E_i \overset{\alpha_i}{\rightarrow} E_i',\ \ i \in I}{E \overset{\alpha}{\rightarrow} E'} \quad \text{in Figure 1}$$

$$\frac{\langle [\![\Delta]\!](X), \Delta\rangle \overset{\alpha}{\underset{\Psi,\Phi}{\mapsto}} \langle E, \Delta\rangle}{\langle X, \Delta\rangle \overset{\alpha}{\underset{\Psi,\Phi}{\mapsto}} \langle E, \Delta\rangle} \quad \text{for } X \in dom(\Delta)$$

$$\frac{\langle [\![\Psi]\!](X), \Delta\rangle \overset{\alpha}{\underset{\Psi,\Phi}{\mapsto}} \langle E, \Delta\rangle}{\langle X, \Delta\rangle \overset{\alpha}{\underset{\Psi,\Phi}{\mapsto}} \langle E, \Delta\rangle} \quad \text{for } X \in dom(\Psi)$$

$$\frac{\langle [\![\Phi]\!](X), \Delta\rangle \overset{\alpha}{\underset{\Psi,\Phi}{\mapsto}} \langle E, \Delta\rangle}{\langle X, \Delta\rangle \overset{\alpha}{\underset{\Psi,\Phi}{\mapsto}} \langle E, \Delta\rangle} \quad \text{for } X \in dom(\Phi)$$

Figure 5. Operational semantics of processes $\mathcal{P}$ affected by both Ψ and Φ.

It is easy to show that the joint semantic effect of faults Ψ and Φ (on processes $\mathcal{P}$) is the same as the effect of the combined fault $\Psi \oplus \Phi$:

Proposition 1
If $\Psi, \Phi \in \mathcal{D}$ and $Q \in \mathcal{P}_{\Psi \oplus \Phi}$
then for all $\alpha \in \mathcal{A}$ we have: $Q \underset{\Psi,\Phi}{\overset{\alpha}{\mapsto}} Q'$ iff $Q \underset{\Psi\oplus\Phi}{\overset{\alpha}{\mapsto}} Q'$.

The proof proceeds by induction on the inference of transitions $Q \underset{\Psi,\Phi}{\overset{\alpha}{\mapsto}} Q'$ and $Q \underset{\Psi\oplus\Phi}{\overset{\alpha}{\mapsto}} Q'$.

Example 4 According to Proposition 1, the joint effect of the three communication faults, creation, omission and permutation of messages, can be specified as $\Psi_\epsilon \oplus \Psi_o \oplus \Psi_p$. Following denotational semantics of $\mathcal{D}$ in Figure 2, $\Psi_\epsilon \oplus \Psi_o \oplus \Psi_p$ equals:

$$[X(s) \mathrel{\hat{=}} \tau.\overline{out}(\surd).X(s) \mid \sharp s = 0]$$
$$[X(s) \mathrel{\hat{=}} \tau.\overline{out}(\surd).X(s) + \tau.X(s') \mid \sharp s = 1]$$
$$[X(s) \mathrel{\hat{=}} \tau.\overline{out}(\surd).X(s) + \tau.X(s') + \tau.\overline{out}((s')_0).X(s_0 : s'') \mid 1 < \sharp s \leq m] \qquad \square$$

3.2. Quantitative Assumptions

Consider $\Psi \in \mathcal{D}$ and suppose that transitions $\underset{\Psi}{\mapsto}$ are assigned types: $\longrightarrow$ having type 0 and $\underset{\Psi}{\dashrightarrow}$ having type 1. In order to specify the quantity of faulty transitions we will use sets $H \subseteq \{0,1\}^*$ of all admissible sequences of transition types. Because H is intended to constrain transitions $\underset{\Psi}{\dashrightarrow}$ only, we assume that $\varepsilon \in H$ and if $h \in H$ then $h : 0 \in H$ (H is closed with respect to the concatenation of 0).

Example 5 Suppose that $n \in \mathbb{N} \cup \{\infty\}$ denotes the maximal number of times transitions $\underset{\Psi}{\dashrightarrow}$ can occur successively (if $n = \infty$ then they can occur at any time; if $n = 0$ then not at all). The set H_n, of all admissible sequences of 0's and 1's, under assumption n, equals:

$$H_n =_{def} \begin{cases} \{0,1\}^* & \text{if } n = \infty \\ \{0,1\}^* - \{h : 1^{n+1} \mid h \in \{0,1\}^*\} & \text{if } n \in [0,\infty) \end{cases} \qquad \square$$

Recall that $Q \underset{\Psi}{\overset{s}{\mapsto}} Q'$ if Q evolves into Q' performing the sequence s of transitions $\longrightarrow$ and $\underset{\Psi}{\dashrightarrow}$. This may be no longer the case if transitions $\underset{\Psi}{\dashrightarrow}$ can only occur under assumption H about admissible sequences of transition types. When this is the case then we use the family $\{\underset{\Psi}{\mapsto}{}^h_{h'}\}_{h,h' \in H}$ of transition relations $\underset{\Psi}{\mapsto}{}^h_{h'} \subseteq \mathcal{P}_\Psi \times \mathcal{A}^* \times \mathcal{P}_\Psi$. If $(Q, s, Q') \in \underset{\Psi}{\mapsto}{}^h_{h'}$ then we write $Q \underset{\Psi}{\overset{s}{\mapsto}}{}^h_{h'} Q'$ and say that Q evolves into Q' by the sequence $s \in \mathcal{A}^*$ of transitions $\underset{\Psi}{\mapsto}$, under assumption H and given that h is the history of transition types before and h' after transition. Formally, $\underset{\Psi}{\mapsto}{}^h_{h'}$ is defined by the following inductive rules:

$$Q \underset{\Psi}{\overset{\varepsilon}{\mapsto}}{}^h_h Q$$
$$Q \underset{\Psi}{\overset{\alpha:s}{\mapsto}}{}^h_{h'} Q'' \text{ iff } \exists_{Q'} \quad Q \overset{\alpha}{\longrightarrow} Q' \underset{\Psi}{\overset{s}{\mapsto}}{}^{h:0}_{h'} Q'' \;\vee \qquad (8)$$
$$Q \underset{\Psi}{\overset{\alpha}{\dashrightarrow}} Q' \underset{\Psi}{\overset{s}{\mapsto}}{}^{h:1}_{h'} Q'' \;\wedge\; h : 1 \in H$$

Observe that the induction is well-defined: the first rule provides the base, for the empty sequence ε, and the second rule decreases the length of the action sequence by one. The family $\{\underset{\Psi}{\mapsto}{}^h_{h'}\}_{h,h' \in H}$ denotes the semantics of processes $\mathcal{P}_\Psi$ when they are affected by Ψ and under assumption H about the quantity of transitions $\underset{\Psi}{\dashrightarrow}$.

4. VERIFICATION: BISIMULATION AND TRANSFORMATIONS

Fault-tolerance is a crucial property for safety-critical systems. Such a system is said to tolerate anticipated faults when its behaviour is 'correct' in an operating environment which contains these faults. As such, fault tolerance depends on the chosen notion of correctness. For verifying correctness we provide the choice of the three relations: trace equivalence, bisimulation equivalence and bisimulation preorder. They give rise to different notions of fault-tolerance which is verified using transformations $\mathcal{T}(\cdot, \Psi)$ to model effects of faults on the semantics of processes. Transformations are also used to verify conditional fault-tolerance, under assumption H about the quantity of faults.

4.1. Functional Correctness

Consider the 'weak' transition relation $\Longrightarrow \subseteq \mathcal{P} \times (\mathcal{L} \cup \{\varepsilon\}) \times \mathcal{P}$ where transitions $\xrightarrow{\tau}$ are ignored and let β range over $\mathcal{L} \cup \{\varepsilon\}$. We define $\stackrel{\varepsilon}{\Longrightarrow}$ as the reflexive and transitive closure of $\xrightarrow{\tau}$ and $\stackrel{a}{\Longrightarrow}$ as the composition $\stackrel{\varepsilon}{\Longrightarrow} \xrightarrow{a} \stackrel{\varepsilon}{\Longrightarrow}$ of relations. If $(P, \beta, P') \in \Longrightarrow$ then we write $P \stackrel{\beta}{\Longrightarrow} P'$. We also write $P \stackrel{s}{\Longrightarrow}$, $s \in \mathcal{L}^*$, if there exists P' such that $P \stackrel{s}{\Longrightarrow} P'$.

Trace equivalence, denoted $\approx_1$, identifies a process with all sequences of (observable) actions that it can perform, like in the standard automata theory:

$$P \approx_1 Q \quad \text{iff} \quad \text{for all } s \in \mathcal{L}^* \;\; P \stackrel{s}{\Longrightarrow} \text{ iff } Q \stackrel{s}{\Longrightarrow} \tag{9}$$

Examples are processes P, Q, R and S in Figure 6 which can all perform the same sequences of actions a and b. Thus $P \approx_1 Q \approx_1 R \approx_1 S$. Trace equivalence however admits a linear-time approach to process executions, ignoring at which execution stages which choices are made. For example $P \approx_1 Q$ but action b is always possible for P and not always for Q. As a result, in the environment which continually demands b, P will always meet this demand and Q will sometimes deadlock. Thus $\approx_1$ is insensitive to deadlock.

This is not the case for bisimulation equivalence, $\approx$ [1, 4], which is defined in terms of relations B such that if $(P, Q) \in B$ then for all $\beta \in \mathcal{L} \cup \{\varepsilon\}$:

$$\begin{array}{llll} \text{whenever } P \stackrel{\beta}{\Longrightarrow} P' & \text{then } \exists_{Q'} & Q \stackrel{\beta}{\Longrightarrow} Q' \;\wedge\; (P', Q') \in B \\ \text{whenever } Q \stackrel{\beta}{\Longrightarrow} Q' & \text{then } \exists_{P'} & P \stackrel{\beta}{\Longrightarrow} P' \;\wedge\; (P', Q') \in B \end{array} \tag{10}$$

Such B is called a bisimulation and we have $P \approx Q$ iff there exists a bisimulation B which contains the pair (P, Q). For processes in Figure 6 we have $P \not\approx Q$ because $Q \stackrel{\varepsilon}{\Longrightarrow} Q'$ and $P \stackrel{\varepsilon}{\Longrightarrow} P$ only, however $P \stackrel{b}{\Longrightarrow} P'$ but $Q' \stackrel{b}{\not\Longrightarrow}$. Also $P \approx R \approx S$ because $\{(P, R), (P', R'), (P', R'')\}$ and $\{(P, S), (P', S'), (P', S'')\}$ are bisimulations. However, only S can engage (after a or b) in the infinite sequence of actions τ so bisimulation equivalence is insensitive to divergence.

The closest to $\approx$, divergence-sensitive relation is bisimulation preorder, $\sqsubseteq$ [6, 7]. Informally, $P \sqsubseteq Q$ iff P and Q are bisimulation equivalent, except perhaps when P diverges, and Q diverges no more than P does. Thus Q is at least as 'good' as P: whenever P converges, Q must converge as well, however if P diverges then Q need not diverge. Consider the predicate $\downarrow \subseteq \mathcal{P}$ where if $P \in \downarrow$ then we write $P \downarrow$ and say that the process P converges. Based on $\downarrow$, we define $P \Downarrow \beta$ if there is no process P' such that $P \stackrel{\beta}{\Longrightarrow} P'$ and

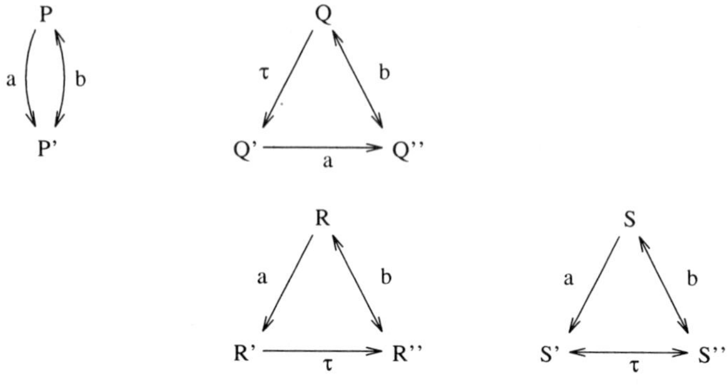

Figure 6. Transition diagrams of P, Q, R and S.

which performs action τ indefinitely:

$$\begin{array}{llll} P \Downarrow \varepsilon & \text{iff} \quad P \downarrow & \text{and whenever} \quad P \xrightarrow{\tau} P' & \text{then} \quad P' \Downarrow \varepsilon \\ P \Downarrow a & \text{iff} \quad P \Downarrow \varepsilon & \text{and whenever} \quad P \stackrel{a}{\Longrightarrow} P' & \text{then} \quad P' \Downarrow \varepsilon \end{array} \tag{11}$$

Bisimulation preorder $\sqsubseteq$ is defined in terms of relations B, called partial bisimulations, such that if $(P, Q) \in B$ then:

$$\begin{array}{lll} \text{whenever} \quad P \stackrel{\beta}{\Longrightarrow} P' & \text{then} & \exists_{Q'} \; Q \stackrel{\beta}{\Longrightarrow} Q' \;\wedge\; (P', Q') \in B \\ \text{whenever} \quad P \Downarrow \beta & \text{then} & Q \Downarrow \beta \;\wedge \\ & & \text{if} \quad Q \stackrel{\beta}{\Longrightarrow} Q' \quad \text{then} \quad \exists_{P'} \; P \stackrel{\beta}{\Longrightarrow} P' \;\wedge\; (P', Q') \in B \end{array} \tag{12}$$

We have $P \sqsubseteq Q$ iff $(P, Q) \in B$ for some partial bisimulation B. For processes in Figure 6 we have $P \sqsubseteq R$ because $\{(P, R), (P', R'), (P', R'')\}$ is a partial bisimulation and $P \not\sqsubseteq S$ because $P \Downarrow a$ and $S \not\Downarrow a$.

4.2. Fault-Tolerance

Suppose that $\preceq$ denotes any one of relations $\approx_1$, $\approx$ or $\sqsubseteq$, and consider the high-level process P. Such P determines the set of admissible implementations Q, $P \preceq Q$, where semantics of both P and Q is defined by transition relation $\longrightarrow$. The situation however is different if we want to ensure that Q is a fault-tolerant implementation of P, according to $\preceq$ and the specification Ψ of faults. Informally, such Q should behave 'properly' in any environment where faults are specified by Ψ. Such a Ψ-affected behaviour of Q is then defined by transition relation $\underset{\Psi}{\longmapsto}$, in contrast to P which still behaves according to $\longrightarrow$. This raises the problem of comparing two processes which behaviour is defined by different transition relations $\longrightarrow$ and $\underset{\Psi}{\longmapsto}$. In order to solve this problem consider the following transformation $\mathcal{T}$ of processes:

$$\mathcal{T}(\langle E, \Delta \rangle, \Psi) \;=_{def}\; \langle E, \Delta \oplus \Psi \rangle \tag{13}$$

It is easy to show that such $\mathcal{T}$ provides an equivalent to $\underset{\Psi}{\mapsto}$, syntactic method of representing effect of faults on the behaviour of processes, i.e. we can show that:

Proposition 2
If $\Psi \in \mathcal{D}$ and $Q \in \mathcal{P}_\Psi$
then for all $\alpha \in \mathcal{A}$ we have: $Q \overset{\alpha}{\underset{\Psi}{\mapsto}} Q'$ iff $\mathcal{T}(Q,\Psi) \xrightarrow{\alpha} \mathcal{T}(Q',\Psi)$.

The proof proceeds by induction on the inference of transitions $\mathcal{T}(Q,\Psi) \xrightarrow{\alpha} \mathcal{T}(Q',\Psi)$ and $Q \overset{\alpha}{\underset{\Psi}{\mapsto}} Q'$. Thus the semantics of Q in the Ψ-affected environment is the same as the semantics of $\mathcal{T}(Q,\Psi)$ in the fault-free environment and in order to prove that Q is a Ψ-tolerant implementation of P (according to $\preceq$) it is enough to show that:

$$P \preceq \mathcal{T}(Q,\Psi) \tag{14}$$

Example 6 Consider the task to ensure a reliable communication (specified by the bounded buffer Buf_{m+1}, Example 2) over a medium of capacity m which creates messages (specified by Ψ_e, Example 3). To this end we can use a process Ret which ignores all $\surd$'s:

$$Ret \ =_{def} \ \langle Z, [Z \mathrel{\hat{=}} in(x).\ \textbf{if}\ x = \surd\ \textbf{then}\ Z\ \textbf{else}\ \overline{out}(x).Z]\rangle$$

Given such Ret, it is easy to prove that $Buf_{m+1} \approx_1 \mathcal{T}(Buf_m \frown Ret, \Psi_e)$. However, we have $Buf_{m+1} \not\approx \mathcal{T}(Buf_m \frown Ret, \Psi_e)$ because after receiving $x \neq \surd$ and before its transmission, Ret does not accept any more $\surd$. The 'better' implementation is Re:

$$\begin{array}{lllll} Re & =_{def} & \langle Z, & [Z & \hat{=} \ in(x).\ \textbf{if}\ x = \surd\ \textbf{then}\ Z\ \textbf{else}\ Z(x)] \\ & & & [Z(x) & \hat{=} \ in(\surd).Z(x) + \overline{out}(x).Z]\rangle \end{array}$$

Then we have $Buf_{m+1} \approx \mathcal{T}(Buf_m \frown Re, \Psi_e)$ but $Buf_{m+1} \not\sqsubseteq \mathcal{T}(Buf_m \frown Re, \Psi_e)$ because $\mathcal{T}(Buf_m \frown Re, \Psi_e)$ but not Buf_{m+1} can diverge, due to arbitrary creation of messages. □

The same approach can be used to verify that Q tolerates multiple faults, say faults specified by Ψ and Φ. Applying Proposition 1 which shows that the joint effect of Ψ and Φ, $\underset{\Psi,\Phi}{\mapsto}$, is the same as the effect of the combined fault $\Psi \oplus \Phi$, $\underset{\Psi\oplus\Phi}{\mapsto}$, and Proposition 2 which allows to express $\underset{\Psi\oplus\Phi}{\mapsto}$ in terms of $\longrightarrow$ and $\mathcal{T}(\cdot, \Psi \oplus \Phi)$, it is enough to prove that:

$$P \preceq \mathcal{T}(Q, \Psi \oplus \Phi) \tag{15}$$

4.3. Conditional Fault-Tolerance

Suppose now that we want to verify $\langle E, \Delta\rangle$ in the presence of faulty transitions $\underset{\Psi}{\dashrightarrow}$ and under assumption $H \subseteq \{0,1\}^*$ about admissible transition type histories. To this end, like before, we will use process transformations. The idea is to use a family $\{X_h\}_{h\in H}$ of the process identifiers, for each identifier X of Δ or Ψ. The index h of X_h denotes the history of transition types. Consider $h \in H$ and the following transformation $\widetilde{\mathcal{T}}_h(\cdot,\Psi,H)$:

$$\widetilde{\mathcal{T}}_h(\langle E,\Delta\rangle,\Psi,H) =_{def} \mathcal{T}(\langle E_h,\Delta_H\rangle,\Psi_H) \tag{16}$$

$$\begin{array}{lll} \text{where } E_h & =_{def} & E\{X_h/X \mid X \in \mathcal{X}(E)\} \\ \Delta_H & =_{def} & [X_h \mathrel{\hat{=}} F\{Y_{h:0}/Y \mid Y \in \mathcal{X}(F)\} \mid [\![\Delta]\!](X) = F \wedge h \in H] \\ \Psi_H & =_{def} & [X_h \mathrel{\hat{=}} F\{Y_{h:1}/Y \mid Y \in \mathcal{X}(F)\} \mid [\![\Psi]\!](X) = F \wedge h \in H \wedge h:1 \in H] \end{array}$$

Thus E_h is obtained from E by replacing all process identifiers $X \in \mathcal{X}(E)$ by X_h; Δ_H is obtained from Δ by replacing each declaration $X \mathrel{\hat=} F$ by the family of declarations $X_h \mathrel{\hat=} F\{Y_{h:0}/Y \mid Y \in \mathcal{X}(F)\}$ for all $h \in H$, and Ψ_H is similar like Δ_H but only taking $h \in H$ such that $h:1 \in H$. Finally, we define $\widetilde{\mathcal{T}}(\langle E, \Delta\rangle, \Psi, H)$ as $\widetilde{\mathcal{T}}_\varepsilon(\langle E, \Delta\rangle, \Psi, H)$.

Recall that the effect of Ψ on the semantics of $\mathcal{P}_\Psi$, under assumption H about the quantity of transitions $\overset{}{\underset{\Psi}{\dashrightarrow}}$, is defined by the family $\{\overset{h}{\underset{\Psi}{\mapsto}}_{h'}\}_{h,h' \in H}$ of transition relations. There are two problems to obtain the same effect using transformations:

1. Consider $X \in dom(\Delta) \cap dom(\Psi)$ and transition $\langle X, \Delta\rangle \overset{\alpha}{\underset{\Psi}{\mapsto}} \langle E, \Delta\rangle$ which can be either inferred from $\langle [\![\Delta]\!](X), \Delta\rangle \overset{\alpha}{\underset{\Psi}{\mapsto}} \langle E, \Delta\rangle$ or from $\langle [\![\Psi]\!](X), \Delta\rangle \overset{\alpha}{\underset{\Psi}{\mapsto}} \langle E, \Delta\rangle$. The problem appears when $\langle X, \Delta\rangle \overset{\alpha}{\underset{\Psi}{\mapsto}} \langle E, \Delta\rangle$ can be inferred from both of them. Then it is regarded as 'normal' by $\overset{h}{\underset{\Psi}{\mapsto}}_{h'}$ but either as 'normal' or 'faulty' by $\widetilde{\mathcal{T}}_h(\cdot, \Psi, H)$. If no such E and α exists then we say that Ψ *has the proper effect on* Δ.

2. The second problem is that in case of $\widetilde{\mathcal{T}}_h(\cdot, \Psi, H)$ (but not $\overset{h}{\underset{\Psi}{\mapsto}}_{h'}$), some transitions do not contribute to the history h of transition types. Suppose that $\Psi =_{def} [X \mathrel{\hat=} \tau.a.X]$ and $\Delta =_{def} [X \mathrel{\hat=} b.X]$. Then we have: $\langle X, \Delta\rangle \overset{\tau}{\underset{\Psi}{\dashrightarrow}} \langle a.X, \Delta\rangle \overset{a}{\longrightarrow} \langle X, \Delta\rangle \overset{b}{\longrightarrow} \langle X, \Delta\rangle$ and thus $\langle X, \Delta\rangle \overset{\tau ab\, h}{\underset{\Psi}{\mapsto}}_{h:100} \langle X, \Delta\rangle$, however $\widetilde{\mathcal{T}}_h(\langle X, \Delta\rangle, \Psi, H) \overset{\tau ab}{\longrightarrow} \widetilde{\mathcal{T}}_{h:10}(\langle X, \Delta\rangle, \Psi, H)$. We can solve this problem assuming that all expressions involved are *linear* i.e. they are of the form $\sum_{i=1}^{k} \alpha_i.X_i$ (Δ is linear if all $F \in ran(\Delta)$ are linear).

Under both conditions it is easy to prove the following proposition:

Proposition 3

If $\quad \Psi \in \mathcal{D}$ and $\langle X, \Delta\rangle \in \mathcal{P}_\Psi$ where

$\quad\quad \Psi$ has the proper effect on Δ and Ψ and Δ are linear

then $\quad \langle X, \Delta\rangle \overset{s\; h}{\underset{\Psi}{\mapsto}}_{h'} \langle Y, \Delta\rangle$ iff $\widetilde{\mathcal{T}}_h(\langle X, \Delta\rangle, \Psi, H) \overset{s}{\longrightarrow} \widetilde{\mathcal{T}}_{h'}(\langle Y, \Delta\rangle, \Psi, H)$

The proof proceeds by induction on the length of s. Thus the 'normal' semantics of $\widetilde{\mathcal{T}}(\langle X, \Delta\rangle, \Psi, H)$ is the same as the Ψ-affected semantics of $\langle X, \Delta\rangle$, under assumption H. As a result, in order to prove that $\langle X, \Delta\rangle$ tolerates Ψ under assumption H (with respect to P and according to $\preceq$), it is enough to show that:

$$P \preceq \widetilde{\mathcal{T}}(\langle X, \Delta\rangle, \Psi, H) \tag{17}$$

Although the linear form of Δ and Ψ is necessary to prove Proposition 3, the meaning of $\widetilde{\mathcal{T}}(\langle E, \Delta\rangle, \Psi, H)$ for non-linear Δ and Ψ is also well-understood. While in the first case all transitions are significant, they all contribute to the history h, in the second case only chosen ones are significant. Thus the main reason for 'mismatch' between $\widetilde{\mathcal{T}}_h(\cdot, \Psi, H)$ and $\overset{h}{\underset{\Psi}{\mapsto}}_{h'}$ for non-linear expressions lies in the restrictive form of the latter. In the sequel we will adopt (17) to verify conditional fault-tolerance for $\langle E, \Delta\rangle$ and Ψ where process expressions of Δ and Ψ are not necessarily linear.

For example we have $Buf_{m+1} \sqsubseteq \widetilde{\mathcal{T}}(Buf_m \frown Re, \Psi_\epsilon, H_n)$ for process Re (Example 6) which ignores all created messages $\surd$ and for H_n (Example 5) where $n \neq \infty$ is the bound on the number of successive occurrences of Ψ_ϵ (Example 3).

Example 7 Consider $n, m > 0$ and the task to ensure a reliable communication (specified by Buf_{m+n+2} and according to $\sqsubseteq$) over a medium of capacity m which permutes messages. To this end we will use two processes: the sender Sp_n and the receiver Rp_n. In order to determine the proper transmission order, messages will be send by Sp_n with their sequence numbers modulo n. The value of n determines the number of parallel components St_i of Rp_n ($i = 0, \ldots, n-1$), each one used to store a message with the sequence value i, received out-of-order. The value of $\perp$ means that no message is stored. Suppose that the summation $i+1$ below is taken modulo n. Then we have:

$$
\begin{array}{lll}
Sp_n & =_{def} & \langle Zs(0), [Zs(i) \triangleq in(x).\overline{out}(x,i).Zs(i+1) \mid 0 \leq i < n] \rangle \\
Rp_n & =_{def} & (Ctr \mid St_0 \mid \cdots \mid St_{n-1}) \backslash \{st_0, \ldots, st_{n-1}\} \\
Ctr & =_{def} & \langle Zr(0),\ [Zr(i) \triangleq in(x,j).\ \mathbf{if}\ i=j\ \mathbf{then}\ \overline{out}(x).Zm(i+1) \\
 & & \qquad\qquad\qquad\qquad\qquad \mathbf{else}\ \overline{st_j}(x).Zr(i) \qquad \mid 0 \leq i < n] \\
 & & \quad [Zm(i) \triangleq st_i(x).\ \ \mathbf{if}\ x = \perp \mathbf{then}\ Zr(i) \\
 & & \qquad\qquad\qquad\qquad\qquad \mathbf{else}\ \overline{out}(x).Zm(i+1) \mid 0 \leq i < n] \rangle \\
St_i & =_{def} & \langle Zm_i,\ \ [Zm_i \triangleq st_i(x).\overline{st_i}(x).Zm_i + \overline{st_i}(\perp).Zm_i] \rangle,\ \ 0 \leq i < n
\end{array}
$$

It can be shown that $Sp_n \frown Buf_m \frown Rp_n$ tolerates Ψ_p provided the number of successive permutations is not greater than n: $Buf_{m+n+2} \sqsubseteq \widetilde{\mathcal{T}}(Sp_n \frown Buf_m \frown Rp_n, \Psi_p, H_n)$. □

5. DEVELOPMENT: STEPWISE TRANSFORMATIONS

So far we were only concerned with how to verify that given a high-level process P, a fault hypothesis Ψ and a low-level, Ψ-affected process Q, Q is an implementation of P (according to some $\preceq$) which tolerates Ψ (perhaps under a certain assumption H about its quantity). The problem of how to design such Q has been completely ignored. This problem is the topic of the current section.

In most cases, Q can be obtained by the transformation $\mathcal{R}(P)$ of P. For example we have: $\mathcal{R}(Buf_m) = Buf_{m-1} \frown Re$ to tolerate Ψ_ϵ according to $\approx$ (Example 6, $m > 1$) and $\mathcal{R}(Buf_m) = Sp_n \frown Buf_{m-n-2} \frown Rp_n$ to tolerate Ψ_p according to $\sqsubseteq$ (Example 7, $m > n+2$). Thus, when no assumption about the quantity of faults is made, our problem is to find a transformation $\mathcal{R}$ such that:

$$P \preceq \mathcal{T}(\mathcal{R}(P), \Psi) \tag{18}$$

To this end, $\mathcal{R}(P)$ will employ various techniques to detect, confine and to recover from erroneous states of P, by introducing some additional, recovery processes. Our task is easier when the recovery processes are assumed not to be affected by Ψ, i.e. when they do not share any process identifiers with Ψ (like Re, Sp_n and Rp_n). It this case, $\mathcal{T}(\mathcal{R}(P), \Psi)$ is identical with $\mathcal{R}(\mathcal{T}(P, \Psi))$ and it is enough to prove that:

$$P \preceq \mathcal{R}(\mathcal{T}(P, \Psi)) \tag{19}$$

For example, in order to verify $Buf_{m+1} \approx \mathcal{T}(Buf_m \frown Re, \Psi_\epsilon)$, it is enough to prove that $Buf_{m+1} \approx \mathcal{T}(Buf_m, \Psi_\epsilon) \frown Re$. It is out of scope of this paper to investigate any particular technique to design such $\mathcal{R}(P)$. Instead, when $\Psi = \Psi_1 \oplus \cdots \oplus \Psi_n$, we would like to propose that $\mathcal{R}(P)$ is obtained from P by the sequence of transformations, one for each component Ψ_i of Ψ. Then, we can expect that the task to tolerate each Ψ_i is easier than the task to tolerate them altogether. This suggests the following development process:

$$P_0 \preceq \mathcal{T}(P_1, \Psi_1) \preceq \mathcal{T}(P_2, \Psi_1 \oplus \Psi_2) \preceq \cdots \preceq \mathcal{T}(P_n, \Psi_1 \oplus \cdots \oplus \Psi_n) \tag{20}$$

where $P_0 \equiv P$ and $P_i \equiv \mathcal{R}_i(P_{i-1})$ for $i = 1 \ldots n$. Such P_i aims to tolerate Ψ_i in the presence of faults specified by $\Psi_1, \ldots, \Psi_{i-1}$ and in general depends on the recovery processes used in the earlier stages of the design. The final transformation $\mathcal{R}(P) = \mathcal{R}_n(\ldots \mathcal{R}_1(P) \ldots)$. Our task can be largely simplified if $\preceq$ is preserved by $\mathcal{R}_i$ i.e. whenever $P \preceq Q$ then $\mathcal{R}_i(P) \preceq \mathcal{R}_i(Q)$ and if recovery processes of $\mathcal{R}_i$ are not affected by $\Psi_1 \oplus \ldots \oplus \Psi_i$ i.e. $\mathcal{T}(\mathcal{R}_i(P_{i-1}), \Psi_1 \oplus \ldots \oplus \Psi_i)$ is identical with $\mathcal{R}_i(\mathcal{T}(P_{i-1}, \Psi_1 \oplus \ldots \oplus \Psi_i))$. Then, in the $i+1$-st stage, it is enough to find $\mathcal{R}_{i+1}$ which transforms P_{i-1}, not P_i:

$$\mathcal{T}(P_{i-1}, \Psi_1 \oplus \ldots \oplus \Psi_i) \preceq \mathcal{T}(\mathcal{R}_{i+1}(P_{i-1}), \Psi_1 \oplus \ldots \oplus \Psi_{i+1}) \tag{21}$$

This means that P_{i+1} does not depend on the recovery processes used in the i-th step, i.e. that both steps are independent. When $\preceq$ is preserved by all transformations $\mathcal{R}_i$ and when none of the recovery processes is affected by faults then all stages are mutually independent and the final transformation $\mathcal{R}(P) = \mathcal{R}_1(\ldots \mathcal{R}_n(P) \ldots)$.

Example 8 Consider the task to ensure a reliable communication (specified by Buf_m and according to $\approx$) over a medium of capacity m which corrupts, creates, omits and replicates messages, i.e. to find a transformation $\mathcal{R}(Buf_m)$ of Buf_m such that:

$$Buf_m \approx \mathcal{T}(\mathcal{R}(Buf_m), \Psi_e \oplus \Psi_c \oplus \Psi_o \oplus \Psi_r)$$

Applying development procedure (20) we will design such $\mathcal{R}(Buf_m)$ in two steps, by first tolerating $\Psi_o \oplus \Psi_r$ and then $\Psi_e \oplus \Psi_c$. To tolerate $\Psi_o \oplus \Psi_r$ we will use a version of the sliding window protocol with the window size m. The protocol consists of two processes, the sender So_m and the receiver Ro_m such that at most m messages are sent by So_m without being acknowledged by Ro_m (we acknowledge messages by actions ack). So_m uses s as the sequence of messages sent but not acknowledged (we have $\sharp s \le m$) and repeatedly retransmits s_0 until acknowledgement for this message is received. Taking all arithmetic operations below modulo $m+1$, we have:

$$\begin{array}{llll}
So_m =_{def} \langle Z(0,\varepsilon), & [Z(i,s) & \hat{=}\ in(x).Z(i,s,x) & \mid 0 \le \sharp s < m \ \wedge\ 0 \le i \le m] \oplus \\
& [Z(i,s) & \hat{=}\ ack.Z(i,s') & \mid 0 < \sharp s \le m \ \wedge\ 0 \le i \le m] \oplus \\
& [Z(i,s) & \hat{=}\ \overline{out}(i-\sharp s, s_0).Z(i,s) & \mid 0 < \sharp s \le m \ \wedge\ 0 \le i \le m] \\
& [Z(i,s,x) & \hat{=}\ \overline{out}(i,x).Z(i+1, s:x) & \mid 0 \le \sharp s < m \ \wedge\ 0 \le i \le m] \oplus \\
& [Z(i,s,x) & \hat{=}\ ack.Z(i,s',x) & \mid 0 < \sharp s < m \ \wedge\ 0 \le i \le m] \oplus \\
& [Z(i,s,x) & \hat{=}\ \overline{out}(i-1,s_0).Z(i,s,x) & \mid 0 < \sharp s < m \ \wedge\ 0 \le i \le m] \rangle \\
Ro_m =_{def} \langle Z(0), & [Z(i) & \hat{=}\ in(j,x).\ \textbf{if}\ i=j\ \ \textbf{then}\ \overline{out}(x).\overline{ack}.Z(i+1) & \\
& & \qquad\qquad\qquad\qquad\quad \textbf{else}\ Z(i)] \rangle &
\end{array}$$

For such So_m and Ro_m we can prove the following equivalence:

$$Buf_m \approx (So_m \frown (\mathcal{T}(Buf_m, \Psi_o \oplus \Psi_r) \frown Buf_1) \frown Ro_m) \backslash \{ack\}$$

Recall a process Re (Example 6) to tolerate Ψ_e by ignoring all messages $\surd$. If we apply Re to the medium which creates, corrupts, omits and replicates messages ($\Psi_e \oplus \Psi_c \oplus \Psi_o \oplus \Psi_r$) then the resulting medium is only affected by the last two faults ($\Psi_o \oplus \Psi_r$):

$$\mathcal{T}(Buf_m, \Psi_o \oplus \Psi_r) \frown Buf_1 \approx \mathcal{T}(Buf_m, \Psi_e \oplus \Psi_c \oplus \Psi_o \oplus \Psi_r) \frown Re$$

Finally, because none of the processes So_m, Ro_m or Re is affected by faults and because $\approx$ is preserved by $\frown$ and $\backslash$, we get the desired transformation:

$$\begin{aligned} \mathcal{R}(Buf_m) &= \mathcal{R}_1(\mathcal{R}_2(Buf_m)) \\ &= \mathcal{R}_1(Buf_m \frown Re) \\ &= (So_m \frown (Buf_m \frown Re) \frown Ro_m) \backslash \{ack\} \end{aligned}$$

The resulting process is a two-layered protocol where the lower layer tolerates $\Psi_e \oplus \Psi_c$ and the higher one tolerates $\Psi_o \oplus \Psi_r$. It is not possible, using only bounded sequence numbers, to extend this protocol to tolerate permutation Ψ_p [8]. □

It is often the case that recovery processes are affected by faults themselves. Even worse, that they introduce new faults (not specified by Ψ), as in case of the sliding window protocol and acknowledgements which are not exchanged by simple synchronisations but using a medium which itself may be faulty. Suppose that Φ specifies 'new' faults, introduced by recovery processes of $\mathcal{R}(P)$. In this case we have to prove that:

$$P \preceq \mathcal{T}(\mathcal{R}(P), \Psi \oplus \Phi) \tag{22}$$

So far, we were not concerned with assumptions H about the quantity of faults. However, such assumptions can be used to support the stepwise procedure (20). To this end, we can first design a process for strong assumptions about the quantity of faults (say H) and then to stepwise transform this process to ensure fault-tolerance for increasingly relaxed assumptions (H' where $H \subseteq H'$). Such a stepwise procedure will be described elsewhere.

6. CONCLUSIONS

Currently, there is a number of methods for specifying and proving correctness of fault-tolerant systems [3, 9–15]. In this paper, we did not aim to provide yet another formalism. Our purpose was to show how the well-established theory of CCS can be extended to reason about fault-tolerance, with emphasis placed on reasoning under weak assumptions about faults. This extension includes two languages (for specifying processes and faults), verification theory based on transformations of processes (and exemplified using bisimulations and partial bisimulations) and the development approach where multiple faults are proposed to be tolerated incrementally, by stepwise transformations.

We plan to continue this work in all three aspects: specification, verification and (first of all) development of fault-tolerant processes. When the last is concerned, we plan to provide some constructive proof rules for certain, well-known techniques for fault-tolerance, e.g. backward recovery and modular redundancy. Also, to utilise assumptions about the quantity of faults for the stepwise design of fault-tolerant processes.

ACKNOWLEDGEMENTS

I am grateful to my supervisor, Mathai Joseph, for many valuable comments on draft versions of this paper, to Zhiming Liu for useful discussions on fault-tolerance, and to David Walker for helpful comments and for putting some literature to my attention. Thanks are also to the referees for their comments.

REFERENCES

1. R. Milner. *Communication and Concurrency.* Prentice-Hall International, 1989.
2. Z. Liu. *Fault-Tolerant Programming by Transformations.* PhD thesis, University of Warwick, 1991.
3. Z. Liu and M. Joseph. Transformations of programs for fault-tolerance. *Formal Aspects of Computing*, 4:442–469, 1991.
4. D. Park. Concurrency and automata on infinite sequences. *LNCS*, 104, 81.
5. G. Plotkin. A structural approach to operational semantics. Technical report, Computer Science Department, Aarhus University, 81.
6. R. Milner. A modal characterisation of observable machine-behaviour. *LNCS*, 112:25–34, 81.
7. D.J. Walker. Bisimulation and divergence. *Information and Computation*, 85:202–241, 90.
8. D. Wang and L. Zuck. Tight bounds for the sequence transmission problem. In *Proc. 8th ACM Symp. on Princ. of Distributed Computing*, pages 73–83, 89.
9. F. Cristian. A rigorous approach to fault-tolerant programming. *IEEE Transactions on Software Engineering*, 11(1):23–31, 1985.
10. He Jifeng and C.A.R. Hoare. Algebraic specification and proof of a distributed recovery algorithm. *Distributed Computing*, 2:1–12, 1987.
11. J. Nordahl. *Specification and Design of Dependable Communicating Systems.* PhD thesis, Technical University of Denmark, 1992.
12. J. Peleska. Design and verification of fault tolerant systems with CSP. *Distributed Computing*, 5:95–106, 1991.
13. D. Peled and M. Joseph. A compositional approach for fault-tolerance using specification transformation. *LNCS*, 694, 1993.
14. K.V.S. Prasad. *Combinators and Bisimulation Proofs for Restartable Systems.* PhD thesis, Department of Computer Science, University of Edinburgh, 1987.
15. H. Schepers. Tracing fault-tolerance. In *Proc. 3rd IFIP Working Conference on Dependable Computing for Critical Applications.* Springer-Verlag, 1993.

INDEX OF CONTRIBUTORS

KEYWORD INDEX